Environmental Soil Science

Environmental Soil Science

Edited by **Henry Wang**

SYRAWOOD
PUBLISHING HOUSE
New York

Published by Syrawood Publishing House,
750 Third Avenue, 9th Floor,
New York, NY 10017, USA
www.syrawoodpublishinghouse.com

Environmental Soil Science
Edited by Henry Wang

International Standard Book Number: 978-1-68286-346-6 (Hardback)

Printed in China.

Contents

Preface

This book was inspired by the evolution of our times; to answer the curiosity of inquisitive minds. Many developments have occurred across the globe in the recent past which has transformed the progress in the field.

Environmental soil science has given new direction to the study of interactions among humans, environment and soil. The book provides a detailed account of the physical, biological and chemical properties of soils, and their subsequent interaction with different layers of environment. This book includes some of the vital pieces of work being conducted across the world, on various topics related to soil stability, soil erosion, nutrient management and land-use management. As this field is emerging at a rapid pace, the contents of this book will help the readers understand the modern concepts and applications of the subject.

This book was developed from a mere concept to drafts to chapters and finally compiled together as a complete text to benefit the readers across all nations. To ensure the quality of the content we instilled two significant steps in our procedure. The first was to appoint an editorial team that would verify the data and statistics provided in the book and also select the most appropriate and valuable contributions from the plentiful contributions we received from authors worldwide. The next step was to appoint an expert of the topic as the Editor-in-Chief, who would head the project and finally make the necessary amendments and modifications to make the text reader-friendly. I was then commissioned to examine all the material to present the topics in the most comprehensible and productive format.

I would like to take this opportunity to thank all the contributing authors who were supportive enough to contribute their time and knowledge to this project. I also wish to convey my regards to my family who have been extremely supportive during the entire project.

Editor

Molecular geochemistry of soil organic matter by pyrolysis gas chromatography/mass spectrometry (GC/MS) technique: A review

Mojtaba Mehrabanian

Department of Soil Science, Faculty of Agricultural Engineering and Technology, Collage of Agriculture and Natural Resources, University of Tehran, Iran.

Soil organic matter (SOM) is arguably the most complex and least understood component of soils. To analyze SOM structure and reduce SOM large structural heterogeneity, different components of SOM need to be separated into entities that differ in terms of source, composition, and turnover. Pyrolysis-gas chromatography/mass spectrometry (Pyrolysis-GC/MS) is to present the most effectual approach in the study of the pedogenesis of SOM; this uses thermal degradation to cleave bonds in the organic macromolecules and enables a sensitive and rapid characterization of organic constituents. Pyrolysis-GC/MS also aids the molecular characterization of microbial and plant-derived biomass and generates valuable data on the degradation and conservation rates of organic debris. This review provides an overview of commonly measured soil organic chemical constituents identified by Pyrolysis-GC/MS, including (i) aliphatics, fatty acids and sterols, (ii) carbohydrates, (iii) lignin, (iv) aromatic compounds and polycyclic aromatic hydrocarbons (PAHs) and (v) N-containing compounds.

Key words: Soil organic matter (SOM), pyrolysis-gas chromatography/mass spectrometry (GC/MS), molecular geochemistry, organic compounds.

INTRODUCTION

Soil organic matter (SOM) occupies a fundamental role in ecological dynamics. It influences a large group of soil chemical, physical and biological attributes, including soil structure, soil fertility and productivity, soil moisture retention and erodability (Kay et al., 1997). These qualities have a tremendous impact on ecosystem roles, such as nutrient cycling, carbon sequestration and pollutant retention. In fact, SOM composition and its changes during biotic and abiotic transformation are key variables in the processes and the size of soil inter-active reactions with the surroundings.

SOM is a combination of all organic materials derived from plants and animals, in different stages of decomposition and degree of association with the mineral matrix (Buurman and Roscoe, 2011). This implies large structural heterogeneity and close linkage to SOM functions. SOM structure analysis helps us to understand chemical processes within soil and to predict changes in the sign and magnitude of terrestrial carbon fluxes in a changing environment. To reduce SOM large structural heterogeneity, different components of SOM need to be separated into entities that differ in terms of source, composition, and turnover. Recent evidence suggests that the classical chemical fractionation into fulvic acids, humic acids, and humin according to solubility characteristics in dilute acid and base (Swift, 1996) is not useful in this respect. Fractionations that are more promising rely mostly on physical fractionations according to particle size or density (Golchin et al., 1994), and the analysis of dissolved organic matter in the soil solution (Guggenberger et al., 1994), primarily to determine labile fraction of carbon.

In order to expand deep appreciation of the origin and function of organic matter, development of modern fractionation methods is vital. The objectives of this review are to present an account of what has been found about SOM characterization by Pyrolysis-gas chromatography/mass spectrometry (GC/MS) technique. The first part of this review will deal with the strengths of this approach, while the second part will focus on more recent data on the identities of constituent compounds of SOM regarding molecular geochemistry.

PYROLYSIS-GC/MS[1] TECHNIQUE AND ITS STRENGTHS

The modern methods used to fractionate SOM into a small number of fractions can be divided into degradative methods involving chemolysis, thermolysis, or thermochemolysis, and non-invasive spectroscopic techniques such as nuclear magnetic resonance (NMR) spectroscopy. Degradative methods provide molecular-level information on specific organic compounds accessible to the degradative step, whereas spectroscopic techniques enlighten about the bulk composition of SOM (Guggenberger, 2006).

In thermolysis, a soil sample is heated according to a heating program, usually with a constant heating rate. A distribution, known as a thermogram, can be produced by measuring the weight loss as a function of temperature. The gases evolved can also be analyzed to determine the amount of carbon or mass lost at different temperatures by different kinds of spectroscopic methods. The thermograms often exhibit different peaks for labile and more resistant SOM (Bruun et al., 2010).

Pyrolysis, coined from the Greek-derived elements pyr "fire" and lysis "separating", along with GC/MS is an effective thermal technique in the study of pedogenetic processes (Abelenda et al., 2011), which uses thermal degradation to cleave bonds in the organic macromolecules. When the heat energy applied to the macromolecule is greater than the energy of specific bonds, those bonds will dissociate in a predictable and reproducible way. The smaller molecules generated by this bond-breaking are identified by the analytical tool and therefore bring an understanding of the original macromolecule and enables a sensitive and rapid characterization of organic constituents.

Pyrolysis-GC/MS is an instrumental technique that enables a reproducible characterization of the intractable and involatile macromolecular complexes found in virtually all materials in the natural ecosystem. It differs from GC/MS in the type of sample analyzed and the method by which it is introduced to the GC/MS system. Instead of the direct injection of a highly refined organic solution, a few amount of the original natural material (for example, soil, sediment etc.) is analyzed directly. The analysis is regularly preceded by an extraction of the sample with an organic solvent to remove any free, unbound components with a low molecular mass that would otherwise obscure analytical data pertaining to the high molecular mass components of interest. Extracted samples are then inserted into a quartz chamber in a pyrolysis unit that is then heated resistively in an oxygen free enivironment at a pre-set temperature for a number of seconds. This results in a heat mediated cleavage of chemical bonds within the macromolecular structures of interest producing a suite of low molecular weight chemical moieties of the composition of which is indicative of specific types of macromolecule (for example, lignin, cellulose, chitin etc.). This mixture of compounds is then swept onto the analytical column of the GC and GC/MS proceeds as normal.

In pyrolysis-gas chromatography (PyGC), the fragments generated by pyrolysis are passed through the GC for separation and identification. Frequently, the major peaks in the resulting chromatogram (pyrogram) are easily identifiable and give direct structural information about the material being pyrolyzed. Many polymers, polystyrene and polymethyl methacrylate for example, generate significant quantities of the monomer used in producing the polymer. For other materials, the pyrograms are more complex and serve as "fingerprints" which may be used to distinguish related materials for identification or for quality control.

Most methods used to identify or quantify individual organic compounds require the target chemical be extracted from a solid or liquid matrix. This is often done using a liquid or supercritical fluid extraction. Solvents, particularly basic solutions, can partially oxidize, or otherwise modify the organic matter being studied. In addition, organic molecules can only be identified by GC/MS if they remain in an inert gas stream at 300 °C or less. Most organic matrices in the environment are composed of material too large to volatilize at 300 °C and cannot be analyzed by GC/MS. Pyrolysis will thermally extract intact molecules or crack large molecules into fragments that can be separated and identified by GC/MS. As such, pyrolysis is an alternative way to "extract" organic matter from complex matrices (White et al., 2004). Pyrolysis is not without its limitations, however. Thermal secondary reaction causes considerable modification of the organic compound, which may bias the interpretation of the pyrolysis products with respect to their mother compounds (Saiz-Jimenez, 1994).

In analytical pyrolysis, low molecular weight compounds and some of the medium molecular weight compounds could be detected at moderate temperatures;

[1] Pyrolysis-Gas Chromatography/Mass Spectrometry

however, non-volatile compounds of low thermal stability, in particular organic compounds of high molecular weight, could be analyzed at higher temperatures of above 500°C (Cabane et al., 2001). Pyrolysis volatile products are chromatographically separated (GC) and identified (MS) based on their ratios of mass to charge (m/z).

The thermal energy required to volatilize a compound indicates bond strength. Thermograms of volatilization, therefore, give useful information on the strengths of chemical bonds within the organic macromolecules[2], or between organic materials and minerals[3] (Balser, 2005). Shifts in such thermograms to higher pyrolysis temperatures indicate stronger bonds (Schulten and Leinweber, 1999).

MAJOR COMPOUNDS IDENTIFIED BY PYROLYSIS-GC/MS

Analysis of the gaseous mixtures that are produced from the pyrolysis of a given organic compound provides a "Fingerprint" [nature and abundance (GC-MS) of released gases], which can be compared to laboratory data bases. Thus, from this fingerprint, we can identify the complex molecules which constitute the pyrolysed samples. Hence, pyrolysis-GC/MS aids the molecular characterization of microbial and plant-derived biomass, and generate valuable data on the degradation and conservation rates of organic debris. Furtheremore, MS of the pyrolysis products can be used to make thermograms of compound classes such as carbohydrates, lignin monomers, lignin dimers, alkyl aromatics, lipids and N-containing compounds, thus, giving rise to a different distribution for each compound class (Bruun et al., 2010).

Based on chemical similarities, the following group of compounds identified by pyrolysis-GC/MS are reviewed in the present paper, including (i) aliphatics, fatty acids and sterols, (ii) carbohydrates, (iii) lignin, (iv) aromatic compounds and polycyclic aromatic hydrocarbons (PAHs) and (v) N-containing compounds.

Aliphatics, fatty acids and sterols

Aliphatics mainly originate from plant waxes and suberin present in tree roots and bark (Lorenz et al., 2007). Aliphatic compounds refer to n-alkanes, n-alkenes, n-methylketones and n-alcohols. The long-chain n-alkanes and n-alkenes are common in fresh litter and in soils with low microbial activity (Abelenda et al., 2011). Short-chain aliphatics can be attributed to microbial products. They are either fragments of microbial lipids, or parts of longer chains, degraded by microorganisms. In addition, long-chain odd C-numbered methylketones probably come from microbially induced oxidation of alkanes (Buurman et al., 2007b).

Fatty acids characterized by pyrolysis are mainly derived from plants, but they can also be of microbial origin (Abelenda et al., 2011). Although C12 fatty acid is probably microbial, fatty acids from plants have a predominance of even over odd chain lengths (Buurman et al., 2007c). The C16 and C18 fatty acids are derived from waxes or lipid biomacromolecules such as cutan and cutin from higher plants. The iso- and anteiso- odd C-numbered fatty acids (mainly C15 and C17), commonly considered to be bacterial biomarkers (Buurman et al., 2007b; Buurman et al., 2007c).

Sterols are part of cell membranes and are generally easily decomposed in aerobic environments. Plant sterols are structures with four rings largely consisting of 27 to 30 C atoms, while microbial sterols (hopanoids) have five rings and up to 35 C atoms. Their identification with pyrolysis-GC/MS is problematic, although quite often a group can be identified (Abelenda et al., 2011).

Carbohydrates

Levoglucosan (anhydroglucosan), cyclopentenone, furan and furaldehyde are some identifiable carbohydrate markers by pyrolysis-GC/MS. Although pyrolysis products of carbohydrates can also have a plant or microbial origin, in most soils polysaccharide products are only major compounds in pyrolysates of fresh plant material.

Levoglucosan and cyclopentenone are largely associated with virtually fresh plant-derived matter, especially when found in combination with C3 lignin and small contributions of N-compounds. On the other hand, the occurrence of furans, furaldehydes and anhydrosugars in many soils suggests that these hardly biodegraded polysaccharides originated from residual plant polysaccharides and newly formed microbial polysaccharides, especially when they dominate over levosugars in a given pyrolysate (Buurman et al., 2007b; Abelenda et al., 2011). Kaal et al. (2007) interpreted low furans to levoglucosans ratio as indicating that the majority of carbohydrates is expected to be well preserved.

Lignin

Groups that mainly represent primary SOM are the aliphatics, fatty acids, ketones and lignins. Lignin is commonly derived from wood and also is the second most abundant component of cellular walls after polysaccharides. Within the lignin group, microbial decay preferentially reduces the contribution of syringols. Syringyl-lignin is known to be less stable than guaiacyl-lignin and there is a relative accumulation of guaiacol fragments (Buurman et al., 2007c).

[2]Organic-organic bonds
[3]Organic-mineral bonds

Lignin degradation includes oxidation of the C3 side-chain, which is recognized in pyrolysates. Consequently, the relative proportion of C3-guaiacols to total guaiacols (C3G/GT) and C3-syringols to total syringols (C3S/ST) is commonly utilized as an index of degradation. In pyrolysates, the vinyl-containing phenols and methoxy-phenols are indicative of fresh lignin (Abelenda et al., 2011). In addition, the pattern of lignin monomers can be used to identify the primary resource of SOM; in contrast to angiosperm lignin, gymnosperm lignin does not contain syringyl units (Guggenberger, 2006).

Phenols originated largely from lignin, perhaps with a small contribution from carbohydrate-derived phenols (Calvelo Pereira et al., 2011). Pyridines can also be formed by microbial decomposition of plant lignins and other phenolics in the presence of NH_3 (Buurman et al., 2007b).

Aromatic compounds and polycyclic aromatic compounds

Aromatic (benzene, toluene, alkylbenzenes, methylbenzenes, dimethylbenzenes, ethylbenzenes and styrenes) and PAH pyrolysis products may be ascribed to a defined origin only in conjunction with other compounds. Thus, aromatic pyrolysates are ascribed to proteins, and specifically to microbial material, when pyridine and toluene are found simultaneously; while a combination of benzene, toluene and PAHs indicates the presence of charred material. PAHs can usually be ascribed to aromatization and condensation reactions upon burning (Abelenda et al., 2011). In other words, the presence of PAHs in SOM pyrolysates is usually considered as testimony of burning (Naafs, 2004; Ross et al., 2005; Rumpel et al., 2007), although small amounts of these may be secondary pyrolysis products of fatty acids (Saiz-Jimenez, 1995).

A significant proportion of the benzenes and PAHs might also originate from charred carbohydrates (Calvelo Pereira et al., 2011). A-type humic acid has a relatively large C-content and large aromatic and carboxyl contents. There is now increasing evidence that much of this aromaticity is due to accumulation of charred plant remains from periodic burning of the vegetation.

Buurman et al. (2007b) conjectured that the naphthalenes Pa4-6 may therefore be pyrolysis products of microbial organic matter. Additionally, series of n-alkylbenzenes (C_4-C_{17}) probably result from incomplete combustion upon burning (Abelenda et al., 2011).

Alkylaromatics are considered a result of the humification of primary plant materials and microbial metabolites to form the skeletons of humic substances (Schulten and Leinweber, 1999). Pyrolysis of fatty acids with one double bond results in traces of toluene and ethylbenzene and such unsaturated fatty acids could be the origin in soil of hydroaromatic and alkylaromatic structures (Buurman et al., 2007b).

Sollins et al. (1996) mentioned similar related behavior in alkyls and aromatics regarding SOM-decomposition. Both fractions are considered relatively recalcitrant to microbial decay (Helfrich et al., 2006; Lorenz et al., 2007).

N-containing compounds

The study of N-containing compounds comprises the identification of specific protein sources as well as the discrimination of N-compounds of either plants or microbial biomass origins.

Although N-compounds (pyrrole, pyridine, benzonitrile, acetobenzonitrile, indole, methylindole and diketodipyrrole) are, in general, most probably drived from amino acid and amino sugar moieties, there is insufficient knowledge of their more specific sources. Furthermore, whether these moieties are still present as intact polypeptides or as modified entities is as yet unknown. Additionally, the presence of rather unusual isomers of some nitrogen-containing products could imply that these are derived from chemically modified biomolecules formed in the soil (Van Bergen et al., 1998).

A large number of N-compounds are from microbial SOM, but others can be both of microbial and plant origin. Large amounts of N-compounds accompanied by carbohydrate pyrolysis products other than levosugars point toward the presence of large amounts of microbial SOM (Abelenda et al., 2011).

According to Schulten and Schnitzer (1998), heterocyclic N-compounds such as pyrroles, pyrrolidines, pyridines, pyranes, and pyrazoles are significant components of the SOM. The pyridines and pyrroles are probably of microbial origin, while the indoles can be predominantly plant derived (Buurman et al., 2007c). Indeed, pyridines can be formed by microbial decomposition of plant lignins and other phenolics in the presence of NH_3. Pyrroles may derive from amino acids such as proline and hydroxyproline. Substituted pyrroles are formed readily upon pyrolysis of porphyrin, which is an essential component of the chlorophyll molecule in terrestrial plants. Indoles are pyrolysis products of tryptophane, an amino acid that occurs in proteins. Furthermore, imidazoles are pyrolysis products of histidine, an amino acid that is essential for tissue growth and repair, but also thermal degradation of grass and soil microorganisms forms imidazoles. Benzonitrile is derived from aromatic amines as well (Buurman et al., 2007b). The work by Buurman and Roscoe (2011) suggests that benzonitrile is presumably mainly microbial.

Acetamides are probably pyrolysis products of chitin. In fact, chitin is a polymer of N-acetylglucosamine[4] that is relatively easy to degrade (Buurman et al., 2007b), and its pyrolysis products are not commonly found in soils. Nonetheless, a large number of chitin pyrolysis products

[4]An N-containing polysaccharide

were obtained from the allophanic soils of the Azores (Buurman and Nierop., 2007a) and Costa Rica (Buurman et al., 2007b).

This implies that in these soils a significant part of the SOM could derive from fungi and from mesofauna such as arthropods and nematodes. In other words, it appears that decay of plant-derived organics is extremely efficient and that microbially and mesofauna-derived moieties represent a large fraction of the SOM.

Diketodipyrrole may be derived from both plant and microbial sources. Several compounds with nitrile functional groups have been also detected. The origin of these compounds remains unknown, but they may be related to charring (Abelenda et al., 2011).

According to Buurman et al. (2007c), 3-methylpyridine, benzeneacetonitrile and indole were associated with fresh litter, while other N-compounds were more associated with strongly degraded SOM, that is, with microbial activity. Nitrogen compounds showed a clear loss of pyridine compounds, benzonitrile, and diketodipyrrole, while a cluster of N-compounds increased. It is likely that the shift indicates a change from predominantly plant-derived nitrogen compounds towards microbial ones.

Low molecular weight amino acids split into small and volatile compounds that generally elute in the first large bulk of the pyrogram. These substances are poorly identifiable and quantifiable (Kaal et al., 2007).

CONCLUSION

In order to achieve a profound understanding of complex organic matrices, applying Pyr-GC/MS technique as an innovative approach in environmental science and engineering is inevitable and indispensable. The superiority of this technique is its capability to help interpret the chemical composition of SOM in terms of sources, decomposition and stability. Consequently, SOM fingerprinting by means of Pyr-GC/MS technique is indubitably the most effectual method to measure SOM chemistry.

The challenge of this technique until recently has been to identify the precursor molecules of the original SOM, which clarifies the limited use of Pyr-GC/MS technique in soil science. However, the fresh rapid progression of the identification of these compounds decipher the genesis of SOM.

Moreover, a sophisticated understanding of SOM chemistry and dynamics will help better modelling of greenhouse gases emissions such as methane and carbon dioxide.

Ultimately, analytical pyrolysis has major limitations that could produce misleading results if disregarded (for example, the effect of the mineral matrix). Nevertheless, within its limitations, analytical pyrolysis has numerous applications and is a useful and valuable tool.

RECOMMENDATIONS

Despite the mentioned uses of Pyr-GC/MS technique, it seems this technique has more relationships with other aspects of soil science. Some recommendations, therefore, are proposed to be the main subject of future studies in this regard.

1) Characterize the type of SOM by pyrolysis-GC/MS that promotes microbial activity and therefore soil denitrification
2) Investigate the effect of SOM on nitrogen cycling especially on N_2O emissions
3) Using the derived data in process-based models such as Daycent to simulate N_2O fluxes with regard to the nature of SOM

REFERENCES

Abelenda MS, Buurman P, Camps Arbestain M, Kaal J, Martinez-Cortizas A, Gartzia-Bengoetxea N, Macías F (2011). Comparing NaOH-extractable organic matter of acid forest soils that differ in their pedogenic trends: A pyrolysis-GC/MS study. Eur. J. Soil Sci. 62:834-848.

Balser TC (2005). Humification, In D. Hillel, et al., eds. Encyclopedia of Soils in the Environment, Vol. 2. Elsevier, Oxford UK. pp. 195-207.

Bruun S, Argren GI, Christensen BT, Jensen LS (2010). Measuring and modeling continuous quality distributions of soil organic matter. Biogeoscience 7:27-41.

Buurman P, Nierop KGJ (2007a). NaOH and Na-$Na_4P_2O_7$-extractable organic matter in two allophanic volcanic ash soils of the Azores Islands - quantified pyrolysis-GC/MS data and factor analysis. In: Soils of volcanic regions in Europe / Stoops, G., Garcia-Rodeja, E., Arnalds, O., Bartoli, F., Buurman, P., Oskarsson, H. Berlin Heidelberg: Springer-Verlag.

Buurman P, Peterse F, Almendros Martin G (2007b). Soil organic matter chemistry in allophanic soils: A pyrolysis-GC/MS study of a Costa Rican Andosol catena. Eur. J. Soil Sci. 58:1330-1347.

Buurman P, Roscoe R (2011). Different chemical composition of free light, occluded light and extractable SOM fractions in soils of Cerrado and tilled and untilled fields, Minas Gerais, Brazil: A pyrolysis-GC/MS study. Eur. J. Soil Sci. 62:253-266.

Buurman P, Schellekens J, Fritze H, Nierop KGJ (2007c). Selective depletion of organic matter in mottled podzol horizons. Soil Biol. Biochem. 39(2):607-621.

Cabane M, Coll P, Rodier C, Israël G, Raulin F, Sternberg R, Niemann H, Mahaffy P, Jambon A, Rannou P (2001). In-situ inorganic and organic analysis (Pyr/CD-GC/MS) of the Martian soil, on the Mars 2005 mission, Planet. Space Sci. 49:523-553.

Calvelo Pereira R, Kaal J, Camps Arbestain M, Pardo Lorenzo R, Aitkenhead W, Hedley M, Macias F, Hindmarsh J, Macia-Agullo JA (2011). Contribution to characterisation of biochar to estimate the labile fraction of carbon. Organic Geochem. 42:1331-1342.

Golchin A, Oades JM, Skjemstad JO, Clarke P (1994). Soil structure and carbon cycling. Aus. J. Soil Res. 32:1043-1068.

Guggenberger G (2006). Organic matter structure and characterization, In R. Lal (Ed.) Encyclopedia of Soil Science, Second Edition. published on-line, Taylor and Francis Group, Boca Raton, FL. pp 1203-1209.

Guggenberger G, Zech W, Schulten HR (1994). Formation and mobilization pathways of dissolved organic carbon: Evidence from chemical structural studies of organic carbon fractions in acid forest floor solutions. Org. Geochem, 21:51-66.

Helfrich M, Ludwig B, Buurman P, Flessa H (2006). Effect of land use on the composition of soil organic matter in density and aggregate fractions as revealed by solid-state C-13 NMR spectroscopy.

Geoderma 136 (1-2):331-341.
Kaal J, Baldock JA, Buurman P, Nierop KGJ, Pontevedra-Pombal X, Martinez-Cortizas A (2007). Evaluating pyrolysis-GC/MS and 13C CPMAS NMR in conjunction with a molecular mixing model of the Penido Vello peat deposit, NW Spain. Organic Geochemistry, 38(7):1097-1111.
Kay BD, Silva da, Baldock AP (1997). Sensitivity of soil structure to changes in organic carbon content: Predictions using pedotransfer functions. Canadian J. Soil Sci. 77(4):655-667.
Lorenz K, Lal R, Preston CM, Nierop KGJ (2007). Strengthening the soil organic carbon pool by increasing contributions from recalcitrant aliphatic bio(macro)molecules. Geoderma 142(1-2):1-10.
Naafs DFW (2004). What are Humic Substances? A Molecular Approach to the Study of Organic Matter in Acid Soils. Ph.D. Thesis. University of Utrecht. pp. 99-106.
Ross AB, Junyapoon S, Jones JM, Williams A, Bartle KD (2005). A study of different soots using pyrolysis-GC-MS and comparison with solvent extractable material. J. Anal. Appl. Pyrol. 74:494-501.
Rumpel C, Gonzalez-Pérez JA, Bardoux G, Largeau C, Gonzalez-Vila FJ, Valentin C (2007). Composition and reactivity of morphologically distinct charred materials left after slash-and-burn practices in agricultural tropical soils. Organic Geochem. 38:911-920.
Saiz-Jimenez C (1994). Analytical pyrolysis of humic substances: pitfalls, limitations and possible solutions. Environ. Sci. Technol, 28:1773-1780.
Saiz-Jimenez C (1995). Production of alkylbenzenes and alkylnaphthalenes upon pyrolysis of unsaturated fatty acids. Naturwissenschaften 81:451-453.
Schulten H R, Leinweber P (1999). Thermal stability and composition of mineral-bound organic matter in density fractions of soil. Eur. J. Soil Sci. 50:237-248.
Schulten HR, Schnitzer M (1998). The chemistry of soil organic nitrogen: A review. Biol Fertil Soils. 26:1-15.
Sollins P. Homann P, Caldwell, BA (1996). Stabilization and destabilization of soil organic matter: Mechanisms and controls. Geoderma 74(1-2):65-105.
Swift RS (1996). Organic matter characterisation. In Methods of Soil Analysis. Part 3. Chemical Methods; Sparks, D.L., Ed.; Soil Science Society of America: Madison, WI. pp. 1011-1069.
Van Bergen PF, Nott CJ, Bull ID, Poulton PR, Evershed RP (1998). Organic geochemical studies of soils from the Rothamsted Classical Experiments - IV. Preliminary results from a study of the effect of soil pH on organic matter decay. Organic Geochem. 29:1779-1795.
White DM, Sarah Garland D, Beyer L, Yoshikawa K (2004). Pyrolysis-GC/MS fingerprinting of environmental samples. J. Anal. Appl. Pyrol. 71:107-118.

2

Effects of three tree species on microclimate and soil amelioration in the central rift valley of Ethiopia

Agena Anjulo Tanga[1] , Tilahun Fromssa Erenso[2] and Bekele Lemma[3]

[1]Department of Plant Science, College of Agricultural Sciences, Arba Minch University, P. O. Box: 21, Ethiopia.
[2]Ziway Soil Laboratory, Oromia Institute of Agricultural Research, Addis Ababa, Ethiopia.
[3]Wondo Genet College of Forestry and Natural Resources, Hawassa University, Ethiopia.

The effect of *Balanites aegyptiaca, Acacia tortilis* and *Acacia seyal* on soil fertility and microclimate was studied in the central rift valley of Ethiopia. Sampling was done in randomized complete blocks at 1/3 radius, 2/3 radius and at crown edge radius from the tree base and control was maintained at an open area. All soil and microclimatic parameters observed under the canopy distances were significantly different from the open control ($P<0.05$) except soil texture. Soil bulk density increased from the trees base to open field. Available Nitrogen under crown of *A. tortilis* in all crown radii was much higher than that of *B. aegyptiaca* and *A. seyal*; available Potassium was significantly higher under *B. aegyptiaca* and *A. tortilis* at their 1/3 and 2/3 crown radii. *B. aegyptiaca* at 1/3 crown radius had significantly higher level of available Phosphorus. Cation exchange capacity (CEC) was significantly higher under the 1/3 and 2/3 crown radii of all the involved trees. Soil moisture decreased whereas relative illumination, soil and below canopy temperatures showed increasing trend towards the crown periphery revealing the significant canopy cooling effect. Inclusion of higher number of these trees can reduce adverse climatic effects in the semi-arid valley and supplement organic source of nutrients.

Key words: *Acacia seyal, Acacia tortilis, Balanites aegyptiaca,* microclimate modification, soil amelioration.

INTRODUCTION

Low soil fertility is the most fundamental cause for low agricultural productivity, food insecurity, low income and poverty in Ethiopia (Abebe, 2006). Maintenance of soil quality is considered essential for ensuring sustainable land use. Hence, land resource management must aim at soil conservation (Parysow, 2001). Loss of soil quality is explained through increased bulk density, reduced inorganic matter content and availability of soil nutrients (Brenner, 1994). Agroforestry can be a viable option to alleviate the degradation and loss of soil fertility from the agricultural fields.

The central rift valley in Ethiopia is being noticed for a shift in the use of land from dense woodland with palatable pasture to a farm land with scattered trees for growing agricultural crops to feed the growing population. The system is described as agroforestry parkland where naturally regenerated and scattered individual trees occur in cultivated fields. The scattered trees provide ecological services such as soil fertility and microclimate amelioration (Tomlinson et al., 1995; Boffa, 1999; Bayala et al., 2002; Teklehaimanot, 2004).

Parkland agfroforestry is an important land use system that has a positive influence on maintaining soil fertility, mainly due to the tree component (Kamara and Haque, 1992; Mulongey and Merck, 1993). The effect of the tree component on soil can be seen by comparing soil properties under individual tree canopies with surrounding tree-less areas. According to Tilahun (2007), *Faidherbia albida* has shown a 50 to 100% increase in organic matter and nitrogen under the canopy. In semi-arid climates, it is common to find higher soil organic matter and nutrient content under tree canopies than in adjacent open land and cropland (Rao et al., 1998). The presence of trees on farmland adds nutrients to the soil through biological nitrogen fixation and efficient nutrient cycling (Tadesse et al., 2002). This is an inexpensive and environmentally safe organic N source that has the potential of meeting nitrogen demand, the most important and scarce agricultural nutrient element (Nair, 1984).

Trees influence microclimate and soil property through organic matter accumulation and canopy produced shade which reduces evaporation from the soil surface and modifies air temperature extremes (Hugues and Philippe, 1998). Tree roots hold soil in place and tree crown reduces the intensive force of rainfall and also acts as physical barrier in controlling runoff. The contribution of *Acacia tortilis* and *Balanites aegyptiaca* to soil fertility improvement in Afar region of Ethiopia was documented by Abule et al. (2005). The tree species (*A. tortilis, B.aegyptiaca* and *Acacia seyal*) taken up for this study, are generally considered valuable species for soil improvement and conservation (Nyberg and Högberg, 1995; Raddad and Luukkanen, 2007). Thus, the present study was initiated with the objective to assess and compare the farm trees potential in site and soil amelioration in the semi-arid environment.

MATERIALS AND METHODS

Description of the study area

The study was conducted in Arsi Negelle district, which is located in West Arsi zone of Oromia, a regional state of Ethiopia. The district is 210 km south of Addis Ababa at 38° 30′ to 38° 40′ E and 7°25′ to 7° 40′N. It is a part of the Ethiopian central rift valley, covering an area of 1400 km^2 (Efrem et al., 2009). The topography of the area is flat to undulating. According to local climatic classification, the altitude ranges from 1610-1680m.a.s.l is categorized as dry Weyna Dega/savannah (Azene et al., 2007). The annual rainfall of the site is between 500 and 835 mm, characterized by high intensity and short duration showers occurring mainly from June to August, sometimes extending up to September. The area is known for its high temperature with little variation between dry and rainy season. The mean monthly maximum temperature is 28°C and minimum temperature is 13.8°C.

The soil is dark brown to dark greyish brown Vitric Andosol and The surface soil, as Murphy (1968) explained, is generally loose (low bulk density, friable-coarse sandy loam). Soil pH ranges from 7.3 to 8.5 (alkaline) while organic matter varies between 2.93 and 3.10%. The soil is deficient in phosphorus (< 12.4 kg ha^{-1}), and to a lesser extent, also in nitrogen (< 272 kg ha^{-1}) as indicated by FAI (1977). The vegetation of the area is classified mainly as *Acacia-Balanities* vegetation type (Nazereth, 1998) with some thorny shrubs of *A. tortilis* and *A. seyal.* Crop production is primarily rain fed and livestock rearing is the mainstay of livelihood of the local population. The major crops grown are maize, wheat, *teff* and barley under parkland agrisilviculture system.

Experimental details

Three tree species namely *B. aegyptiaca* (L.) Del., *A. tortilis* (Forsk.) and *A. seyal* (Del.) were taken.

Distances from tree base were: at 1/3 of tree crown radius; 2/3 of tree crown radius; crown edge; and open control away from canopy/tree root influence. Soil physico-chemical properties and some microclimatic changes under tree crown were then studied at 0 at 20 cm soil depth at three distances from tree base and one open control common for all the three tree species. The experiment was conducted in randomized complete blocks with three replications.

Tree selection and soil sampling

Tree selection

Nine scattered trees, three from each species were randomly selected in the blocked area based on their relative similarities in diameter at breast height (DBH) and total height. There were dissimilarity in crown diameter and crown area measured (Tables 1 and 2) for the tree species involved, hence, variables were studied at 1/3, 2/3 and crown edge of the individual trees so as to enable comparison between species.

Soil sampling

Three concentric circles were drawn under each tree crown at 1/3 crown radius, 2/3 crown radius and edge of the crown. Additionally, samples were also taken from the blocks in the open control area away from tree effect. Aiming for a representative soil samples, four sub-composites from each direction under nine trees and four distances including the open control area were collected. Hence, thirty six samples which consisted of 144 sub-composites were analysed for soil fertility.

Data collection

Data on physico-chemical properties of soil

The distribution of different particle size fractions for soil texture evaluation was determined by the modified Bouyoucos method using a hydrometer after destruction of organic matter with 30% of H_2O_2 and using 5% sodium hexametaphosphate (Calgon) as a dispersing agent (Day, 1965; Gee and Bauder, 1989). Soil bulk density (g/cm^3) was determined using the non-destructive core-volume method by dividing the weight of oven dried soil in the core (g) to the volume of the dry soil (cm^3). Soil pH was determined in water at a soil-water suspension ratio of 1:2.5 after stirring for 2 h; a reading was made using combined electrode method (Jackson, 1973). Similarly, electrical conductivity was measured from the same soil-water suspension prepared for pH determination, using a conductivity meter at 25°C (Jackson, 1973).

Table 1. Mean height, DBH, and crown area distribution of the selected *Balanites aegyptiaca, Acacia tortilis* and *Acacia seyal* tree species.

Tree species	Height (m)	Std. error	DBH (cm)	Std. error	Crown area (m^2)	Std. error
B. aegyptiaca	8.33	0.33	37.44	0.46	90.05	11.54
A. tortilis	12	0.58	39.23	0.30	236.55	17.79
A. seyal	8.35	0.43	30.89	0.55	126.65	13.61

Table 2. Mean distance in meter of crown radii of the studied tree species.

Tree species	Crown radius	Mean distance in meter	Std. error
B. aegyptica		3.07	0.19
A. tortilis	1/3 crown radius	5.01	0.20
A. seyal		3.65	0.19
B. aegyptica		4.35	0.27
A. tortilis	2/3 crown radius	7.08	0.28
A. seyal		5.17	0.27
B. aegyptica		5.33	0.33
A. tortilis	Crown edge	8.67	0.35
A. seyal		6.33	0.33

Available nitrogen was assessed by Alkaline Potassium Permanganate method by using the Kjeldhal distillation unit where the digest was distilled in the presence of sodium hydroxide and distillate trapped in 0.02N H_2SO_4. The ammonia liberated was titrated with 0.02 N NaOH to assess the available nitrogen content of the samples (Subbiah and Asija, 1956).

Available phosphorus was determined by the Olsen method in which samples were extracted with sodium bicarbonate solution at pH 8.5 and the P content in the extracted samples was determined calorimetrically after they developed the blue colour with ammonium molybdate in the presence of ascorbic acid as a reducing agent (Olsen et al., 1954).

Available potassium was measured by the Neutral Ammonium Acetate extraction method (Merwin and Peech, 1951). The extracted sample was estimated by flame-photometer. Soil organic carbon was determined by Rapid Titration method in which the organic matter was oxidised by potassium dichromate in the presence of concentrated H_2SO_4.The excess potassium dichromate was back titrated with ferrous ammonium sulphate using diphenylamine indicator (Walkley and Black, 1934).

The cation exchange capacity (CEC) of the samples was estimated by 1 M Ammonium Acetate extraction method as outlined by Reeuwijk (2002).

Modification of microclimatic conditions

Sampling and all the measurements in the microclimate of the area were carried out in the midday of the study period. Soil moisture was determined gravimetrically where the samples of fresh soil were weighed and oven-dried at 105°C for 24 h. The percentage of water present in the soil was calculated as the weight difference between field soil and oven dried soil, divided by the weight of the oven dried soil, multiplied by 100. Soil temperature was measured using a Mercury Soil Thermometer whereas below canopy temperature by using Digital Infrared Thermometer (Model AG–42). The relative illumination for the respective distances under canopy was calculated in relation to open field by using Lux meter. The data was collected for three months during the main rainy season, that is, in the months of May, June and July; reason being foliage regrowth and full crown cover is achieved during this time.

Data analysis

The data collected from laboratory analysis on soil physico-chemical properties and field observation on microclimatic changes was subjected to two way analysis of variance (ANOVA). The significant differences between means were determined by LSD at 5% probability level.

The values for microclimatic study were the averages of the daily observations at midday of the particular month.

RESULTS AND DISCUSSION

Evaluation of the tree species potential in soil amelioration

Soil texture

The difference in the treatment effects of the trees on the mean values of soil particle fractions of sand, silt and clay at different distances from the trees base at 0 to 20 cm soil layer was not significant (Table 3). As a result, the particle size distribution of the soils at the measured soil layer of all treatments was categorized under sandy loam textural class indicating no significant effect of the tree

Table 3. Particle size distribution as affected by tree species and distance from tree bole.

Species	Crown radius	Soil particle distribution			Textural class
		Sand%	Silt%	Clay%	
	1/3 crown radius	63.00	22.33	14.66	Sandy loam
B. aegyptiaca	2/3 crown radius	61.00	24.66	14.33	Sandy loam
	Crown edge	63.00	24.33	12.66	Sandy loam
	1/3 crown radius	66.33	21.00	12.66	Sandy loam
A. tortilis	2/3 crown radius	66.33	21.00	12.66	Sandy loam
	Crown edge	65.66	21.00	12.66	Sandy loam
	1/3 crown radius	60.33	23.00	16.66	Sandy loam
A. seyal	2/3 crown radius	59.66	25.00	15.33	Sandy loam
	Crown edge	59.00	24.33	16.66	Sandy loam
	Open control	63.33	22.33	13.66	Sandy loam
LSD 0.05		7.29 (NS)	6.21 (NS)	5.19 (NS)	

species involved in soil texture. The finding is in agreement with Akpo et al. (2005) who reported no significant influence of tree species on soil texture modification. Soil texture is mainly dependent on parent materials of the soil. It affects soil quality; influences the nutrient supplying ability of soil solids, soil moisture and air relations, and root development (Spur and Barnes, 1980).

Soil pH

Analysis of soil pH revealed that it was not significantly affected by the different treatment combinations of tree species and the distances from tree base (Table 4). It was observed to be slightly above neutral in all the treatments; however, in semi-arid climates such as the area under study, pH at this range is expected and normal. For the growth of the most predominant crop of the area under study which is maize, the pH value is slightly higher; yet, would not be deleterious and need gypsum application. Agbede (2008) and Hailemariam et al. (2010) reported lower pH values under *B. aegyptiaca* at Goblel and Korbebite sites from northern Ethiopia. Higher organic matter and clay content coupled with sufficient soil moisture may have led to the lower pH values.

Electrical conductivity

The maximum and significantly higher mean value of electrical conductivity was recorded under *B. eagyptica* at 1/3 and 2/3 crown radii whereas the mean values of electrical conductivity recorded under the other tree crown radii were much lower and statistically at par with the open control and one another (Table 4). The higher values at the inner and middle crown radii of *B. aegyptiaca* may be caused due to the relatively higher leaf biomass and fruit drop which upon decomposition release soluble nutrients to the soil. Hailemariam et al. (2010) also reported higher EC value under canopy than the open field of *B. aegyptiaca* at Limat site in northern Ethiopia.

Organic carbon in soil

Soil organic carbon under *A. tortilis* at 1/3 crown radius was recorded the highest; it was also statistically at par with and significantly high under tree crown of other treatments as compared to the open control area. However, *B. aegyptiaca* at the crown edge had lower value of organic carbon due to the narrower crown radius and its sparse crown edge cover which possibly yield less leaf litter at that distance (Table 4). Rao et al. (1998) argues that it is common to find higher soil organic matter and nutrient content under tree canopies than in adjacent open land and cropland in semi-arid climates. For soil under cultivation, mineralization is favoured over humification (Brinson *et* al., 1980; Toky et al., 1989). The major avenues for addition of organic matter to the soil where trees are growing, is through litter that is, leaves, twigs, branches, fruits, and so on (Quideau and Bockheim, 1996).

Available Nitrogen

Available nitrogen was significantly influenced by the combined treatments of tree species and distances where the maximum value was recorded under *A. tortilis* at 1/3 crown radius which however, was statistically at par with its mean value at 2/3 crown radius and crown

Table 4. Bulk density, pH, electrical conductivity and organic carbon as influenced by tree species and distance from tree bole.

Species	Crown radius	Bd gcm^{-3}	pH	EC (dS/m)	%OC
	1/3 crown radius	0.94[a]	7.68[a]	0.13[a]	2.49[a]
A. aegyptiaca	2/3 crown radius	1.02[b]	7.31[a]	0.13[a]	1.62[a]
	Crown edge	1.10[b]	7.58[a]	0.11[b]	1.35[b]
	1/3 crown radius	0.87[a]	7.63[a]	0.10[b]	2.83[a]
A. tortilis	2/3 crown radius	0.95[a]	7.66[a]	0.08[b]	2.24[a]
	Crown edge	1.02[b]	7.6[a]	0.08[b]	2.17[a]
	1/3 crown radius	0.97[a]	7.69[a]	0.12[b]	2.73[a]
A. seyal	2/3 crown radius	1.10[b]	7.74[a]	0.11[b]	2.14[a]
	Crown edge	1.11[b]	7.73[a]	0.10[b]	2.16[a]
	Open control	1.24[b]	7.66[a]	0.09[b]	1.32[b]
$LSD_{0.05}$		0.23	0.47(NS)	0.03	0.55

Different superscript alphabets within a column at a particular treatment represent significant difference from the control at P < 0.05

Table 5. Available Macronutrients and CEC as influenced by tree species and distance from tree bole.

Species	Crown radius	Soil variables			
		Av. N (kg/ha)	Av. P (kg/ha)	*Av. K (kg/ha)*	CEC (cmol Kg^{-1})
	1/3 crown radius	159.93[a]	14.34[a]	211.61[a]	34.31[a]
B. aegyptiaca	2/3 crown radius	139.02[b]	12.63[b]	188.26[a]	32.69[a]
	Crown edge	129.62[b]	10.29[b]	171.5[b]	31.69[b]
	1/3 crown radius	185.02[a]	12.86[b]	216.96[a]	37.55[a]
A. tortilis	2/3 crown radius	169.34[a]	12.63[b]	198.62[a]	33.29[a]
	Crown edge	164.11[a]	11.41[b]	189.09[a]	29.75[b]
	1/3 crown radius	157.84[a]	13.11[b]	153.96[b,bc]	34.65[a]
A. seyal	2/3 crown radius	139.02[b]	11.33[b]	145.26[b,bc]	32.69[a]
	Crown edge	137.02[b]	11.22[b]	136.99[c]	28.56[b]
	Open control	111.15[b]	9.06[b]	106.38[d]	23.83[c]
$LSD_{0.05}$		33.35	4.32	29.21	5.03

Different superscript alphabets within a column at a particular treatment represent significant difference from the control at P < 0.05

edge, *B. aegyptiaca* and *A. seyal* at their 1/3 crown radius (Table 5). The lowest value of available nitrogen was recorded at the open control which was preceded but statistically alike to the values under *B. aegyptiaca* and *A. seyal* at their 2/3 crown radius and crown edge. Despite the higher contribution of the tree species in available nitrogen content, the observed values in both the soil layers were categorized as "low fertility class" as per the ratings of FAI (1977). Total nitrogen and the mineral nitrogen flux had progressive decrease from the trunk to the canopy margin under the tree of *Acacia Senegal, B. aegyptiaca and Adansonia digitata* in semi-arid zones of northern Senegal (Young, 1989). Higher soil nitrogen content under tree canopies than away from tree influence was reported for all distances and soil depths under *F. albida*, *B. aegyptaica* and *Cordia africana* (Kamara and Haque, 1992; Akpo et al., 2005; Abebe Yadessa et al., 2009).

Available Phosphorus

Significantly higher amount of available phosphorus was recorded under *B. aegyptiaca* at 1/3 crown radius, which was followed by statistically different values under *A. seyal* and *A. tortilis* at their 1/3 crown radius. On the other hand, minimum value of available phosphorus was recorded at the open control away from tree influence which however, was statistically same to the values in the rest of the trees distance treatments (Table 5). Higher level of available phosphorus at the closest distance to the tree base can be ascribed to organic phosphorus input from leaf litter deposition and release at mineralization, higher microbial population stimulated by organic matter input which supported phosphorus solubilisation from fixation. However, the observed values near the tree base for available phosphorus put the soil under category of "medium fertility class" (FAI, 1977). The soil in the open control away from tree influence is deficient in available P content is in agreement with the findings of Fikru (1998). Soil phosphorus concentration decreased as the distance increased from the bole to the outside of the canopy (Kamara and Haque, 1992).

Available Potassium

The analysis indicated that the mean value of available potassium content was significantly higher under *A. tortilis* at 1/3 crown radius, which was followed by statistically same values under *B. aegyptiaca* at 1/3 and 2/3 0crown radii; and *A. tortilis* at all crown distances (Table 5). Lowest available potassium content was recorded at the open control which was preceded by the value under *A. seyal* at the crown edge (Table 5). The observed values under the trees' crown radii for available potassium are in the medium range (FAI, 1977). In soil amelioration study, high total nitrogen and potassium concentration was observed under canopy of *Parkia biglobosa* (Jacq.) Benth. in West Africa (Tomlinson et al., 1995). The reason behind the high concentration of available potassium under canopy than the open area might be high organic matter accumulation, decomposition and release in the soil (Brady and Weil, 2002).

Cation exchange capacity

The treatment combinations involving tree species and distances from tree base have significantly influenced the mean values of cation exchange capacity of the soil. Cation exchange capacity (CEC) was maximum (37.55 cmol kg^{-1}) under *A. tortilis* at 1/3 crown radius which however, was statistically similar to the mean values at its 2/3 crown radius; *B. aegyptiaca* and *A. seyal* at 1/2 and 2/3 crown radii. The minimum value of 23.83 cmol kg^{-1} cation exchange capacity was recorded at the open control which was preceded, but statistically similar to the values under *A. tortilis* and *A. seyal* at their crown edge (Table 5). Soil with CEC values above 25 cmol kg^{-1} is commonly taken as having a good nutrient holding capacity. The results depict the trees' influence with added organic matter to attain higher CEC values at their inner crown radii. Similar findings by Taddese et al. (2002) state the concentration of CEC under canopy of *Millettia ferruginea* to be higher than the open field in south Ethiopia. In the middle awash rift valley of Ethiopia, concentration of CEC under canopy of *Prosopis juliflora* was higher as compared to the open area with no clear difference in soil layers (Ameha, 2006).

Tree species relative effect on microclimate during the main cropping season

Soil moisture

Soil moisture is one of the factors that hinder crop success in the semi-arid rift valley. The valley has history of several crop failures due to erratic rains. Under canopy of the studied tree species it appeared to be significantly higher than the open control. The highest soil moisture recorded in the month of May was 11.3% under *B aegyptiaca* at the 1/3 crown radius which increased to 12.87% in June (Figure 1). Then due to the denser canopy cover, *A. tortilis* took over the maximum in July which was recorded 18.99%. The differences in soil moisture from the maximum at 1/3 crown radius of the tree species over the open control were 4.13, 4.55 and 7.87% in May, June and July respectively. The soil moisture, from the beginning of the first month of cropping season to the second and third month increased under each tree species. The reasons for this change were increased rainfall, decreased evapotranspiration and increased relative humidity in the valley. Similarly, Kessler (1992) reported that improved availability of water under tree canopies was due to decrease in actual evapotranspiration as well as better water infiltration. On the other hand, Akpo et al. (2005) observed that soil moisture under shade was not significantly different under *A. tortilis* and *B. aegyptiaca*, suggesting that there was no species effect on soil moisture content.

Soil temperature

The tree species had cooling effect with their aboveground influence on the soil under crown cover. The lowest temperature recorded was 30°C under *B. aegyptiaca* at 1/3 crown radius in the month of May. Later on it decreased under 1/3 canopy radius of *A. tortilis* to 28 and 25°C in the month of June and July respectively. About 6.11, 4.44 and 6.0°C reduction in soil temperature was achieved (Figure 2) over the maximum at the open control in May, June and July respectively. In a similar microclimate modification study, Lin (2007) observed that the amount of shade cover is directly related to the

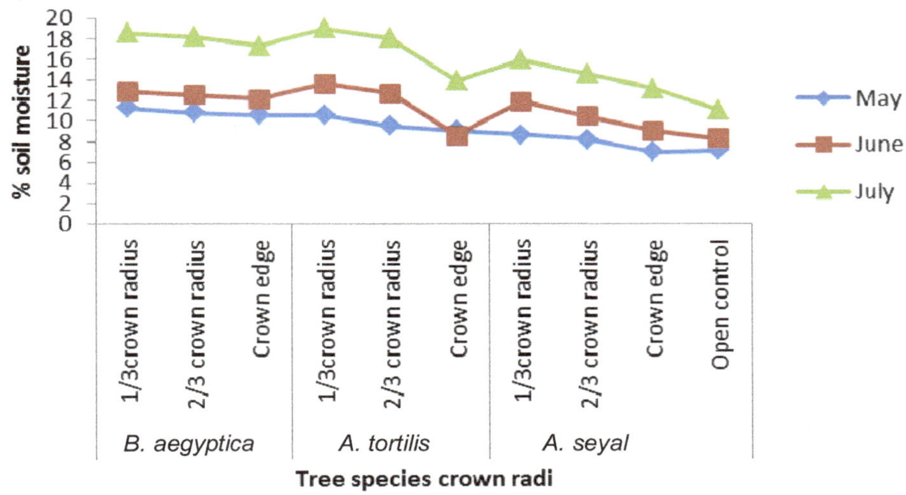

Figure 1. Soil moisture as influenced by tree species and distance away from tree base.

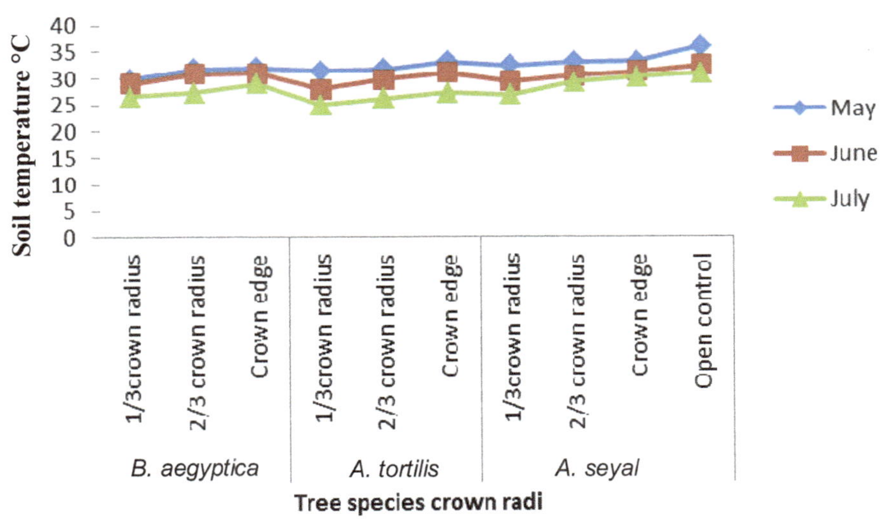

Figure 2. Soil temperature in °C as influenced by tree species and distance from tree base.

mitigation of variability in microclimate for the crop of interest. The use of agroforestry systems is an economically feasible way to protect crop plants from extremes in microclimate and should be considered a potential adaptive strategy for farmers in areas that will suffer from extremes in climate.

Below-canopy temperature

In the month of May erratic rains stimulated the regrowth of new flush leaves on the trees thereby influencing the below canopy temperature of the trees. *B. aegyptiaca* do not shed all the leaves in the dry season and moreover it develops edible fruits of pecan nut size during the dry season. Therefore, *A. tortilis* and *B. aegyptiaca* had significantly lower below canopy temperature in the studied months over the open control. Accordingly, mean temperature reductions over the open control during May, June and July were 6.3, 6.7 and 7.3°C at 2/3 crown radius of *A. tortilis*, 1/3 crown radius of *B. aegyptiaca* and 1/3 crown radius of *A. tortilis* respectively (Figure 3). However, below canopy temperatures at crown edge of *B. aegyptiaca*, at 2/3 crown radius and crown edge of *A. seyal* were statistically same to the open control. Narrower were the below canopy temperature differences at middle and peripheral distances of *A. seyal* to that of the open control. From the data, it is evident that *A. tortilis* is superior in reducing below canopy temperature at its nearest and middle canopy distances as compared

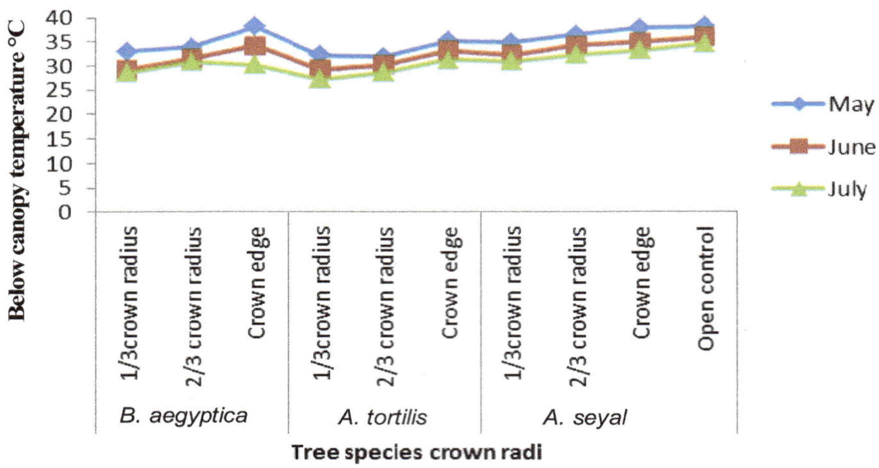

Figure 3. Below canopy temperature in °C as influenced by tree species and distance from tree base.

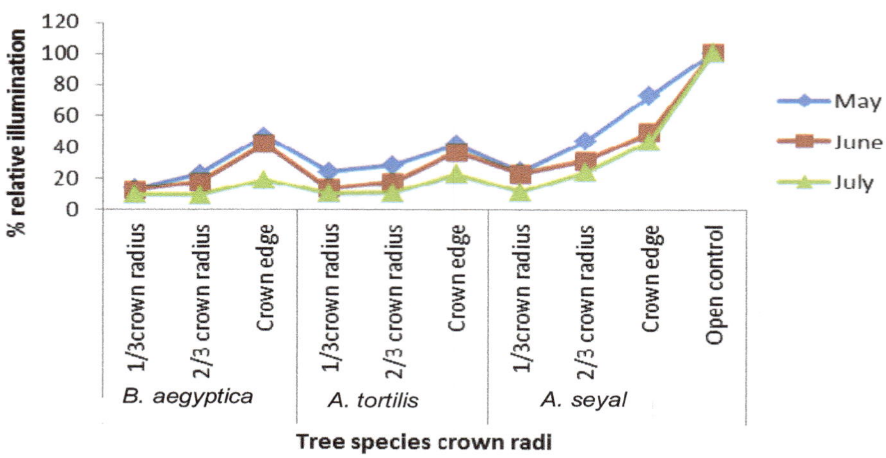

Figure 4. Percent relative illumination as influenced by tree species and distance from tree base.

to the other tree species. Akpo et al. (2005) recorded mean difference of 5°C between the shaded, that is, 1 m away from the tree trunk and in the open, that is, 16m away from trunk areas in the midday. The authors interestingly noted that early in the morning temperature under the shade was higher than in the open by about 1.5°C. The results indicated that trees reduce below canopy temperature fluctuations by decreasing the maxima and increasing the minima.

Percent relative illumination

The tree canopies during the month of May had transmitted much of the light received from the sun. When the deciduous trees regrow new leaves stimulated by the first rains, the canopy started to cover reducing the light transmission ratio in the following months. Relative illumination followed decreasing trend from May through June to July. In all the studied months, light transmission was the lowest through the crown of *B. aegyptiaca* at the 1/3 crown radius followed by *A. tortilis*. Percent relative illumination received at 1/3 crown radius under *B. aegyptiaca* was reduced by 86% in May, 87.9% in June and 91.44% in July over the open area. However, much higher intensity of relative illumination was received at the middle and peripheral distances of the trees species under study (Figure 4). Trees and shrubs may reduce solar radiation under canopies by 45 to 60% in the Sahelian zone and 85 to 95% in the Sudan savannah. The transmission of direct solar radiation through tree canopies at midday differed a little across species, that is, *A. tortilis* 21.3%, *B. aegyptiaca* 16.8% and *Ziziphus Mauritania* 20.1% (Akpo et al., 2005).

Conclusion

From the findings of the study of the effect of the trees on soil and microclimate, it can be concluded that (i) the tree species studied had no effect on soil texture; (ii) there is an improvement in the physical properties of the soil and also an increase in the content of plant available nutrients in the soil below tree canopies. *A. tortilis* contribute much higher level of available N than that of *B. aegyptiaca* and *A. seyal* in all crown radii; however, it was found to be in lower range to satisfy growth of associated crops and requires inorganic amendments. Available K was significantly higher under *B. aegyptiaca* and *A. tortilis*. *B. aegyptiaca* at 1/3 crown radius had significantly higher and superior level of available phosphorus. Similarly, CEC was significantly higher under the 1/3 and 2/3 crown radii of all the involved trees than in the open field. (iii) There is favourable effect on the microclimate as shown by a reduction in relative illumination, soil and below canopy temperatures as well as higher level of soil moisture below tree canopies; (iv) The semi-arid rift valley demands higher tree density than it currently has, to have broader area canopy cover to avert crop failure and ensure crop success, reduce livestock heat stress and improve food security in the region.

Conflict of Interest

The authors have not declared any conflict of interests.

REFERENCES

Abebe N (2006). Soil fertility status under indigenous tree canopies on farmland in selected areas of Eastern and Western Hararghe zone. M.Sc. Thesis, Haramya University, Ethiopia.

Abebe Y, Fisseha I, Olsson M (2009). Scattered trees as modifiers of agricultural landscapes: the role of Waddeessa (*Cordia africana* Lam.) trees in Bako area, Oromia, Ethiopia. Afr. J. Ecol. 47:78-83. http://dx.doi.org/10.1111/j.1365-2028.2008.01053.x

Abule E, Smit GN, Snyman HA (2005). The influence of woody plants and livestock grazing on grass species composition, yield and soil nutrients in the Middle Awash valley of Ethiopia. J. Arid Environ. 60:343-358. http://dx.doi.org/10.1016/j.jaridenv.2004.04.006.

Agbede OO (2008). Soil husbandry: Life for national food security and economic empowerment. An inaugural delivered on March 19th, 2008 at Nasarawa State University, Keffi, Nigeria.

Akpo LE, Goudiaby VA, Grouzis M, Houerou HN (2005). Tree shade effects on soil and environmental factors in a Savanna of Senegal. West Afr. J. Appl. Ecol. 7:41-52.

Ameha T (2006). Impact of Prosopis juliflora (SW.DC.): Invasion of plant biodiversity and soil properties in the middle Awash, rift valley, Ethiopia. M.Sc. Thesis, Hawassa University, Ethiopia.

Azene B, Ensermu K, Sebsibe D, Maundu P, Tegnas B (2007). Useful trees and shrubs in Ethiopia. Identification and management for 17 agro climatic zones. World Agroforestry Centre, Eastern Africa Region, Nairobi, Kenya.

Bayala J, Teklehaimanot Z, Ouedraogo SJ (2002). Millet production under pruned tree crowns in a parkland system in Burkina Faso. Agroforest. Syst. 54:203–214. http://dx.doi.org/10.1023/A:1016058906682

Boffa JM (1999). Agroforestry parklands in sub-Saharan Africa. FAO Conservation Guide. Rome. 34:230.

Brady NC, Weil RR (2002). The nature and properties of soils, 13th Ed. Prentice-Hall Inc., New Jersey, USA. P. 960.

Brenner T (1994). Accelerated soil erosion in watersheds of Yunnan province, China. J. Soil. Water Conserv. 1(49):67-68.

Brinson M, Bradshaw HD, Holmes RN, Elkins JB (1980). Litter fall, stem flow and through fall nutrient fluxes in an alluvial swamp forest. Ecology 61:827-835. http://dx.doi.org/10.2307/1936753

Day PR (1965). Hydrometre method of particle size analysis. In: methods of soil analysis (Black CA ed.) American Society of Agronomy, Madison, Wisconsin.

Efrem G, Sandewall M, Soderberg U, Campbell BM (2009). Land-use and land-cover dynamics in the central rift valley of Ethiopia. Environ. Manage. 44:683-694. http://dx.doi.org/10.1007/s00267-009-9355-zPMid:19688359

FAI (1977). Hand book of fertilizer usage. Fertilizer Association of India, New Delhi, India.

Tilahun G (2007). Soil fertility status as influenced by different land uses in Maybar areas of South Wollo zone, North Ethiopia. MSc. Thesis. Haramaya University, Ethiopia.

Gee GW, Bauder JW (1986). Particle size analysis. In: methods of soil analysis. Part 1, (2nd ed). (Klute A ed.). Agron. Monogr. 9, ASA and SSSA Madison, WI. pp. 383-411.

Hailemariam K, Kindeya G, Charles Y (2010). Balanites aegyptiaca, a potential tree for parkland agrofrorestry systems with sorghum in Northern Ethiopia. J. Soil Sci. Environ. Manage. 1(6):107-114.

Hugues D, Philippe DL (1998). Trees and multi-storey agriculture in Africa, Brussels, Belgium.

Jackson ML (1973). Soil chemical analysis. Prentice Hall of India Pvt. Ltd., New Delhi. P. 498.

Kamara CS, Haque I (1992). Faidherbia albida and its effects on Ethiopian highland vertisoils. Agroforest. Syst. 18:17-29. http://dx.doi.org/10.1007/BF00114814

Kessler JJ (1992). The influence of Karite (*Vitellaria paradox*) and Nere (Parkia biglobosa) trees on sorghum production in Burkina Faso. Agroforest. Syst. 17:97-134.http://dx.doi.org/10.1007/BF00053116

Lin BB (2007). Agroforestry management as an adaptive strategy against potential microclimate extremes in coffee agriculture. Agricul. Forest Meteorol. 144:85–94. http://dx.doi.org/10.1016/j.agrformet.2006.12.009

Merwin HD, Peech M (1951). Exchangeability of soil potassium in sand, silt and clay fraction as influenced by the nature of complementary exchangeable cations. Soil Sci. Soc. Am. Proc. 15:125-128. http://dx.doi.org/10.2136/sssaj1951.036159950015000C0026x

Mulongey K, Merck R (1993). Soil organic matter dynamics and sustainability of tropical agriculture. John Wiley and Sons, Inc., New York, P. 392.

Murphy HF (1968). A report on the fertility status and other data on soils of Ethiopia. Experimental Bulletin Haile Sellassie I university, Dire Dawa. 44:294-297.

Nair PKR (1984). Soil productivity aspects of agroforestry. International Centre for Research in Agroforestry, Nairobi, Kenya. P. 85.

Nazereth F (1998). Study of farmers species preference and evaluation of some fuel wood qualities. A case study in Abijata Shalla national park in Arsi-Negelle district, Ethiopia. M.Sc. Thesis, SLU, Sweden.

Nyberg G, Högberg P (1995). Effects of young agroforestry trees on soils in on-farm situations in Western Kenya. Agroforest. Syst. 32:45–52. http://dx.doi.org/10.1007/BF00713847

Olsen SR, Cole CV, Watanabe FS, Dean LA (1954). Estimation of available phosphorus in soil by extraction with $NaHCO_3$; U.S. Department of agriculture circular, U.S. Government Printing Office, Washington, D.C. P. 939.

Parysow ET (2001). Assessing uncertainty of erodibility factors in national cooperative soil surveys: A case study at Fort Hood, Texas. J. Soil. Water Conserv. 56(3):207-208.

Quideau SA, Bockheim JG (1996). Vegetation and cropping effects on pedogenic processes in a sandy prairie soil. Soil Sci. Soc. Am, 60:536-545. http://dx.doi.org/10.2136/sssaj1996.03615995006000020028x

Raddad EY, Luukkanen O (2007). The influence of different Acacia Senegal agroforestry systems on soil water and crop yields in clay soils of the Blue Nile region. Agric. Water Manage. 87:61-72. http://dx.doi.org/10.1016/j.agwat.2006.06.001.

Rao MR, Nair PKR, Ong CK (1998). Biophysical interactions in tropical agroforestry. Agroforest. Syst. 38:3–50. http://dx.doi.org/10.1023/A:1005971525590
Reeuwijk LPV (2002). Procedures for soil analysis. 6th edition. Technical paper/international soil reference and information centre, Wageningen, The Netherlands.
Spur SH, Barnes BV (1980). Forest ecology (3rd ed). John Wiley and Sons. P. 679.
Subbiah BV, Asija GL (1956). Rapid procedure for the estimation of available nitrogen in soil. Current Sci. 25:259-260.
Tadesse H, Lemma N, Olsson M (2002). *Millettia ferruginea* from southern Ethiopia: Impacts on soil fertility and growth of maize. Agroforest. Syst. 48:9-24.
Teklehaimanot Z (2004). Exploiting the potential of indigenous agroforestry trees: *Parkia biglobosa* and *Vitellaria paradoxa* in sub-Saharan Africa. Agroforest. Syst. 61:207-220. http://dx.doi.org/10.1023/B:AGFO.0000029000.22293.d1
Toky OP, Kumar P, Khosla PK 1(989). Structure and function of traditional agroforestry systems in the Western Himalayan.II. Nutrient Cycling Agroforest. Syst. 9:71-89.
Tomlinson H, Teklehaimanot Zewuge, Traore A, Olapade E (1995). Soil amelioration and root symbioses of *Parkia biglobosa* (Jacq.) Benth. in West Africa. Agroforest. Syst. 30:145 159. http://dx.doi.org/10.1007/BF00708918
Walkley AJ, Black IA (1934). Estimation of soil organic carbon by chromic acid titration method. Soil Sci. 37:29-38.http://dx.doi.org/10.1097/00010694-193401000-00003
Young A (1989). Agroforestry for soil conservation, CAB International, Netherlands.

3

Composition and structure of humic substances in long-term fertilization experimental soils of southern China

Bocar Ahamadou[1], Qiaoyun Huang[2], Lin Yaping[2] and Javed Iqbal[3]

[1]Unit of Soil Science, Department of Agronomy, Katibougou Rural Polytechnical Institute of Training and Applied Research, P. O. Box 06 Koulikoro, Mali.
[2]Key Laboratory of Subtropical Agriculture and Environment, Ministry of Agriculture, Huazhong Agricultural University, Wuhan, 430070, P.R. China.
[3]Department of Plant and Soil Science, 2207 Agronomy Hall, Iowa State University, Aames IA, 50010, USA.

Humic substances are extremely important attributes of soil quality but the impacts of long-term fertilization on their composition and structure remain controversial and are still not fully understood. A study, employing various methods was conducted to elucidate the influence of long-term fertilization practices on the composition and structure of humic substances. Soil samples were collected from eight treatments at the Red Soil Experimental Station of the Chinese Academy of Agricultural Sciences. Data were subjected to analysis of variance (ANOVA), using the MSTAT-C software package. In general, the Fourier transform-infrared (FT-IR) spectra discriminated bands between humic fractions. The humic acid (HA) from the NPK plus manure (NPKM) and manure alone (M) treatments showed higher values of the O/R ratios (1.24 and 1.18, respectively) as compared to that from the 1.5 NPKM and fallow (H) treatments, which decreased these ratios (0.66 and 0.85, respectively). Except for NPKM plus wheat-soybean-sweet potato between the lines (NPKMR), the O/R ratios of the fulvic acid (FA) did not differ among the treatments. However, the humin (HM) from H and the control (CK) displayed the highest value of O/R ratios (about 3.23 for both). Humic substances from organic treatments showed more aliphatic nature, whereas more condensed-alkali soluble humic substances were formed in the unfertilized and mineral treatments. Among fractions, the HA formed more phenolic or aromatic structures than did the FA or HM. In addition, the C, H, N and S contents of humic substances increased significantly, whereas their oxygen content decreased.

Key words: Long-term fertilization, soil quality, humic substances, Fourier transform-infrared (FT-IR) spectroscopy, elemental composition.

INTRODUCTION

Concerns regarding progressive soil degradation and the long-term sustainability of current agricultural practices have led to the study and assessment of soil quality (Haynes, 2008). Humic substances (HS) are the major components of soil organic matter (SOM) and extremely important attributes of soil quality since they profoundly influence the physical, chemical and biological properties and processes of soils. Humic substances are source of energy and nutrients for soil biota. They are also charge carriers and play vital roles in complexing organic compounds and adsorbents in controlling the toxicity, transport and fate of a variety of pollutants. Humic substances are involved in soil formation. They affect aggregate stability, trafficability, water retention and hydraulic properties. As a result, soil organic matter content and quality are now recognized as key factors in the evaluation of the sustainability of soil management practices.

Agricultural management practices, namely tillage, crop rotation and fertilization are known to affect SOM strongly with respect to the amount and composition at shorter and longer time scales. By evaluating the chemical changes of SOM in a sandy clay loam in North-east Viterbo city of Central Italy, Marinari et al. (2007) found that the humic-C content in the upper soil layer was 55% greater under organic than conventional management. They also noticed significantly higher humification degree and less condensed humic substances under the organic management. After 11 years of different tillage operations in Chinese Mollisols, Liu et al. (2005) reported higher levels of soil organic C and N in the upper soil layer. Ding et al. (2002) studied the changes of SOM quantity under conventional and conservation tillage in a Norfolk loamy sand soil of South Carolina, USA. They found that the aliphatic-C content of HA declined with the increase of soil depth under both systems, whereas the reverse was true for the aromatic- C content. Their results further revealed that HA contained more recalcitrant functional groups in conventional than in conservation tillage system. Recently, Liu et al. (2006) reported that the inclusion of legume crops in a long-term rotation trial of Pellic and Chromic vertisols in Australia resulted in an increase of liable carbon concentrations compared with continuous wheat or a long fallow period. They also observed that during a 9-year crop rotation experiment in the Chinese Mollisol, the soil organic carbon in the treatments of the wheat-sweet clover and wheat-soybean with addition of pig manure or wheat straw was significantly (22%) greater than that of the commonly used wheat-soybean alone. In a Pullman soil of Texas in USA, Acosta-Martinez et al. (2004) found that organic C, microbial biomass C, N and the enzymes activities were higher in perennial pasture compared with continuous cotton.

Although humic substances represent the most microbially recalcitrant and stable reservoir of organic carbon in soil, the application of organic matter can influence their amount and structural characteristics (Dou et al., 2008). These authors reported for example, that maize stalk and wheat straw applications increased the E_4/E_6 ratio and the content of carboxylic and aliphatic carbon in humic substances from a Dark brown soil of Heihe City, China. In a fluvi-calcaric cambisol of the long-term rotation experiment at Legnaro in Italy, Nardi et al. (2004) demonstrated that farmyard fertilizations improved the production of humus with a high degree of polycondensation, whereas the absence of organic fertilizer inputs led to a higher percentage of non-complex and light-weight humus.

Das and Nand (2006) studied the nature of humic substances under long-term manuring and fertilization in a rice-wheat-cowpea sequence on a mollisol at Pantnagar (India). They found that in general, in terms of quantity, the fulvic acid fraction dominated about 2 to 5 times over humic acid fraction, possibly due to biochemical transformations of open-chain compounds of plant residues into soil humus. They further noticed that regarding the elemental composition, the humic acid showed comparatively wider C/H and C/N but narrower O/H than fulvic acid isolated from the same treatment due to the stronger aromatization of humic acid in comparison with fulvic acid. According to these authors, the absorbance of humic substances measured at 465 (E_4) and 665 (E_6) revealed that humic acid absorbed more light than fulvic acid and this was attributed to accelerated mobility of electrons over unsaturated structures conjugated with humic substance nuclei.

During the past few decades, there has been much research on HS, but their chemical structure and composition remain controversial and not fully understood (Baldock et al., 1992; Dong et al., 2006). Previous long-term studies mostly focused on temporal changes of the amount of SOM (Kaiser et al., 2007) and changes induced by anthropogenic factors such as long-term fertilization practices on humic substances are still being researched. There is a crucial need in a more fundamental understanding of HS functions in the stabilization of soil aggregates, soil formation and quality, the binding of anthropogenic organic chemicals, and the sequestration of C from atmospheric CO_2 among other functions. To our knowledge most studies conducted on these issues have dealt predominantly with the HA fraction. However, the characterization and evaluation of long-term fertilization effects on the composition and structure of all the three fractions of humic substances, using integrative approaches are still scarce. We hence hypothesized that: (1) a better understanding of the changes occurring in HA, FA and HM fractions might give additional and comprehensive information on the effects of fertilisation practices and (2) reliable and complementary techniques are an essential precondition for more accurate interpretations of these structural changes in HS. The objectives of the present study were (1) to elucidate the influence of different long-term fertilization practices on the composition and structure of the isolated HAs, FAs and Humin fractions; (2) to discriminate between HAs, FAs and Humin fractions based on their characteristics that are influenced by the long-term fertilization conditions; and (3) to compare nondegradative spectroscopic (FT-IR) data with those produced by the elemental analyses to evaluate the suitability and specificity of individual results.

Table 1. Basic data of the long-term fertilization experimental soils (0-20 cm) (2008).

Treatment	Available nutrients(mg.kg^{-1})				Total Nutrients (g.kg^{-1})			Al (%)		Fe (%)		C_{org} (g.kg^{1})	pH	Moisture (%)
	NO_3^-	NH_4^+	P	K	N	P	K	DCB	PYRO	DCB	PYRO			
CK	7.07^f	3.82^f	3.41^g	81.5^g	0.86^e	0.52^g	12.6^e	9.06^b	3.10^b	13.6^b	4.73^b	0.48h	4.91	10.8^g
NPK	12.9^e	9.96^d	10.3^e	138.7^e	1.55^d	1.16^f	14.2^d	8.41^c	2.85^c	12.6^c	4.27^c	1.20^g	5.11^f	11.8^e
NPKM	19.6^a	13.6	17.9^a	254.8^a	2.98^a	2.82^a	17.7^a	6.51^e	2.06^e	9.75^e	3.10^e	2.65^a	5.87^c	12.9^c
1.5NPKM	14.6^d	10.2	13.1^c	179.4^c	2.08^c	1.81^d	14.9^c	6.16^f	1.88^f	9.24^f	2.82^f	1.65^e	5.58^d	12.2^d
NPKMR	16.2^c	11.1^c	15.3^b	157.6^d	1.16^e	2.09^c	14.1^d	7.58^d	2.47^d	11.4^d	3.71^d	2.06^c	5.32^e	11.5^f
NPKS	17.0^b	12.8	7.06^f	205.6^b	1.69^d	2.47^b	16.4^b	4.65h	0.88h	6.97h	1.33h	1.89^d	4.34h	13.1^b
M	14.2^d	8.18^e	11.3^d	121.2^f	1.86^{cd}	1.52^e	14.8^c	9.53^a	3.62^a	14.3^a	5.53^a	1.40^f	6.33	11.5^f
H	7.14^f	10.0	3.44^g	65.7^h	2.54^b	0.93^f	16.6^b	5.64^g	1.56^g	8.46^g	2.34^g	2.38^b	6.75^a	14.2^a

DCB, Dithionate-citrate bicarbonate; Pyro, pyrophosphate; C_{org}, organic carbon in soil; Different letters in the same column indicate a significant difference at $P<0.05$ (ANOVA).

MATERIALS AND METHODS

Experimental site and design

A long-term fertilization experiment with double cropping wheat and corn was begun in September 1990 at the Red Soil Experimental Station (RSES) of the Chinese Academy of Agricultural Sciences (CAAS), Southern Hunan Province (26.8°N, 111.9°E, 120 m altitude). The soil was originated from the Quaternary Red Clay and classified as Ultisols. Under average climate conditions, the area receives 1290 mm of annual precipitation, about 70 to 80% of which occurs from April to October. The mean annual temperature, annual evaporation; annual frost-free days and sunshine hours are 18°C, 1470 mm, 300 day and 1610 h, respectively. Prior to the experiment, the field had been under wheat-corn rotation for several years.

Based on the organic matter contents, eight plots (27 m × 10 m - 270 m^2) with different fertilization treatments were chosen for this study, that is, (1) CK, (2) NPK, (3) NPKM, (4) NPKMR, (5) 1.5 NPKM, (6) NPKS, (7) M and (8) H. The N, P and K fertilizers for corn and wheat were provided as urea at 456 and 195 kg N hm^{-2} ; P as single superphosphate at 699 and 300 kg P hm^{-2}, and K as KCl at 140 and 60 kg K hm^{-2} , respectively. Chemical fertilizer N was yearly applied as top dressing (40%) and base dressing (60%); chemical P and K fertilizers and manure were fertilized only as base dressings. Thirty percent of total N was used as chemical N and 70% as manure for the NPKM treatment; the same amount of chemical N was applied for other treatments. Thirty percent (30%) of the total amount of individual fertilizer applied each year was used for wheat and 70% was used for corn.

Soil sampling and analyses

Soil samples from the top red soil (0-20 cm) were collected in September 2008 at nine randomly selected points in each plot and mixed to form a bulk sample. Samples were sieved (<2 mm) and kept in polyethylene bags at 4°C. The basic data of soils were determined conventionally and summarized in Table 1. In brief, total organic carbon was determined by the method of Walkely and Black (1934) which consisted to oxidize the sample with potassium dichromate and sulphuric acid. The excess potassium dichromate was titrated against ferrous ammonium sulphate. Soil total nutrients were determined in the Kjeldahl acid digest using FIASTAR 5000 (total N and P) and flame photometer (for total K). The available N, P and K were measured by FIASTAR 5000 after their extraction with 2 M KCl, 0.5 M $NaHCO_3$ (pH 8.5) and 1 M NH_4OAC (pH 7), respectively (Page et al. 1982). The determination of dithionite extractable iron and aluminum was performed by the procedure of Mehra and Jackson (1960). The principle of the method consisted to heat the sample in a complexing buffer of sodium citrate/bicarbonate to which solid sodium dithionite was added as a reducing agent. Iron and aluminium were then measured in the extract by AAS. For the determination of the pyrophosphate extractable Al and Fe, 44.6 g of $Na_4P_2O_7.10H_2O$ (0.1 M) were dissolved in distilled water and brought to 1 l and the pH was adjusted to 10.0. One gram soil was placed into a 250 ml polyethylene tube (with screw stopper) and 100 ml of 0.1 M sodium pyrophosphate reagent were added and agitated on a rotary shaker for 16 h at ambient temperature (20°C). The mixture was centrifuged at 20,000 g. The supernatant was retained in test tubes and two blanks (with only reagents) and two reference samples in each series were also performed. The concentrations of Al and Fe were measured by atomic absorption spectrometry with standards diluted in the extraction matrix.

Extractions and separation of humic fractions

Method "Schnitzer (1982)" was used with some modifications to maximize the extraction of humic compounds and to minimize their degradation. Soil humic substances were extracted with a solution containing 0.1 M NaOH and 0.1 M $Na_4P_2O_7$ (60 ml per 15 g of soil) with permanent shaking for 18 h at 25°C and the suspension was centrifuged at 9000 x g for 15 min. The supernatant was carefully decanted into a clean centrifuge bottle and retained for separation into humic acid (HA) and fulvic acid (FA) fractions. The residue (humin fraction) was mixed with 50 ml of deionized water, sonicated for 10 min and the suspension was centrifuged. The supernatant was removed and discarded, while the residue was dried and transferred into polyethylene bags for storage and analysis.

Table 2. Some characteristics of humic fractions in long-term fertilization experimental soils (0-20cm) (2008).

Treatment	Elemental composition (%)														
	Humic acids (HA)					Fulvic acids (FA)					Humin				
	C	H	N	S	O	C	H	N	S	O	C	H	N	S	O
CK	40.7^{g}	0.60^{h}	0.90^{h}	0.20^{g}	57.6^{a}	34.6^{h}	0.40^{g}	0.70^{g}	0.60^{e}	63.7^{a}	41.0^{h}	0.90	1.20^{g}	0.70^{f}	56.2^{a}
NPK	42.6^{fg}	1.30^{g}	1.60^{g}	1.00^{f}	53.5^{b}	39.4^{g}	1.00^{fg}	1.30^{fg}	1.70^{d}	56.6^{b}	43.7^{g}	1.70^{f}	2.30^{f}	1.20^{f}	51.1^{b}
NPKM	53.5^{ab}	5.60^{b}	6.40^{b}	1.70^{ef}	32.8^{g}	51.0^{b}	3.70^{bc}	5.00^{b}	2.30^{d}	38.0^{e}	54.2^{b}	6.40^{b}	6.90^{a}	1.90^{e}	30.6^{g}
1.5NPKM	48.6^{cd}	3.20^{d}	4.60^{d}	2.60^{cd}	40.7^{e}	44.9^{d}	2.10^{de}	3.60^{c}	3.73^{c}	45.6^{d}	49.5^{d}	3.90^{d}	5.20^{c}	2.50^{d}	38.9^{e}
NPKMR	56.3^{a}	7.40^{a}	8.20^{a}	3.40^{b}	25.0^{h}	53.2^{a}	4.80^{a}	6.20^{a}	4.50^{bc}	31.3^{f}	56.9^{a}	8.20^{a}	7.40^{a}	3.60^{c}	23.9^{h}
NPKS	51.4^{bc}	4.20^{c}	5.30^{c}	4.80^{a}	34.4^{f}	48.7^{c}	3.00^{cd}	3.10^{cd}	5.90^{a}	39.3^{e}	52.0^{c}	5.00^{c}	6.00^{b}	5.10^{a}	31.9^{f}
M	44.2^{ef}	2.60^{e}	2.60^{f}	2.00^{de}	48.6^{c}	41.3^{f}	1.50^{ef}	2.00^{ef}	4.00^{c}	52.2^{c}	45.3^{f}	3.40^{de}	3.50^{e}	3.00^{d}	44.8^{c}
H	46.4^{de}	2.00^{f}	3.40^{e}	3.00^{bc}	45.2^{d}	42.7^{e}	4.20^{ab}	2.60^{de}	5.20^{ab}	45.3^{d}	48.1^{e}	2.70^{e}	4.20^{d}	4.20^{b}	41.5^{d}

Different letters in the same column indicate a significant difference at P<0.05.

Using a dropping tube and a pH-meter, 6 M HCl was added to the decanted supernatant in the centrifuge bottle until a pH of 1.5 was attained. To separate the humic fractions, the dark brown to black precipitated portion and that of yellowish in color dissolved in the solution were centrifuged and the supernatant (FA) was removed and retained in 50-ml centrifuge bottle for storage and analysis. 50 ml of 0.1 M NaOH were added to the centrifuge bottle containing the precipitate (HA) to redissolve it. The HA was then transferred into 50-ml centrifuge bottle for storage and analysis. In a further step, the obtained HA and FA were freeze-dried and the dried samples were stored. The extractions were carried out three times for each sample.

Fourier-transform infrared spectroscopy

FT-IR spectra of humic fraction samples were collected with a Nicolet 380 spectrophotometer, in the 400 to 4000 cm^{-1} range with a 4 cm^{-1} resolution and 32 scans per sample. Pellets were prepared by pressing a mixture of 0.8 mg of sample and 80 mg of spectroscopy grade KBr.

Elemental analysis

Elemental analysis of humic fractions was determined by using CHNS analyzer (Perkin Elmer EA 2400). Oxygen was calculated by difference:

O%= 100−(C+H+N+S)%.

Statistical analysis

All results for basic soil properties and the other parameters were expressed on an oven-dried soil weight basis (105°C, 24 h). The data were subjected to analysis of variance (ANOVA), using the MSTAT-C software package (Michigan, USA). The means and standard deviations for triplicates were calculated. Significant differences of means for all treatments were judged by least significant difference (LSD) multiple-comparison tests and expressed at P<0.05 and/or P<0.01.

RESULTS AND DISCUSSION

Elemental and atomic ratios analyses of humic fractions

Elemental analysis of humic substances is often used to establish their nature and origin (Baddi et al., 2004). The contents of C, H, N and S among fertilization treatments significantly increased in all humic fractions as compared to the control (CK) (Table 2). In contrast, the oxygen content decreased in these fractions. C, H, N and S concentrations of humic fractions reached maximum values in 1.5 NPKM followed by NPKM, while their minimum values were recorded in CK followed by NPK. Among the treatments, the mean carbon content was higher in HA fraction (47.96%) than that in the FA fraction (44.47%) but remained lower if compared to the mean C content in the HM fraction (48.83%). The mean content of H and N in the treatments revealed a similar behavior, though to a lesser extent. Conversely, the mean content of S was higher in the FA fraction (3.49%) than that in the HA (2.34%) and HM (2.77%) fractions. These results showed that the carbon, hydrogen, and sulphur contents of humic fractions were relatively higher in the organic treatments (1.5 NPKM, NPKM, NPKMR, NPKS, M and H), although their C/N ratios were low as compared to the unfertilized (CK) and mineral (NPK) treatments. For example the C/N ratios of the mentioned organic treatments for the HA fraction were 8.01, 9.75, 12.3, 11.3, 19.9 and 15.9, respectively while that of the control and the NPK treatments recorded higher values (52.8 and 31.1,

Table 3. Atomic ratios of humic substances in long-term fertilization experimental soils (0-20 cm).

Treatments	Humic acids			Fulvic acids			Humin		
	H/C	O/C	C/N	H/C	O/C	C/N	H/C	O/C	C/N
CK	0.18	1.06	52.7	0.14	1.38	5.77	0.26	1.03	39.9
NPK	0.37	0.94	31.1	0.30	1.08	35.4	0.47	0.88	22.2
NPKM	1.26	0.46	9.75	0.87	0.56	11.9	1.42	0.42	9.16
NPKMR	0.79	0.63	12.3	0.56	0.76	14.5	0.94	0.59	11.1
1.5 NPKM	1.58	0.33	8.01	1.08	0.44	10.0	1.73	0.31	8.97
NPKS	0.98	0.50	11.3	0.74	0.60	18.3	1.15	0.46	10.1
M	0.70	0.82	19.8	0.43	0.95	24.1	0.90	0.74	15.1
H	0.52	0.73	15.9	1.18	0.80	19.2	0.67	0.65	13.3

respectively). This decreasing trend of the values of the C/N ratios in organic treatments as compared to the mineral ones was also true for the FA and humin fractions. However, for the FA fraction, the control recorded the lowest value of the C/N ratio (5.77), whereas its highest value was observed in the NPK treatment (35.4). The low value of the C/N ratio in organic treatments could probably be attributed to the high levels of products arising from protein decomposition and hence may indicate a greater level of humification. According to Baddi et al. (2004), high values of nitrogen and sulphur are indicative of high contents of non-humified biomolecules (polysaccharides and polypeptides) in the material. A high nitrogen values could also be attributed to the incomplete hydrolysis of proteinaceous constituents.

The decomposition of the organic material in soil tends to lead to the formation of phenolic structures deriving from lignin. These structures are hardly decomposed with respect to sugars and proteins (Rosa et al., 2005). Therefore, the C/H, C/N, and C/O atomic ratios have been used as important indicators of the aromaticity and level of organic material decomposition. The highest H/C ratios of HA and humin fractions were observed in 1.5 NPKM, followed by NPKM, whereas that of FA was recorded in H, followed by 1.5 NPKM. The lowest H/C ratios of all the three fractions were found in CK, followed by NPK (Table 3). According to the literature (Stevenson, 1994), higher C/H, C/O, and C/N atomic ratios are associated with higher degrees of humification due to decreased acid, carbohydrate, and amino acid/protein content.

For the different humic fractions, our results generally indicate that the H/C ratios in organic treatments were higher than those observed in the unfertilized and mineral plots. This suggests that the humic fractions from organic treatments may be more aliphatic than those from the unfertilized and mineral ones, which should contain more aromatic carbon. In contrast to our findings, Dou et al. (2008) have reported that the C/H ratio decreased after organic matter application and consequently reduced the degree of condensation. The values of the O/C atomic ratio appeared higher in CK and lower in 1.5 NPKM for all humic fractions. While the C/N atomic ratio of the HA and humin fractions displayed a similar trend, that of the FA fraction showed higher values in NPK and lower ones in CK. Moreover, the mean values of the C/N ratio for the different humic fractions did not follow the changes in C and N content in these fractions. These values were relatively higher in the HA fraction (20.12) than those in FA (17.40) and HM (16.23) fractions, suggesting the formation of more phenolic or aromatic structure in HA fraction as compared to that in FA and HM fractions.

The distribution of carbon in different humic fractions (Table 4) revealed that the mean bulk of it was present in humin (48.83%), followed by HA (47.96%) and FA (44.47%) fractions in all treatments. The HA/FA ratio showed a predominance of HA in all treatments. This may be attributed to the supply of raw humus through manure, straw and the incorporated detritus of fallow causing incomplete decomposition of humic acids. Despite the lower expected biological activity in CK and NPK, it is interesting to observe that the HA/FA ratios of these treatments displayed relatively higher values. This probably indicates an encouragement for the formation of condensed alkali-soluble humic substances with greater stability in CK and NPK. The reasons for such a controversy should be further investigated because they are still unclear.

FT-IR spectra

Based on values from FT-IR spectroscopy (Appendix1), the spectra for humic acid, fulvic acid and humin from each treatment of the long-term fertilization soils were presented in Figure 1. In general, the FT-IR spectra showed similar bands in all treatments but different bands occurred between the humic fractions (HA, FA and HU). The broad intense bands at 3400 and 3700 cm^{-1} are due to the stretching vibration of bonded and non-bonded hydroxyl groups (Solomon et al., 2007; Meccozi et al., 2009). Small contributions of the bands at 2280 and 2360 cm^{-1}come from, NH_2^+, $CO\text{-}NH_2$, CO-NH-, S-H, and P-H (Tatzber et al., 2007). According to Mecozzi et al. (2009),

Table 4. Total organic carbon content and its distribution in humic fractions from long-term fertilization experimental soils (0-20 cm)*.

Treatment	TOC (%)	TN(%)	C/N ratio	HAC (%)	FAC (%)	HUC (%)	HAC/FAC ratio
CK	0.48	0.09	5.58	40.7	34.6	41.0	1.18
NPK	1.20	0.15	7.74	42.6	39.4	43.7	1.08
NPKM	2.65	0.30	8.89	53.5	51.0	54.2	1.05
NPKMR	1.65	0.21	7.93	48.6	44.9	49.5	1.08
1.5 NPKM	2.06	0.12	17.8	56.3	53.2	56.9	1.06
NPKS	1.89	0.17	11.2	51.4	48.7	52.0	1.05
M	1.40	0.19	7.53	44.2	41.3	45.3	1.07
H	2.38	0.25	9.37	46.4	42.7	48.1	1.09
Mean	1.71	0.18	9.50	47.9	44.5	48.8	1.08

*TOC, Total organic carbon; TN, total nitrogen; HAC, FAC and HUC are the contents of elemental carbon in humic acid, fulvic acid and humin fractions, respectively.

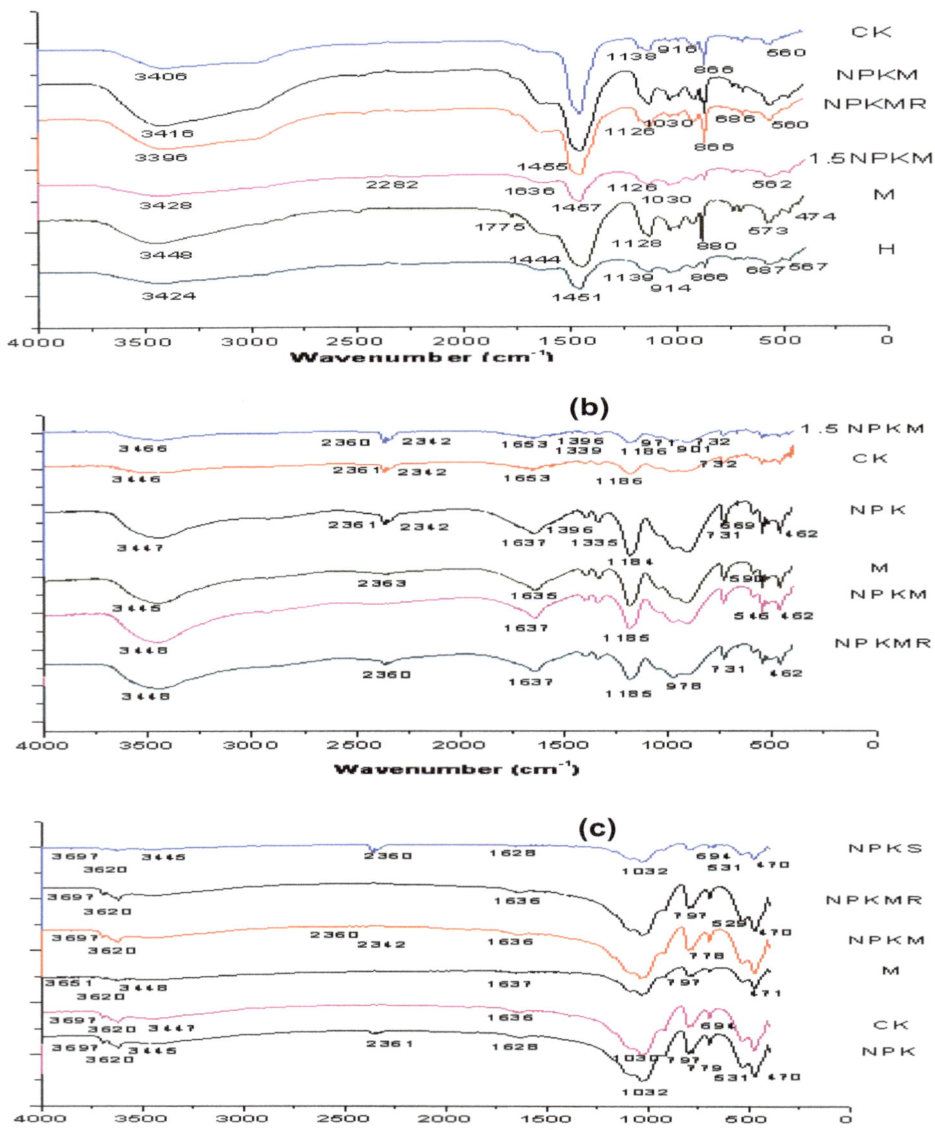

Figure 1. FT-IR spectra of humic substances in long-term fertilization soils (0-20 cm): humic acid (a), Fulvic acid (b), and Humin (c).

Table 5. The O/R ratios of humic substances from FT-IR spectra in the long-term fertilization soils (0-20 cm)*.

Treatments	HA[a]	FA[b]	HM[c]
CK	1.01	0.48	3.22
NPK	1.02	0.46	1.74
NPKM	1.24	0.46	2.28
NPKMR	1.04	0.54	0.91
1.5 NPKM	0.66	0.40	2.21
NPKS	1.01	0.46	0.92
M	1.18	0.46	2.21
H	0.85	0.46	3.23

*O/R ratios are the intensities of oxygen-containing functional groups vs. aliphatic and aromatic groups; HA is humic acid; FA is fulvic acid; and HM stands for humin. [a]HA O/R ratio was generated from HA FT-IR spectrum and calculated at (3406 +1138)/(1454 + 916 +866+ 686+ 560); [b]FA O/R ratio was generated from FA FT-IR spectrum and calculated as: (3446+1186)/(2361+1653+1397+1338+974+732+592+545+463); [c]HM O/R ratio was generated from HM FT-IR spectrum and calculated at (3697+3620+3447+1636)/(1030+797+694+530+470).

the band at 1775 cm-1 originates from the C=O stretch of ester fatty acids. The broad band at 1630 cm^{-1} can be assigned to the aromatic C=C vibrations, symmetric stretching of COO^- groups, and H-bonded C=O of conjugated ketones (Mao et al., 2008). The bands at 1440 to 1457 cm^{-1} can be attributed to the CH and NH (amide II) bending motions, molecule skeleton vibrations and CO-bond vibration (Antil et al., 2005). These bands can also be due to the carbonate, C-O and C=C stretches as well as to the aliphatic C-H stretches (Piccolo et al., 1992, Richard et al., 2007, Hardie et al., 2009). Evidence for the presence of COOH groups was indicated by the peaks at 1120 to 1150 cm^{-1} which were attributed to the C-O stretches. Sulfone functional groups and C-O-C polysaccharides can also be associated with these peaks (Mao et al., 2008, Mecozzi et al., 2009). Si-O vibrations of clay minerals were assigned at around 1030 cm^{-1}(Artz et al., 2008, D'Orazio and Senesi, 2009). The bands at 900 to 460 cm^{-1} were indicative of inorganic materials, such as clay and quartz minerals (Ding et al., 2002; Filip and Demnerova, 2007).

By comparing the spectra among fractions, strong alkyl halide stretches (C-F at1000-1400 cm^{-1}; C-CL at 600-800 cm^{-1}; C-Br at 500-600 cm^{-1}; and amine N-H stretch at 3300-3500 cm^{-1}) were common to all the three fractions. The humin fraction was distinguished by the presence of a strong and sharp free OH stretch (3620-3700 cm^{-1}) whereas, a strong carbonyl C=O stretch (1670-1820 cm^{-1}) appeared in the HA fraction only. A C-N stretch (1080-1360 cm^{-1}) was present in both the HA and FA fractions but the humin fraction lacked this type of vibration.

It was a common practice to draw conclusions on factors affecting SOM composition by spectral peak positions and intensity changes. However, this method does not often provide the kind of detailed information required to ascertain fertilization effects on SOM.

Therefore, detailed insight on the reactivity of HA, FA and HM was provided by calculation of O/R ratios, which were the intensities of oxygen-containing functional groups vs. aliphatic and aromatic groups (Ding et al., 2006). The reactivity of SOM was related with organic oxygen because it was present in all major functional groups, e.g., carboxyl, phenolic, hydroxyl, alcohol, and carbonyl and could be associated with SOM binding characteristics. The impact of fertilization on spectral composition was summarized in the O/R ratio (Table 5).

With different fertilization applications, the HA fraction isolated from the NPKM treatment had the highest O/R ratio (1.24) followed by the manure alone treatment (1.18), while that from 1.5 NPKM and H exhibited the lowest values (0.66 and 0.85, respectively). The relatively high O/R ratios in the NPKM and M treatments may indicate that HA was chemically and biologically more active in these treatments if compared to the unfertilized and other plots. This higher biological activity is probably related to the intrinsic accumulation of fatty acids in animal manure, corroborated in our study by the presence of a pronounced carbonyl C=O stretch at 1775 cm^{-1}. In addition, the cellular residues from increased soil microbial growth that is often associated with animal manure application may also explain the observed high biological activity of HA in the NPKM and M treatments. It is interesting to note that 1.5 NPKM and H treatments significantly decreased the O/R ratio of the HA, whereas no impact was observed for NPK, NPKS and NPKMR as compared to CK. This phenomenon is difficult to explain. However, either an inhibition of microbial activity or a formation of more recalcitrant materials is supposed to be occurred in 1.5 NPKM and H. It can also be speculated that the type, quality and quantity of the incorporated amendments are key factors in the potential structural changes in the HA fraction. In this regard, tannins and lignins could be formed as main contributors of aromatic and phenolic structures in these treatments. Such a structure can be linked with the E_4/E_5 ratio which is independent of the concentration of humic substances but varies with source and origin of incorporated residues.

The O/R ratios of the FA fraction did not differ among the treatments. Interestingly the HM fraction isolated from H and CK demonstrated high but similar values of O/R ratios (3.23 and 3.22, respectively), whereas that from NPKMR and NPKS had the lowest but also very close values of this ratio (0.91 and 0.92, respectively). NPKM, 1.5 NPKM and M treatments displayed little differences in the O/R ratio although these values remained higher than that for the NPK treatment. This probably indicated that the intrinsic accumulation of fatty acids in animal manure and the cellular residues from increased soil microbial growth were key factors for the high chemical and biological activities in organic treatments if compared to the control or mineral fertilizers alone. Among the humic substances, our results showed higher O/R ratios in the HM fraction followed by those in the HA fraction, whereas lower values were recorded in the FA fraction. It can be deduced that the degree of complexity of the humic fraction controls its biological activity.

Conclusion

The results of this study demonstrated that organic fertilizers, namely 1.5 NPKM, NPKM, NPKMR, NPKS, M and H were the best treatments in increasing the elemental composition of the humic substances and in decreasing the C/N ratios which may indicate their greater level of humification as compared to the control or the mineral fertilizers. In addition these results were corroborated by the FT-IR analysis through the O/R ratios which revealed higher chemical and biological activities in the aforementioned organic treatments in comparison to the control or the mineral fertilizers. However, the type, quantity and quality of the incorporated amendments have to be considered for fertilization. The elemental analysis, the atomic and O/R ratios as well as the spectral characterization are important indicators to evaluating the impacts of fertilization on the structure and composition of humic substances. However, the in-depth understanding of the functional and structural features of soil humic substances appeals to more cross-site collaborative experiments and more specific techniques such as ^{13}C-NMR and fluorescence resonance. Foreseeing the future in this area can also be achieved through approach of coupling long-term experiments and modeling.

ACKNOWLEDGEMENTS

The authors express their thanks to Hong Zhineng, Cao Yuanyuan and Wei Xing for their suggestions and advices in the analysis and interpretation of the FT-IR spectra. The study was financially supported by the National Natural Science Foundation of China (40825002).

REFERENCES

Acosta-Martínez V, Zobeck TM, Allen V., (2004). Soil microbial, chemical and physical properties in continuous cotton and integrated crop–livestock systems. Soil Sci. Soc. Am. J. 68:1875–1884.

Antil RS, Gerzabek MH, Haberhauer G, Eder G (2005). Long-term effects of cropped vs. fallow and fertilizer amendments on soil organic matter I. Organic carbon. J. Plant Nutr. Soil Sci. 168:108-116.

Artz RRE, Chapman SJ, Robertson AH. J, Potts JM, Laggoun-Defarge F, Gogo S, Comont L, Disnar J-R, Francez A-J (2008). FTIR spectroscopy can be used as a screening tool for organic matter quality in regenerating cutover peatlands. Soil Biol. Biochem. 40:515–527.

Baddi GA, Hafidi M, Cegarra J, Alburquerque JA, Gonzalvez J, Gilard V, Revel J-C (2004). Characterization of fulvic acids by elemental and spectroscopic (FTIR and ^{13}C-NMR) analyses during composting of olive mill wastes plus straw. Bioresource Technol. 93:285–290.

Baldock JA, Oades JM, Waters AG, Peng, X, Vassallo AM, Wilson MA (1992). Aspects of the chemical structure of soil organic materials as revealed by solid-state ^{13}C NMR spectroscopy. Biogeochemistry 16:1-42.

Das DK, Ram N (2006). The nature of humic substances under long-term manuring and fertilization in a rice-wheat system. Int. Rice Res. Notes 31(1):29-31.

D'Orazio V, Senesi N (2009). Spectroscopic properties of humic acids isolated from the rhizosphere and bulk soil compartments and fractionated by size-exclusion chromatography. Soil Biol. Biochem. 41:1775–1781.

Ding G, Novak JM, Amarasiriwardena D, Hunt PG, Xing B (2002). Soil organic matter characteristics as affected by tillage management. Soil Sci. Soc. Am. J. 66:421–429.

Ding G, Liu X, Herbert S, Novak J, Amarasiriwardena D, Xing B (2006). Effect of cover crop management on soil organic matter. Geoderma 130:229–239.

Dong LH, Yuan Q, Yuan HL (2006). Changes of chemical properties of humic acids from crude and fugal transformed lignite. Fuel 85:2402–2407.

Dou S, Zhang JJ, Li K (2008). Effect of organic matter applications on ^{13}C-NMR spectra of humic acids of soil. Eur. J. Soil Sci. 59:532–539.

Filip Z, Demnerova K (2007). Humic substances as a natural factor lowering ecological risk in estuaries. In: Linkov, I., Sahay, S. (Eds.), Environmental Security in Harbors and Coastal Areas. Springer, Netherlands, pp. 343–353.

Hardie AG, Dynes JJ, Kozak LM, Huang PM (2009).The role of glucose in abiotic humification pathways as catalyzed by birnessite. J. Mol. Catal A-Chem. 308:114-126.

Haynes RJ (2008). Soil organic matter quality and the size and activity of the microbial biomass: Their significance to the quality of agricultural soils. In: Huang Q, Huang, P. M., Violante, A. (Eds.), Soil Mineral–Microbe-Organic Interactions Theories and Applications. Springer-Verlag, Berlin Heidelberg, pp. 201-221.

Kaiser M, Ellerbrock RH, Gerke HH (2007). Long-term effects of crop rotation and fertilization on soil organic matter composition. Eur. J. Soil Sci. 58:1460–1470.

Liu X, Liu J, Xing B, Herbert SJ, Meng K, Han X, Zhang X (2005). Effects of long-term continuous cropping, tillage, and fertilization on soil organic carbon and nitrogen of black soils in china. Commun. Soil Sci. Plant Anal. 36:1229–1239.

Liu X, Herbert SJ, Hashemi AM, Zhang, X, Ding G (2006). Effects of agricultural management on soil organic matter and carbon transformation – a review. Plant Soil Environ. *52*:531–543.

Mao J, Olk DC, Fang X, He Z, Schmidt-Rohr K (2008). Influence of animal manure application on the chemical structures of soil organic matter as investigated by advanced solid-state NMR and FT-IR spectroscopy. Geoderma 146:353–362.

Marinari S, Liburdi K, Masciandaro G, Ceccanti B, Grego S (2007). Humification mineralization pyrolytic indices and carbon fractions of soil under organic and conventional management in central Italy. Soil Till. Res. 92:10–17.

Mecozzi M, Pietrantonio E, Pietroletti M (2009). The roles of carbohydrates, proteins and lipids in the process of aggregation of natural marine organic matter investigated by means of 2D correlation spectroscopy applied to infrared spectra. Spectrochim. Acta A 71:1877–1884.

Nardi S, Morari F, Berti A, Tosoni M, Giardini L, (2004). Soil organic matter properties after 40 years of different use of organic and mineral fertilizers. Eur. J. Agron. 21:357–367.

Piccolo A, Zaccheo P, Genevini, PG (1992). Chemical characterization of humic substances extracted from organic –waste amended soils. Bioresour. Technol. 40:275–282.

Richard C, Guyot G, Rivaton A, Trubetskaya O, Trubetskoj O, Cavani L, Ciavatta C (2007). Spectroscopic approach for elucidation of structural peculiarities of Andisol soil humic acid fractionated by SEC-PAGE setup. Geoderma 142:210-216.

Rosa AH, Simões ML, Camargo de Oliveira L, Rochac JC, Neto LM, Milori DMBP (2005). Multimethod study of the degree of humification of humic substances extracted from different tropical soil profiles in Brazil's Amazonian region. Geoderma 127:1-10.

Schnitzer M (1982). Organic matter characterization. In: Page, A.L.,Miller, R.H., Keeney, D.R. (Eds.), Methods of Soil Analysis, Part 2. Chemical and microbiological properties. Am. Soc. Agron. Madison, Wisconsin, USA, pp. 581-594.

Solomon D, Lehmann J, Thies J, Schäfer T, Liang B, Kinyangi J, Neves E, Petersen J, Luizão F, Skjemstad J (2007). Molecular signature and sources of biochemical recalcitrance of organic C in Amazonian Dark Earths. Geochim. Cosmochim. Acta 71:2285–2298.

Stevenson FJ (1994). Humus Chemistry: Genesis, Composition, Reactions. Wiley, New York.

Tatzber M, Stemmer M, Spiegel H, Katzlberger C, Haberhauer G, Mentler A, Gerzabek MH (2007). FTIR-spectroscopic characterization of humic acids and humin fractions obtained by advanced NaOH, $Na_4P_2O_7$, and Na_2CO_3 extraction procedures.

APPENDIX

Appendix 1. FT-IR spectral values of humic substances in different long-term fertilization treatments (2008).

CK	NPK	NPKM	NPKMR	1.5NPKM	NPKS	M	H
			Humic acids (HA)				
3406	3452	3416	3396	3428	3399	3448	3424
1454	1452	1452	1455	1457	1451	1444	1451
1138	1137	1127	1126	1126	1140	1128	1139
916	914	915	917	920	913	926	914
866	866	866	866	867	866	-	866
686	686	686	686	-	686	-	687
560	560	559	560	562	562	573	567
-	-	1030	1030	1030	-	1030	-
-	-	880	-	-	-	880	880
-	-	-	-	2282	-	-	-
-	-	-	-	1636	-	-	-
-	-	-	-	735	-	735	-
-	-	-	-	474	-	474	-
-	-	-	-	-	-	1775	-
-	-	-	-	-	-	989	-
-	-	-	-	-	-	702	-
			Fulvic acids (FA)				
CK	NPK	NPKM	NPKMR	1.5NPKM	NPKS	M	H
3446	3447	3448	3448	3466	3448	3445	3458
2361	2361	-	2360	2360	2365	2363	2362
1653	1637	1637	1637	1653	1637	1635	1636
1397	1396	1397	1397	1396	1396	1395	1396
1338	1335	1335	1336	1339	1334	1335	1335
1186	1184	1185	1185	1186	1185	1185	1185
974	-	974	978	971	-	-	-
732	731	731	731	732	731	731	731
592	591	591	591	592	591	590	591
545	545	546	545	544	546	546	545
463	-	462	462	463	463	462	462
-	522	522	521	521	523	523	522
-	907	909	-	901	908	909	907
-	-	-	-	669	-	-	-
			Humin				
3697	3697	3697	3697	-	3697	-	3697
3620	3620	3620	3620	3620	3620	3620	3620
3447	3445	-	-	3433	-	3448	3446
1636	1628	1636	1636	1636	1654	1637	1636
1030	1032	1032	1032	1032	1031	1032	1032
797	797	797	797	797	796	797	797
694	694	694	694	694	694	694	694
530	531	528	529	529	531	530	529
470	470	471	470	470	470	471	470
-	2361	2360	-	-	2361	-	-
-	779	778	-	-	-	-	-
-	-	2342	-	-	2342	-	-
-	-	-	-	-	669	-	-

Organic carbon distribution in alluvial soils according to different flood risk zones

Vernhar Gervais-Beaulac[1], Diane Saint-Laurent[1,2], Jean Sébastien Berthelot[1] and Mhamed Mesfioui[3]

[1]Département des sciences de l'environnement, 3351, boul. des Forges, C.P. 500, Trois-Rivières, Québec, G9A 5H7, Canada.
[2]Géographie et Laboratoire de recherche en géomorphologie fluviale et sols, Université du Québec à Trois-Rivières, 3351, boul. des Forges, C.P. 500, Trois-Rivières, Québec, G9A 5H7, Canada.
[3]Département de Mathématique et informatique, Université du Québec à Trois-Rivières, 3351, boul. des Fortes, C.P. 500, Trois-Rivières, Québec, G9A 5H7, Canada.

This study examines the spatial distribution of organic carbon in alluvial soils subjected to frequent flooding according to different flood risk zones, that is, interval recurrences of 0-20 years (FFz) and 20-100 years (MFz). Sites located outside of flood zones (NFz) were also selected to compare the soil organic carbon (SOC) in different zones. The selected sites are located in floodplains covered by forest dominated by silver maple (*Acer saccharinum* L.) and green ash (*Fraxinus pennsylvanica* Marsh.) in southern Québec. These floodplains are affected by frequent flooding, especially in the last decades, which has a direct impact on pedogenic processes, particularly in terms of in situ soil biomass and organic matter. The soil samples (0-20 cm depth) collected in a frequent flood zone (FFz), generally show a lower content of soil organic carbon (SOC%) ranging from 1.74 to 2.59% (median values), and mean values between 1.79 and 2.83%, respectively. In areas not affected by the floods, levels of SOC (%) are generally higher, with values ranging between 2.86 and 3.73% (mean), and mean values between 3.18 and 5.17%. Loss of biomass (litter) during the flood recession causes a net loss of organic matter to the subsurface soils. Successive flooding leads to an impoverishment of alluvial soils and undermining of the pedogenic processes and soil development. This confirms the trends observed in our previous work on soil depletion in active floodplains in the study area.

Key words: Alluvial soils, soil organic carbon (SOC), floods, spatial variability, climate change.

INTRODUCTION

The number of studies on global warming and their impact on river systems have grown over the past few years (Monirul et al., 2003; Alcamo et al., 2007; Kay et al., 2008; Whitehead et al., 2009; Wilby and Keenan, 2012). Hydroclimatic changes have had a direct impact on river flow and play a key role in the homeostasis of riverside environments (Tockner et al., 1999; Steiger and Gurnell, 2003; Rokosh et al., 2009). The effect of floods on riverside ecosystems, in particular on microflora and microfauna, have been widely documented (Heimann and Roell, 2000; Clinton et al., 2006, Schilling et al., 2009; Kayranli et al., 2010). Conversely, there are fairly few studies on the pedogenetic development of riverside soils and their physical and chemical properties regarding the increase in flood recurrence (Daniels, 2003; Bailey and Guimond, 2009; Ruiz-Sinoga et al., 2012). Frequent flooding may result in soil depletion with a loss of litter and reduced input in organic matter (Saint-Laurent et al.,

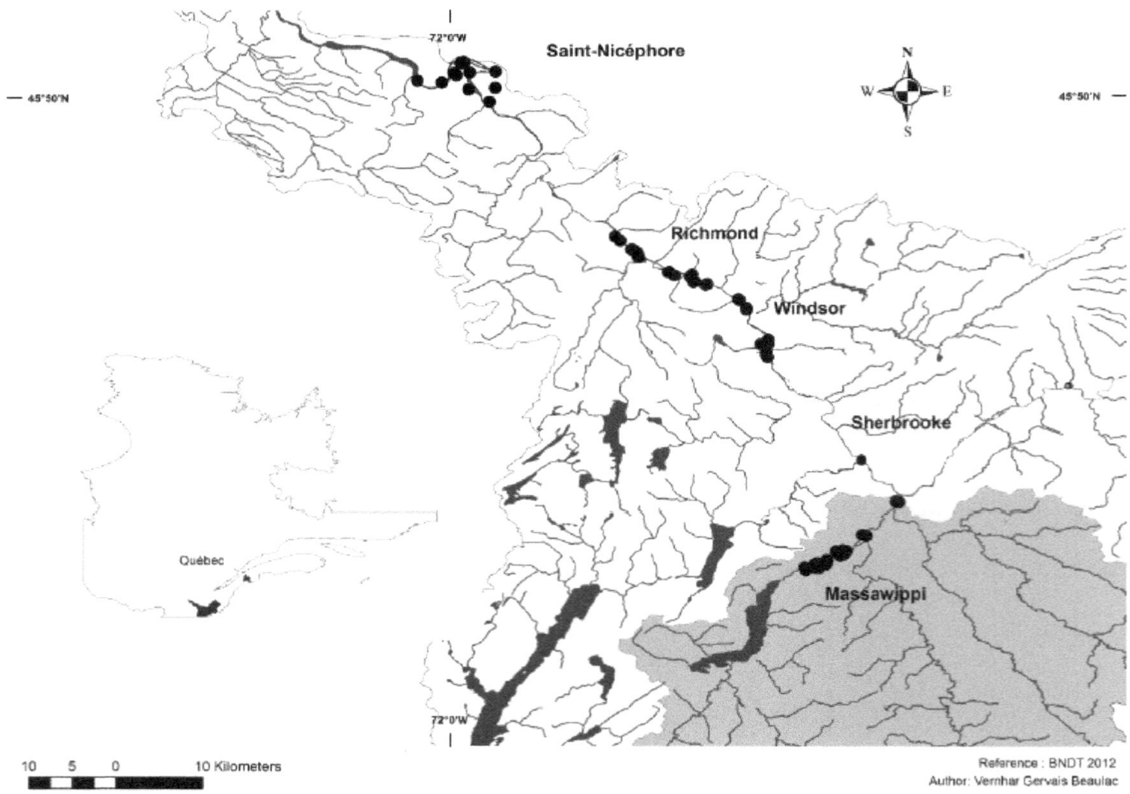

Figure 1. Location of sampling sites along the Massawippi and Saint-François rivers (Southern Québec, Canada).

2010; Drouin et al., 2011). During flood recessions, the litter (annual or biennial) is transported further downstream by the current, which ultimately depletes organic matter sinks and causes a decrease in organic carbon for the soil.

Organic carbon is known to play a key role in pedogenetic processes and soil fertility (Schilling et al., 2009; Dai et al., 2011; Ngailo and Vieira, 2012). Furthermore, soil is known to be a natural storehouse for organic carbon and to contribute to the various biogeochemical exchanges (sinks and sources) between the atmosphere and other environmental components (Zhang and Mitsch, 2007; Kayranli et al., 2010). Given that soil plays a key role in the conservation of sinks of organic carbon and their impact on variations in atmospheric CO_2, one can see the importance of measuring this pedological parameter on different spatial scales and in various physical environments. It appears in fact, important to understand the spatial variability of soil organic carbon in environments subject to constant fluctuations such as river environments, especially those subject to periodic flooding. What would be the long-term impact of successive floods on the soil organic content of floodplains, for instance?

This study aims at measuring the soil organic carbon of subsurface soils based on the various flood risk zones (recurrence intervals of 0-20 years and 20-100 years). Our preliminary work (Drouin et al., 2011; Saint-Laurent et al., 2010) conducted in the same study areas, shows that frequent river overflowing results in a depletion of SOC% in active sedimentation zones (recurrence of 0-20 years). The zones less affected by periodic flooding (recurrence of 20-100 years) generally show higher SOC%, while no-flood zones show higher SOC% compared to the flood risk zones. For this study, the number of sampling sites was increased in order to assess and compare the SOC% in the various flood and no-flood zones.

MATERIALS AND METHODS

Description of the study area

The study was conducted along the Massawippi (MAS) and Saint-François (STF) rivers located in southern Québec (Figure 1). These rivers cross through wooded or partially wooded areas, farmland, and urban areas with a moderate to high population density (Drummondville and Sherbrooke areas). The middle section of the Saint-François River is characterized by low floodplains (1-3 m in height) covered mainly by fluvial deposits (recent and ancient). Under these fluvial deposits are mainly found glaciolacustrine sediments, glaciofluvial outwash materials and rocky outcrop (Lavoie et al., 2006; Saint-Laurent et al., 2008). The middle section of the Massawippi River is also characterized by low alluvial terraces (1 to 2 m) made up of recent fluvial deposits based on

Table 1. Hydrological characteristics and river discharges of the Massawippi and Saint-François sectors

Rivers	Mean channel width (m)	Mean channel depth (m)	Mean channel height (m)	Gauging station number	Location	Period recorded	Mean annual discharge (m^3/s)	Peak discharge (m^3/s)
Massawippi	30	2.2	1.50	02OE019	45°17'03''N 71°57'45''W	1955-2011	10	67
Saint-François (Richmond)	180	16.0	1-2	02OF001	45°39'32''N 72°08'37''W	1915-1965	183	2080
Saint-François (Drummondville)	260	8.0	1-2.50	02OF002	45°51'42''N 72°27'11''W	1925-2002	189	2719
Saint-François (Windsor)	180	16.0	1-2	02OF004	45°33'50''N 72°00'23''W	1936-1972	165	2080

Sources: Environment Canada/CD-HYDAT (2011) ; MDDEP/CEHQ (2013a).

various quaternary deposits (e.g. moraines, glaciofluvial). The peak discharge registered in 1925-2002 in the Saint-François River (middle section/ station no. 030203) is 2719 m^3/s and the mean annual discharge is 189 m^3/s (Table 1) (Environment Canada, 2011; MDDEP, 2013a).

Southern Québec is characterized by a cool and humid climate with mean annual temperatures ranging from -10.2 to 20.8°C, with a mean annual temperature of 6.06°C and annual precipitation ranging from 68.7 to 107.1 mm, with total annual precipitation of 1107 mm (1981-2010) (EC, 2012; MDDEP, 2013b/Drummondville station no. 7022160). The mean and maximum flow rates obtained for the two rivers differ significantly due to the depth and width of the river sections. For the Massawippi River, the mean flow rates recorded from 1955 to 2010 are 10 to 67 m^3/s, respectively, which is 12 to 20 times lower than those recorded in the medium section of the Saint-François River (Table 1).

The Massawippi and Saint-François rivers are frequently affected by flooding (Saint-Laurent et al., 2009), owing to the number of tributaries and shallow riverbeds that favour overflowing during spring floods or periods of heavy rain. An increase in flood frequency was noted especially in the last decades (Saint-Laurent et al., 2008; 2009), which favours the formation of large alluvial plains and the aggradations of river terraces, along with affecting the development process of riverside soils (Saint-Laurent et al., 2010; Drouin et al., 2011). For instance, the sedimentation and erosion processes observed in the river corridor take the form of a loss of fine materials at the foot of the riverbanks during strong currents, especially spring floods; however, during flood recessions, the fine sediment that is transported along the river is accumulated on the alluvial plain bench and can be up to 4-5 cm thick. For example, a mean sedimentation rate of 36.8 mm was recorded over a decade in the medium section of the Saint-François River between Windsor and Richmond (Saint-Laurent et al., 2010).

Data gathering

The sites selected for the soil sampling were spread out based on the flood risk zones, including the flood zones with a recurrence interval of 0-20 years (FFz: Frequent Flood zone) and 20-100 years (MFz: Moderate Flood zone), along with an adjacent zone located outside the floodplains (NFz: No Flood zone). The delimitation of the flood zones was established based on flood risk maps (scale of 1:10,000 or 1:15,000) produced by Environment Canada and Québec's environment ministry (EC and MENV, 1982), as well as flood risk maps created in the municipal development plans in the areas under study. A total of 90 quadrats (20 x 10 m) were selected along the MAS and STF rivers (middle section), including 42 quadrats in Frequent Flood zones (FFz), 17 quadrats in Moderate Flood zones (MFz) and 31 quadrats in No Flood zones (NFz). The soil samples were collected at each end of the quadrat using a cylindrical manual auger (Eijkelkamp model) at a depth of 0-20 cm. In addition, other samples were taken every 20 cm in depth (0-20, 20-40, 40-60, 60-80 and 80-100 cm) at the end of the quadrat. More than 600 soil samples were collected for all the quadrats in the two study areas (MAS and STF). Samples were collected according to the criteria established by the Canadian System of Soil Classification (McKeague, 1978; CSCW, 1998).

Data analysis

A total of 180 soil samples was selected (2 samples at 0 to 20 cm in depth for all the quadrats) for the physical and chemical analysis (particle size, soil organic carbon (SOC %), and pH) in accordance with the analytical standards of the Soil Sampling and Methods of Analysis (McKeague, 1978; Carter and Gregorich, 2006). Soil samples were air-dried (2 to 3 weeks) and sieved at 2 mm (copper sieve, U.S. No. 10). The sandy fraction (<2 to 0.05 mm) was obtained by sieving (dry materials), and the finer fractions (silt/<0.05-0.002 mm and clay/<0.002 mm) were obtained using laser diffractometry techniques (Fritsch/Analysette 22, Micro Tec Plus) with a measurement range of 0.08 to 2,000 µm. The protocol developed by Yeomans and Bremner (1988) was used for the total organic carbon. Soil samples were placed in a digestion tube to which was added 5 ml of acidified dichromate solution (K2Cr2O7-H2SO4) for 30 m. The tube was placed in a preheated digestion block at 170°C for 30 m. The titration was performed with an ammonium ferrous sulfate solution at 0.05 mol L^{-1} (Yeomans and Bremner, 1988). Soil nitrogen was determined with the Kjeldahl method (Quikchem Method, 1996). The samples were placed in glass tubes with a reactive solution (sulfuric acid) and placed on a block digester for 2 h. The residues were analyzed using a Flow Injection Analyser based on Lachat Method (no. 13-107-06-2-D). A 1:2 soil-solution ratio (CaCl2, 0.01M) was used to determine the pH of the soil samples and the liquid solution was measured with pH meter electrodes (Carter and Gregorich, 2006).

Table 2. Chemical and physical properties of subsurface soils along the Massawippi (MAS) riverbanks in different flood zones (FFz and MFz) and no flood zones (NFz)

Soil samples (0-20 cm depth)	pH ($CaCl_2$)	Organic carbon (%)	Sand (%)	Silt (%)	Clay (%)	Texture[a]
Frequent Floods (FFz) (*n* = 13)						
Mean	4.69	2.83	36.60	60.24	3.15	Silt loam
Maximum	5.68	5.13	52.60	81.55	4.64	
Minimum	3.69	1.08	14.56	45.18	1.71	
Median	4.68	2.59	39.00	58.20	3.13	
Standard deviation	0.57	1.29	12.86	12.06	0.93	
Moderate floods (MFz) (*n* = 8)						
Mean	4.30	3.13	32.43	63.77	3.80	Silt loam
Maximum	5.12	5.21	46.74	82.03	6.20	
Minimum	3.47	1.75	11.76	51.16	2.10	
Median	4.27	2.73	34.23	63.18	3.32	
Standard deviation	0.65	1.17	10.31	9.18	1.54	
No floods (NFz) (*n* = 10)						
Mean	4.05	3.18	38.11	57.32	3.65	Silt loam
Maximum	5.19	5.28	70.69	85.58	8.39	
Minimum	3.18	1.76	6.01	27.72	1.37	
Median	3.98	2.86	43.30	54.15	2.56	
Standard deviation	0.69	1.09	20.23	17.27	2.45	

[a] Textural classes are based on the Canadian System of Soil Classification (CSCW, 1998).

Statistical analysis

Soil properties were treated with standard statistics and other statistical tests (Shapiro-Wilk and Mann-Whitney U) were performed to evaluate the distribution and normality of the soil data (soil organic carbon (SOC), pH, and texture). The student's *t*-test was also used to test the significance of mean differences in the SOC and the two recurrence interval flood zones (FFz and MFz) and no flood zone (NFz). Before analyzing the Student's *t*-test, we checked the distribution of the data with the Shapiro-Wilk's test for the series of three different zones (FFz, MFz and NFz). Next, we used the Mann-Whitney U test since two of the three groups (FFz and NFz) did not have a normal distribution. Pearson's and Spearman's correlation analyses were also done between the two SOC (%) and Silt (%) variables. The correlations were done on all the data, after which a second series of correlation analyses was done by removing a value deemed marginal from the lot, that is, the value with a concentration of 30.67% (SOC). All the statistical analyses were done with a significance level of $p = 0.05$ using SPSS PASW Statistics v18 software.

RESULTS AND DISCUSSION

Soil properties in different flood-risk zones

Tables 2 and 3 provide the mean, maximum, minimum and median values for the different soil properties for the two study areas (MAS and STF) based on the different zones (FFz, MFz and NFz). The pH values obtained for the subsurface soils (0-20 cm in depth) are relatively comparable between the two study areas, although greater acidity is noted for the STF area soils (Figure 2). In general, the soils located in the no-flood zone (NFz) had the highest acidity levels, with median values of 3.98 and 3.77, compared to 4.68 and 5.31 for the soils in the frequent-flood zones (FFz), and 4.27 and 4.15 for the soils in the moderate-flood zones (MFz). Based on the vertical distribution of the pH values in the soil profiles (0-100 cm in depth), the most acidic soils are found in no-flood zones. The litter, which is substantially greater in no-flood zones, may be contributing to acidifying the soils through a greater input of organic matter (e.g. fulvic and humic acids) which comes from the decomposition of the organic matter in the litter (Duchaufour, 2001; Brady and Weil, 2008).

Regarding the textural properties of the soils, differences were noted between the two study areas (Tables 2 and 3; Figure 3). The surface soils (0-20 cm) in the MAS area in fact, had higher proportions of silt than the STF area, with mean values of 60.24, 63.77 and 57.32%, compared to 49.98, 46.15 and 42.37% (FFz, MFz and NFz), respectively.

These textural variations are most likely due to the type of materials that make up the riverbanks of the two study areas (MAS and STF). The higher proportion of silt in the MAS area soils is due to a dominance of fine fluvial deposits which provide a finer matrix. For instance, the Massawippi riverbanks extend over a total of 37 km and fluvial deposits account for 66% of the channel banks in this sector (Lavoie et al., 2006; Saint-Laurent et al., 2008). The riverbanks between Sherbrooke and Drummondville (Saint-Nicéphore area) extend over 80 km

Table 3. Chemical and physical properties of subsurface soils along the Saint-François (STF) riverbanks in different flood zones (FFz and MFz) and no flood zones (NFz).

Soil samples (0-20 cm depth)	pH ($CaCl_2$)	Organic carbon (%)	Sand (%)	Silt (%)	Clay (%)	Texture[b]
Frequent Floods (FFz) (*n* = 29)						
Mean	5.12	1.79	47.95	49.98	2.07	Silt loam
Maximum	6.07	4.56	76.84	67.29	3.72	
Minimum	3.53	0.63	29.43	21.85	1.05	
Median	5.31	1.74	42.58	55.44	2.15	
Standard deviation	0.76	0.83	13.25	12.81	0.53	
Moderate floods (MFz) (*n* = 9)						
Mean	4.49	2.78	51.89	46.15	1.95	Sandy loam
Maximum	5.36	5.39	72.80	65.39	3.74	
Minimum	3.95	1.54	32.19	26.00	1.10	
Median	4.15	2.22	49.40	49.01	1.63	
Standard deviation	0.56	1.31	13.58	13.04	0.78	
No floods (NFz) (*n* = 21)						
Mean	3.84	5.17	56.35	42.37	1.72	Sandy loam
Maximum	5.78	30.67	81.07	74.44	2.70	
Minimum	2.79	0.85	30.47	18.29	0.63	
Median	3.77	3.73	58.48	39.91	1.63	
Standard deviation	0.74	6.12	14.50	14.81	0.68	

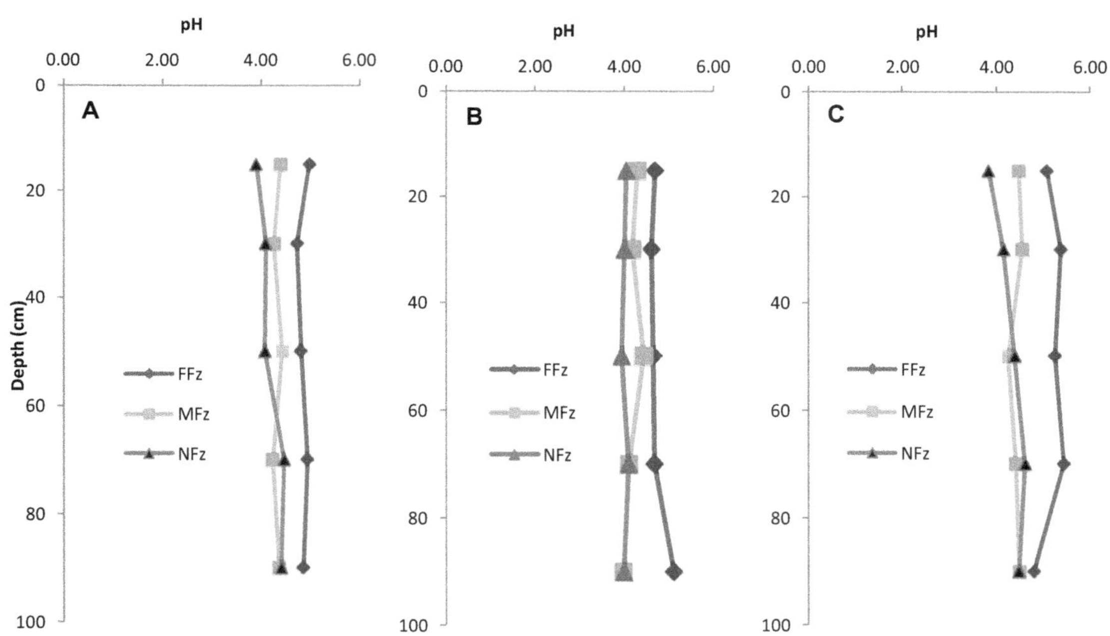

Figure 2. Vertical distribution of mean values of the pH in soil profiles in the three study zones (FFz, MFz and NFz). A: Total of soil samples (n = 90); B: Soil samples of the Massawippi sector (n = 31); C: Soil samples of the Saint-François sector (n = 59).

and are composed of fluvial deposits and other materials (e.g. glacial, fluvioglacial, glaciolacustrine). Between Windsor-Richmond specifically, the banks consist predominantly of fluvial deposits (42%) and glaciolacustrine deposits (22%), along with a large proportion of glacial materials (18%) (Lavoie et al., 2006). Lastly, the hydrological characteristics, which differ from one area to another, must also be considered. The flow rate of the Massawippi River differs substantially from that of the Saint-François River, which can explain why,

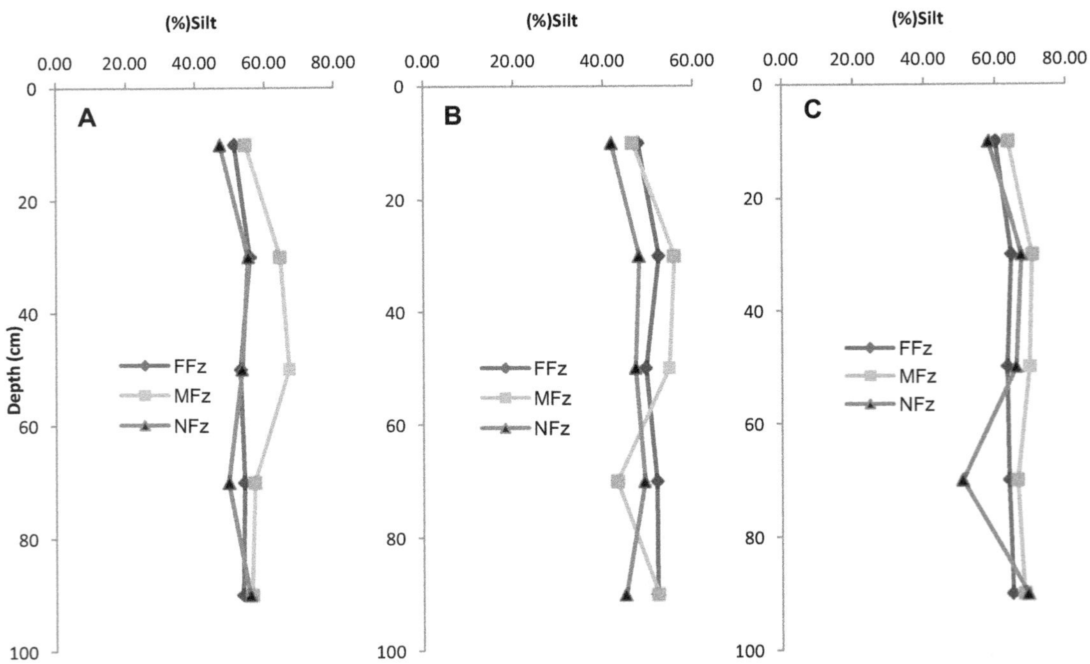

Figure 3. Distribution of silt content (%) in soil profiles in the three study zones (FFz, MFz and NFz). A: Total of soil samples (n = 90); B: Soil samples of the Massawippi sector (n = 31); C: Soil samples of the Saint-François sector (n = 59).

Table 4. Results of student *t*-test and Mann-Whitney U test for the comparison of average (SOC%) of the three groups (zones FFz, MFz and NFz).

Comparison between the three groups[a]	Group 1 NFz	Group 2 MFz	Group 1 NFz	Group 3 FFz	Group 2 MFz	Group 3 FFz
MAS sector (student *t*-test analysis)	0.919*		0.494*		0.601*	
STF sector (Mann-Whitney U test)	0.099*		0.000**		0.040**	
MAS and STF sectors (Mann-Whitney U test)	0.095*		0.000**		0.011**	

[a] H0: *No statistical difference between groups ($p > 0.05$) ; H1: **Statistical difference between groups ($p < 0.05$)

during overflowing in periods of flooding, the deposits left by the Massawippi River contain more fine particulate matter than in the soils in the Saint-François riverbanks.

Organic carbon contents in soils

Soil organic carbon content (SOC %) are relatively different based on the flood zones and the no flood zones. The mean concentrations of SOC% in surface soils (0-20 cm in depth) ranges from 2.83±1.29% and 1.79±0.83% (FFz) to 3.13±1.17 and 2.78±1.31% (MFz) for soils in the flood zones, and for the soils in the no flood zones (NFz), the mean SOC values are 3.18±1.09% and 5.17±6.12% (Tables 2 and 3). These mean SOC values in the no-flood zones are significantly higher than those observed in the flood zones. The results of the Student's *t*-test and Mann-Whitney U test (Table 4) show that there is a significant difference ($p <$ 0.05) between the three zones regarding the SOC% content of the surface horizons (0-20 cm). These concentrations are higher in the NFz zone than in the FFz zone, and are comparable between the MFz and NFz zones. The FFz zone basically consists of the zone with the lowest soil organic carbon content in the surface horizons. The values obtained are statistically significant (p = 0.000), confirming that the soils in frequent flood zones have a lower SOC% content.

Lastly, note that the soils in the MAS area compared to those in the STF area show slightly higher levels in the frequent flood zones (FFz), with mean values of 2.83% versus 1.79% (Tables 2 and 3). However, in the no-flood zones (NFz), the soils in the Saint-François River area have the highest SOC concentrations, with values of 5.17% compared to 3.18%. These variations in SOC content may partly depend on the hydrogeomorphological conditions specific to each area. The riverbanks of the MAS area are less flooded than the STF area (Jones,

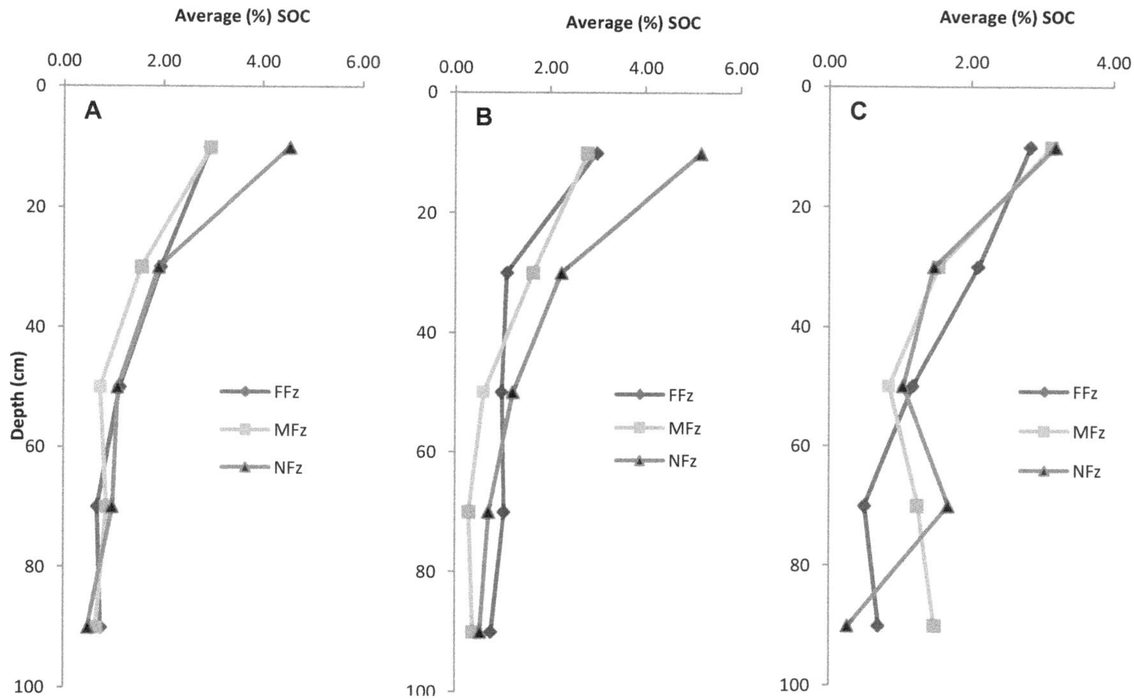

Figure 4. Distribution of SOC (%) in soil profiles in the three study zones (FFz, MFz and NFz). A: Total of soil samples (n = 90); B: Soil samples of the Massawippi sector (n = 31); C: Soil samples of the Saint-François sector (n = 59).

2008; Saint-Laurent et al., 2009; 2010), which is more favourable to maintaining litter and, as a result, provides a more regular organic matter input for the soils. The flood frequency is more than double (8 vs 3 floods) for the STF area (Sherbrooke-Drummondville areas) compared to the MAS area from 2000 to 2013.

It is known that several factors may be involved in the contents of organic matter in the soil, including texture, drainage, soil biomass, pH and microbial activity which itself depends on the acidity of the soil and the nature of the organic materials (e.g. labile C proportion vs. aromatic C) (Lützow et al., 2006; Brady and Weil, 2008). For the soils studied, aside drainage and litter thickness, the other soil properties (e.g. texture, acidity) does not appear to be involved in the content of organic carbon. Soil biomass (litter) is much greater in thickness for soils in the no flood zones (see next sections), which seems to be the determining factor in the explanation of the highest content of SOC (%) in these soils.

Vertical variation of SOC% in soil profiles

Figure 4 shows the vertical distribution of the SOC% based on the depth and the different flood zones (FFz, MFz and NFz) for each area (MAS and STF). The soils not affected by flooding show a more regular pattern of SOC distribution, with a higher concentration in the surface horizons (0-20 cm) with a mean value of 4.53% and a decrease of SOC at the base of the profile (mean = 0.49%) at all the sites. Soils in the flood zones (FFz et MFz) show a more vertical distribution of the SOC content in the surface soil at the base of the profile, in addition to having significantly lower values than the soils in the NFz zones (Figure 4). The mean SOC values in soils in the frequent flood zones (FFz) are 2.93% (0-20 cm), 1.13% (60-80 cm) and 0.75% (80-100 cm), a difference of ±2.18% between the surface and the base of the profile, whereas the average difference is ±4.04% from the surface to the base of the profile for soils in the no-flood zones (NFz). This variation is even more pronounced for the STF area, with a difference of 2.21% and 4.64%, respectively (Figure 4c). For the MFz zones, the soils have values equivalent to those in the FFz zones.

Organic carbon content vs soil texture

Soil texture does not appear to account for the variability of SOC content (0-20 cm depth) observed in the various zones (FFz, MFz and NFz). It is known that fine fractions (clay and silt) retain more organic particles than coarser fractions (Nadeu et al., 2011; Ruiz-Sinoga et al., 2012). Even though the soils in the FFz zone have generally finer matrices (e.g. fine or very fine sand and silt, with low clay particles), their sand and silt proportion is comparable to that determined in the other zones (MFz and NFz).

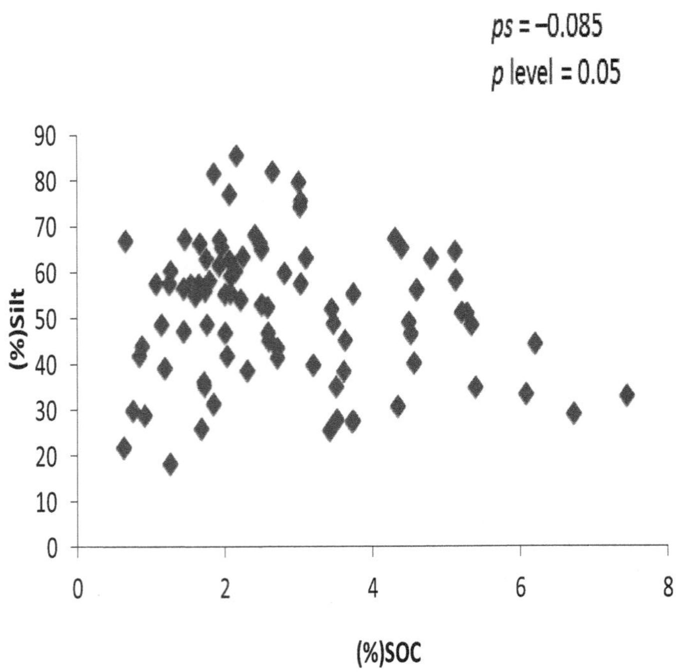

Figure 5. Correlation (Spearman's test) between SOC (%) and Silt (%) contents for all soil samples (n = 89).The significance level considered for statistical tests are 0.05 (confidence interval of 95%).

If these soils contained a greater proportion of clay particles (e.g. 20 to 30%), the concentration of SOC (%) may be higher due to the formation of organo-clay complexes that promote the retention of fine organic particles in the soil horizons. To measure the degree of correlation between these two variables (SOC% and Silt %), two standard statistical tests were used. The Pearson and Spearman correlation that was done for the SOC% concentrations and Silt (%) percentage did not show any relationship between the two variables in all the zones (Figure 5). The statistical values obtained are $rp = -0.113$ and $p = 0.292$ (Pearson's test) and $p_s = -0.097$ and $p = 0.368$ (Spearman's test). The marginal value (30.67 SOC %) was excluded from the correlation analysis, and the correlations remain low ($rp = -0.140$, and $p = 0.193$, and $p_s = -0.085$, and $p = 0.431$ (Spearman's test) respectively (Figure 5). The significance level considered for all statistical tests is 0.05 (confidence interval of 95%). The low proportion of clay in the soils (low microaggregate particles) that were analyzed may account for the lack of correlation between the two variables (SOC% and Silt%).

Microtopography, drainage and soil biomass

Figure 6 provides additional elements on edaphic conditions (e.g. microtopography, drainage, soil biomass) for the various areas under study (FFZ, MFz and NFZ). In terms of topography, the floodplain zones (FFz and MFz) are characterized by flat (75%) or subhorizontal (10%) landforms that may have irregular surfaces (hollows and bumps), while the outlying areas (NFz) are generally sloping (with slight to moderate or abrupt slopes), and are often characterized by the presence of micro mounds, which indicate old buried stumps. Internal drainage (moisture) is largely determined by the microtopography of the site as well as texture and stoniness. The flat surfaces of the flooded areas (Figure 6a and b) are often poorly (28 to 37%) or moderately drained (15%), although soil mainly consisting of sand or silty sand can facilitate the drainage of water in the soil profiles. The sites with abrupt or moderate slopes have better drainage (63%) than the flat areas, and the presence of gravel in some profiles facilitates internal drainage. The soil outside the flood zones (Figure 6c) has the largest quantity of litter (%), namely three to four times more (29%) than the soil subject to frequent flooding (7%). In fact, these sites (NFz) have the highest concentrations of SOC (%) and nitrogen (%), that is, close to twice the levels found in the other zones (FFz and MFz). The soil pH values were also found to be lower (92%) for soil with more soil biomass (Figure 6c). It is known that litter decomposition and humification releases acidifying products (e.g. fulvic and humic acids), which can increase soil acidity, especially in the surface horizons (Brady and Weil, 2008). High levels of nitrogen (N %) were also found to be correlated with high concentrations of SOC (%).

Conclusion

Soils in the frequent flood zones show lower values of SOC% content compared to the no flood zones. During successive floods, the litter can be transported by currents, which limits long-term enrichment of soil through organic matter. In zones less affected by floods (MFz) or in no-flood zones (NFz), the SOC% content is generally higher. For other soil properties, there are minor differences between the zones and sectors. Soils not affected by the floods are generally more acidic, which can be explained by the decomposition of organic matter that provides organic acid substances (e.g. fulvic and humic acids). The soils in the MAS area generally contain a higher proportion of silt, which is likely attributable to the hydrogeomorphic conditions and dominance of fluvial deposits in this area. However, no positive correlation was noted between the proportion of silt and the organic carbon content for all the soils that were analyzed.

ACKNOWLEDGEMENTS

The authors would like to thank the Natural Sciences and Engineering Research Council of Canada (NSERC) for their financial support. We thank sincerely the editor and referees for their constructive remarks. We would like to also acknowledge Rachelle Bellerive for her support in statistical analysis, and I. Bazier, C. Ouellet and L. Pelletier for their assistance during field works.

REFERENCES

Alcamo J, Flörke M, Märker M (2007). Future long-term changes in global water resources driven by socio-economic and climatic change. Hydrol. Sci. J. 52(2):247–275.

Bailey SE, Guimond JK (2009). Aboveground biomass and nutrient limitation in relation to river connectivity in Montane Floodplain Marshes. Wetlands 29:1243–1254.

Brady NC, Weil RR (2008). The Nature and properties of soils, fourteenth edition. Pearson, Columbus, OH, USA.

Carter MR, Gregorich EG (2006). Soil Sampling and Methods of Analysis. Canadian Society of Soil Science, 2e edition, Lewis Publishers, Boca Raton, Fl. 2006.

Clinton SM, Power M, Swenson R (2006). Impacts of inundation regime, floodplain vegetation, and burrowing animals on the incorporation of carbon into floodplain soils, CALFED Science Fellows Program, Project Number : R/SF-1, 2006, pp. 1-9.

CSCW (Canadian Soil Classification Working Group) (1998). The Canadian System of Soil Classification, 3rd ed. Agriculture and Agri-Food Canada Publication 1646. P. 187.

Dai F, Su Z, Liu S, Liu G (2011). Temporal variation of soil organic matter content and potential determinants in Tibet, China Catena, 85:288-294. DOI: 10.1016/j.catena.2011.01.015.

Daniels JM (2003). Floodplain aggradation and pedogenesis in a semiarid environment. Geomorphology 56:225–242.

Drouin A, Saint-Laurent D, Lavoie L, Ouellet C (2011). High-Precision Digital Elevation Model to evaluate the spatial distribution of Soil organic carbon in active floodplains. Wetlands 31:1151-1164.

Duchaufour P (2001). Introduction à la science du sol : sol, végétation, environnement. 6e édition, Dunod, Paris, P. 331.

Environment Canada (2012). National Climate Data and Information Archive, Canadian Climate Normals 1971-2000. http://climate.weatheroffice.gc.ca/ climate_ normals/index_e.html.

Environment Canada (2011). Hydrological Data, Hydat CD-Rom, version 2.02, Canada.

Environment Canada (EC) and Ministère de l'Environnement (MENV), (1982). Flood risk map: Saint-François River Basin (scales 1: 15,000, 1:10,000). Québec, Canada.

Heimann DC, Roell MJ (2000). Sediment loads and accumulation in a small riparian wetland system in northern Missouri. Wetlands 20(2):219-231.

Jones NK (2008). An Overview of Recent Climate Change and Streamflow in the Massawippi River Basin. J. Eastern Townships Stud. 32-33:23-38.

Kay AL, Jones RG, Reynard NC (2008). RCM rainfall for UK flood frequency estimation. II. Climate change results. J. Hydrol. 318(1-4):163-172.

Kayranli B, Scholz M, Mustafa A, Hedmark A (2010). Carbon storage and fluxes within freshwater wetlands: a critical review. Wetlands 30:111–124.

Lavoie L, Saint-Laurent D, St-Laurent J (2006). Analyse pédologique et sédimentologique des sols alluviaux et paléosols des terrasses d'inondation. Can. J. Soil Sci. 86:1-14.

Lützow MV, Kögel-Knabner I, Ekschmitt K, Matzner E, Guggenberger G, Marschner B, Flessa H (2006). Stabilization of organic matter in temperate soils: mechanisms and their relevance under different soil conditions - a review. Eur. J. Soil Sci. 57:426-445.

McKeague JA (1978). Manual on Soil Sampling and Methods of Analysis. Canada Soil Survey, Subcommittee on Methods of Analysis. (2th ed.), Ottawa, pp. 212-211.

MDDEP (2013a). Centre d'expertise hydriques du Québec (CEHQ). Ministère du Développement durable et des Parcs, gouvernement du Québec. http://www.cehq.gouv.qc.ca/hydrometrie/index.htm

MDDEP (2013b). Normales climatiques du Québec 1981-2010. Ministère du Développement durable et des Parcs, gouvernement du Québec. http://www.mddep.gouv.qc.ca/climat/normales/sommaire.asp.

Monirul QM, Warrick RA, Ericksen NJ (2003). The Implications of Climate Change on Floods of the Ganges, Brahmaputra and Meghna Rivers in Bangladesh. Climatic Change 57(3):287-318.

Nadeu E, De Vente J, Martinez-Mena M, Boix-Fayos C (2011). Exploring particle size distribution and organic carbon pools mobilized by different erosion processes at the catchment scale. J Soil Sedim. 11(4):667-678.

Ngailo JA, Vieira SR (2012). Spatial patterns and correlation of soil properties of a lowland soil J. Soil Sci. Environm. Manag. 12:287-296.

Rokosh AE, Bouchard V, Fennessy S, Dick R (2009). The use of soil parameters as indicators of quality in forested depressional wetlands. Wetlands 29(2):666-677.

Ruiz-Sinoga JD, Pariente S, Diaz AR, Martinez Murillo JF (2012). Variability of relationships between soil organic carbon and some soil properties in Mediterranean rangelands under different climatic conditions (South of Spain). Catena 94:17–25.

Saint-Laurent D, St-Laurent J, Lavoie L, Ghaleb B (2008). Use geopedological methods for the evaluation of sedimentation rates on river floodplains. Catena 73:321-337.

Saint-Laurent D, Mesfioui M, Evin G (2009). Hydroclimatic variability and relation with flood events (Southern Québec, Canada). Water Resources, 36(1):43-56. DOI:10.1134/S0097807809010047.

Saint-Laurent D, St-Laurent J, Lavoie L, Drouin A, Ghaleb B (2010). Floodplain sedimentation rates, soil properties and recent flood history in Southern Québec. Global. Planetary Change 70:76-91.

Schilling KE, Palmer JA, Bettis III EA, Jacobson P, Schultz RC, Isenhart TM (2009). Vertical distribution of total carbon, nitrogen and phosphorus in riparian soils of Walnut Creek, southern Iowa. Catena 77:266-273.

Steiger J, Gurnell AM (2003). Spatial hydrogeomorphological influences on sediment and nutrient deposition in riparian zones: observation from Garonne River, France. Geomorphology, 49(1-2):1-23. DOI: org/10.1016/S0169-555X(02)00144-7.

Tockner K, Pennetzdorfer D, Reiner N, Schiemer F, Ward JV (1999). Hydrological connectivity, and the exchange of organic matter and nutrients in a dynamic river – floodplain system (Danube, Austria), Freshwater Biol, 41:521-535.

Whitehead PG, Wilby RJ, Battarbee RW, Kernan M, Wade AJ (2009). A review of the potential impacts of climate change on surface water quality. Hydrol. Sci. J. 54(1):101–124.

Wilby RL, Keenan R (2012). Adapting to flood risk under climate change. Progr. Phys. Geogr. 36(3):348–378.

Yeomans JC, Bremner JM (1988). A rapid and precise method for routine determination of organic carbon in soil. Soil Sci. Plant Ann. 19:1467-1476.

Zhang L, Mitsch WJ (2007). Sediment chemistry and nutrient influx in a hydrologically restored bottomland hardwood forest in Midwestern USA. River. Res. Applic. 23(9):1026-1037.

5

Soil sulfur availability due to mineralization: Soil amended with biogas residues

Mewael Kiros Assefa[1], S. von Tucher[2] and Urs Schmidhalter[2]

[1]Department of Plant Sciences, Aksum University, Shire- campus P. O. Box 314, Shire-endaslassie, Ethiopia.
[2]Plant Nutrition, Technische Universität München, Emil-Ramann-Straße 2, 85350 Freising-Weihenstephan, Germany.

Sulfur, an essential element for plant growth, has received lesser attention than it deserves. Current inputs of sulfur to agricultural soils from atmospheric deposition have reduced to less than the amount sulfur required by most crops. In such soils the release of sulfur from organic matter is vital for the supply of sulfur. A pot experiment without plant (incubation study) in a green house was conducted at the experimental station Dürnast TU Weihenstephan to investigate the potential of biogas residues as sources of available sulfur. The pot experiment comprised 20 different fertilization variants; 16 biogas residues and 3 mineral sulfur fertilizer variants, that is, S- 30, S-60, and S- 90 mg/pot, and a control (S-0), replicated trice. Pots were arranged in completely randomized design. Soil SO_4^{2-} -S and NO_3^--N contents upon were measured trice during the study period. Soil SO_4^{2-}-S and NO_3^--N content were significantly influenced by fertilizer treatments at all times. Generally, liquid biogas residues tend to show higher soil SO_4^{2-}-S and NO_3^--N content. Soil SO_4^{2-}-S content varied from 26.1 to 100.20 mg pot^{-1} (625-S, S-90); 30.4 to 100.4 mg pot^{-1} (621-S, 626-L) and 31.4 to 98.2 mg pot^{-1} (S-0, 626-L), in first, second and third sampling times, respectively. Soil NO_3^--N content also varied from 114.5 to 526.5 mg pot^{-1} (629-S, 616-L); 99.0 to 1054 mg pot^{-1} (620-S, 616-L); 114.8 to 1045.4 mg pot^{-1} (620-S, 616-L), in first, second and third sampling times, respectively. Biogas residues containing more than 0.1% S_t in fresh matter and C_{org}: S ratio lower than 30 could replace short term sulfate limitation.

Key words: Biogas residues, sulfur mineralization, sulfur availability, C_{org}:S ratio.

INTRODUCTION

Sulfur, an essential element for plant growth, has received lesser attention than it deserves. In recent years, deficiency of sulfur in crops has increased worldwide and has been recognized as a constraint in crop production (Eriksen et al., 2004; Girma et al., 2005; Mascagni et al., 2008). The main reasons are: (1) the environmental control of sulfur dioxide emissions in industrial areas, (2) the increasing use of high-analysis, sulfur free fertilizers, (3) the decreasing use of sulfur-containing pesticides and fungicides, (4) the adoption of high-yielding crop cultivars which demand a high fertility level and results in greater exploitation of soil reserve nutrients and removal of much larger quantities of nutrients in the harvested crop, and (5) the increased cropping intensity (intensive cultivation) (Scherer, 2001). microfauna, have been widely documented (Heimann Current inputs of sulfur from atmospheric deposition are less than 10 kg ha^{-1} in most Western European countries (Hu et al., 2005), which is less than the amounts of sulfur required by most crops (McGrath et al., 2002 as cited in

Scherer, 2009). In Western Europe, incidence of sulfur deficiency has increasingly been reported in *Brassica* over the last decade (Scherer, 2001). In Bavaria, in the last 15 years, sulfur emissions were reduced from 30 to 60 kg ha^{-1} (Guster and Tucher, 2001). Therefore, demand based sulfur fertilization for specific crops (20-70 kg ha^{-1}) is becoming increasingly important.

In agricultural systems, where sulfur inputs from fertilizer and atmospheric deposition are low the release of sulfur from organic forms is important for the supply of sulfur to plants and a valuable parameter is the measurement of the potential contribution that the organic sulfur pool makes to plant-available sulfur, especially following the addition of organic matter to the soil. According to Bermuth (2008), despite the steadily growing number of biogas plants which produce biogas residues which are heterogeneous in their composition and used as organic fertilizers, the effect of the biogas residues as a source of available sulfur is largely unknown. Moreover, biogas residues are often mechanically separated into a liquid and a solid fraction. The effect of separation on the availability of S is largely unknown. This study was, therefore, conducted to determine available sulfur in an incubation experiment of soil amended with 16 different biogas residues by way of mineralization.

MATERIALS AND METHODS

Pot experiment in a green house was conducted at the experimental station Dürnast TU Weihenstephan to investigate the potential of biogas residues as sources of available sulfur.

Experimental set up

Top soil was collected from a permanent arable land near the experimental station Dürnast TU Weihenstephan. The soil was sandy loam (58% sand, 27% silt and 15% clay), pH 5.7, and contained C_t 0.80%; N_t 0.09%; P CAL 3.5 mg 100 g^{-1}; kcal 9.6 mg 100 g^{-1}; Mg $CaCl_2$ 11.2 mg 100 g^{-1}, which was percolated to wash out SO_4^{2-} -S and was air dried and mixed to create homogenous soil conditions which was then filled up to a 6 plus kg pot containing holes that ensure free drainage to impose treatments. The experiment comprised 20 different fertilization variants replicated trice setup in a Completely Randomized Design (CRD) in a green house. Sixteen of which included biogas residues from different provenances. Three of the variants were mineral sulfur fertilizer different rates *vis*., S- 30, S-60, and S- 90 mg/pot.

Biogas residues and fertilization

Biogas residues (BGR) with a broad spectrum of substrates composition were collected from different provenances. BGR were either used directly after the fermenter, that is, raw non-separated (NS) (AD7=628-NS) or mechanically separated by a screw separator into a liquid fraction (614-619+627) and a solid fraction (620-625+629). Biogas residue 626 was derived from the liquid fraction 627 after reverse osmosis. In all biogas plants the purification of the biogas from SO_2 was performed by the addition of iron chloride and/or the injection of air.

BGR were analyzed for dry matter content, S_t, SO_4^{-2}-S, N_t, NH_4-N, P, K, Mg and Cl on fresh matter basis. Since BGR contained very different amounts of sulfur and had different dry matter, accurate adjustments were necessary to 90 mg S per pot application.BGR already contained certain amounts of N, P, K and Mg, which varied with residue types and thus based on pak choi requirement (experimental plant of parallel experiment- result not presented here) all pots received adequate amount of NH_4NO_3, Ca $(H_2PO_4)_2.H_2O$, KNO_3 and Mg $(NO_3)_2.6H_2O$ to level out N (900 mg/pot), P (100 mg/pot), K (800 mg/pot), and Mg (100 mg/pot).

In order to impose mineral sulfur fertilization variant treatments (S-30, S-60 and S-90, mg of sulfur), sulfur was applied in the form of $MgSO_4$ along with KNO_3 as a compensatory fertilization of nitrogen for shooting. Each of the sixty pots were filled with approximately 6 kg of soil portion in such a way that the bottom layer contained 2.5 kg of soil, mechanically shaked to adjust the bulk density. The next layer (middle layer) was filled with another 2.5 kg of soil to which the fertilizer treatments were added and thoroughly mixed and was again shaked. The top layer was only 1 kg of soil, which was not shaked, to cover the fertilized layer.

Pots were regularly watered with distilled H_2O according to 60 to 70% of the maximum water holding capacity (100% water holding capacity: 29.3% gravimetric soil water content) corresponding to a gravimetric soil water content of 17.6 to 20.5%. Pots were monitored for their SO_4^{2-}-S and NO_3^- content.

Laboratory analysis

Analysis of the biogas residues

Liquid, solid and raw BGR were analyzed for dry matter content (drying at 100°C after the addition of sand for the liquid residues), for pH value (with pH sensitive electrode after the addition of $CaCl_2$ for the solid residues), for total contents of S, P, K and Mg after wet digestion with HNO_3 and perchloric acid with an ICP-OES (Liberty, Varian, Mulgrave, Australia) using the lines of 182.034 nm for S, 213.618 nm for P, 769.896 nm for K, 285.213 for Mg), for total N after Kjeldahl digestion, steam distillation and titration with H_2SO_4, for NH_4-N by steam distillation and titration with H_2SO_4, for Cl^- and SO_4^{2-}-S after extraction with H_2O dist. by ion chromatography (for further details see "soil analysis"). Organic carbon was determined by subtracting carbonate-C (measured following the Scheibler procedure, that is gas-volumetric measurement of CO_2 release after the addition of HCl) from total C which was measured after freeze-drying using an isotopic ratio mass spectrometer (Anca, SL 20-20, Europa Scientific, Crewe, UK).

Soil analysis

Soil samples were air dried and ground to pass a 2 mm sieve. All samples were extracted in 100 ml of $CaCl_2$ (0.01mol/l) by shaking for 60 min on an end-over-end shaker. Extracts were filtered through a filter paper and were stored at -15°C prior to analysis. Thawed extracts were taken directly for SO_4^{2-} -S and NO_3^--N determination by ion chromatography. Chromatographic conditions were: column AS4A; eluent: 1.8 mmol L^{-1} Na_2CO_3 / 1.7 mmol L^{-1} $NaHCO_3$ pH 8-9; flow: 2 ml min^{-1}; detection by conductivity.

Statistical analysis

Statistical significance of differences between treatments were determined by Tukey HSDa,b mean separation test for multiple comparisons after analysis of variance (ANOVA), using IBM SPSS 19. All tests were performed at a significance level of $P < 0.05$. Correlation analysis was also employed.

Table 1. Some selected characteristics of biogas residues variants and their composition.

BGR fertilizer variants	Organic C (% in FM)	Total N (% in FM)	Total S (% in FM)	C:N ratio	C_{org}:S_t ratio	N:S ratio
Liquids						
614	2.28	0.57	0.038	4.4	60.4	15.0
615	3.21	0.75	0.039	4.6	81.9	19.2
616	1.59	0.54	0.032	3.7	49.5	16.8
617	2.75	0.63	0.043	4.8	64.5	14.9
618	3.46	0.76	0.046	4.9	75.6	16.6
619	2.17	0.53	0.032	4.4	68.8	16.8
626	0.23	0.42	0.166	0.6	1.4	2.5
627	1.77	0.78	0.120	2.9	14.8	6.5
Solids						
620	10.25	0.56	0.080	18.4	128.4	7.0
621	9.70	0.64	0.060	15.3	161.4	10.7
622	9.45	0.77	0.082	12.4	115.3	9.4
623	10.28	0.57	0.044	18.0	233.6	13.0
624	11.41	0.74	0.068	15.7	168.3	10.9
625	9.31	0.70	0.063	13.5	148.3	11.1
628NS*	3.53	0.81	0.122	4.9	28.9	6.6
629	7.00	0.79	0.251	9.7	27.9	3.1

NS*, Raw non-separated BGR.

RESULTS

Soil sulfate (SO_4^{2-}-S) content

Determination of available soil sulfur from biogas residue fertilizers is very important to gain better understanding of their mineral equivalence capability. SO_4^{2-}-S content was significantly influenced by fertilizer treatments at all times. Generally, liquid biogas residues tend to show higher SO_4^{2-}-S content.

At first time sampling, that is 4 days after fertilizer application, SO_4^{2-}-S content was significantly influenced by fertilizer treatments. Significantly highest SO_4^{2-}-S content (100.2 mg pot^{-1}) was recorded with S-90 mineral fertilization, which was on par with pots treated with 626-L and S-60 followed by pots treated with 627-L, 629-S, 628-NS, and S-30 where as the lowest SO_4^{2-}-S content (26.0 mg pot^{-1}) was recorded in pots fertilized with 625-S, which was on par with pots treated with 621-S , S-0, 624-S, 620-S, 623-S, and others.

Similarly, at second time sampling, that is, 22 days after fertilizer application, SO_4^{2-}-S content was significantly influenced by fertilizer treatments. Significantly highest SO_4^{2-}-S content (101.4 mg pot^{-1}) was recorded in pots fertilized with 626-L which was on par with pots treated with S-90, S-60, 627-L, 628- NS and 629-S followed by pots treated with 616-L, 617-L, and 614-L where as significantly lowest SO_4^{2-}-S content (30.4 mg pot^{-1}) was observed in pots treated with 621-S which was on par with pots treated with S-0, 620-S, 625-S, 623-S, 624-S, 619-L and others.

Also at the third sampling, that is, 30 days after fertilizer application, SO_4^{2-}-S content was significantly influenced by fertilizer treatments. Significantly highest SO_4^{2-}-S content (98.2 mg pot^{-1}) was recorded in pots fertilized with 626-L which was on par with pots treated with S-90, S-60, 629-S, 627-L and 628-NS followed by pots treated with 617-L, 616-L and 622-S where as significantly lowest SO_4^{2-}-S content (31.4 mg pot^{-1}) was recorded in pots treated with S-0 which was on par with pots treated with 621-S, 620-S, and others.

The relatively higher $CaCl_2$- extractable SO_4^{2-}-S status in 626-L and 627-L fertilized pots could be attributed to higher totals sulfur and lowers C_{org}:S ratio of biogas residues. Table 2 was worked out from Table 1.

At each sampling time there was highly significant positive correlation between SO_4^{2-} S contents in soil and the S_t contents of the biogas residues and a highly significant negative correlation to the C_{org}: S_t ratio of the biogas residues.

Figure 1 depicts that changes in the soil SO_4^{2-} -S content did not show clear pattern of mineralization in the course of the study period. However, 67% (622-S) up to 95% (615-L) of the SO_4^{2-} -S found at the 3rd sampling 30 days after fertilizer application were already present at the 1st sampling 4 days after fertilizer application. The increase in soil SO_4^{2-} -S content from the 2nd (22 days after fertilizer application) to the 3rd sampling was low and seemed to be more pronounced in the group of the solid biogas residue fractions (e.g. 620-S, 621-S, 623-S and

Table 2. Correlation coefficient (r) between biogas residue properties and $CaCl_2$ extractable soil SO_4^{2-}-S.

Biogas residue properties	$CaCl_2$ extractable soil SO_4^{2-} content		
	1st sampling	2nd sampling	3rd sampling
S_t	0.68**	0.70**	0.74**
C_{org}: S_t	-0.83**	-0.82**	-0.79**
N:S	-0.548*	-0.575*	-0.654**

*Significant at 5% level; **Significant at 1% level; BGR substrate composition (Fouda, 2011) is - 614 + 620: Turkey manure 2%, Maize 98%; 615+621: Clover 15%, Triticale 25%, Maize 60%; 616+622: Triticale meal 5%, Corn grain meal 7%, Green rye 13%, Maize 75 %; 617+623: Cereals 6%, Sudan grass, Green grass 8%, Maize 86%; 618+624: Maize 20%, Grass silage 80%; 619+625: Cracked wheat 5%, Sudan grass 48%, maize 47%; and 626, 627 +628,629: Grass silage 10%, Grass meal 5%, whole plant silage (rye, barley, sorghum) 5%, maize silage 15%, Poultry manure 15%.

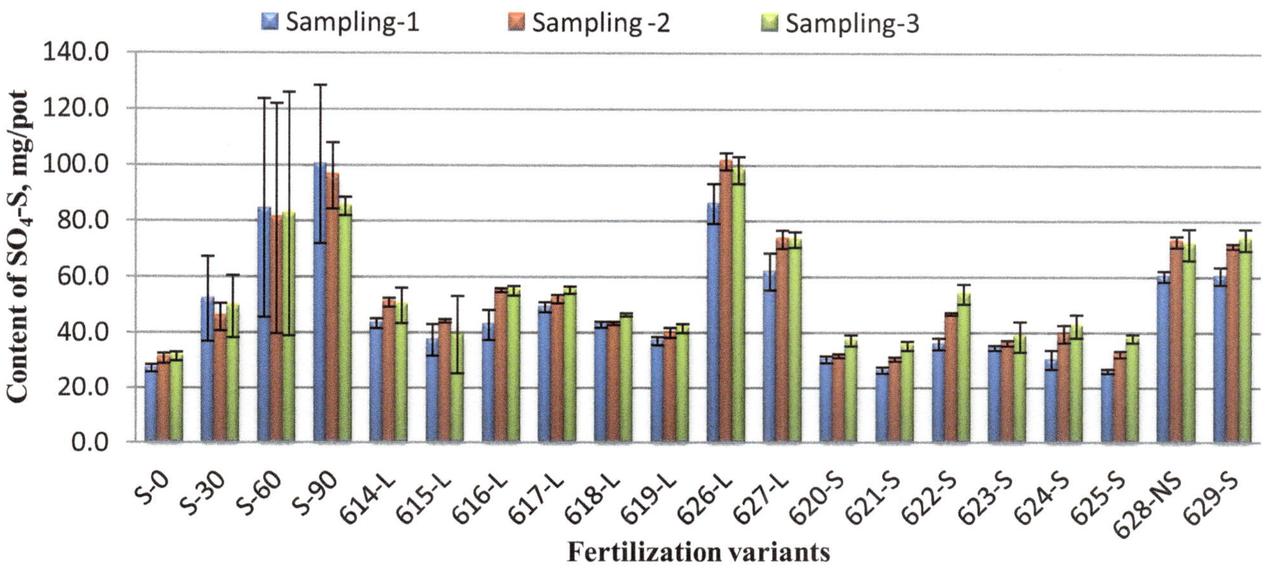

Figure 1. Time course of SO_4^{2-}-S release from biogas residues. Error bars represent standard deviations.

625-S).

Figure 2 depicts the amounts of SO_4^{2-}-S recovered as the proportion of S present as SO_4^{2-}-S from the biogas residues (that is, minus the unfertilized control) in relation to the S applied as fertilizer (90 mg) from biogas treatments variants using 10 mM $CaCl_2$ in relation to C_{org}:S_t ratio. It is obvious that the relation was non-linear. None of the samples showed values significantly lower than zero.

Soil nitrate (NO_3^--N) content

Soil nitrate (NO_3^--N) content was significantly influenced by fertilizer treatments at all times. Similar to SO_4^{2-}-S content, liquid biogas residues tend to show higher NO_3^--N content.

At first time sampling, NO_3^--N content was significantly influenced by fertilizer treatments. Significantly highest NO_3^--N content (526.6 mg pot^{-1}) was recorded with 619-L fertilization, which was on par with pots treated with 618-L and 617-L followed by pots treated with 616-L where as the lowest NO_3^--N content was recorded in pots fertilized with 629-S, which was on par with pots treated with 620-S, 621-S, 626-L, 622-S, and 623-S.

Also during second time sampling, NO_3^--N content was significantly influenced by the fertilizer treatments. Significantly highest NO_3^--N content was recorded with 616-L fertilization, which was on par with pots treated with 618-L, 619-L, 614-, 615-L and 617-L followed by pots treated with 627-L where as the lowest NO_3^--N content (99.0 mg pot^{-1}) was recorded in pots fertilized with 620-S, which was on par with pots treated with 621-S, 629-S, 623-S, and others.

Similarly, in third time sampling, NO_3^--N content was significantly influenced by the fertilizer treatments. Significantly highest NO_3^--N content (1045.4 mg pot^{-1}) was recorded with 616-L fertilization, which was on par with pots treated with 618-L, 619-L, 617-L and 614-L (followed by pots treated with 615-L where as the lowest NO_3^--N content (114.81 mg pot^{-1}) was recorded in pots fertilized with 620-S, which was on par with pots treated

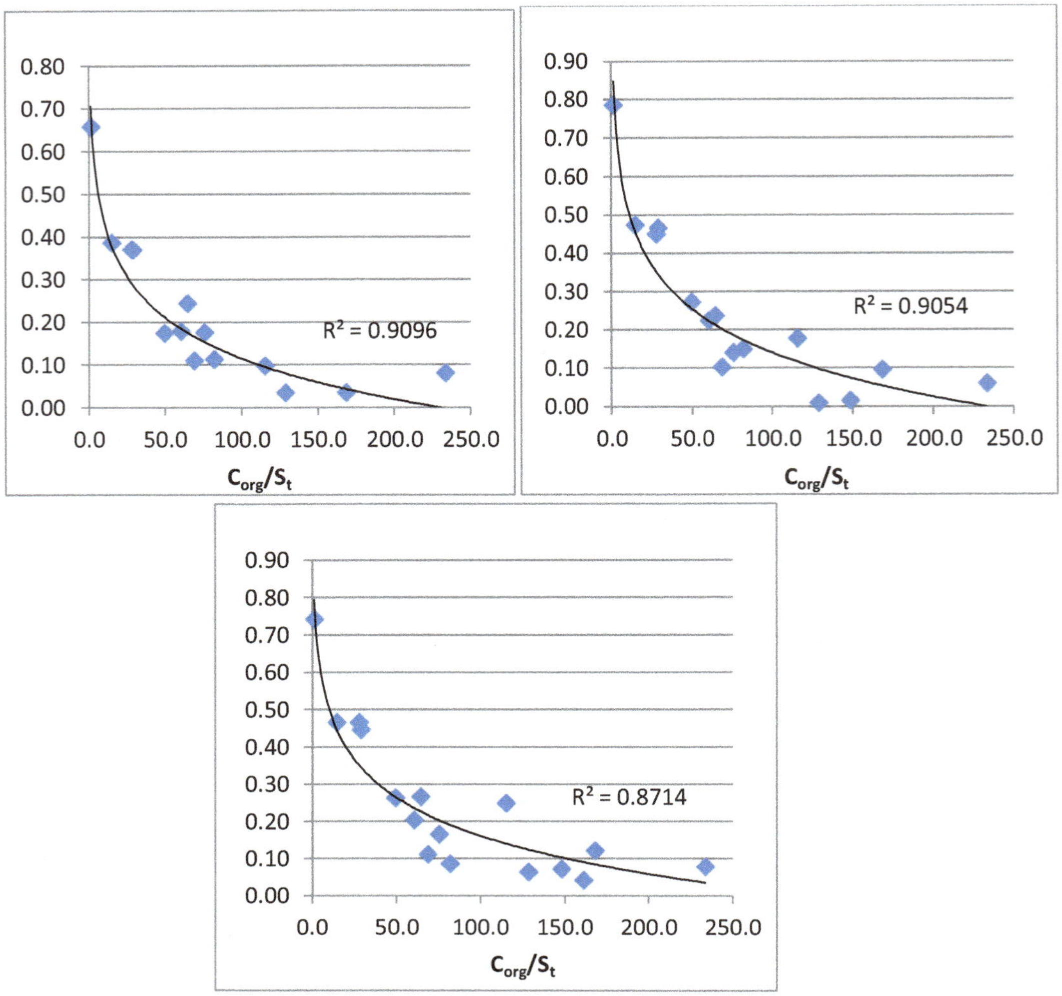

Figure 2. Relation of S-recovery to C_{org}:S_t ratio.

with 621-S, 629-S, 623-S, 626-L.

DISCUSSION

The study of sulfur transformation is important for sulfur fertilizer optimization. It has been suggested that, the mineralization of sulfur involves both biological and biochemical mineralization (McGill and Cole, 1981). Biological mineralization is driven by the microbial need for organic C to provide energy, and S released as sulfate is a by-product when microorganisms break down organic compounds to acquire C for their energy metabolism. Biochemical mineralization is the release of sulfate from sulfate-esters through enzymatic hydrolysis. Whereas the mineralization of C-bonded S is strictly dependent on microbial activity, sulfate-esters can be readily hydrolyzed by sulfatase enzymes in the soil, and biochemical mineralization is, therefore, controlled by the supply of S rather than the turnover of organic matter.

The experiment showed biogas residues' clear differences presumably attributed to their different substrate composition and/or the efficiency of the biogas process. The substrate compositions were, actually, subject to different degradation rates during the fermentation process depending on the length of time, temperature and mixing ratio in the fermenter. The C_{org}:S ratio of various biogas residues fertilizers ranged from 1 to 234.

SO_4^{2-}-S content measured in the unplanted soil was significantly influenced by the fertilizer treatments at all times. Generally, liquid biogas residues tended to show higher SO_4^{2-}-S content. Among the biogas residue variants the mean available sulfur content varied from 26 to 86.3 mg pot^{-1} in first sampling, 30 to 101 mg pot^{-1} in second sampling, and 31-98 mg pot^{-1} in the third sampling.

About 67 to 95% of the SO_4^{2-}-S measured in the third sampling were already present 4 days after the application of the biogas residues. As the SO_4^{2-}-S applied by biogas residues was marginal (up to about 1 mg SO_4^{2-}-S per pot), the rapid increase in soil SO_4^{2-}-S may be due to oxidation of S^0 or S^{2-} derived from the biogas purification process (Abatzoglou and Boivin, 2009). In soils under aerobic conditions these S forms will be readily oxidized to sulfate (Janzen and Ellert, 1998; Scherer, 2009). Although there is no information on the S forms present in the biogas residues tested here, it is interesting that sulfate formation seemed to be completed within 3 to 4 weeks (Figure 1).

The overall mineralization of SO_4^{2-} -S 4 days, 22 days and 30 days after fertilizer application was strongly correlated with the organic C and total S (S_t) of the biogas residues (BGRs). Table 2 shows the extent to which available sulfur mineralized was strongly correlated with the total sulfur, that is, 0.68**, 0.70**, 0.74**; and Corg:S, that is, -0.83**, -0.82**, and -0.79** at first, second and third sampling, respectively. The fact that there exist significant negative correlation between the short term S availability and C: S ratio of the organic fertilizers agrees with the finding by Guster and Tucher (2001).

The biogas residue (BGR) L-626 happened to consistently record highest SO_4^{2-} -S content (when compared with rest of BGRs) which was on par with the apparently highest SO_4^{2-} -S recorded in the S-90 treated pots (86.3, 100.2; 101.4, 96.2 and 98.2, 85.4 mg/pot of BGR and S-90 SO_4^{2-} -S content, respectively across the three samplings). The relatively higher S mineralization of the BGR L-626 was presumably attributed to its low C_{org} content and high S_t content (1.4 g/100 g, 0.169 g/100 g) (Table 1). L-627, S-628 and S-629 also showed consistent and comparable high SO_4^{2-} -S content throughout the study period.

The BGRs to record low $CaCl_2$- extractable SO_4^{2-}-S consistently were S-625, S-621 and S-620 which are characterized by one of the highest C_{org} (9.31, , 9.70, 10.25 g/100 g) and least S_t content (0.064, 0.067, 0,095 g/100g), respectively.

Moreover, data for the amounts of SO_4^{2-} -S recovered from all of the treatments using 10 mM $CaCl_2$ in relation to C_{org}:S_t ratio depicted in Figure 2 reveals that S-recovery increased with smaller C_{org}:S_t ratio at all times. This again substantiated the fact that mineralization of organic fertilizers increase as its C_{org}:S ratio decreases. None of the samples showed values significantly lower than zero which implies there was no immobilization. Taking the third sampling time into consideration, recovery was from as high as 74% in L-626 to as low as only 4 % in 621-S. When compared against S-90 (81, 73 and 60%), except in the first sampling, S-recovery was higher in 626-L (66, 79 and 74.5%) which implies some organic biogas residue fertilizers could replace mineral fertilizer equivalent for sulfur nutrition or that at least no immobilization could indicate a potential source of available sulfur. S-recovery showed most of the solid BGRs almost did not release SO_4^{2-} -S.

Earlier studies have shown that the incorporation of organic residues with narrow C_{org}:S ratio (< 200:1) can increase the contents of SO_4^{2-} -S in the soil, whereas the incorporation of those organic residues which have wide C_{org}:S ratio (>400: 1) may result in a net immobilization of soil inorganic-S (Barrow, 1960; Stevenson, 1986). In this study, despite the non-equivalency of the above reported cut off points for mineralization and immobilization the tendency matches. In the treatments using biogas residues, the C_{org}:S ratio in the pot 626-L which recorded highest SO_4^{2-} -S had a very low C_{org}:S ratio (that is, organic carbon was by far lower in 626-L which is the biogas residue which was improved by reverse osmosis after the separation of the liquid fraction) whereas the pot with the least SO_4^{2-} -S record had C_{org}:S ratio of 234 (623-S) followed by 168 (624-S). Moreover, the N/S ratio showed significant inverse relationship with sulfur mineralization of $r= -0.548^*$, $r= -0.575^*$, $r= -0.654^{**}$ at first sampling, second sampling and third sampling, respectively. In agreement to this result, Kowalenko and Lowe (1975) reported that a high N: S ratio resulted in a decrease in mineralization of sulfur in soil sample during incubation.

Mineralized sulfur probably originates from both the sulfate-ester and the C-bonded S pool (Freney et al., 1975; McLaren et al., 1985) although C-O-S is considered more labile than C-S. The pot experiment without plants indicated mineralization-immobilization processes were occurring simultaneously resulting in minimal net changes in $CaCl_2$-extractable SO_4^{2-} -S concentrations. This is presumably because short term sulfur cycling evaluation involves small fraction of potentially available sulfur when mineralization may need longer time to show its effect. This goes in line to the finding by Saggar et al. (1981) that the short term supply of available sulfur upon mineralization is low.

There was significant difference in NO_3^--N content among fertilizer treatments at all times. BGRs L-619, L-618, L-617 and L-616 showed consistently higher NO_3^--N content. This, however, did not seem to relate to the S-fertilization variation. Higher NO_3^--N records tend to show strong positive correlation ($r=0.851^{**}$) to the amount of supplemental mineral NH_4-N fertilizer applied to bring treatments to equal bases. And, thus, the increased changes in NO_3^--N content with time in many cases is seemingly due to nitrification rather than mineralization.

In conclusion it gives the impression sulfur limitation from less atmospheric deposition can in the short term be replaced by some biogas residues, given that the biogas residues are high in total sulfur content *i.e.*, higher than about 0.1% S_t in fresh matter and in C_{org}:S ratio lower than 30.

ACKNOWLEDGMENT

This work was supported by the Chair of Plant Nutrition, Technische Universität München. The authors would like to express their deep gratitude to the greenhouse and laboratory technicians in the experimental station Dürnast and the Institute of Plant Nutrtion TU Weihenstephan, Germany.

REFERENCES

Abatzoglou N, St. Boivin (2009). A review of biogas purification processes. Biofuels Bioprod. Bioref. 3:42-71.

Barrow, N. J. (1960). A comparison of the mineralization of nitrogen and of sulfur from decomposing organic materials. Aust. J. Agric. Res. 11:960–969.

Bermuth, S. (2008): Effect sulfur on from biodigest on yield compostion and quality of wheat. *Master Thesis*. Technical University of Munich, Germany.

Eriksen, J., K. Thorup-Christensen, M. Askegard (2004). Plant availability of catch crop sulfur following spring incorporation. J. Plant Nutr. Soil Sci. 167: 609–615.

Freney JR, Melville GE, Williams CH (1975). Soil organic matter fractions as sources of plant-available sulfur. Soil Biol. Biochem. 7:217–221.

Girma K, Mosali J, Freeman KW, Raun WR, Martin KL, Thomason WE (2005). Forage and grain yield response to applied sulfur in winter wheat as influenced by source and rate. J. Plant Nutr. 28:1541–1553.

Guster R, Tucher SV (2001). Plant availability of sulfur from organic fertilizers. *In*: Horst W.J. *et al.* (eds.) Plant Nutrition-Food Security and sustainability of agro-ecosystems through basic and applied research. XIV International Plant Nutrition Colloquium. Kluwer Academic Publishers, Dordrecht, Developments in Plant and Soil Sciences 92:844-845.

Hu ZY, Zhao FJ, McGrath SP (2005). Sulfur fractionation in calcareous soils and bioavailability to plants. Plant Soil 268:103–109.

Janzen, H.H., B.H. Ellert (1998): Sulfur dynamics in cultivated, temperate agro-ecosystems. In: Maynard, D.G. (ed.), Sulfur in the Environment. Marcel Dekker, Inc., New York, pp. 11–43.

Kowalenko CG, Lowe LE (1975). Mineralization of sulfur from four soils and its relationship to soil carbon, nitrogen and phosphorus. Can. J. Soil Sci. 55:9-14.

Mascagni H J Jr., Harrison SA, Padgett GB (2008). Influence of sulfur fertility on wheat yield performance on alluvial and upland soils. *Commun.* Soil Sci. Plant Anal. 39:2133–2145.

McGill, W.B., C.V. Cole (1981): Comparative aspects of cycling of organic C, N, S and P through soil organic matter. Geoderma 26:267–286.

McGrath, S. P., F.J. Zhao, M. M. A. Blake-Kalff (2002): History and outlook for sulfur fertilizers in Europe. Proceedings No. 497. International Fertiliser Society, York, UK.

McLaren, R. G., J.I. Keer, R.S. Swift (1985): Sulfur transformations in soils using sullfur-35 labelling. Soil Biol. Biochem. 17:73–79.

Saggar, S, Bettany JR, Stewart JWB (1981). Sulfur transformations in relation to carbon and nitrogen in incubated soils. Soil Biol. Biochem. 13:499–511.

Scherer HW (2001). Sulfur in crop production- invited paper. Eur. J. Agron. 14:81-111.

Scherer HW (2009). Sulfur in soils. J. Plant Nutr. Soil Sci. 172:326-335.

Stevenson FJ (1986). Cycles of soil: Carbon, Nitrogen, Phosphorus, Sulfur, Micronutrients. John Wiley & Sons, New York.

Isolation of microorganism from soil contaminated with degraded paper in Jharna village

Sharma Richa , Chandra Subhash and Singh Amrita

Department of Biotechnology and Allied Sciences, Jayoti Vidyapeeth Women`s University, Jaipur, Rajasthan, India.

Pulp and paper mills are categorized as a core sector industry and are the fifth largest contributor to industrial water pollution. The pollutant, discharge from the pulp and paper industry, affect all aspects of the environment such as water, air and land. Various studies have reported detrimental effects of pulp and paper mill effluent on animals living in water bodies receiving the effluent. The effects are in form of respiratory stress, oxidative stress, liver damage and genotoxicity. This study aimed to determine the species of fungi present in the soil environment of Jharna area. Different types of fungi were isolated from soil samples contaminated with degraded papers. Identification of fungi was carried out by culturing on Potato dextrose and Sabouraud's Dextrose agar media and microscopic method. In our study, dominant fungal flora, that is, *Aspergillus niger*, *Aspergillus flavus, Aspergillus fumigatus, Aspergillus terrus, Penicillium* sp., *Trichoderma* sp. and yeast were isolated. These fungi play important role in the paper and pulp industry in degradation of paper waste.

Key words: Pulp, paper, fungi, *Aspergillus* sp.

INTRODUCTION

The production, use and recycling of paper can have a number of adverse effects on the environment, that is, pollution. Pulp mills can emit various waste products to the air, water, and land. Discarded paper is a major component of many landfills sites, accounting for about 35% by weight of municipal solid waste before recycling (US EPA, 2005). Pulp and paper production is a major industry in India, with a total capacity of over 3 million tons per annum (CPCB, 2001). Most paper is made from cellulose pulp from tree logs. Cellulose is the most common organic crystalline polymer. Microbial degradation of lignocellulosic waste and the downstream products resulting from it is accomplished by a concerted action of several enzymes, the most prominent of which are the cellulases, which are produced by a number of microorganisms (Sukuraman et al., 2005). Cellulases are a group of hydrolytic enzymes capable of hydrolyzing cellulose to smaller sugar components like glucose units. Cellulolytic enzymes play an important role in nature's biodegradation processes where plant lignocellulosic material is efficiently degraded by cellulolytic fungi and bacteria. In industry, these cellulolytic enzymes have found novel application in the production and processing of chemicals, food and manufactured goods such as paper, rayon and cellophane (Mandels, 1985; Kader et al., 1999). Microorganisms play an important role in natural biodegradation processes in which plant lignocellulosic materials are efficiently degraded by cellulolytic fungi, bacteria, actinomycetes and even protozoa. Most commonly studied cellulolytic organism include fungal species are *Trichoderma, Humicola, Penicillium, Aspergillus*; Bacteria-*Bacilli, Pseudomonas, Cellulomonas*; and Actinomycetes-*Streptomyces* and *Actinomucor. Trichoderma .reesei*, *Humicola*, *Penicillium*

Table 1. Physiochemical properties of soil samples.

Physiochemical property of soil	Readings
pH	6.0 ±1.4
Temperature	31 °C
Humidity	36% Fair
Air pressure	749 mb

Table 2. Fungal growth in the soil environment of Jharna area using two different media.

Fungi	Growth of fungi on PDA	Growth of fungi on SDA
Aspergillus niger	+4	+4
Aspergillus flavus	+3	+2
Aspergillus fumigatus	+3	+2
Aspergillus terrus	+4	+4
Aspergillus nidulans	+2	+2
Penicillium chrysogenum	+4	+4
Trichoderma sp.	+2	+2
Mucor sp.	+3	+4
Yeast	+3	+3

Growth was scored in the following manner: +4: Excellent growth; +3: Good growth; +2: Fair growth; +1 poor growth, 0: no growth.

and *Aspergillus* have the ability to yield extracellular cellulases (Sukuraman et al., 2005). Generally, fungi produce three major types of cellulolytic enzyme: Endoglucanase, exoglucanase and cellobiohydrolase (Klyosov, 1990). These enzymes are extracellular and inductive in nature (Enari, 1983; Kader et al., 1999). The ability to produce cellulose is widespread among fungi and this has become the subject of extensive investigation. In the pulp and paper industry, cellulases and hemicellulases have been employed for biochemical pulping for modification of the coarse mechanical pulp and hand sheet strength properties, de-inking of recycled fibers and for improving drainage and runnability of paper mills (Bedford et al., 2003; Akhtar, 1994; Pere et al., 1995; Prasad et al., 1992). The objective of this work was to isolate microorganisms from soil contaminated with degraded paper.

MATERIALS AND METHODS

Study area

The study was carried out in the soil environment of Vedant Gyan Valley, Jharna village, Jaipur, Rajasthan State, India. Jharna is 48 km away from Jaipur and located on east longitude-75°27'38" and north latitude-26°49'34" and situated on altitude 450 to 500 masl.

Samples collections

Soil samples of 200 g were collected from 5 different sites in Jharna village. The samples were collected with small sterile shovels into sterile plastic containers. The soil samples were sent to the laboratory within 30 min for analysis. The pH and temperature of soil samples were determined using digital pH meter and thermometer respectively. Humidity and air pressure was determined by hygrometer and barometer respectively.

Isolation and identification of fungal isolates

Potato dextrose (PDA) and Sabouraud's dextrose agar (SDA) media were employed for the isolation of fungi by spread plate method using serial dilution technique. All the plates were incubated at 30°C for 7 days. Fungal isolates were identified by cultural and microscopic features.

RESULTS

In our results, the soil of Jharna village was slightly acidic, the humidity and temperature was 36% and 31°C respectively. The physiochemical properties of soil play an important role in the growth of microorganism. Tables 1 and 2 show physiochemical properties of soil and isolated fungi. In our findings, *Aspergillus niger*, *Aspergillus flavus, Aspergillus fumigatus, Aspergillus terrus, Penicillium* sp., *Trichoderma* sp. and yeast were found (Figure 1). *A. niger* and *Penicllium* sp. were dominating species, they show vigorous growth and are found in large numbers. *A. niger* has black colour colony with conidial production on potato dextrose media. *A. niger* is biseriate, conidiophores are long, smooth and hyaline, becoming darker at apex and terminating globose vesicle, metulae and phialides cover the entire

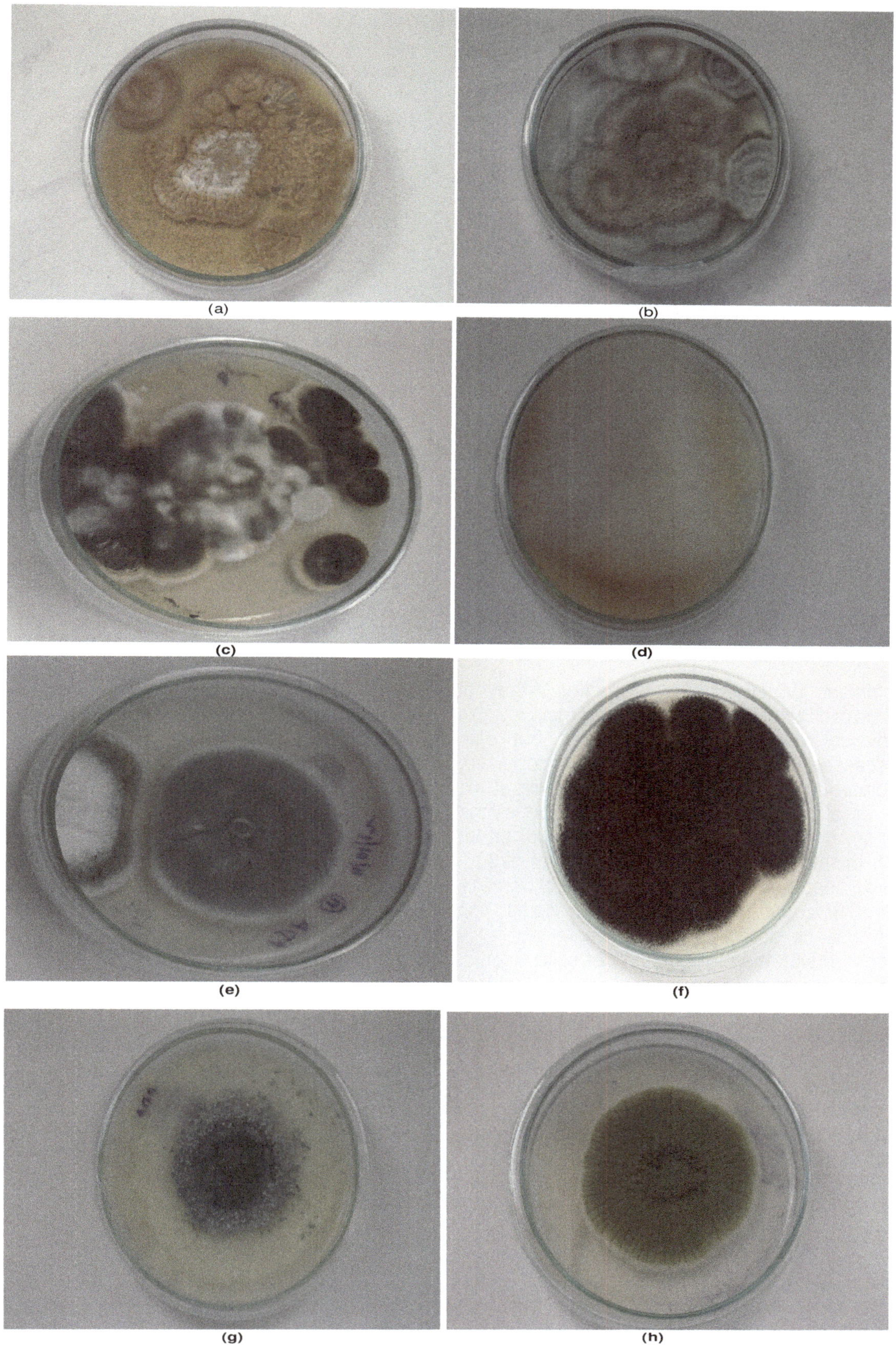

Figure 1. The growth of fungi on potato dextrose agar (a) *A. terrus;* (b) *A. nidulans;* (c) *P. chrysogenum;* (d) *Mucor* sp. (e) *A. fumigates;* (f) *A. niger;* (g) *Trichoderma* sp.*;* (f) *A. flavus.*

vesicle, conidia are brown to black very rough and globose and measures 4 to 5 µm in diameter. *A. terrus* have a cinnamon colour colony on potato dextrose agar media, hyphae are septate and hyaline, and conidial heads are biseriate; conodiophore are smooth walled and hyaline, terminating mostly globose vesicle, conidia are small globose and smooth and measure 2 to 6 µm in diameter. *A. flavus* have a green color colony on potato dextrose media, hyphae are septate and hyaline, conidiophores are coarsely roughened, globose vesicle, metulae covering the entire vesicle, conidia are smooth walled and 3 to 6 µm in diameter. *A. fumigates* have gray-green colony in potato dextrose agar. Hyphae are septate and hyaline, conidiophores are smooth walled, uniseriate with closely compact phialides occurring only on the upper portion of the columner vesicle. Conidia are smooth to finely roughed and 2 to 3 µm in diameter. *Mucor* show white cottony growth on potato dextrose agar media, nonseptate hyphae, having sporangiphore, sporangia and spores. *Trichoderma* colony is transparent at first on pda, after seven days of incubation at 25 to 28°C, green colour tufts conidia form. Conidiophores are highly branched.

DISCUSSION

Fungi have unique bioremediation properties. Many of the remediation technologies currently being used involve not only physical and chemical treatment, but also biological processes. A. flavus, A. niger and Rhizopus sp. were also isolated from paper mill effluents (Bajwa et al., 2010). Raju et al. (2007) also isolated A. flavus and *Trichoderma* sp. from the polluted soil samples of paper mills of South Indian paper mills. Jahangeer et al. (2005) also isolated *Aspergillus, Alternaria, Trichoderma, Penicillium* and *Rhizopus* sp. from the soil samples. The results would also be useful to environmental agencies that are concerned with the management of solid wastes. A number of species of bacteria and fungi were isolated and partially characterized from enrichment cultures containing the fungicide thiram (Sahin and Tamer, 2000). Our findings were also similar with Luis Henerique et al. (2010); they isolated *A. niger, A. flavus,* and Mucor from the soil samples. *Fusarium* spp., Penicillium spp., *Trichoderma* spp., *Paecilomyces* spp., *Epicoccum* sp. *Diheterospora* spp. and *Verticillium* spp. were capable of hydrolyzing EPTC from a soil previously exposed to the herbicide EPTC (Lee, 1984). Aspergillus sp., Fusarium sp., Mortierella sp. and Penicillium sp. were able to defluorinate activity 1080 from soils previously exposed to Namonofluoracetate (1080) (Wong et al., 1992). Mishra and Dadhich (2010) also isolated fungi A. niger, Penicillium and Trichoderma from soil samples of rajasthan having amylase and xylanase activity. Environmental agencies can take advantage of the result of this investigation and then invest in further studies which could lead to the isolation and characterization of specific high-yielding fungal strains found in the local soil environments. This will go a long way in reducing the cost of production in such industries as well as the cost of management of solid waste by environmental agencies.

Conclusion

A. niger, A. fumigatus, A. flavus, Trichoderma sp., *Rhizopus* sp. and *Penicillium* sp. were isolated from the soil contaminated with degraded paper. This investigation may lead to the development of strains of soil fungi that would be used locally for the biodegradation of cellulose materials. These organisms are so recommended for management of solid wastes containing cellulose.

ACKNOWLEDGEMENT

The authors are grateful to Department of Biotechnology and Allied Sciences, Jayoti Vidyapeeth Women`s University, Jaipur (Rajasthan) for providing the laboratory facilities used in this study.

REFERENCES

Akhtar M (1994). Biochemical pulping of aspen wood chips with three strains of *Ceriporiopsis subvermispora, Holzforschung.* 48:199-202.

Bajwa R, Khokhar I, Mukhtar I, Mushtaq S (2010). Isoaltion and identification of filamentous fungi from different Industrial Effluents. Plant Prod. Res. J. 14:32-35.

Bedford MR, Morgan AJ, Fowler T, Clarkson KA, Ward MA, Collier KD, Larenas EA (2003). Enzyme feed additive and animal feed including it. US Pal6562340 (to Genecore International Inc, USA).

CPCB (2001). Comprehensive industry document for large pulp and paper industry. COINDS/36/2000-2001.

Enari TM (1983). Microbial cellulases. In Fogarty, W.F.(ed.) Microbial enzymes and biotechnology. Applied Sciences Publishers, London. pp. 183-223.

Jahangeer S, Khan N, Jahangeer S, Sohail M, Shahzad S, Ahmad A and Khan SK (2005). Screening and characterization of fungal cellulases isolated from the native environmental source. Pak. J. Bot. 37(3):739-748.

Kader AJ, Omar O, Feng LS (1999). Isolation of cellulolytic fungi from the Barrio Highlands, Srawak. ASEAN Review of Biodiversity and Environmental Conservation (ARBEC).

Klyosov AA (1990). Trends in biochemistry and enzymology of cellulose degradation. Biochemistry 29:10577-10585.

Lee A (1984). EPTC (S-ethyl N, N-dipropylthiocarbamate) degrading microorganisms isolated from a soil previously exposed to EPTC. Soil Biol. Biochem. 16(5):529-531.

Luis Henerique SG, Simone CPN, Michele M, Ana CSR, Valeria CS, Fabiana FZ,Ana Carla MMA, Altino BJ (2010). Sreening of filamentous fungi for the production of enzymes for biotechnological interest. Brazilian J. Microbiol. 37:474-480.

Mandels M (1985). Application of cellulases. Biochemical Society transaction. 13:414-416.

Mishra BK, Dadhich SK (2010).Production of amylase and xyalanse enzyme from soil fungi of rajasthan.J. Adv. Dev. Res. 1(1):21-23.

Pere J, Siika-Aho M, Buchert J, Viikari L (1995). Effects of purified *T. reesei* cellulases on the fibre properties of kraft pulp. Taapi J. 78:71-78.

Prasad DY, Heitmann JA, Joyce TW (1992). Enzyme de-inking of black

and white letterpress printed newsprint waste. Prog Paper Recycl. 1:21-30.
Raju NS, Venkataramana GV, Girish ST, Raghavendra VB, Shivashankar P (2007). Isolation and evaluation of indigenous Soil fungi for decolourization of Textile dyes. J. Appl. Sci. 7(2):298-301.
Sahin NÜ, Tamer A (2000). Isolation, Characterization and Identification of Thiram degrading Microorganisms from Soil Enrichment Cultures. Turk J. Biol. 24:353–363.
Sukuraman RK, Shingania RR, Pandey A (2005). Microbial cellulases-production, application and challenges. J. Scientific. Ind. Res. 64:832-844.
US EPA (2005). Executive summary: Municipal solid waste in the united states; 2005 facts and figures. US Environmental Protection Agency. 2005. Retrived on 2008-05-06.
Wong DH, Kirkpatrick WE, King DR, Kinnear JE (1992). Defluorination of sodium monofluoroacetate (1080) by microorganisms isolated from western Australian soils. Soil Biol. Biochem. 24(9):833-838.

7

Pedological characteristics and classification of soils along landscapes at Abobo, southwestern lowlands of Ethiopia

Yacob Alemayehu[1], Heluf Gebrekidan[2] and Sheleme Beyene[3]

[1]Soil and Water Research Section, Gambella Agricultural Research Center, P. O. Box 62, Gambella, Ethiopia.
[2]School of Natural Resources Management and Environmental Sciences, Haramaya University, P. O. Box 138, Dire Dawa, Ethiopia.
[3]School of Plant and Horticultural Sciences, Hawassa University, P. O. Box 5, Hawassa, Ethiopia.

Soils of the lowlands in southwestern Ethiopia are under intensive cultivation, however, very little is known about them. The objective of this study is to provide basic information on how pedogenic properties along the landscape relate to their use, which could assist better management to establish sustainable crop production system. Four soil profiles from varying slope positions and land management were described *in-situ*, and sampled horizon-wise for physico-chemical analysis. Accordingly, soils at upper-slope position (Pedons 2 and 4) were characterized by shallow depth with the upper boundary of C-horizon 47/52 cm from the soil surface, whereas those at lower-slope position (Pedon-3) exhibited thick sola with a deep C-horizon (>200 cm). The C-horizon of all Pedons, except Pedon-3, had high coarse fragments, ranging from gravelly (48%) to extremely gravelly (69%). The soil consistence (wet) showed various grades of stickiness and plasticity along the profile depth. Soil organic carbon (OC), total N, exchangeable bases (Ca, Mg, K) and cation exchange capacity (CEC) generally decreased with profile depth, whereas available P content showed increasing trend with depth of Pedons and along the slope, ranging from 8.7 to 33.7 mg kg^{-1}. According to WRB legend, Pedon-1 and 2 with *cambic* horizons were classified as *Haplic Cambisols* and *Fluvi-Mollic Cambisols*, respectively, whereas Pedon-3 was classified as *Leptic Regosols* soil unit. Overall, this study revealed that the effects of land management on soil properties as a function of landscape is profound, suggesting the need for developing appropriate management options for sustainable use of the land.

Key words: Landscape, Pedon, slope, soil characteristics, soil classification.

INTRODUCTION

Soil types and characteristics show great variations across the regions of Ethiopia, because of the country's wide range of topographic, geologic and climatic features (Mesfin, 1998). Various studies of soil properties at watershed level (Shimeles, 2006; Alemayehu, 2007; Ashenafi et al., 2010; Sheleme, 2011) also confirm that topographic position largely governs the change in types and characteristics of soils. Such variability in types and characteristics influence not only the pedogenesis and development, but also the use and productivity of soils (Mesfin, 1998). On the other hand, increased pressures on soils caused by demographic increase: Such as deforestation and expansion of croplands, and continuous cultivation including sloppy landscapes and marginal areas (Sanchez et al., 1997; Bewket and Stroosnijder, 2003), resulting in land degradation and reduction in the productive capacities of soils in many areas of the country. Thus, study and understanding of soil properties and their distribution over a landscape is crucial for developing soil management practices that will maintain the productive potential of a soil.

The capacity of soil to function can be reflected by measured soil physical and chemical properties (Buol et al., 2003). These properties, however, are largely affected by land use/ management practices. Land management/use for crop production in landscapes influence runoff, soil erosion and drainage, which result in varying physical and chemical properties (Mishra et al., 2004). Taking into cognizance that soils differ with physiographic position, various researches have been carried out on impacts of land management on the properties of soils, but results vary due to soil type, landuse scenario (cropping systems, residue management) and climate of the area. Marked differences in profile depth and OC due to slope gradient were reported by Mebit (2006). Moreover, studies from northern (Gebeyaw, 2007), central (Wakene and Heluf, 2003; Attah, 2010) and southern Ethiopia (Alemayehu, 2007; Fentaw and Abdu, 2011) reported severe deterioration of soil quality due to long-term cultivation, which call for serious emphasis for fine-tune the ongoing soil fertility degradation.

Similarly, large areas of land at Abobo, southwestern Ethiopia, are under a prolonged agricultural pressure on variable physiographic positions for more than four decades. This area is center for huge farms, including private investors engaged in cotton and oil crop production. However, comprehensive characterization of the soils of the area has not been carried out so far. Alike, classification of soils in southwestern is still unclear, except small scale soil survey study by Yeshi Ber Consultant (2003) which was based on limited soil profile data derived from small scale. In the past, lack of detail soil information the area has been a serious drawback not only for improving productivity of land, but also for the transfer of agricultural technologies to the area.

In light of the above, it is of paramount importance to have an in-depth knowledge on the characteristics of soils of the area, in order to manage and use the resources based on their potentials and limitations and thereby maximize crop production to the allowable genetic potential limits and conserve the soils for future use. It will also be equally important to classify the soils using their inherent characteristics and in a manner that will ease communication and transfer of knowledge about such soils to farmers, stakeholders and soil scientists. This study, therefore, was involved at the research field and its surrounding farms at Abobo area with the objectives: (i) To characterize the morphological, physical and chemical properties of the soil profiles along the landscape under different land management systems; (ii) To classify the soils following the major soil classification schemes, and (iii) Provide comparisons of the properties of soil profiles in a landscape for sustainable crop production of the area.

MATERIALS AND METHODS

Site descriptions

The study site is located at Abobo district, in Gambella region, southwest Ethiopia, at 7° 51′ 18″ to 7° 54′ 23″ N latitude and 28° 31′ 40″ to 28° 33′ 35″ E longitude, and 480 to 550 m a.s.l altitude (Figure 1). The area falls within hot to warm humid lowland agro-ecological zone (Yeshi Ber Consultant, 2003). Long term rainfall records (1986 and 2012) show a uni-modal pattern with the annual average of 1039.4 mm (Figure 2). The rainy season starts at the end of April and lasts in October with maximum rainfall in July. The mean annual maximum and minimum temperatures are 35.3 and 17.4°C, respectively.

The site includes about 76% of land under cultivation, which has been cultivated for approximately 40 years with annual mechanized plow. The remaining part of the site is Open woodland (OWL) occupied by Sudan grass (*Sorghum bicolor* subsp. *drummondii*) with Acacia (*Acacia sp.*) and Neem (*Azadirachta indica*) trees, located at the southern part of the site. The field under cultivation has been cropped predominantly to maize (*Zea mays* L.), sorghum (*Sorghum bicolor* L.), cotton (*Gossypium hirsutum L.*), haricot bean (*Phaseolus vulgaris* L.), groundnut (*Arachis hypogae*) and sesame (*Sesamum astivum*).

Field survey and site setup

Soil investigation was carried out in two successive steps (auger observation and profile sampling) based on major land/soil characteristics: Slope, soil depth and surface soil texture, as described in Food and Agriculture Organization (FAO) (2006a) guidelines for soil description. In this case, fixed-grid survey technique (100 by 50 m interval) was employed. The grid system was oriented by a baseline which traverses the site along the slope gradient. Soil auger observations were implemented, using 'Edelman auger', at every 100 m by 50 m distance to identify variations in soil depth and texture characteristics along the slope gradient. The depth of augering was 110 cm, unless the soil depth limited or lithic contact prevents the operation.

A total of 105 auger observations were made for soil depth and surface texture, and each were grouped into the FAO (2006a) slope and depth classes. Individual soil characteristics were recorded on standard format. Points with the same soil depth class and surface soil texture in a given slope class were considered as a single mapping unit. Accordingly, three mapping units namely: (a) Back-slope (b) Upper-slope and (c) Lower-slope, were identified along the landscape (Table 1).

Soil profile description and sampling

Following the identification of the mapping units, further soil

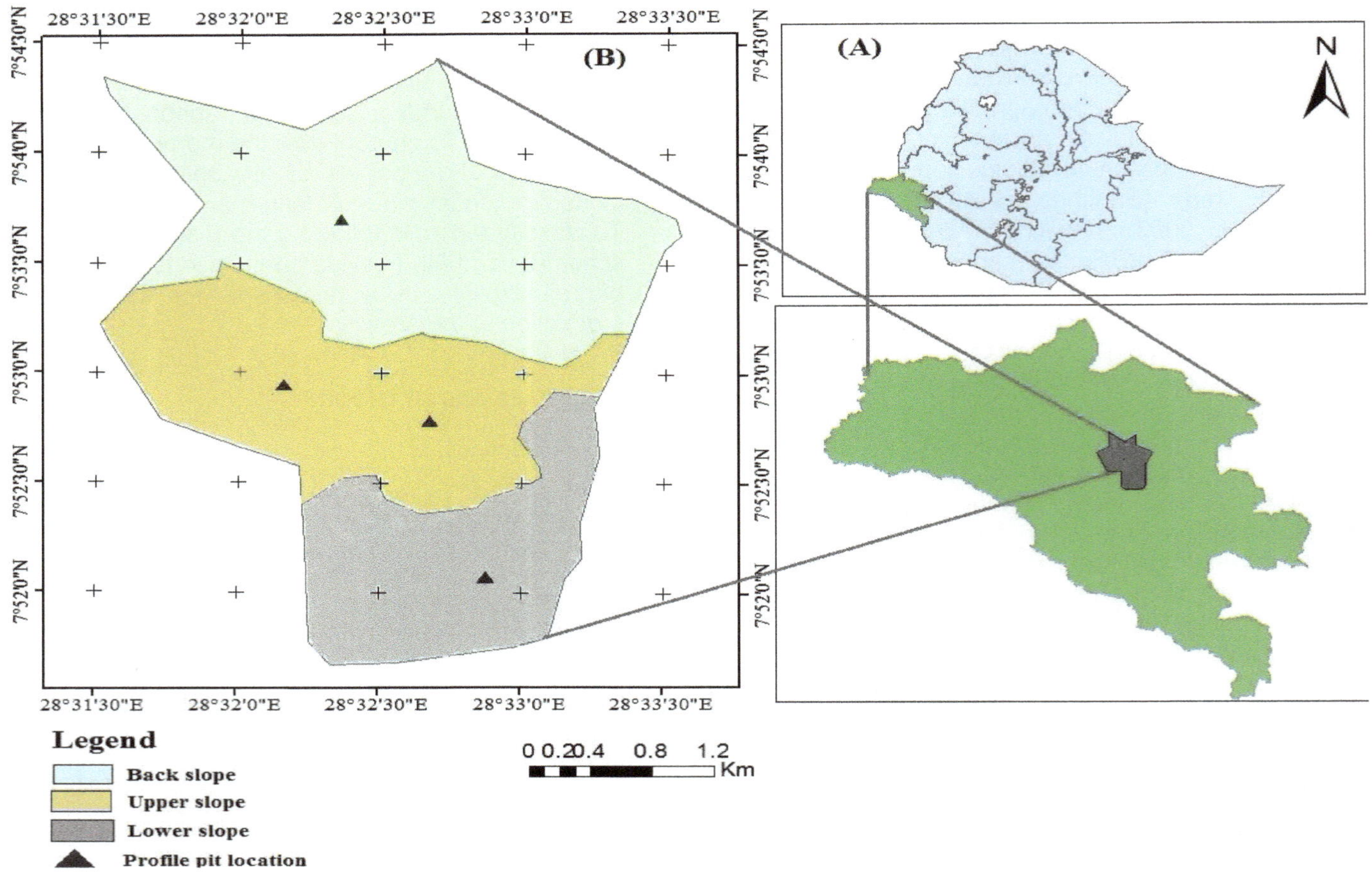

Figure 1. Map of Ethiopia (A) showing location of the study area (Abobo), and (B) soil mapping units of the study site with locations of soil profile pits sampled.

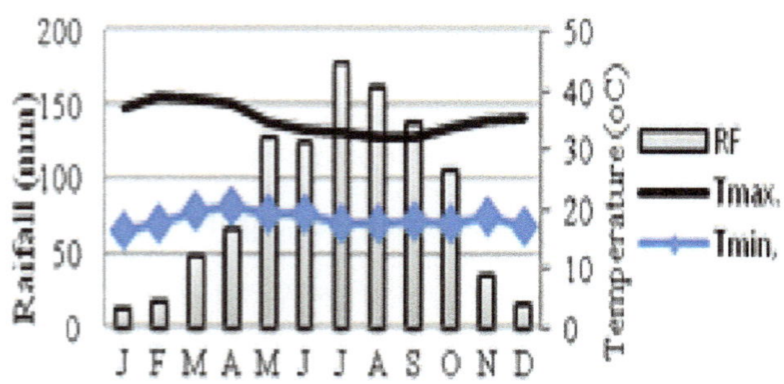

Figure 2. Monthly mean temperature and rain fall of the study area (1986-2012).

characterization was done using soil profile pits which were excavated at the representative sites (Table 2). Four soil profiles (Pedon-1 at back-slope; Pedon-2 and 4 at upper-slope; Pedon-3 at lower-slope) were described *in-situ,* for morphological characteristics, and samples were taken from every identified horizon. Both disturbed (bulk) and undisturbed (core) soil samples were collected from each horizon designated for laboratory determination of soil properties. Coarser material (>2 mm) contents were determined separately.

In general, soil profile horizonation and description, and sampling were carried out following the standard methods outlined by FAO (2006a) guidelines. During the site mapping, important landmarks such as altitude, slope and geographic coordination of the land unit boundaries were recorded using altimeter, clinometers and global positioning system (GPS) devices, respectively.

Determinations of soil properties

Disturbed soil samples were air-dried and sieved (<2 mm) for the analysis of all soil properties, while core samples were used for bulk density and soil-water retention determinations. For determinations of OC and total N, the samples were passed through 0.5 mm sieve. Particle size distribution was determined by hydrometer method (Bouyoucos, 1962). Solid particle density (ρ_s) was measured by pycnometer method (Black and Hartge, 1986), while bulk density (ρ_b) was determined using core sampling method, as described by Gupta et al. (1979). Total porosity was then estimated from the values of ρ_b and ρ_s. Soil water retention at Field Capacity (FC = 0.33 bars) and Permanent Wilting Point (PWP = 15 bars) was determined using pressure plate apparatus (Van Reeuwijk, 2006), then Available Water Holding Capacity (AWHC) computed from FC and PWP. Soil pH was measured in 1:2.5 soil liquid ratio where the liquids were water and 1 *M* KCl solution. EC was measured by conductivity meter in soil to water ratio of 1:2.5 (Van Reeuwijk, 2006). Soil OC was analyzed by wet digestion method (Walkley and Black 1934), and total N was determined by a modified micro-Kjeldahl digestion procedure. Plant available P quantified by Olsen method (Olsen et al., 1954), as the method was recommended for

Table 1. Distinguishing features of the mapping units based on major land/soil characteristics of the Site.

Mapping unit	Slope gradient		Soil depth		Surface texture	Area coverage
	%	Class	Depth (cm)	Class	(0-30cm)	(ha)
Back-slope	0.5 - 2.0	Nearly level	50-100	Moderately deep	Clay	49.83
Upper-slope	7.0 - 9.6	Sloping	<50	Shallow	Clay	25.16
Lower-slope	2.0 - 5.0	Gently sloping	$50\text{-}100^{+}$	Moderately deep to deep	Clay loam	23.01

The mapping units were named after their landscape position.

Table 2. Location and characteristics of Pedons under each mapping unit.

Pedon	Mapping unit	Geographic location		Altitude	Slope	Landscape	Parent	Land	Erosion
		Latitude	Longitude	(m asl)	(%)	position	material	use[1]	status
1	Back-slope	7° 53' 45.8"	28° 32' 20.7"	510	1.2	Nearly level	basalt	CL	Slight
2	Upper-slope	7° 52' 56.7"	28° 32' 09.0"	480	7.2	Sloping	basalt	CL	Moderate
3	Lower-slope	7° 52' 05.5"	28° 32' 53.2"	470	3.0	Gently sloping	basalt	OWL	Nil
4	Upper-slope	7° 52' 45.5"	28° 33' 41.8"	476	8.3	Sloping	basalt	CL	Moderate

[1] OWL = Open Woodland; CL= Cultivated land.

all types of Ethiopian soils (Tekalign and Haque, 1991). Cation Exchange Capacity (CEC) was determined after extracting the soil samples by ammonium acetate at pH 7.0 (Chapman, 1965). Exchangeable Ca and Mg in the extracts were determined using atomic absorption spectrophotometer (AAS), whereas Na and K were quantified by flame photometer (Van Reeuwijk, 2006). Total porosity (%), Base saturation (PBS), exchangeable sodium percentage (ESP) and CEC of clay were also computed, using their respective equation as follows:

$$TP = \left(1 - \frac{pb}{ps}\right) x\ 100 \quad (1)$$

$$PBS = \frac{Ca + Mg + K + Na}{CEC} \times 100 \quad (2)$$

$$ESP = \frac{Exchangeable\ Na}{CEC} x\ 100 \quad (3)$$

$$CEC\ of\ clay = \frac{CEC\ of\ soil - (\%OM\ x\ 200)}{\%\ Clay} \quad (4)$$

Soil classification was carried out following *World Reference Base for Soil Resources* [WRB] (IUSS Working Group, 2006) and *Soil Taxonomy* (Soil Survey Staff, 2006) classification schemes.

RESULTS AND DISCUSSION

Depth and morphological features of soils

The soil profiles described at the back-slope (Pedon-1), upper-slope (Pedon-2 and 4) and lower-slope (Pedon-3) positions were identified by Ap-B-CB-C, Ap-AC-C-R/Ap-AC-C-R and A_1-A_2-B-BC-Cr_1-Cr_2 horizon sequences, respectively (Table 3). Depth and horizon features of the profiles were highly variable along the landscape (Figure 3). Pedon-2 and 4 (at upper-slope) were characterized by truncated surface layer (16/19 cm), a shallow depth to the C-horizon with the upper boundary 47/62 cm from the soil surface, whereas Pedon-3 (at lower-slope) was exhibited a thick A-horizon (38 cm), moderately developed B-horizon and a deep C-horizon with relatively weathered *saprolites (soft rock)* up to more than 200 cm depth. On the other hand, the C-horizons of Pedon-1, 2 and 4 had high amount of coarser materials, ranging from 48 to 69% of the total volume of the horizons (Table 3). Soils containing >60% gravels can be considered as extremely gravelly soils (Buol et al., 2003). Accordingly, the C-horizon of investigated pedons varies from gravelly (Pedon-1) to extremely gravelly (Pedon 2 and 4). In contrast, relatively deeper solum and fewer amounts of coarser materials in Pedon-3 suggest that the rate of weathering process were increasing toward the lower-slope area.

At upper-slope (Pedon 2 and 4), where cultivation pressure is higher on sloppy position, the depth of A-horizon has shown reduction by > 50%, as compared to that under woodland area of lower-slope (Pedon-3) position. This is because prolonged intensive tillage on sloping site has influenced erosion that accelerated the removal of topsoil from this position to the adjacent lower-slope landscapes. Conversely, thicker surface layer of Pedon-3 is partly linked with the gradual accumulation of soil materials from upper-slope, and the role of vegetation cover that promotes better soil development at low-slope position.

Soil color (dry, moist) *value* and *chroma* was invariably increased with soil depth, ranging from dark reddish brown (5YR 3/4) to dark brown (10YR 5/6) owing to

Table 3. Selected morphological properties of the soil profiles.

Depth (cm)	Horizon	Color		Structure[1]	Consistency[2]			Texture (Field)[3]	HB[4]	CF[5]
		Dry	Moist		Dry	moist	wet			(vol %)
Pedon-1: Back-slope										
0-22	Ap	7.5YR 4/4	7.5YR 3/4	MO FI SB	HA	FR	ST PL	C	CS	-
22-59	B	7.5YR 4/4	7.5YR 4/6	MO ME SB	HA	FR	ST PL	C	GS	-
59-93	CB	5YR 3/4	5YR 4/6	WE ME SB	HA	FI	ST PL	CL	GW	-
93-176^{+}	C	5YR 5/6	5YR 4/4	SG	HA	FI	ST PL	CL	-	48
Pedon-2: Upper-slope										
0-19	Ap	7.5YR 5/4	7.5YR 3/4	WE VFI SB	SHA	FR	VST PL	C	CS	-
19-32	AC	7.5YR 4/4	7.5YR 4/6	WE ME SB	HA	FI	ST PL	C	GS	-
35-62	C	7.5YR 5/6	7.5YR 4/6	SG	VHA	FI	ST PL	C	-	69
62^{+}	R	-	-	-	-	-	-	-	-	-
Pedon-4: Upper-slope**										
0-16	Ap	7.5YR 5/4	7.5YR 4/3	WE FI SB	SHA	FR	VST PL	C	GS	-
16-24	AC	7.5YR 5/4	7.5YR 3/3	WE FI SB	HA	FR	VST PL	C	CW	32
24-47	C	7.5YR 5/6	7.5YR 4/4	SG	VHA	FI	SST SPL	C	-	63
47^{+}	R	-	-	-	-	-	-	-	-	-
Pedon-3: Lower-slope										
0-27	A_1	10YR 5/5	10YR 3/6	MO ME SB	SHA	FR	ST PL	C	CS	-
27-38	A_2	10YR 4/5	7.5YR 4/6	MO ME SB	SHA	FR	ST PL	CL	GS	-
38-77	B	5YR 4/4	5YR 4/6	MO ME SB	SHA	FR	ST PL	CL	GS	-
77-90	BC	5YR 4/6	5YR 4/4	WE ME SB	HA	FR	ST PL	C	DS	-
90-125	Cr_1	5YR 4/6	5YR 4/6	WE ME SB	HA	FR	SST SPL	C	GS	13
125-200^{+}	Cr_2	7.5YR 3/4	7.5YR 3/4	MA	HA	FR	SST SPL	CL	---	12

[1]WE=Weak; MO=Moderate; VF =Very fine; FI=Fine; ME=Medium; SB=Sub-angular blocky; SG=Single grain, MA=Massive; [2]SHA=Slightly hard; HA=Hard; FR= Friable; FM=Firm; ST=Sticky; SST= Slightly sticky; VST=Very sticky, PL=Plastic; SPL= Slightly plastic. [3]Texture field feel method: C=Clay; CL=Clay loam.[4]Horizon boundary: CS=Clear and smooth; GS=Gradual and smooth; DS=Diffuse smooth; CW= Clear and wavy. [5]CF = Coarse fragments (>2 cm diameter); ** = Only the morphological data has taken.

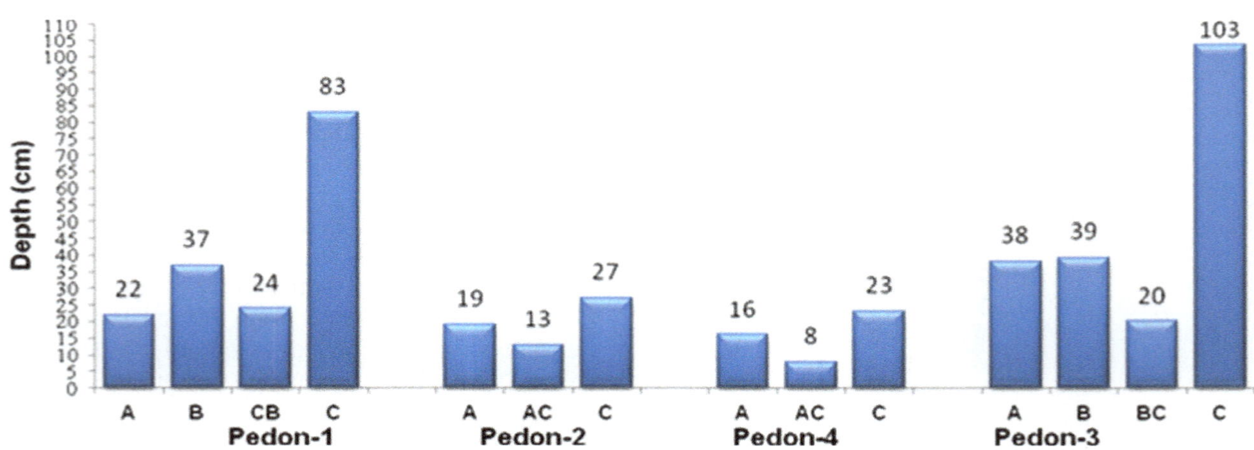

Figure 3. Horizon thickness of the profiles (cm) described at the back-slope (Pedon-1), up-slope (Pedon-2 & Pedon-4) and low-slope (Pedon-3) landscape positions.

decrease in OC with depth. Surface layer of Pedon-3 (at lower-slope) has a darker *hue* (10YR) due to relatively higher OC compared to other profiles (Table 3), but it became redder (5YR) with depth. This could be due to

Table 4. Selected physical properties of the soils profiles.

Depth (cm)	Horizon	Fine earth (%) Sand	Silt	Clay	Textural class	ρs* (g cm^{-3})	ρb* (g cm^{-3})	TP* (%)	FC (vol %)	PWP (vol %)	PAW (m^3 m^{-3})
Pedon-1: Back-slope											
0-22	Ap	21.4	34.4	44.2	C	2.69	1.45	46.84	46.30	29.74	0.202
22-59	B	16.4	36.2	47.4	C	2.64	1.48	44.32	48.57	26.22	
59-93	CB	19.1	35.3	45.6	C	2.58	-	-	-	-	
93-176$^+$	C	32.4	37.3	30.3	CL	2.73	-	-	-	-	
Pedon-2: Upper-slope											
0-19	Ap	14.4	36.1	49.6	C	2.67	1.47	44.94	49.72	32.05	0.178
19-32	AC	20.4	29.0	51.6	C	2.68	1.50	44.40	46.22	29.44	
35-62	C	43.2	38.3	18.5	L	2.68	-	-	-	-	
62$^+$	R	-	-	-	-	-	-	-	-	-	
Pedon-3: Lower-slope											
0-27	A_1	22.9	38.2	37.9	CL	2.68	1.39	49.14	43.16	26.03	0.168
27-38	A_2	23.6	38.1	38.3	CL	2.66	1.45	46.49	43.60	27.68	
38-77	B	20.4	35.1	44.5	C	2.75	-	-	-	-	
77-90	BC	26.4	32.2	41.4	C	2.66	-	-	-	-	
90-125	Cr_1	33.4	37.0	30.6	CL	2.65	-	-	-	-	
125-200$^+$	Cr_2	28.4	36.0	35.6	C	2.66	-	-	-	-	

*ρs = Particle density; ρb = Bulk density; TP = Total porosity; PWP = Permanent wilting point; FC = Field capacity; PAW = Plant available water content; - = Not determined.

the oxidation of iron oxides, responsible for the reddish colour in subsoil horizons (Buol et al., 2003). Overall, the determination of colour, higher *value* and *chroma* indicates absence of soil moisture. This can be explained by the fact that light reflection and purity of the spectral colour increases when the soil is dry resulting in higher *chroma* and *value* of soil colour. Besides, absence of mottles in subsoil layers of all Pedons confirms that the soils were in well drained condition.

Soil structure of the surface horizons of Pedons was generally sub-angular blocky with some variations in size of peds. Weak grade of peds was observed at upper-slope (Pedon 2 and 4) indicating structure deterioration attributed to prolonged cultivation that fragments soil aggregates. Similarly, soil consistence (dry) remained hard throughout the horizons of Pedon-1, while it changed from slightly hard to extremely hard with the depth of Pedon-2. Consistence (wet) also showed various grades of stickiness and plasticity along the profile depth. This could be attributed to a change in particle size distribution of soil.

Soil physical properties

Soil texture determined in the field (feel method) and laboratory (Table 3 and 4) were similar in most cases. Results on the particle-size distribution of the pedons indicated that the clay content slightly increased from A- to B-horizon and then decreased with depth, while sand contents increased in all Pedons (Table 4). Across the profiles, the highest (52%) and lowest (18%) clay contents were recorded in Pedon-2. Although the increase in clay content in the B-horizons of Pedon 1 and 3, clay skins (cutans) were not found on the sides of ped faces, implying that clay illuviation did not occur, and high clay contents of B-horizons are due to *in-situ* weathering of the parent material. The silt/clay ratios increased with depth of profiles (1 and 2) that followed opposite pattern of variation with clay particles. High silt/clay ratios (2.1) in the subsoil layers along with resistant skeletal composition of parent material (Pedon 2) reflect that the soils at upper-slope position are at young stage of development. In line with this, Shimelis (2006) reported similar silt/clay ratio relationship with depth for *gleyic Cambisols* of Tenocha catchment, central Ethiopia.

The bulk densities of plough layers of Pedon-1 and 2 were higher than the *'optimum'* value (1.40 g cm^{-3}) for clayey soils (USDA, 2008). This could be attributed to soil compaction as a result of continuous and intensive cultivation with heavy farm machinery for many years. The total porosity of soils in all pedons falls below the *'ideal porosity'* value (> 50%) for healthy root growth (Lawrence, 1977). Poor soil structure (less interped spaces), low soil OM and perhaps soil compaction associated with prolonged cultivation practices might have contributed to lower porosity of soils at back-slope and upper-slope positions as compared to that of

Table 5. Soil reaction, electrical conductivity organic carbon and total N contents of soil profiles.

Depth (cm)	Horizon	pH_{H2O} (1:2.5)	pH_{KCl} (1:2.5)	ΔpH	EC ($dS\ m^{-1}$)	$CaCO_3$	OC (%)	Total N	C/N	P ($mg\ kg^{-1}$)
Pedon-1: Back-slope										
0-22	Ap	6.27	5.43	0.84	0.13	0.38	0.80	0.09	8.8	12.06
22-59	B	6.52	5.34	1.18	0.16	0.38	0.83	0.05	17.3	8.44
59-93	BC	6.48	5.70	0.78	0.12	0.27	0.26	0.04	7.3	8.71
93-176$^+$	C	6.45	5.91	0.54	0.15	0.41	0.20	0.02	10.0	10.08
Pedon-2: Upper-slope										
0-19	Ap	6.42	5.39	1.03	0.10	0.25	0.72	0.08	8.9	13.40
19-32	AC	6.46	5.20	1.26	0.10	0.19	0.39	0.05	7.8	13.67
35-62	C	6.12	5.50	0.96	0.07	0.22	0.23	0.03	7.6	11.06
62$^+$	R	-	-	-	-	-	-	-	-	-
Pedon-3: Lower-slope										
0-27	A_1	6.84	5.69	1.15	0.13	0.26	2.28	0.13	17.5	29.65
27-38	A_2	6.80	5.52	1.28	0.15	0.15	1.52	0.09	16.8	15.40
38-77	B	6.35	5.60	0.75	0.17	0.15	0.70	0.04	17.4	12.06
77-90	BC	5.96	5.42	0.54	0.11	0.38	0.80	0.03	26.6	19.43
90-125	Cr_1	6.21	5.25	1.06	0.03	0.28	0.55	0.03	16.7	33.74
125-200$^+$	Cr_2	6.26	5.40	0.86	0.09	0.37	0.27	0.04	6.7	26.41

EC = Electrical conductivity; OC=Organic carbon.

woodland area (Pedon-3) at lower-slope position.

The volumetric soil water content at FC (0.33 bar) and at PWP (15 bars) had shown a slight variation among the studied pedons (Table 4). Across the pedons, the highest FC (49.7%) and PWP (32.1%) were recorded for the surface horizon of Pedon-2 that contained higher clay content, regardless of its high bulk density and low porosity. Likewise, the computed PAW values for root depth (0.5 m soil depth) of all Pedons fell within the *'ideal'* range (0.15 - 0.20 $m^3\ m^{-3}$) for fine-textured soils (Cockroft and Olsson, 1997). This was probably due largely to the high clay and silt contents in the upper horizons, which could, therefore, facilitate a longer period of soil water retention for crop utilization.

Chemical properties

Soil pH

The results of soil reaction (pH) determined in a 1: 2.5 soil to water/KCl suspension generally showed increasing trends (Pedon-1), while inconsistently decreased with increasing depth of Pedon-2 and 3 (Table 5). On the other hand, ΔpH, (pH_{H2O} - pH_{KCl}) values were positive in all Pedons, ranging in between 0.5 and 1.3 units. The value for ΔpH can be positive, zero, or negative, depending on the net surface charge at the time of sampling, and a positive ΔpH indicates the presence of negatively charged clay colloids (Soil Survey Staff, 2006); hence, higher ΔpH (1.3) for A-horizon of Pedon-3 indicates the presence of appreciable amount of negatively charged clay colloids. Following the soil pH rating set by Jones (2003), the pH_{H2O} values throughout horizons of all Pedons were within slightly acidic and neutral range.

The electrical conductivity (EC) of soils in all Pedons was found to be well below unity (Table 5). Similarly, there was no significant difference in the $CaCO_3$ values within soil horizons and among pedons. Field determination of carbonates with 10% HCl also confirmed that there was no audible and/or visible effervescence throughout the depth. Low values (<0.50%) of $CaCO_3$ indicate non-calcareous class, as explained by Landon (1991).

Soil organic carbon, total nitrogen and available P

Apparent change in OC content was observed in the surface layers of Pedons (Table 5), ranging from 'very low' (Pedon 1 and 2) to 'medium' (Pedon-3) based on the ratings by Tekalign and Haque (1991). This is due to the fact, long-term cultivation coupled with frequent burning of crop residues (common land clearing practice for cultivation) have accelerated the rapid turnover rates of organic materials of soils under cultivation. Studies from northwest Ethiopia (Habtamu et al., 2009) have also noticed that burning and removal of crop residue had significantly reduced soil OC and total N contents under

Table 6. Exchangeable bases, CEC and base saturation of the soil profiles.

Horizon	Depth (cm)	Exchangeable bases and acidity ($cmol_c\ kg^{-1}$)					Ca/Mg	CEC ($cmol_c\ kg^{-1}$)		ESP[1]	BS[2]
		Na	K	Ca	Mg	Al+H		Soil	Clay	(%)	(%)
Pedon-1: Back-slope											
Ap	0-22	1.14	1.01	12.20	8.69	1.09	1.4	32.8	68.0	3.5	70.2
B	22-59	0.92	1.70	14.20	9.25	0.80	1.6	35.2	69.9	2.6	74.0
CB	59-93	1.58	0.96	5.44	9.24	0.96	0.6	35.1	75.2	4.5	49.1
C	93-176+	0.94	1.14	7.80	8.82	0.58	1.1	31.6	98.4	3.2	57.9
Pedon-2: Upper-slope											
Ap	0-19	0.81	2.04	8.41	10.62	0.14	0.8	34.4	64.6	2.4	63.6
AC	19-32	0.68	1.65	8.83	11.48	0.31	0.8	35.9	67.0	1.9	63.2
C	35-62	0.80	0.05	6.34	8.68	0.21	0.7	24.7	85.0	2.5	62.3
R	62+	-	-	-	-	-	-	-	-	-	-
Pedon-3: Lower-slope											
A_1	0-27	1.34	1.12	9.70	15.81	0.53	0.6	39.1	82.5	3.4	71.5
A_2	27-38	0.86	1.02	10.43	12.20	0.40	0.8	39.4	93.7	2.2	62.1
B	38-77	1.90	2.93	11.72	13.44	0.61	0.9	36.3	73.5	5.2	80.3
BC	77-90	1.10	1.04	9.80	19.60	0.91	0.5	36.3	85.6	3.9	86.6
Cr_1	90-125	0.94	1.82	11.20	18.91	0.66	0.6	37.6	82.2	3.3	87.4
Cr_2	125-200+	0.66	0.42	11.28	16.09	0.69	0.5	32.3	78.3	2.0	78.1

ESP= Exchangeable sodium percentage; BS=Percentage base saturation.

cultivated land as compared to the uncultivated (woodland) counterpart of the site at low-slope (Pedon-3) position. The total N content of soils was directly associated with OC content along the profile depth, that the highest (0.13%) value was recorded in A-horizon of Pedon-3, where the highest (2.28%) OC content was found.

The C:N ratio of soils varied from the narrow (7:3) to wider range (27:1) and showed an irregular decrease with depth of all Pedons (Table 5). The C:N ratio of the layers (Pedon-1 and 2) is low below the *'optimum'* range (10-12:1) for arable soils (Havlin et al., 1999) as compared to the uncultivated (woodland) area of lower-slope (Pedon-3) position, indicating oxidation and loss of OM is higher in soils under long-term cultivation.

Olsen extractable available P has shown increasing trend along a landscape of the site, and became higher with depth particularly in Pedon-3 (Table 5). The values for most horizons of Pedons were above the 'high' range (Olsen et al., 1954); hence, P is not a limiting nutrient for crop production. As soil pH ranges 6.0 to 7.5 is ideal for P availability (Havlin et al., 1999), high concentration of available P is a reflection of slightly acidic to neutral soil reaction (Table 5) and low contents of Al (Table 6). Moreover, soils became increasingly deeper and more developed at lower-slope (Pedon-3) position; this could be the possible reason for the increasing concentrations of available P with profile depth (Pedon-3). In favour of this, comparable results and trends along a landscape were also reported by Mebit (2006) for soils of Woreta area, northern Ethiopia.

Base saturation and Cation exchange capacity

The analytical values of exchangeable bases indicated that exchangeable Mg and Ca were dominant cations in the exchange complex (Table 6), accounting for more than 85% of exchangeable bases and 44 to 69% of total cations of the exchange sites. The abundance in decreasing order was: $Mg^{++} > Ca^{++} > K^{+} > Na^{+}$, except for Pedon-1.

According to the nutrient ratings by FAO (2006b), exchangeable Ca and Mg contents of the soils were in medium and high classes, respectively. However, Ca:Mg ratio of soils (< 1.6:1) fell below the *'optimum'* (2 - 4:1) value as proposed by Landon (1991) for most crops. Ca:Mg ratio below 3:1 results unavailability of Ca (Landon, 1991); hence, low ratio indicates the probable limitation of Ca uptake due to excess amount of Mg. Higher Mg content, particularly in the subsoil horizons of Pedon-3 suggests that the dominance of Mg-bearing minerals in the weathering environment. The Ca Mg trend in this study is comparable to a study report by Sheleme (2011) for soils along toposequence of Gununo area, southern Ethiopia.

Exchangeable K occupied 2 to 9% of total exchangeable bases in the exchange site. The values in most horizons can be rated as very high (>1.2 $cmol_c\ kg^{-1}$) according to FAO (2006b), indicating K sufficient for soils under study. The result supports the common belief that Ethiopian soils contain sufficient K for crop production. On the other hand, the ESP value throughout the pedons was far below the critical level (> 15%) that causes deterioration of soil structure and Na toxicity (Landon, 1991); hence, the soils are free from the adverse effects.

Overall, base saturation (BS) of the soils high (> 60%) to be considered as fertile soil (Landon, 1991), is related to the basic nature of soils developed from basaltic parent materials (Table 2). As BS indicates the degree of leaching of basic cations, the studied pedons were weakly leached soils (60-80%), based on the ratings set by Hazelton and Murphy (2007). Relatively lower BS values in the surface horizons of Pedons at upper-slope (cultivated field), as compared to that under woodland (at lower-slope), might be due to depletion of basic cations in the crop harvest and their removal through erosion with topsoil from the sloping position. However, there is an increase in levels of K, Ca and Mg, P and clay contents in the subsoil horizons of Pedon-3 (at lower-slope), indicating a richer *saprolites* of the profile, compared to the other Pedons.

The CEC of soils across the landscape varied between 24.7 and 39.3 $cmol_c\ kg^{-1}$. Variations in CEC at the surface layers is largely connected with limited recycling and depletion of OM as a result of residue mismanagement (burning), that have attributed to lower the CEC of soils (Pedon-1 and 2) compared to the uncultivated (woodland) counterpart of the site (Pedon-3) at lower-slope position. These values are comparable to those obtained by Alemayehu (2007) and Fentaw and Abdu (2011) for soils under varying management from southern Ethiopia. Moreover, the CEC of soil profiles followed the trend exhibited by the exchangeable basic cations, reflecting that basic cations are the main ion contributors in the exchange complexes.

The CEC of clay fraction can be used as an indication of type of clay mineralogy (Buol et al., 2003). Accordingly, most of the horizons of Pedon-3 and C-horizon of Pedon 1 and 2 were expected to have *montmorillonitic* minerals. As shown in Table 3, very sticky and plastic consistency characteristics of soils also confirmed that the occurrence of 2:1 clay minerals in the studied soils.

Soil classification

The studied Pedons were classified at subgroup/subunit level, based on the identified diagnostic horizons, properties and qualifiers to the employed classification systems (*WRB* and *Soil Taxonomy*).

Pedon-1 described at back-slope was characterized by the beginning stage of subsoil horizon differentiation, evident from changes in colour and structure, absence of *illuviated* clay and OC with increasing depth. Bw-horizon had redder hue, moderately developed soil structure, friable consistence and higher clay content than the underlying horizon, and the horizon thickness met the requirements of *cambic* diagnostic horizon (IUSS Working Group, 2006). In addition, higher (48%) coarse fragments at lower depth have ascertained the Pedon to be classified as *Haplic Cambisols (Endoskeletic)* subunit (IUSS Working Group, 2006).

Pedon-3 (at lower-slope) had a surface layer of strong and well developed soil structure; darker colour (value < 3); high OC (>0.6%); PBS (71.9%) and horizon thickness (27 cm) that meet the necessities of *mollic* diagnostic horizon (IUSS Working Group, 2006). The subsoil (B) layer below *mollic* horizon also identified to have a *Cambic* horizon. The Pedon had *fluvic* material, affirmed by stratification and irregular decreasing in OC content with depth. Thus, the pedon can be classified as *Fluvi-Mollic Cambisols*.

Pedon-2 and 4 (at upper-slope) had shown very weak profile development, shallow solum depth along with higher quantities of weatherable minerals overlying the consolidated rocks. The Pedons had no diagnostic horizons other than an *ochric* epipedon (Soil Survey Staff, 2006), and the observed properties could not qualify for any other Reference Soil Group (RSG), except Regosols (IUSS Working Group, 2006). The Pedons are, hence, classified as *Leptic Regosols (skeletic)* subunit.

The findings are in harmony with several research results in Ethiopia (Mesfin, 1998; Yihenew 2002; Shimeles, 2006; Alemayehu, 2007; Ashenafi et al., 2010) that *Cambisols* are characterized by high CEC and PBS, neutral pH, and low OC and total N contents. However, our results are lower than those reported by Yeshi Ber Consultant (2003) for *Cambisols* under the savanna woodlands of the nearby district.

USDA soil taxonomy equivalent

The soil moisture and temperature regimes in the study site are *ustic* and *isohyperthermic*, respectively. Considering this, Pedon 1 and 3 had *cambic* horizon, show lack of free carbonates, but Pedon 1 contains low PBS (<60%) within 59 to 93 cm depth. Besides, Pedon-2 showed irregular decrease in OC with depth and absence of *lithic* contact throughout was classified as *Fluventic Haplustepts,* while Pedon-1 lacks further characteristics and was grouped under *Typic Dystrustepts* at subgroup level (Soil Survey Staff, 2006). On the other hand, Pedon-2 and 4 had *ochric* epipedon and *lithic* contact within 50 cm profile depth, and higher (69%) coarse fragments, were classified as *Lithic Ustorthents* at subgroup level.

Conclusions

Results of the study showed that differences in slope gradient and land management across the landscape had influence on the characteristics of the soils in the studied site. The four studied Pedons across the landscape were variable in their profile depth, horizon features and some physicochemical properties of soils. Pedons 2 and 4 (*Leptic Regosols*) located on upper-slope, where intensive cultivation prevailed, were characterized by shallow profile depth with truncated A-horizon, which resulted from the removal of topsoil from the sloping position of the site. Conversely, Pedon-3 (*Fluvi-Mollic Cambisols*) placed on lower-slope (woodland area) had darker colour, thick A-horizon and deep soil profile >200 cm, indicating better soil development; this is mainly linked with the gradual accumulation of soil materials from upper-slope as well as the role of vegetation cover in promoting soil development at low-slope position. Besides, severe depletion in soil OM and total N contents was observed in the soil profiles subject to long-term cultivation (Pedons, 1, 2 and 4) as compared to the wood land area of lower-slope (Pedon-3) position. Such a drastic loss of soil OM has also brought to reduction in soil aggregates, porosity, CEC and nutrient losses. Therefore, care must be taken concerning agricultural pressure (including residue burring practices) that deteriorates the quality of soils at upper-slope position. Proper residue management and soil conservation measures highly needed to combat the ongoing land degradation and improve soil health of the study area.

Conflict of Interest

The authors have not declared any conflict of interest.

ACKNOWLEDGEMENTS

The authors gratefully acknowledge Dr. Taye Bekele and Dr. Dhyan Singh for their valuable comments on the earlier manuscript. Their sincere thanks are also due to the anonymous assistants of the soil laboratory of both *Hawassa University* and *Debrezeit Agricultural Research Center*, for their unreserved collaboration in analyzing the soil parameters. Financial support to the study was obtained from Rural Capacity Building Project (RCBP) for training and research in Ethiopia.

REFERENCES

Alemayehu K (2007). Effects of land use and topography on soil properties at Delbo watershed. MSc Thesis, Hawassa University, Ethiopia. P. 64.

Ashenafi A, Abayneh E, Sheleme B (2010). Characterizing soils of Delbo Wegene watershed, southern Ethiopia. J Soil Sci. Envron. Manage. 1(8):184-199.

Attah L (2010). Physicochemical characteristics of the rhizosphere soils of some cereal crops in Ambo area, Ethiopia. Int. J. Sci. Tech. 4(1):93-100.

Black G, Hartge K (1986). Particle density. In: Klute A (ed.). Methods of soil analysis. Madison. Am. Soc. Agron.1:377-382.

Bouyoucos G (1962). Hydrometer method improvement for making particle size analysis of soils. Agron. J. 54:179-186. http://dx.doi.org/10.2134/agronj1962.00021962005400050028x

Buol S, Hole F, McCracken J, Southard R (2003). Soil genesis and classification, 5th ed. Iowa State Press, USA. pp. 126-327.

Chapman H (1965). Cation exchange capacity. In: Black C, Ensminger L, Clark F, (eds.). Methods of soil analysis. Am. Soc. Agron. 9:891-901.

Cockroft B, Olsson K (1997). Management of soil structure for roots in duplex soils. Dev. Soil Sci. 25:339-350. http://dx.doi.org/10.1016/S0166-2481(97)80043-8

Fantaw Y, Yimer A (2011). The effect of crop land fall owing on soil nutrient restoration in the Bale Mountain. Sci. Dev. J. 1(1):43-51.

FAO (2006a). Guidelines for soil description. 4th ed. Viale delle Terme di Caracalla. Rome, Italy.

FAO (2006b). Plant nutrition for food security: A guide for integrated nutrient management. Fertilizer & Plant Nutrition Bulletin 16, Rome.

Gebeyaw T (2007). Soil fertility status as influenced by different land uses in Maybar, north Ethiopia. MSc Thesis. Haramaya University. P. 56.

Habtamu K, Husien O, Haimanote B, Tegenu E, Charles F, Amy S, Tammo S (2009). The effect of land use on plant nutrient availability and carbon sequestration. pp 208-219. Proc. 10th conf., March 25-27. Addis Ababa.

Havlin J, Tisdale S, Nelson W, Beaton J (1999). Soil Fertility and Fertilizers, 6th ed. Macmillan Publishing. USA. pp. 85-196.

Hazelton P, Murphy B (2007). Interpreting soil test results. 2nd ed. CSIRO Publisher. P. 52.

IUSS Working Group (2006). World Reference base for Soil Resources 2006. 2nd ed. World Soil Resources Reports P. 103. FAO, Rome.

Jones J (2003). Agronomic Handbook: management of crops, soils, and their fertility. CRC Press, Boca Raton, Florida, USA. P. 482.

Landon J (1991). Booker tropical soil manual: A handbook for soil survey and agricultural land evaluation in the tropics and subtropics. Longman, NY.

Lawrence G (1977). Measurements of pore sizes in fine-textured soils: A review of existing techniques. Journal of Soil Science, 28: 527-540. http://dx.doi.org/10.1111/j.1365-2389.1977.tb02261.x

Mebit K (2006). Physicochemical properties and classification of soils at Woreta ATEVT research farm. Ethiopia. MSc Thesis. Alemaya University, P. 53.

Mesfin A (1998). Nature and management of Ethiopian soils. Alemaya University, Ethiopia. P. 272.

Mishra B, Heluf G, Kibebew K (2004). Soils of Ethiopia: Constraints in Relation to Food Security. Food, Agric. Environ. J. 2(3):269-279.

Olsen S, Cole C, Watanabe F, Dean L (1954). Estimation of available phosphorus in soils by extraction with sodium bicarbonate. USDA circular 939. U.S. Govt. Printing, Washington D.C. pp.1-19.

Sanchez P, Sheperd K, Soule F, Place R, Buresh A, Izac A, Mokwunye F (1997). Soil Fertility Replenishment in Africa. An Investment in Natural Resources Capital. In: Buresh R, Sanchez P, Calnoun F (eds.). No. 51. Am. Soc. Agron. Madison, Wisconsin, USA pp. 1-46.

Sheleme B (2011). Characterization of soils along a toposequence in Gununo area, southern Ethiopia. J. Sci. Dev. 1(1):31-41.

Shimeles D (2006). Characterization and classification of soils at Tenocha catchment. MSc Thesis. Alemaya University. P. 39.

Soil Survey Staff (2006). Keys to Soil Taxonomy, 10th ed. USDA, NRCS. Washington DC, United States Govt Printing Office. pp. 161-323.

Tekalign M, Haque I (1991). Phosphorus status of some Ethiopian soils. III. Evaluation of soil test methods. Trop. Agric. 68:51-56.

USDA (2008). Bulk density, Soil quality indicators. http://www.usda.gov/sqi/publications/files.

Van Reeuwijk L (2006). Procedure for soil analysis, 7th ed. Technical Report 9. Wageningen, Netherlands, ISRIC – World Soil Information.

Wakene N, Heluf G (2003). Influence of land management on morphological, physical and chemical properties of some soils of Bako, Western Ethiopia. Agropedology 13(2):1-9.

Walkley A, Black C (1934). An examination of the Degtjareff method for determining soil organic matter and modifying the chromic acid method. Soil Sci. 37:29-38.

http://dx.doi.org/10.1097/00010694-193401000-00003
Woldeamlak B, Stroosnijder L (2003). Effects of agro-ecological land use succession on soil properties in Blue Nile basin. Geoderma 111:85-98.http://dx.doi.org/10.1016/S0016-7061(02)00255-0
Yeshi Ber Consultant (2003). Gambella Peoples Regional State land use/land allotment study. Amended draft final report. Addis Ababa. Ethiopia. (Unpublished document)
Yihenew G (2002). Selected physicochemical characteristics of soils at Adet Research Center, Ethiopia. Ethiop. J. Nat. Resour. 4(2):199-215.

8

Assessing the Input and Output Flows and Nutrients Balance Analysis at Catchment Level in Northern Ethiopia

Gebremedhin Kiros[1] , Mitiku Haile[2] and Girmay Gebresamuel[2]

[1]Department of Soil Resources and Watershed Management, Aksum University, Shire- campus: P. O. Box -314, Shire - Endaselassie, Ethiopia.
[2]Department of Land Resources Management and Environmental Protection, Mekelle University: P. O. Box -231, Mekelle, Ethiopia.

Soil fertility depletion in smallholder farms is one of the fundamental biophysical causes for declining use per capita food production in Ethiopia. In the present study, resource flow analysis was made at catchment scale in northern Ethiopia, analyzing nutrient balances. JMP 5, a computer program for statistics and used in applications such as design experiment and scientific research was used to process and to analyze the data of different resources flow into and out of the watershed. Nutrient balances for N, P and K from four nutrient fluxes entering and four nutrient fluxes leaving the watershed were calculated. Some of the fluxes (e.g. wet deposition and gaseous losses) were estimated using transfer functions. At the catchment level, full nutrient balance results indicate a depletion rate of -42.5 to 13.1 kg N ha^{-1}year^{-1} (for the rich) and -32.7 kg N ha^{-1} year^{-1}, -11.8 kg K ha^{-1}year^{-1} (for the medium) in the upper landscape, -49 kg N ha^{-1} year^{-1}, -14.3 kg K ha^{-1} year^{-1} (for the rich) and -28.5 kg N ha^{-1} year^{-1}, -11.8 kg K ha^{-1} year^{-1} (for the medium) in the middle landscape, -57 kg N ha^{-1} year^{-1}, -16.1 kg K ha^{-1} year^{-1} (for the rich) and -33.5 kg N ha^{-1} year^{-1}, -12.1 kg K ha^{-1} year^{-1} (for the medium) in the lower landscapes. Less negative value of nutrient balances of the poor socio-economic groups across the three landscapes shows N depletion in the poor socio-economic groups with -17.7kg N ha^{-1} year^{-1} and -5.59 kg K ha^{-1} year^{-1}, -21 kg N ha^{-1} year^{-1} and -12.8 kg K ha^{-1} year^{-1} and -19.7 kg N ha^{-1} year^{-1}, and -7.52 kg K ha^{-1} year^{-1} in the upper, middle and lower landscapes respectively. Soil nutrient stocks in all the landscapes were decreasing with the exception of phosphorus which is positive in all the landscapes. In the analysis, harvested crops (OUT1) and crop residues (OUT2) were the major cause for nutrients depletion. It was calculated that the contribution of harvested crops (OUT1) and crop residues (OUT2) to N losses was 80.5%, and its contribution to K losses was 65.8%, therefore, application of organic and inorganic inputs, proper management of crop residues and sustaining soil conservation measures are very crucial.

Key words: Resources flow, nutrient balances, soil fertility, landscapes, wealth status.

INTRODUCTION

Agriculture is the main economic activity of Ethiopia. It is dominated by smallholder farming (MoFED, 2002; CSA, 2009). The contribution of smallholder agriculture to the country is very high. It accounts for about 45% of the GDP, 85% of the exports and 80% of the total employment (EPA, 1997). However, agricultural productivity is continuously challenged by land degradation, which is manifested in various ways. For example, footpaths develop into gullies, soils become thin and stony, topsoil is gone due to accelerated soil erosion etc (Stocking and Murnaghan, 2001).

Over 50% of the highlands in general and cropped areas of Ethiopia are in an advanced stage of land degradation (Elias, 2002). This is because of the continuous cultivation at least since the 13th century (Tewolde Berhan, 2006). Therefore, soil organic matter (SOM) content and nutrients are generally lower, where land degradation is more severe (Elias, 2002) it also leads to poor soil structure consequently to water erosion. Therefore, soils in many areas of the country especially in Tigray become shallow and stony (Stocking and Murnaghan, 2001).

The decline in soil physical, chemical and biological properties is revealed in many parts of the densely populated highlands of Ethiopia. For example, in Tigray nitrogen and phosphorus are highly deficient (Mitiku et al., 2003). Nitrogen in the cultivated surface soils was between 0.07-0.13% in Melbe area of Tigray (Tegene, 1996). Moreover, the soil depth in many areas of Ethiopia is less than 20 to 30 cm, this means that it is reaching the lower limits of productivity of the arable land and has lost much of its capacity to retain moisture; with consequent decline in agricultural yield (Stocking and Murnaghan, 2001; Elias, 2002; World Bank, 2007). The annual grain production loss estimate due to burning of dung as fuel than using for soil fertility improvement is estimated at 50,000 tons per annum, while the loss due to accelerated erosion is around 40,000 t by 1990. This will be accelerated into 170,000 t in 2010, if not controlled (EPA, 1997).

Soil nutrient depletion is rarely directly linked to food shortage, as it is a gradual process unlike natural disasters (e.g. drought, flooding and, etc.). Nevertheless, several studies have revealed that lack of plant nutrients is one of the principal causes for low agricultural productivity and food insecurity in Africa (Smaling, 1993; Sanchez, 2002; Haileselassie, 2005). Agriculture in Ethiopia is no exception: More soil nutrients are exported compared to natural and anthropogenic inputs (Elias et al., 1998; Okumu, 2000). Stoorvogel and Smaling (1990) predicted for Ethiopia that the national nutrient balances would be on average: -47 kg N ha^{-1}, -15 kg P_2O_5 ha^{-1} and -38 kg K_2O ha^{-1} for the year 2000. This prediction was twice as high as the average value for sub Saharan Africa and indicates the severity of nutrient depletion in Ethiopia.

Previous efforts to quantify nutrient balances were at sub-continental and national scales without addressing the large spatial variability of nutrient balances within a region and catchment level. Hence, there is a need for assessment of resource flow analysis at catchment level (scale) in different landscapes and under different socio-economic conditions, as a basis for the design of technically feasible, ecologically non-degrading, and economically viable nutrient management strategies. The study described in this paper responds to that need of reporting on assessment of macro-nutrient balances at catchment scale in May-leba, Northern Ethiopia.

The purpose of this study is to investigate the potential of resources flow analysis to tackle the increasingly severe soil fertility deterioration in May-leba catchment, Northern Ethiopia. Based on primary, available secondary data and orders-of-magnitude estimations, the resources (nutrient) flows through the catchment system are analyzed, traced and quantified, and the key material sources identified. Scenario calculations allow then to determine most effective mitigation measures which are either favorable for nutrient conservation or which are responsible for high nutrient losses under current management practices. This kind of information is critical for decision makers to plan and implement integrated nutrient management strategies at a regional and catchment level.

MATERIALS AND METHODS

Description of the study area

Location

The May- Leba catchment lies (longitude: 13°14'35" and latitude: 39°15'53") is located at 35 km West of Mekelle, along the road from Mekelle to Abi-Adi, about 10 km east of Hagere-Selam town (Figure 1) . It is part of the woreda of Dogua Tembien. The lower edge of the study area is the May-Leba dam at 2290 m.a.s.l. (Van de Wauw et al., 2008).

Soils and climate

The major soils of May-leba catchment are Leptosol, Cambisol, Vertisol, Regosol and Fluvisol (Gebresamuel et al., 2010). They contain 1.1-1.6 percent organic carbon, 0.09-0.16 percent total nitrogen and 39.1-57.9 cmol (+) kg^{-1} soil Cation Exchange Capacity (Van de Wauw et al., 2008). The climate is characterized by an ustic (intermediate between the aridic and the udic) moisture regime implying dryness with a distinct rainy season in July and August. The average rainfall is 716mm. The site is located in the semi-arid agro-ecological zone. Rainfall is distributed between the short rainfall season (March to April) and the main rainy season (June to September).The average monthly temperature varies between 12 and 19°C (Van de Wauw et al., 2008).

Land use

The dominant land use in the May-Leba catchment is cropland (about 50%). The rest is used for grazing (30%), housing (15%) and a small part (5%) is closed for grazing and cropping (Van de Wauw et al., 2008). The major crops grown in the study area are barley (*Hordeum vulgare*), wheat (*Triticum sp.*), teff (*Eragrostis tef*) and grass pea (*Lathyrus sativus*) where the total area of the study area is 1733.6ha and 50% of this is cropland therefore, the relative share of the total cultivated crop land among these crops is 797.456ha (Gebresamuel et al., 2010).

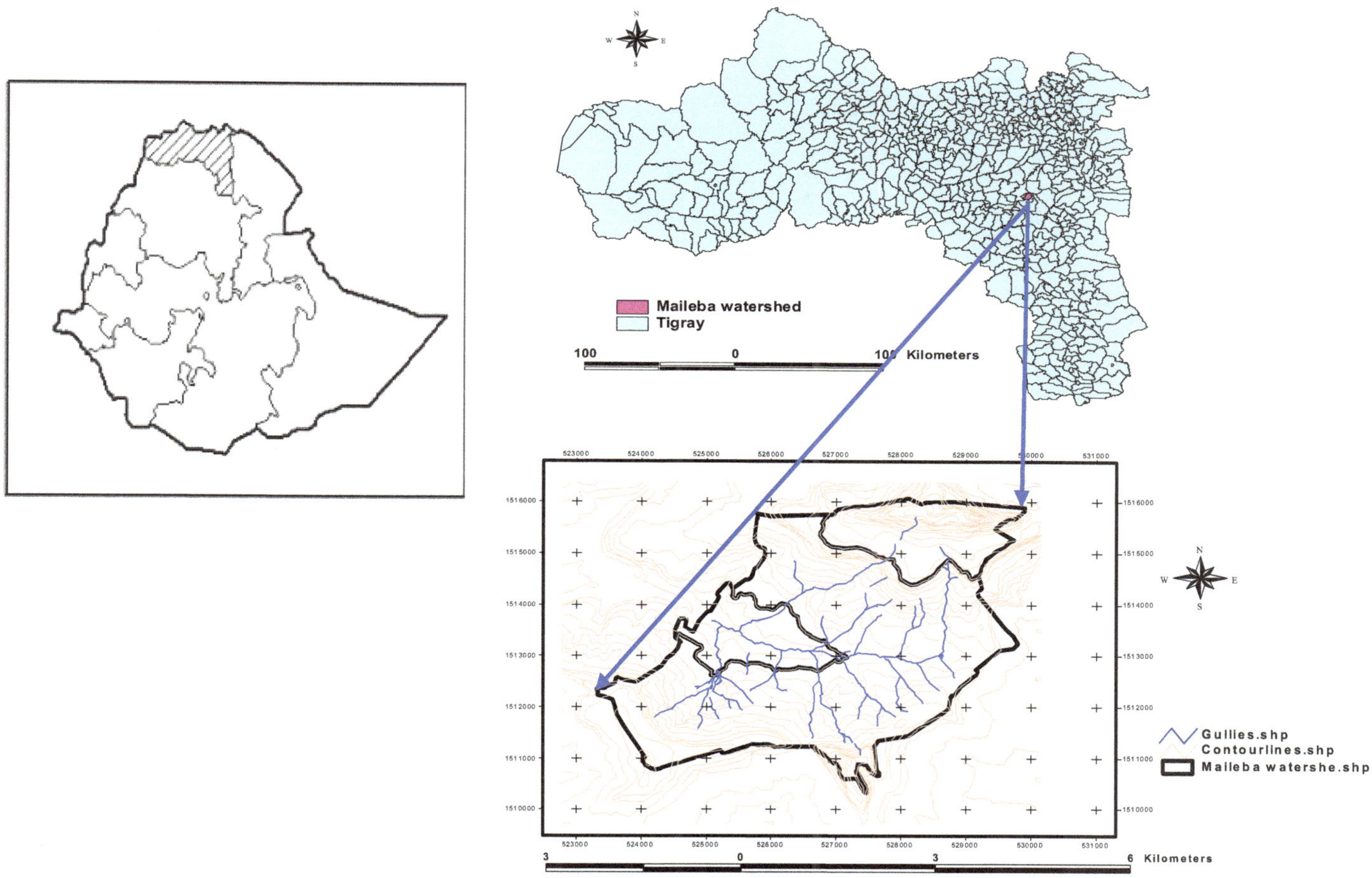

Figure 1. Location map of the May-Leba catchment.

Quantifying of resources flow

The N, P and K balances were calculated from a combination of four input and four output flows using the nutrient/material balance model developed by Stoorvogel and Smaling (1990). The input flows monitored were mineral fertilizer, manure, compost, ash and atmospheric deposition. The major output flows quantified were harvested products (like straw, crop grain), crop residues, hay and gaseous losses. Quantification of the N, P and K in the input and output flows was achieved through the combination of different methods: field measurement, use of empirical quantitative relations (i.e. transfer functions) and assumptions based on secondary data from a variety of sources. The type of data required and the method of quantification for each of the input and output flows are summarized in Tables 1 and 2.

Quantifying Inputs

For N, P, K and OC primary data were obtained on type and amount of mineral fertilizer, organic inputs (manure and compost), ash whereas atmospheric deposition was found by transfer function. Field observation was made on type and amount of resources input and output, type of cereal crop grown and their grain and residue yield were recorded. Routine field measurement of the transfer of resources in and out of each land use type was monitored by maintaining a book keeping of material flows. The quantity of nutrient inputs via wet deposition (IN4) depends on the amount of precipitation and proximity to a city (Mengel and Kirkby, 1996). The method of Stoorvogel and Smaling (1990) and Smaling (1993) were used to estimate wet deposition as the square root of average annual rainfall (in mm) using coefficients of 0.014 for N, 0.053 for P and 0.11 for K (Stoorvogel and Smaling, 1990; Smaling, 1993).

IN4 (N) = $0.14 \times (\text{rainfall})^{1/2}$, IN4 ($P_2O_5$) = $0.053 \times (\text{rainfall})^{1/2}$ and IN4 (K_2O) = $0.11 \times (\text{rainfall})^{1/2}$

Where: IN4 is expressed in kilograms per hectare per year; and rainfall in millimeters per year.

Quantifying outputs

Removal of harvested products (OUT1), crop residues (OUT2) and hay (OUT3) are usually the major pathways of nutrient losses from agricultural soils. The amount varies considerably depending on crop type, soil type, agronomic practices and plant nutrient uptake (Brady and Weil, 2002). Reliable information on crop nutrient composition is, therefore, essential for quantifying this important nutrient export. Biomasses, grain, residue yields of crops were determined by harvesting from a **12 m^2** areas at ground level from randomly selected fields. The harvested area was randomly demarcated in each field and the plot was harvested manually at crop maturity on dates preferred by farmers. Samples were air dried

Table 1. Type of data required and method of quantification of the four input processes employed in the calculation of N, P and K balances.

Input process	Code and nutrients	Data required	Method of quantification
Mineral fertilizer	IN1	Type of fertilizes applied	Field measurement
	N and P	Amount of fertilizers applied	Field measurement
Manure	IN2a	Amount of manure applied	Field measurement
	N, P and K	Nutrient content of manure	Laboratory analysis
Compost	IN2b	Amount of compost applied	Field measurement
	N, P and K	Nutrient content of compost	Laboratory analysis
Ash	IN3	Amount of ash applied	Field measurement
	N, P and K	Nutrient content of ash	Laboratory analysis
Atmospheric	IN4	Average annual rainfall	Rainfall records
Deposition	N, P and K	N, P and K deposition in the rainfall	Transfer function

Table 2. Type of data required and method of quantification of the four output processes employed in the calculation of N, P and K balances.

Output process	Code and nutrients	Data required	Method of quantification
Harvested products	OUT1	Crop yield	Field measurement
	N, P and K	Nutrient content of products	Laboratory analysis
Crop residues	OUT2	Residue yield	Field measurement
	N, P and K	Destination of residue	Field observation
		Nutrient content of residue	Laboratory analysis
Hay	OUT3	Hay yield	Field measurement
	N, P and K	Destination of hay	Field observation
		Nutrient content of hay	Laboratory analysis
Gaseous losses	OUT4 N	Average annual rainfall	Rainfall records
		Nutrient content in applied manure	Field measurement
		Nutrient content in applied fertilizer	Field measurement
		N- losses	Transfer functions

and weighed using a hand held scale or sensitive balance to give values for total biomass, grain, and residue yields and hay.

$OUT1_N = Y * N$ = kg N/ha, $OUT1_P = Y * P$ = kg P/ha and $OUT1_K = Y * K$ = kg K/ha

Y = yield (kilograms/ha), N = N content of crops (kg N/ kilograms harvested product); P = P content of crops (kg P/ kilograms harvested product), and K = K content of crops (kg K/ kilograms harvested product).

The quantity of crop residues (OUT2) and hay (OUT3) were calculated from residues/products ratio given in FAO (1986). It is assumed that in most parts of Ethiopia 85% of total residues are removed from arable lands being used either as energy sources or for animal feed.

$OUT2_N = R * N * F$ = (kg N/ha), $OUT2_P = R * P * F$ = (kg N/ha) and $OUT2_K = R * K * F$ = (kg N/ha)

R = amount of residues (kilograms/ha); N = N content of crops residues (kg N/ kilograms harvested product); P = P content of crops residues (kg P/ kilograms harvested product); K = K content of crops residues (kg K/ kilograms harvested product), and F = removal factor of crop residues

Gaseous losses (OUT4) can be an important pathway of N fluxes in many agricultural production systems. Stoorvogel and Smaling (1990) linked a limited number of reliable data on denitrification of sub Saharan African soils using multiple regression analysis. Some other studies estimate gaseous losses as function of soil N, fertilizer N, precipitation and soil clay contents (Van den Bosch et al., 1998; Smaling, 1993). More recent studies (FAO, 2003) includes volatilization losses in OUT4 estimation as given by the equation below in which OC is organic C content (in percent), R is mean annual precipitation (mm), Nf is mineral N fertilizer and organic fertilizer N (kg N ha^{-1}).

$$OUT4 = (0.025 + 0.000855R + 0.01725Nf + 0.117OC) + 0.113Nf$$

Nutrient export in the harvested products and residues was derived from the yield data for each crop. Sub-samples of harvested products and residues of wheat, barley, teff and grass pea were taken for determination of nutrient content at the Tigray agricultural Research Institute (TARI) laboratory.

Chemical analysis of the resources flow

The resources flow samples were processed for laboratory test including proper registration, air-drying, grinding, sieving through 2 mm sieve and storage (Sahlemedhin and Taye, 2000). The analysis was conducted on ground and sieved (< 2 mm) samples. The material samples were used for chemical property analysis. The following resources flow parameters have been analyzed: organic carbon, total nitrogen, available phosphorus, and available potassium.

The percentage of organic carbon was determined by a modified Walkley-Black procedure (Smith and Weldon, 1940). The organic matter was obtained by wet oxidation technique, that is, calculation of 1.724 × percentage of carbon (Black, 1965). It is based on the assumption that organic matter contains 58% organic carbon.

The wet digestion of the Kjeldahl procedure (Bremmer and Mulvaney, 1982) was used to test for total nitrogen while Olsen's method was employed to determine available phosphorus (Olsen and Sommers, 1982; Anderson and Ingram, 1993). Available K was extracted by ammonium acetate extraction method (Sahlemedhin and Taye, 2000).

Chemical analysis of plant biomass

The N, P and K contents of the plants of different crop resources were analyzed in the soil laboratory of the Tigray Agriculture Research Institute (TARI). The preparation of the plant material was through drying the green material at a maximum of 60 to 70°C, grinded to pass through a 0.15 mm mesh and 10 g is taken for analysis (Anderson and Ingram, 1993). The concentration of the total nitrogen in plant was determined by the Kjeldahl method. The organic nitrogen is oxidized into ammonium by acid hydrolysis with H_2SO_4 together with the reagent potassium sulfate to raise temperature and to hasten the rate of decomposition, copper sulfate and selenium powder were used as catalyst. The plant analysis procedure for Phosphorous and Potassium concentration was done following ashing method. About 1.0 g of ground plant sample was dried out in aluminum dish over night at 105°C in an oven. The ash was dissolved in concentrated HCl and diluted with de-ionized water. After addition of color reagent (molybdate-vandate-solution), the phosphorous concentration was measured by spectrophotometer and K by flame photometry.

Data analysis

The types of analysis that were carried out were the analysis of variance (ANOVA) to evaluate how resources flow (treatments) varies with landscapes. Regression and correlation analyses among the wealth ranking, landscape and resources flow was conducted.

RESULT AND DISCUSSION

Major source of nutrients flow and their contribution to soil fertility

The major sources of N, P, K and OC for the homestead fields was animal manure along with homestead refuse, leftover feed of animals from around the houses. Household refuse and animal manure from around the hut were used on homestead fields to improve the physical properties and nutrient retention of the soils. In contrast, mineral fertilizer (DAP and Urea) were the main sources of nutrient addition to the distance fields.

Nutrient inputs to the catchment

Farmers use specific soil fertility management strategies for different parts of their farms. They grow mainly permanent and vegetable crops near the homestead and others (cereals, pulses and oil crops) on more distant fields. Applications of manure and inorganic fertilizers reflect this differentiation (Tilahun et al., 2001). Currently, DAP and Urea are the only inorganic fertilizers applied by smallholders. Potassium application from inorganic fertilizer (IN1) is not reported in Ethiopia (NFIA, 2001). Fertilizer trials conducted on major cereal crops (for N, P and K) also indicated that cereal crops were not responsive to K. But the importance of long term K fertilization must not be overlooked, as leaching, erosion and depletion via crop harvest may deplete K stocks (Haileselassie, 2005). For all the Socio-economic groups, inflows into the catchment are in the form of chemical fertilizer, organic fertilizer like compost and manure and ash. Inflows of the inorganic fertilizers urea and diammonium phosphate (DAP) were small i.e. differences in the magnitude of IN1 across the socio-economic groups may have similar underlying causes these are like less awareness about using inputs by the farmers, approach of the agricultural extension situations (Table 3).

Nitrogen and phosphorus inputs from inorganic fertilizer were (3.29 kg N ha^{-1} $year^{-1}$, 0.56 kg P ha^{-1} $year^{-1}$) and (23.01 kg N ha^{-1} $year^{-1}$, 4.18 kg P ha^{-1} $year^{-1}$), (10.5 kg N ha^{-1} $year^{-1}$, 1.93 kg P ha^{-1} $year^{-1}$) and (28.39 kg N ha^{-1} $year^{-1}$, 5.21), (5.2 kg N ha^{-1} $year^{-1}$, 0.93 kg P ha^{-1} $year^{-1}$) and (19.6 kg N ha^{-1} $year^{-1}$, 0.28 kg P ha^{-1} $year^{-1}$) higher in the resource endowed households than the medium and poor households in the upper, middle and lower landscape respectively. Hence, the chemical fertilizer inflows of N and P were higher for the rich farmers than for the medium and the poor in all the landscapes. The quantity of fertilizer used by different groups of farmers and the area of application are presented in Table 3. The data revealed a highly significant difference ($p<0.01$) for N among households in the quantity of fertilizer used across all socio-economic groups in all the landscapes. Richer farmers purchase significantly larger amounts of

Table 3. Resources inflow (kg/ha/yr) across the socio-economic groups in the three landscapes in May-leba catchment, Northern Ethiopia.

Landscape	Wealth	Chemical fertilizer		Organic Inputs			Ash			Atmospheric Deposition		
		N	P	N	P	K	N	P	K	N	P	K
	Rich	37.4	6.8	16.3	0.24	12.56	1.12	0.231	11.34	3.75	1.42	2.94
Upper	Medium	34.13	6.26	11.5	0.16	11.58	1.24	0.221	10.14	3.75	1.42	2.94
	Poor	14.4	2.64	9.65	0.12	9.83	1.38	0.219	9.12	3.75	1.42	2.94
	Rich	42.8	7.85	17.6	0.28	14.41	1.19	0.209	10.55	3.75	1.42	2.94
Middle	Medium	32.3	5.92	10.5	0.2	12.1	1.03	0.231	8.55	3.75	1.42	2.94
	Poor	14.4	2.64	6.93	0.14	7.2	1.12	0.790	6.40	3.75	1.42	2.94
	Rich	35.6	6.52	15.9	0.26	11.68	1.9	0.232	10.66	3.75	1.42	2.94
Lower	Medium	30.4	5.87	12.2	0.2	9.85	1.26	0.223	9.64	3.75	1.42	2.94
	Poor	16	5.59	11.5	0.14	7.51	0.7	0.238	9.42	3.75	1.42	2.94

Table 4. The different resources inputs (kg/ha/year) into the catchment across the socio-economic groups (wealth status) in May-leba catchment, Northern Ethiopia.

Wealth status	Chemical fertilizer (IN1)		Compost (IN2a)			Manure (IN2b)			Ash (IN3)		
	N	P	N	P	K	N	P	K	N	P	K
Rich	(38.6± 2.18)[a]	(7.06± 0.41)[a]	(13.7± 0.28)[a]	(0.27± 0.06)[a]	(11.11 ± 0.86)[a]	(19.4± 1.14)[a]	(0.25 ± 0.02)[a]	(14.65 ± 0.92)[a]	(1.00 ± 0.15)[a]	(0.22 ± 0.01)[a]	(2.12± 0.79)[a]
Medium	32.2± (1.07[b]	(5.92 ± 0.19)[a]	(9.45± 1.45)[a]	(0.20± 0.02)[b]	(9.22 ± 0.39)[a]	(13.3± 0.75)[b]	(0.16 ± 0.01)[a]	(13.13 ± 0.96)[a]	(1.39 ± 0.26)[a]	(0.23 ± 0.00)[a]	(1.10± 0.21)[a]
Poor	(14.9± 0.53)[c]	(3.72± 1.08)[ab]	(8.19± 2.99)[a]	(0.11± 0.02)[c]	(5.94 ± 1.07)[ab]	(10.5± 0.63)[c]	(0.14 ± 0.01)[ab]	(10.39 ± 1.25)[b]	(1.25 ± 0.10)[a]	(0.42 ± 0.19)[a]	(2.82± 1.79)[a]

Mean values along the columns with different letters indicate significant difference at $P<0.05$ level of confidence.

fertilizer compared to the poor farmers in all the landscapes. This is because there is no relatively financial constraint in the richer farmers in all the landscapes ($p<0.01$) and also there are a sizeable number of livestock that could secure the rich farmers than the poor from the debts of the chemical fertilizer. There is no significant difference across the landscapes in the quantity of fertilizers used. This finding in agreement with the findings by Elias (2002) who observed no statistically significant difference among socio-economic groups in the rate of fertilizer application

The national scale nutrient inputs from urea and DAP were estimated at 12.24 kg N ha^{-1} $year^{-1}$ and 12.87 kg P ha^{-1} $year^{-1}$ (Haileselassie et al., 2005). Demeke et al. (1998) reported a similar trend of inorganic fertilizer applications across all regional states of Ethiopia. In addition to soil fertility, intensity of crop cultivation, cropping pattern, livestock management system (open grazing, confined management) and differences in other competitive uses of manure (e.g. household energy) were the major causes of variation in the application of IN2 across the different socio-economic groups (Table 4). The rich endowed households had remarkably higher amount of IN2 compared to the other socio-economic groups in all the landscapes. This can be accounted for homestead farms that require regular application of manure. The rich famers also have better livestock management practices that contributed access to dung and increased IN2 in this socio-economic group. Compared to the medium and poor farmers the rich farmers apply a high amount of nutrient inputs from organic fertilizer (IN_2). Nitrogen, Phosphorus and Potassium inputs from organic fertilizer were higher in the rich endowed households that is, (4.8 kg N $ha^{-1}year^{-1}$, 0.1 kg P $ha^{-1}year^{-1}$ and 1 kg K $ha^{-1}year^{-1}$) higher than the medium, (6.7 kg N $ha^{-1}year^{-1}$, 0.12 kg P $ha^{-1}year^{-1}$ and 2.73 kg K $ha^{-1}year^{-1}$) and the poor in the upper landscape and (7.1 kg N $ha^{-1}year^{-1}$, 0.1 kg P $ha^{-1}year^{-1}$ and 2.31 kg K $ha^{-1}year^{-1}$) higher than the medium, (10.7 kg N $ha^{-1}year^{-1}$, 0.2 kg P $ha^{-1}year^{-1}$ and 7.21 kg K $ha^{-1}year^{-1}$) and the poor in the middle landscape and (3.7 kg N $ha^{-1}year^{-1}$, 0.1 kg P $ha^{-1}year^{-1}$ and 1.85 kg K $ha^{-1}year^{-1}$) higher than the medium, (4.4 kg N $ha^{-1}year^{-1}$, 0.12 kg P $ha^{-1}year^{-1}$ and 4.2 kg K $ha^{-1}year^{-1}$) and the poor in the lower

Table 5. The different resource inputs (kg/ha/year) into the catchment across the Landscapes in May-leba catchment, Northern Ethiopia.

Land scape	Chemical fertilizer (IN1)		Compost (IN2a)			Manure (IN2b)			Ash (IN3)		
	N	P	N	P	K	N	P	K	N	P	K
Upper	(28.6 ±7.18)^{a}	(5.23 ±1.31)^{a}	(9.82 ±1.75)^{a}	(0.1 ±0.06)^{a}	(8.87 ±0.88)^{a}	(15.12± 2.23)^{a}	(0.17 ±0.02)^{a}	(13.7 ±0.74)^{a}	(1.25 ±0.08)^{a}	(0.22 ±0.00)^{a}	(1.42 ±1.16)^{a}
Middle	(29.83 ±8.29)^{a}	(5.47 ±1.52)^{a}	(8.19 ±3.04)^{a}	(0.22 ±0.04)^{a}	(8.81 ±2.65)^{a}	(15.16 ± 3.28)^{a}	(0.19 ±0.05)^{a}	(13.63 ± 1.63)^{a}	(1.11 ±0.05)^{a}	(0.41 ±0.19)^{a}	(2.77 ±1.81)^{a}
Lower	(27.32 ± 5.85)^{a}	(5.99 ±0.28)^{a}	(13.39 ± 0.57)^{a}	(0.20 ±0.04)^{a}	(8.59 ±1.04)^{a}	(13.00 ±2.42)^{a}	(0.19 ±0.03)^{a}	(10.77 ± 1.39)^{a}	(1.29 ±0.35)^{a}	(0.23 ±0.01)^{a}	(1.85 ±0.90)^{a}

Mean values along the columns with the same letter are not significantly different at P<0.05 confidence interval.

landscape. Hence, the organic inflows of N, P and K were higher for the rich farmers than for medium and poor in all of landscapes. Thus, the data revealed a significant difference ($p<0.01$ for N, $p<0.05$ for P and $p<0.01$ for K) in the amount of organic inputs (manure and compost) moved (used) to fields for soil fertility maintenance among households of different socio-economic groups (Table 4). Richer farmers used significantly ($p<0.01$) larger volumes of manure compared to poor farmers in the three landscapes. This is because there are a sizeable number of livestock in the rich endowed households than the poor farmers.

As would be expected, the statistical correlation test showed a strong and positive correlation ($r=0.992$, $p<0.01$) between the number of livestock held per household and the amount of manure for soil fertility. This is in agreement with the findings of Elias (2002). The variation in the amount of manure used among households is largely explained by differences in the herd size. Nonetheless, in societies (Burkina Faso and Northern Benin) where animal dung is used for household energy supply (fuel), a significantly negative relationship would occur (Millennium Ecosystem Assessment, 2005).

No significant variation was observed in the amount of chemical fertilizer, compost, manure and ash transferred to the fields for soil fertility maintenance across landscape positions (Table 5). The total national estimate of IN2 application rates for all cropping systems was 29 kg N ha^{-1} year^{-1}, 7.2 kg P ha^{-1} year^{-1} and 34.3 kg K ha^{-1} year^{-1} (Haileselassie et al., 2005). Ash (IN3) and atmospheric deposition (IN4) were the other nutrient inputs considered in this study. Globally, N deposition rates are estimated to be 5 kg N ha^{-1} year^{-1} for less densely populated and non-industrial countries and range from 20 kg N ha^{-1} year^{-1} to 50 kg N ha^{-1} year^{-1} in countries of Western Europe and parts of China (Sheldrick et al., 2003). Mengel and Kirkby (1996) also discussed that atmospheric deposition for N can be as high as 60 kg ha^{-1} year^{-1} depending on proximity to a city and amount of precipitation. Atmospheric deposition estimates showed no clear difference between socio-economic groups.

Nutrient outputs from the catchment

Removals in crops (OUT1) and (OUT2) were the major causes of export of nutrients from the soil followed by hay (OUT3) and gaseous (OUT4) losses in most of the socio-economic groups (Table 6).

Differences in the landscapes and corresponding primary productivities were the main cause of variation (OUT1) among the socio-economic groups (Table 6). Notably, socio-economic groups with a high IN1 level had also a high OUT1 (e.g. the rich in the three landscapes). This imply for nutrient management is assessing the effects of future land use change on nutrients; and soil management practice influence the soil nutrients related soil processes, such as erosion, oxidation, mineralization, and leaching, etc. and consequently modify the processes of transport and re-distribution of nutrients. At the catchment level, the losses of N, P and K via harvested crops (OUT1) varied from (28.69 to 82.64 kg ha^{-1}year^{-1}, 0.12 to 0.32 kg ha^{-1}year^{-1} and 9.37 to 18.02 kg ha^{-1}year^{-1}) across all the landscapes.

As was the case for products harvested, there were clear differences in OUT2 across socio-economic groups (Table 6). This was caused by differences in yields and cropping pattern this indicates that the yearly sequence and spatial arrangement of crops and fallow on a given area where selection of crops and their varieties is to be made depending on the soil and rain fail situation in the rained areas are Teff-Barley/Wheat- Teff; Barley-Teff-Wheat-Grass pea/Bean; Barley/Wheat -Bean-Wheat/Barley. The losses of N, P and K via crop residues (OUT2) varied from (5.1 to 13.3 kg ha^{-1}year^{-1}, 0.611 to 1.21 kg ha^{-1}year^{-1} and 6.29 to 13.55 kg ha^{-1}year^{-1}) along with the landscapes. High crop residue losses were observed in the rich farmers compared to the medium and the poor, because the rich farmers have a sizeable number of livestock so that they do not leave the crop residues in their farms for soil fertility maintenance. This finding is in agreement to the study at a national scale where export through residues (OUT2) was estimated at 9.8 kg N ha^{-1} year^{-1}, 1.5 kg P ha^{-1} year^{-1}, but contrasts for K which is 18.5 kg ha^{-1} year^{-1} (Haileselassie et al., 2005)

Table 6. Resources outflow (kg/ha/yr) across the socio-economic groups in the three landscapes in May-leba catchment, Northern Ethiopia.

Land scape	Wealth	Harvested products			Crop residues			Hay			Gaseous losses
		N	P	K	N	P	K	N	P	K	N
Upper	Rich	70.89	0.19	13.43	12.7	1.21	13.55	12.64	0.16	12.92	4.8
	Medium	60.24	0.15	11.66	8.6	0.89	11.57	9.24	0.15	11.25	4
	Poor	34.73	0.15	9.37	5.1	0.61	8.86	3.78	0.14	9.15	3.3
Middle	Rich	82.64	0.32	18.02	13.3	1.11	10.06	13.79	0.17	14.11	4.7
	Medium	52.02	0.16	15.36	9.7	0.73	8.08	10.33	0.16	11.93	4
	Poor	28.69	0.16	13.01	5.3	0.71	6.82	9.96	0.14	9.46	3.3
Lower	Rich	75.60	0.20	16.19	10.5	0.92	9.84	22.17	0.18	15.38	4.7
	Medium	56.74	0.16	13.80	7.75	0.76	8.16	12.74	0.16	12.56	4.5
	Poor	46.37	0.12	11.12	5.6	0.61	6.29	8.6	0.15	9.98	4

The results suggest that denitrification (OUT4) may be the least important cause of N loss. Indeed, averages are of the magnitude of 4.73 kg ha^{-1} year^{-1}, 4.17 kg ha^{-1} year^{-1}and 4 kg ha^{-1} year^{-1} for the rich, medium and the poor in the study area, values which are close to the result of Van den Bosch et al. (1998). Denitrification losses predicted by Stoorvogel and Smaling (1990) were higher than the present study's estimate (7 kg ha^{-1} year^{-1}). Rich farmers of the study areas had medium level of attention. Of course there are very limited numbers of farmers were found to have a high attention of integrated soil fertility and nutrient management efficiency for sustainable crop production.

The input-output nutrients balance

Partial nutrient balance

The partial nutrient balance of the study considered only the most important inputs and outputs. The inputs include mineral fertilizer (IN1), organic inputs (IN2) while the output harvest products or grain (OUT1) and residues removed (OUT2). The partial nutrient balances are calculated as the difference between sum of inputs (IN1 and IN2) and sum of outputs (OUT1 and OUT2). Based on the partial input-output nutrient balance the study area shows negative balance for Nitrogen and Potassium regardless of the wealth status and landscape.

Marked variations were observed among the socio-economic groups (Figure 2). The rich with a higher rate of fertilizer or dung input (in the three landscapes) have a higher negative balance even though harvests were also higher. It is also observed apparently that strong positive balances of P in all the socio-economic groups exist in all the three landscape positions. Higher inorganic and organic nutrient inputs and relatively strong negative partial nutrient balances in the rich households might be due to intensively cultivated farms, however nutrient-saving techniques, such as soil and water conservation practices, were practiced more frequently by the rich group. In all the socio-economic groups and the landscapes, the level of inputs (IN1 and IN2) was clearly lower than the outputs (OUT1 and OUT2), resulting in a negative partial balance for all nutrients except phosphorus (P). The losses of N and K (-29.91 kg N ha^{-1} year^{-1}, -14.420 kg K ha^{-1} year^{-1}), (-23.20 kg N ha^{-1} year^{-1}, -11.65 kg K ha^{-1} year^{-1}) and (-15.77 kg N ha^{-1} year^{-1}, -8.40 kg K ha^{-1} year^{-1}) in the upper, (-35.48 kg N ha^{-1} year^{-1}, -13.67 kg K ha^{-1} year^{-1}), (-18.99 kg N ha^{-1} year^{-1}, -11.34 kg K ha^{-1} year^{-1}) and (-12.65 kg N ha^{-1} year^{-1}, -12.63 kg K ha^{-1} year^{-1}) in the middle and (-34.61 kg N ha^{-1} year^{-1}, -14.35 kg K ha^{-1} year^{-1}), (-21.90 kg N ha^{-1} year^{-1}, -12.11 kg K ha^{-1} year^{-1}) and (-24.49 kg N ha^{-1} year^{-1}, -9.90 kg K ha^{-1} year^{-1}) in the lower landscapes of the rich, medium and poor socio-economic groups respectively. Higher negative balance was greater in lower, middle and upper landscapes respectively. The average nutrient depletion from the catchment of the study area were -33.32 kg N ha^{-1} year^{-1}, -14.15 kg K ha^{-1} year^{-1} for the rich, -21.36 kg N ha^{-1} year^{-1}, -11.7 kg K ha^{-1} year^{-1} for the medium and -17.64 kg N ha^{-1} year^{-1}, -10.31 kg K ha^{-1} year^{-1} for the poor; this relation shows nutrient depletion on the partial nutrient balance for different agro-ecological conditions of the central highlands of Ethiopia where similar to the study area was also reported by Dechert et al. (2005) as -33 kg N ha^{-1} year^{-1}, -39 kg K ha^{-1} year^{-1} for the rich; -29 kg N ha^{-1} year^{-1}, -39 kg K ha^{-1} year^{-1} for the medium and -16 kg N ha^{-1} year^{-1}, -35 kg K ha^{-1} year^{-1} for the poor socio economic groups in the entire catchment.

At the national scale, the partial balances were positive for all nutrients considered. These values were comparable with the mean partial balance of 26 farms in the Kisii, Kakamega and Embu districts in Kenya where the agroecological zones are similar (De Jager et al., 1998)

In many studies partial nutrient balance at land use level are negative for the Teff based farming system in the central high lands of Ethiopia -28 kg N ha^{-1} year^{-1}, -87 kg K ha^{-1} year^{-1} for Barley; -21 kg N ha^{-1} year^{-1}, -23 kg K

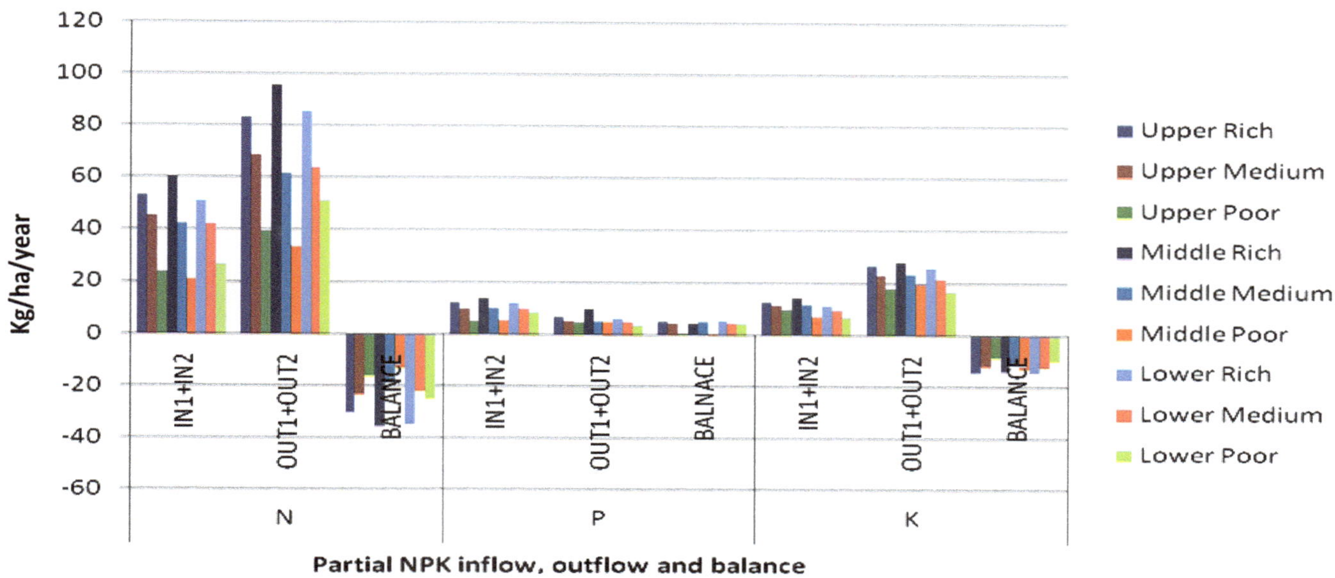

Figure 2. Partial nutrient balance (Kg/ha/year) for different resources inflow and outflow across the socio-economic groups in the landscape positions in the catchment, northern Ethiopia.

ha^{-1} year^{-1} for Wheat; -9 kg N ha^{-1} year^{-1}, -11 kg K ha^{-1} year^{-1} for Teff; -71 kg N ha^{-1} year^{-1}, -81 kg K ha^{-1} year^{-1} for Faba bean and -11 kg N ha^{-1} year^{-1} , -51 kg K ha^{-1} year^{-1} for Maize (Dechert et al., 2005; Van Dung et al., 2008)

However, the study by Haileselassie (2005) reported that the partial nutrient balance is showing positive for the Tigray Region (+10 N, +6 P, +10 K kg ha^{-1} year^{-1}) and (+10 N, +11 P, +7 K kg ha^{-1}year^{-1}) for Ethiopia at national level. So far there is no detailed study conducted in the study area except the plot level study in the region by Hengsdijk et al. (2005) who reported that a negative balance -27 N ha^{-1}.year^{-1}. This report shows a lower estimate as compared to the national level. The lower estimates reported in this study and by others might be due to the severe level of degradation of the region (Hagos et al., 2002; Mitiku et al., 2003; Tewolde Berhan, 2006), as compared to the other well-endowed areas of Ethiopia (Elias et al., 1998), undermining the soils' poor in Nitrogen.

On the other hand the study by Abegaz (2005) in Teghane Atsbi, Tigray Region, reported nutrient depletion between -56.5 to (-115) kg N ha^{-1}year^{-1}, 0 to (-5.8) kg P ha^{-1}year^{-1} and -34.6 to (-112) kg K ha^{-1}year^{-1}. The high nutrient depletion in the country in general and Tigray Region in particular are because of limited applications of organic fertilizers like crop residues and manure, and the socio-economic problems in the mineral fertilizer (Abegaz, 2005).

The results of the partial nutrient balance showed that the nutrient removals by the crop harvest and crop residues were the greatest contributors to the N and K negative balances except for the phosphorous in the three landscapes and in all the socio-economic groups, particularly in the case of rich farmers in the middle landscape. These differences are associated with the differences in crop production, that is, higher crop production by rich farm group.

Full nutrient balance

The full nutrient balance results also demonstrate a large variability in the socio-economic groups (Figure 3). Variations in the high rates of nutrient mining across the socio-economic groups were mainly explained by land holding, family size, herds (with cattle dung mainly used for fuel) and use of external inputs. The balances of N and K for the rich, medium and poor farmers in the lower, middle and upper landscapes were all negative. Whereas the balances of P showed positive values for all the socio-economic groups and in all landscape positions.

The full nutrient balances results were similar across the three landscapes. Higher negative nutrient balances were found highest in the rich farmers followed by the medium and the poor as the result of chemical and physical soil fertility of the cultivated fields of the rich is generally higher and they implement more frequently nutrient-saving techniques, such as soil and water conservation practices and apply more external and internal inputs. The N and K depletion in rich and medium socio-economic groups with high intensive management of their farms -42.5 kg N ha^{-1} year^{-1}, -13.1 kg K ha^{-1} year^{-1} (for the rich) and -32.7 kg N ha^{-1} year^{-1}, -11.8 kg K ha^{-1} year^{-1} (for the medium) in the upper landscape, -49 kg N ha^{-1} year^{-1}, -14.3 kg K ha^{-1} year^{-1} (for the rich) and -28.5

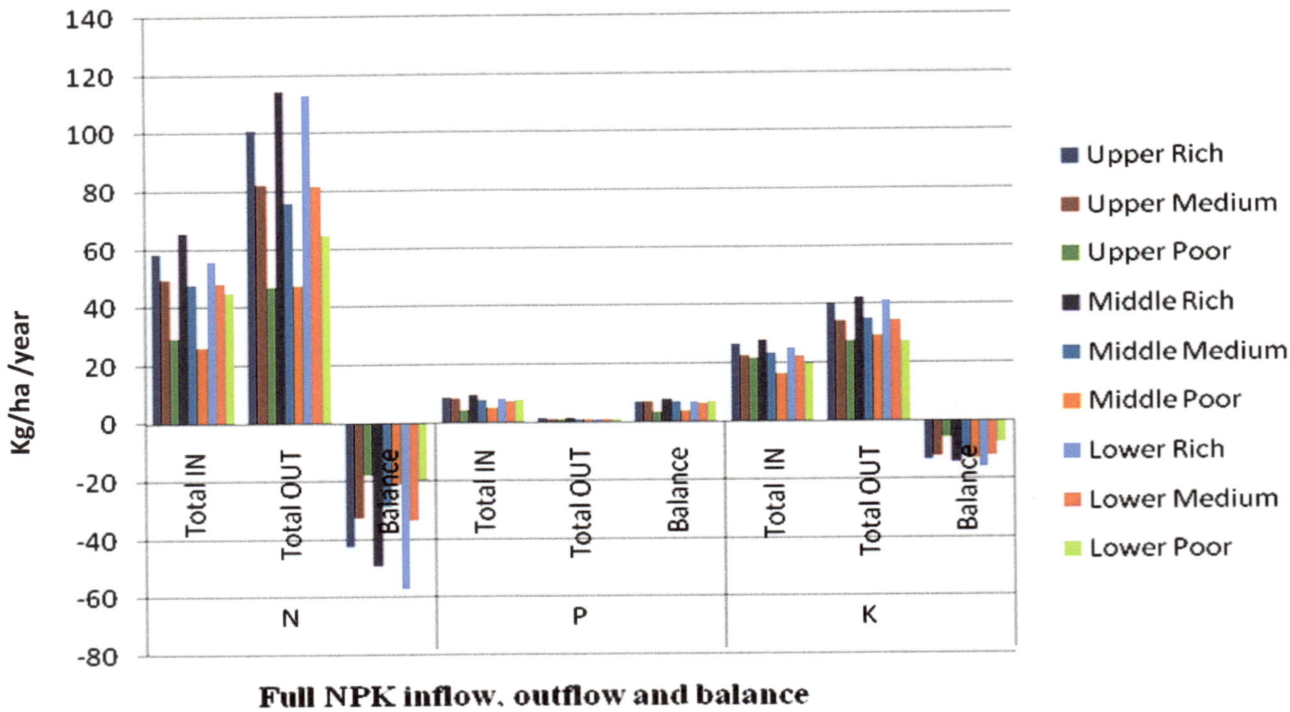

Figure 3. Full nutrient balance (Kg/ha/year) for different resources inflow and outflow across the socio-economic groups in the landscape positions in the catchment, northern Ethiopia.

kg N ha^{-1} $year^{-1}$, -11.8 kg K ha^{-1} $year^{-1}$ (for the medium) in the middle landscape, -57 kg N ha^{-1} $year^{-1}$, -16.1 kg K ha^{-1} $year^{-1}$ (for the rich) and -33.5 kg N ha^{-1} $year^{-1}$, -12.1 kg K ha^{-1} $year^{-1}$ (for the medium) in the lower landscapes.

Comparably nutrient imbalances were low in the case of the poor socio-economic groups across the three landscapes. N depletion in the poor socio-economic groups which apply less intensive farm management were (-17.7 kg N ha^{-1} $year^{-1}$, and -5.59 kg K ha^{-1} $year^{-1}$), (-21 kg N ha^{-1} $year^{-1}$, and -12.8 kg K ha^{-1} $year^{-}$1) and (-19.7 kg N ha^{-1} $year^{-1}$, and -7.52 kg K ha^{-1} $year^{-1}$) in the upper, middle and lower landscapes respectively (Figure 3). This was largely due to, lower rainfall, lower biomass production and lower soil fertility. N and K balances for all the socio-economic groups across the three landscapes were negative (Figure 3). In all the socio-economic groups and the landscapes, positive full balances for P was obtained as 7.132 kg P ha^{-1} $year^{-1}$, 8 kg P ha^{-1} $year^{-1}$ and 7.13 kg P ha^{-1} $year^{-1}$ for the rich, 6.86 kg P ha^{-1} $year^{-1}$, 6.71 kg P ha^{-1} $year^{-1}$ and 6.32 kg P ha^{-1} $year^{-1}$ for the medium and 3.48 kg P ha^{-1} $year^{-1}$, 3.97 kg P ha^{-1} $year^{-1}$ and 6.77 kg P ha^{-1} $year^{-1}$ for the poor in the upper, middle and lower landscapes respectively. P depletion rates generally were not as high as N and K rates. This can be explained by the low quantity of P in plant and soil systems, and because its ability to fix in the soil. Similar trends of P depletion were reported by Sheldrick et al. (2003). This is mainly the impact of higher IN1. As was the case for N and K, farms of the poor households with less nutrient management showed a lower degree of P depletion, while the farms of rich with high farm management show comparatively high amounts of P depletion.

Many nutrient balance studies results in Ethiopia reported negative values. The present study reveal lower nutrient depletion than reported in previous studies where the full nutrient balance is negative for Tigray Region with the depletion rate of -41 N, -1 P, -36 K kg ha^{-1} $year^{-1}$; the national level depletion rate for N, P and K was calculated at -122 kg ha^{-1} $year^{-1}$, -13 kg ha^{-1} $year^{-1}$ and -82 kg ha^{-1} $year^{-1}$, respectively. It may be because other factors, like inputs through sedimentation and outputs like leaching, erosion are not calculated (Dechert et al., 2005; Haileselassie, 2005, 2007). Moreover farmers in the study areas are introducing different extension approaches like using the chemical fertilizers, organic inputs to replenish their farm lands so that the depletion rate of the macro nutrients are less than the national depletion rate.

CONCLUSION AND RECOMMENDATION

From this study we can conclude that there is higher nutrient mining rate, particularly for N in all the landscapes primarily due to large removals of nutrients in harvested output and residue as a result of poor nutrient management by these farmders, leading to low nutrient use efficiency, especially N. The use and management of specific source of soil nutrients like application of mineral fertilizer and organic inputs and nutrient depletion rates differ significantly among socio-economic groups. The highest rates were recorded for the rich farm group, followed by the medium and poor farm groups. These differences are associated with the differences in crop production as the result of chemical and physical soil fertility of the cultivated fields of the rich is generally higher and these rich farmers implement more frequently nutrient-saving techniques, such as soil and water conservation practices and apply more external and internal inputs. On the other hand, the low rates of depletion in the poor farm group are partly associated with low crop production, due to low indigenous soil fertility, low inputs and poor crop management.

Under the prevailing natural and socio-economic conditions, farmers' current strategies (practices) are inadequate to cope with the decline in soil fertility entirely at farm (catchment) level in all the landscapes. Fertilizer use is seen as the only way to maintain or improve soil fertility, and soil management is particularly lacking in the return of organic matter to the soil. However, despite all the extension efforts, fertilizer consumption in the study area as well as in the region level is extremely low. Therefore, there is a need for targeted agronomic interventions, improvements in nutrient use efficiency from different inputs, awareness creation through integrated nutrient management to mitigate nutrient losses and there by improve the sustainability of crop production at farm (catchment) level.

Finally this study recommends that farmers should have a tendency to give more weight for the application of organic and inorganic inputs, proper management of crop residues and sustaining soil conservation measures. In addition, further research about the quantification and analysis of resources flow through the catchment scale and their effect on soils fertility management, landscape positions and farm management practices is required. This can be done by conducting researches on improving the quality and quantity of organic inputs enhancing the macro nutrients at catchment level and further long-term and short-term research on socio-economic conditions and effects on soil fertility, especially on the relationship between social/wealth status and the use of organic and inorganic inputs.

REFERENCES

Abegaz A (2005). Farm management in mixed crop-livestock systems in the Northern Highlands of Ethiopia. Wageningen University and Research Center, PhD Thesis.

Anderson JM, Ingram SJ (1993). Tropical biology and fertility: A handbook of methods. CAB Int., Wallingford, England.

Black CA (1965). Methods of Soil Analysis. America Society of Agronomy, Madison.

Brady NC, Weil RR (2002). The nature and properties of soils. 13th ed. Pearson Education. Inc. USA.

Bremmer JM, Mulvaney CS (1982). Nitrogen total. In Page AL, Miller RH, Keeney DR (eds.). Methods of Soil Analysis. Part 2 Chemical and Microbial Properties. Agronomy, Am. Soc. Agro. Madison, Wis. 9(2):595-624.

CSA (Central Statistics Authority) (2009). Agricultural sample survey 2008/ 2009 (2001E.C.) (September– December 2008). Volume I - Report on area and production of crops (private peasant holdings, Meher season), Statistical Bulletin CSA, Addis Ababa. P. 446.

Dechert G, Veldkamp E, Brumme R (2005). Are partial nutrient balances suitable to evaluate nutrient sutainability of land use systems? Results from a case study in Central Sulawesi, Indonesia. Nutr. Cycl. Agroecosyst, 72:201–212.

De Jager A, Kariuku I, Matiri FM, Odendo M, Wanyama JM (1998). Monitoring nutrient flows and economic performance in African farming systems (NUTMON) IV.Linking nutrient balances and economic performance in three districts in Kenya. Agriculture, Ecosyst. Environ. 71:81-92.

Demeke M, Kelly V, Jayne TS, Said A, Le Vallee JC, Chen H (1998). Agricultural Market Performance and Determinants of Fertilizer Use in Ethiopia. Working paper number P. 10. Ministry of Economic Development and Cooperation, Addis Ababa.

Elias E (2002). Farmers' perceptions of soil fertility change and management. ISD and SOS-Sahel International (UK). EDM Printing Press. Addis Ababa, Ethiopia.

Elias E, Morse S, Belshaw DGR (1998). Nitrogen and phosphorus balance of Kindo-Koisha farms in Southern Ethiopia. Agric. Ecosyst. Environ. 71:93–113.

EPA (Environmental Protection Authority) (1997). The conservation strategy of the federal democratic republic of Ethiopia: Volume II. April 1997, Addis Ababa.

FAO (2003). Ethiopian Highlands Reclamation Study, vols. 1 and 2. FAO, Rome.

Gebresamuel G, Singh BR, Dick, Øystein (2010). Land-use changes and their impacts on soil degradation and surface runoff of two catchments of Northern Ethiopia', Acta Agricultural Scandinavica, Section B - Plant Soil Sci. 60(3):211-226.

Hagos F, Pender J, Gebreslassie N (2002). Land degradation and strategies for sustainable land management in the Ethiopian Highlands, Tigray Region. Socioeconomics and Policy Research (Working P. 25). ILRI, Nairobi, Kenya.

Haileselassie A, Priess JA, Veldkamp E, Teketay D, Lesschen JP (2005). Assessment of soil nutrient depletion and its spatial variability on smallholders' mixed farming systems in Ethiopia using partial versus full nutrient balances. Agriculture, Ecosyst. Environ. 108:1–16.

Haileselassie A, Priess JA, Veldkamp E, Teketay D, Lesschen JP (2007). Nutrient flows and balances at the field and farm scale: Exploring effects of land-use strategies and access to resources. Agric. Syst. 94:459–470.

Mengel K, Kirkby EA (1996). Principles of Plant Nutrition. Panima Publishing Corporation, New Delhi.

Millennium Ecosystem Assessment (MEA) (2005). Ecosystems and human well-being: synthesis, Washington, DC, Island Press..

Mitiku H, Berhanu G, Amare B (2003). The status of soil fertility in Tigray. In: Proceeding the "Policies for Sustainable Land Management in the Highlands of Tigray, Northern Ethiopia." Socio-economic and Policy Research (Working. 54): 28-29 March 2002.

MoFED (Ministry of Finance and Economic Development) (2002). Ethiopia: Sustainable development and poverty reduction program. MoFED. Addis Ababa, Ethiopia.

National Fertilizer Industry Agency (NFIA) (2001). Agronomic and Environmental Aspects of Fertilizer.

Olsen SR, Sommers LE (1982). Phosphorus. In A.L. Page, R.H. Miller, & D.R. Keeny (Eds.), Method of Soil Analysis. II. Chemical and Microbiological Properties, Agronomic Monograph P. 9 (2nd edn). SSSA, Madison, Wisconsin.

Okumu BN (2000). Bio Economic Modeling Analysis of Watershed Conservation in the Ethiopian Highland. Ph.D. Thesis, University of Manchester, UK, P. 257.

Sahlemedhin S, Taye B (Editors) (2000). Procedures for soil and plant analysis: Technical P. 74. National Soil Research Center and Ethiopian Agricultural Research Organization, Addis Ababa, Ethiopia.

Sanchez PA (2002). Soil fertility and hunger in Africa. Sci. Compass 129:2019–2020.

Sheldrick WF, Syers JK, Lingard J (2003). Soil nutrient audits for China to estimate nutrient balances and output/input relationships. Agric.

Ecosyst. Environ. 94:341–354.
Smaling EMA (1993). An Agro-ecological Framework for Integrated Nutrient Management, with Special Reference to Kenya. Ph.D. Thesis, Agricultural University of Wageningen, Wageningen, P. 250.
Smith HW, Weldon MD (1940). A comparison of some methods for the determination of soil organic matter. Proc. Soil Sci. Soc. Am. 5:177-182.
Stocking M, Murnaghan N (2001). A handbook for the field assessment of land degradation. Earthscan, London.
Stoorvogel JJ, Smaling EMA (1990). Assessments of Soil Nutrient Depletion in sub Saharan Africa 1983–2000. Report 28. The Winand Staring Centere for Integrated Land, Soil and Water Research (SC-DLO), Wageningen.
Tegene B (1996). Characteristics and landscape relationships of vertisols and vertic luvisols of Melbe, Tigray, Ethiopia. SINET: Ethiopia J. Sci. 19(1):93-115.
Tewolde-Berhan GE (2006). The role of forest rehabilitation for poverty alleviation in dry lands. J. Dry lands. 1(1):3-7.
Tilahun A, Takele B, Endrias G (2001). Reversing the degradation of arable land in the Ethiopian Highlands. Managing African Soils, London, pp. 23-24.
Van de Wauw J, Baert G, Moeyersons J, Nyssen J, De Geyndt K, Nurhussein T, Amanuel Z, Poesen J, Deckers J (2008). Soil–landscape relationships in the Basalt dominated highlands of Tigray, Ethiopia. Catena 75:117–127.
Van den Bosch H, Gitari JN, Ogaro VN, Maobe SA, Vlaming J (1998). Monitoring nutrient flows and economic performance in African farming system (NUTMON). Monitoring nutrient flows and balances in three districts in Kenya. Agric. Ecosyst. Environ. 71:63–80.
Van Dung N, Duc V, Thanh L, Nguyen, Manh T, Tran, Cadisch G (2008). Analysis of the sustainability within the composite swidden agroecosystem in northern Vietnam. 1. Partial nutrient balances and recovery times of upland fields. Agriculture, Ecosyst. Environ.128:37–51.
World Bank (2007). Ethiopia: Accelerating equitable growth country economic memorandum. Part II Thematic Chapters-Report No. 38662-ET. World Bank Africa Region Poverty Reduction and Econ. Manage. Unit, Washington DC.

9

Intermittent leaching of copper from copper based fungicide through a saturated soil profile

J. A. Osunbitan[1] , K. O. Adekalu[1] and P. O. Aina[2]

[1]Department of Agricultural Engineering, Obafemi Awolowo University, Ile - Ife, Nigeria.
[2]Department of Soil Science, Obafemi Awolowo University, Ile - Ife, Nigeria.

Copper is a micronutrient required for plant growth. However, it can have negative effect on plant growth if the concentration is too high. The aim of this study was to examine copper concentrations of leachate from soil columns treated with copper based fungicide. A 2 × 4 factorial arrangements consisting of two soil types (0 to 60 cm sample depth) and four rates of copper based fungicide concentrations (0, 1.88, 3.75 and 7.50 g/L) was used with three replications. The column leaching experiment was conducted by saturating the soil profile with water at 5 mm/h rate of water flow for 1 day. After saturation, the fungicides was applied and the soil columns kept completely saturated with water by maintaining ponding by continuously applying de-ionized water at a rate of 10 mm/h throughout the experimental period. The process was repeated three more times. Leachate samples were collected every 6 h for 7 days and analyzed for copper using Atomic Absorption Spectrophotometer. The results indicated that drainage time, soil type and fungicide concentrations significantly ($p < 0.01$) effected copper concentration in the leachate. Highest metal concentration in the leachate was obtained between 2.5 to 3.25 days. On the average, the concentration of copper in the leachate during the first fungicide application and leaching was 1.94 mg/L. This concentration reduced to about 1.12 mg/L③ during the second application. It was also observed during the experiment that the flow rate of water decreases during the leaching of the second fungicide application. The concentration then increases to 1.85 and 2.83 mg/L for the third and fourth applications, respectively. Repeated application of the metal through the fungicide leads to the accumulation of the metal in the soil.

Key words: Leaching, copper, saturated soil profile.

INTRODUCTION

Pesticide application is essential for pest control in agricultural production in order for pesticides to be effective; especially pre-emergence herbicides must be able to move within the soil profile to reach the germinating weed seeds. However, too much movement can move a pesticide away from the target pest and this could lead to reduced pest control, injury of non-target species including humans, and contamination of surface and groundwater (Arnold and Briggs, 1990). Pesticides can be transferred in five ways: volatilization, runoff, crop removal, absorption, and leaching (Arnold and Briggs, 1990). Volatilization is the conversion of a solid or liquid into a gas. Once volatized, a pesticide can move away from the treated soil surface. Runoff occurs when the rate of water application exceeds percolations or bound to eroding soil particles.

Pesticide leaching depends in parts on the chemical and physical properties of the pesticide (Kordel and Kleim, 1992). A pesticide strongly adsorbed to the soil particles is less likely to be leached. Solubility is another factor that affects the leaching of pesticides (Kordel and Kleim, 1992). A pesticide that dissolves in water can move with water in the soil. The pesticide longevity influenced the likelihood of it being leached down the soil profile (Kordel and Kleim, 1992). A pesticide that is rapidly broken down by a degradation process is less likely to be leached because it may remain in the soil only a short time. Soil factors that influence pesticide leaching include texture, organic matter content, soil moisture content and availability, soil pH, and microbial community. These factors influence leaching because of their effect on pesticide adsorption (Arnold and Briggs, 1990). Soil permeability, or how readily water moves through the soil may also be important. The more permeable a soil is the greater potential for pesticide to leach down the soil profile. Therefore, pesticides that are highly water soluble, not readily adsorbed and not rapidly degraded are the most likely to leach. Pesticide leaching down the soil profile is of great concern both in relation to environmental contamination and pesticide efficiency (Peoples et al., 1980; Weaver et al., 1983). Low pesticide concentrations in the soil effectively control pest. On the other hand, high pesticide concentration due to excess application is wasteful and could potentially lead to environmental pollution.

The use of metal based fungicides such as copper sulphate could be more detrimental if not properly managed. This is due to increased potential risk of pollution from the chemical and the heavy metal. Unlike organic chemicals, which are eventually broken down to simple non-toxic compounds, the addition of metals and other potentially toxic elements to soil leaves a residue, which is permanent unless leaching occurs (Logan, 1990). Contamination of agricultural soils by toxic heavy metals is one form of chemical degradation of soils (Logan, 1990). Environmental hazards derived from heavy metals are closely link to metal mobility, and therefore concentrations of the metals in soil solution. The mobility of heavy metals in terms of leachability depends not only on the total concentration in soil but also on soil properties and environmental factors (Li and Shuman, 1997). Most of them are associated with pollution of soil and water environment, though some of them are essential to plant and animals at low levels.

The movement of heavy metals in soil profiles has received considerable attention, since even a slow transport through soil and subsoil materials may result in an increased content of heavy metal in the groundwater. Dowdy and Volk (1983) suggested that the movement of heavy metals in soils would occur in sandy, acid, and low - organic matter soil, subjected to heavy rainfall or irrigation. However, some investigators have reported that metal movement was low in soil treated with sludge (Williams et al., 1987; Dowdy et al., 1991). The low mobility of metal in sludge - treated soils was attributed to the retention of organic matter on the soil in the upper layers (Giusquiani et al., 1992) because most of the heavy metals in the organic wastes were associated with colloidal organic matter (Sposito et al., 1982; Sims and Kline, 1991).

Generally, the metal content of soils is low, at least in soils that are not adjacent to main roads or zones of high industrial or mining activity (Greppi and Preti, 1999). On agricultural land, however, the application of agrochemicals, wastewater, sewage sludge or compost (for disposal and/or fertilization) can introduce significant amount of metals. Most metal mobility studies have concentrated on pollution and movement of waste - borne heavy metals and not pollution associated with pesticide application especially copper based fungicides. Copper is one of the heavy metals which can be detrimental to plant growth if excessively present in the soil. Therefore, it is important to understand the impact copper based fungicides applied in agriculture lands, could potentially have in the soil and environment. This study was established to determine the copper concentrations in leachate from soil columns treated with copper fungicide.

MATERIALS AND METHODS

Soil description

The experiment was conducted using soils collected from the Obafemi Awolowo University Teaching and Research Farm at 0 to 60 cm soil depth. The soils were air-dried at 25°C, gently crushed to pass through a 2 mm diameter sieve and stored. All soil had been under secondary forest and did not have any history of fungicide application. The soils are classified at the series level as *Apomu* and *Egbeda* series (Smyth and Montgomery, 1962). The Egbeda series classified as Oxic Tropudalf (Ojanuga, 1975; Soil Survey Staff, 1992) was derived from granite and gneiss parent rock, while the Apomu series classified as Oxisol (Soil Survey Staff, 1992) was derived from sedimentary sandstone parent rock. According to the FAO/UNESCO terminology, *Egbeda* and Apomu soils are classified as Nitisol and Luvisol, respectively (FAO, 1988). Some physico-chemical properties of the two soils are shown in Table 1. For metal mobility experiment, the columns were prepared in three replicates with disturbed soil samples.

Leaching experiment and management

The soil column method used in this study represents a simplification of the natural processes. The movement of the metals in the field will be much slower than in the soil columns due to the heterogeneity of soils. Collected soils were homogenized, air-dried, stems and roots removed by hand and sieved through a 2 mm sieve. Plastic cylinders (40 cm internal diameter and 60 cm long) were used. The columns were perforated at the base for drainage of leachate and wire mesh screen placed at the base of the

Table 1. Some physical and chemical properties of the soils.

Description	Apomu series		Egbeda series	
	0 - 15 cm	15 - 30 cm	0 - 15 cm	15 - 30 cm
Textural classification	**Loamy sand**	**Loamy sand**	**Sandy clay loam**	**Sandy slay loam**
pH	4.98	4.65	5.82	5.64
Cation exchange capacity (cmol/kg)	7.75	5.75	13.25	7.00
Electrical conductivity (mmhos/cm)	0.058	0.029	0.04	0.034
Organic carbon content (%)	1.72	1.16	2.10	1.85
Sand (%)	78.5	82.5	68.0	70.0
Silt (%)	12.0	7.0	10.0	10.0
Clay	9.5	10.5	22.0	20.0
Copper ion (mg/kg)	0.008	ND	0.020	ND
Manganese ion (mg/kg)	34.29	23.56	87.98	54.87
Lead ion (mg/kg)	ND	ND	27.21	6.83
Zinc ion (mg/kg)	0.178	0.170	0.037	0.244
Calcium ion (cmol/kg)	1.45	0.88	2.12	1.64
Magnesium ion (cmol/kg)	1.21	1.19	1.58	0.72
Potassium ion (cmol/kg)	0.35	0.22	0.41	0.33
Sodium ion (cmol/kg)	0.34	0.31	0.39	0.39

columns to prevent the blockage of the drainage holes and to prevent large clumps of soil from falling off the perforated holes of the soil columns. For each soil type, four soil columns (soil packed to 1.5 g/cm^3 bulk density) with 0, 1.88, 3.75 and 7.50 g/L fungicides concentration rates replicated three times was established. All the columns were packed with 103.67 kg of the two soil types to a height of 55 cm. The columns were then raised of about 30 cm from the ground surface. Basins were put at the base of the columns for water samples collection during the leaching experiment (Figure 1). This led to a total of 24 soil columns for the entire experiment, 12 columns for each soil type.

The soil columns were then saturated with water at the rate of about 5 mm/h using perforated bowl placed directly above the soil columns for 1 day (A preliminary experiment was conducted before the commencement of the experiment to determine the flow rate from the perforated bowl). After saturation, the fungicides (at the above mentioned rates) were applied at the top of the soil columns. The columns were then kept completely saturated with water by maintaining ponding through continuous application of de-ionized water at a rate of about 10 mm/h for the entire duration of the experiment.

The volume of outflow was measured as a function of time during the experiment to confirm the steady flow conditions in all the columns. The outflow rate remains relatively constant for each of the two soils. The experiment was conducted for 7 days after which the bowl supplying the columns with the de-ionized water was removed. Leaching activities were then allowed to cease for 24 h to permit adsorption of the metals in the fungicides left in the columns to the soils. During this time all columns were covered with plastic to prevent surface evaporation. After the 24 h "adsorption" period, the plastic cover was removed, another 250 ml of fungicide solution at the same concentration was added to the surface of the column and the leaching process resumed by placing the bowls back at the top of each of the soil columns and the de-ionized was poured into them for another round of 7 days period of leaching. This process of displacement experiment was repeated three times and for each of the experiments, leachate samples were collected every 6 h for next 7 days and the leachate transferred to the Laboratory for analysis of copper. The concentration of copper in the leachate solution was determined using Atomic Absorption Spectrophotometer. Analysis of variance (SAS, 1987) was used to test the variation of metal concentrations in the leachate as affected by the fungicide application and time after the commencement of drainage, while Duncan multiple range test was used for the separation of means.

RESULTS AND DISCUSSION

The concentration of copper measured in the leachate after leaching started with de-ionized water from the columns treated with the three different concentrations of the copper fungicide (Kocide 101) are shown in Figures 2, 3 , 4 and 5 for the four series of fungicide applications. The concentration of copper in the leachate for the four series of fungicide applications increased with sampling time and reached maximum at 60 h after the commencement of leaching then dropped as the time increases. Soil types and initial fungicide concentration had significant ($p < 0.01$) effects on the concentration of copper in the leachate for the four series of fungicide application. As expected, the concentration of copper in the leachate was at the maximum when 7. 5 g/L fungicide concentration was applied for the four series of fungicide applications. On the average, for the two soil types considered in this study, the concentration of copper in the leachate during the first fungicide application and leaching was 1.94 mg/L. Interestingly, this concentration reduces to about 1.12 mg/L on the average during the second application and increased to an average concentration of 1.85 and 2.83 mg/L during the third and fourth applications, respectively. It was also observed during the experiment that the flow rate of water decreases during the leaching of the second fungicide application and increased with the third and fourth fungicide applications.

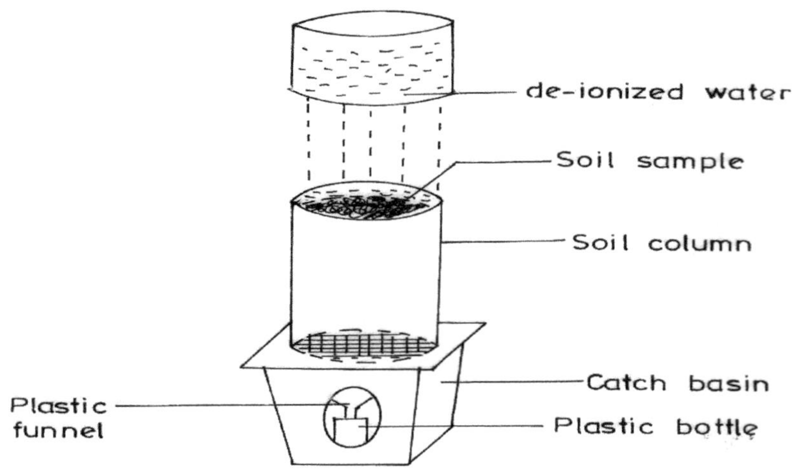

Figure 1. Experimental setup of the soil columns in the laboratory.

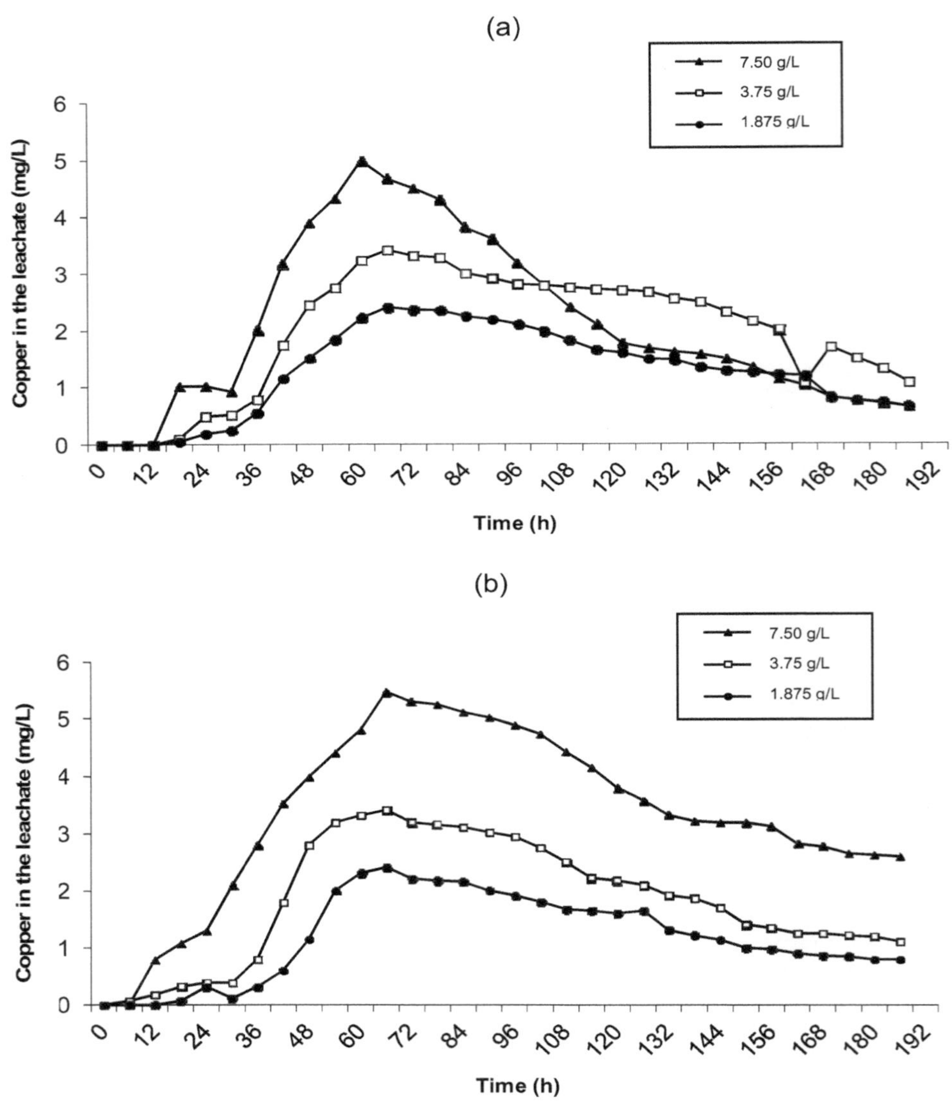

Figure 2. Leaching of copper from Kocide 101 with concentrations of 7.50, 3.75 and 1.875 g/L in (a) Egbeda soil and (b) Apomu soil after the first fungicide application.

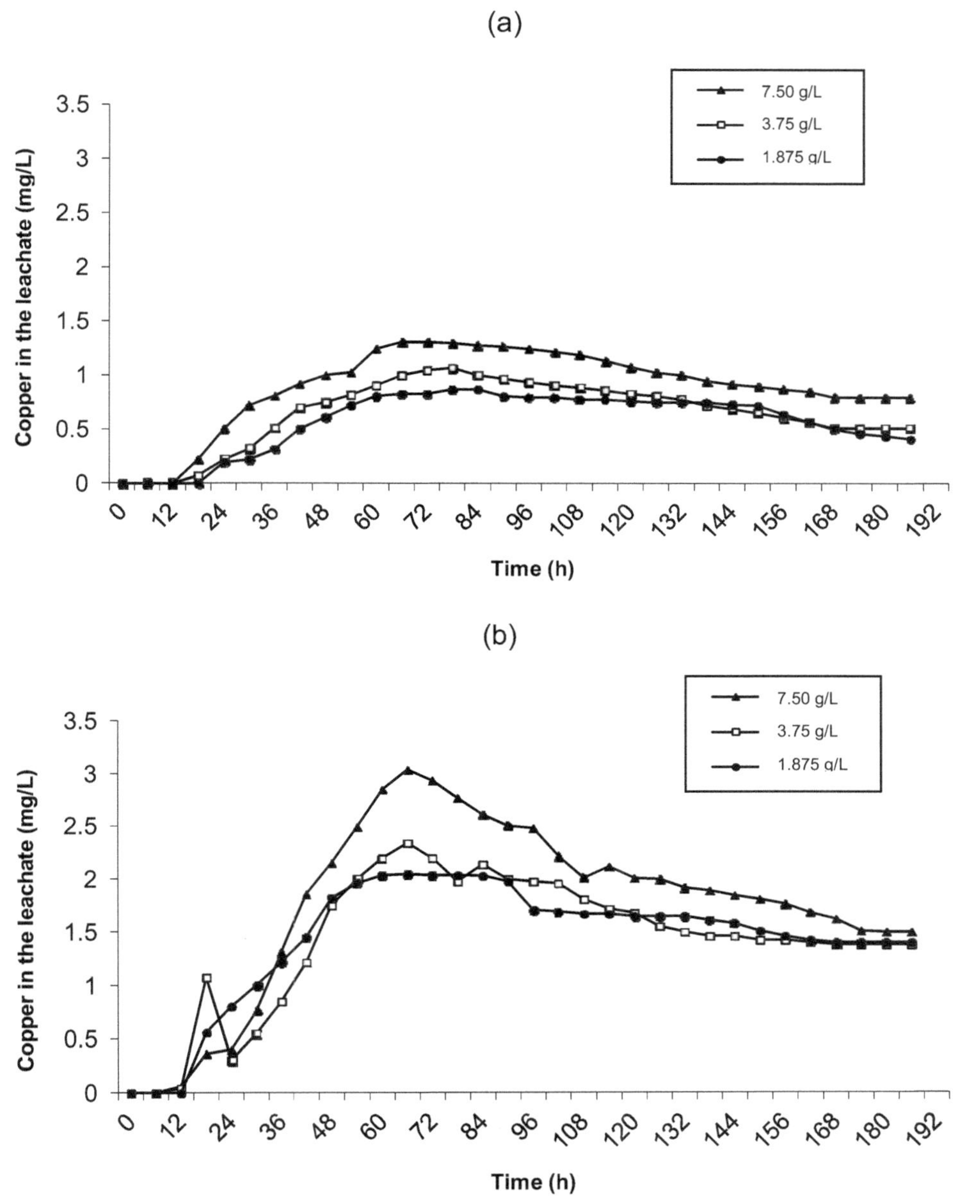

Figure 3. Leaching of copper from Kocide 101 with concentrations of 7.50, 3.75 and 1.875 g/L in (a) Egbeda soil and (b) Apomu soil after the second fungicide application.

Leaching of the columns with first fungicide application

The concentration of copper in the leachate when the soil columns were leached with the three different fungicide concentrations increases from zero at time 0 h to maximum of 3.63 mg/L at the sampling time of 66 h after which the concentration of copper in the leachate begins to decline. Figure 2 shows the concentration of copper in the leachate as a function of time after the commencement of drainage for the first fungicide application. From Figure 2, the concentration of copper in the leachate from both soils increase from 0 mg/L to the maximum of 3.49 and 3.76 mg/L for *Egbeda* and *Apomu* soils, respectively.

The three factors considered that is, time after the commencement of drainage, soil and fungicide concentrations were all significant in their effects on the copper concentration of the leachate ($p < 0.01$). On the average, the concentration of copper in the leachate from soil columns containing Egbeda soil was 1.77 mg/L, while the concentration was 2.11 mg/L for Apomu soil. This is expected because Apomu soil has more macro pores that will permit mobility of water and solute than Egbeda soil. On the average, the concentrations of copper in the leachate were 1.19, 2.82 and 3.31 mg/L for fungicide

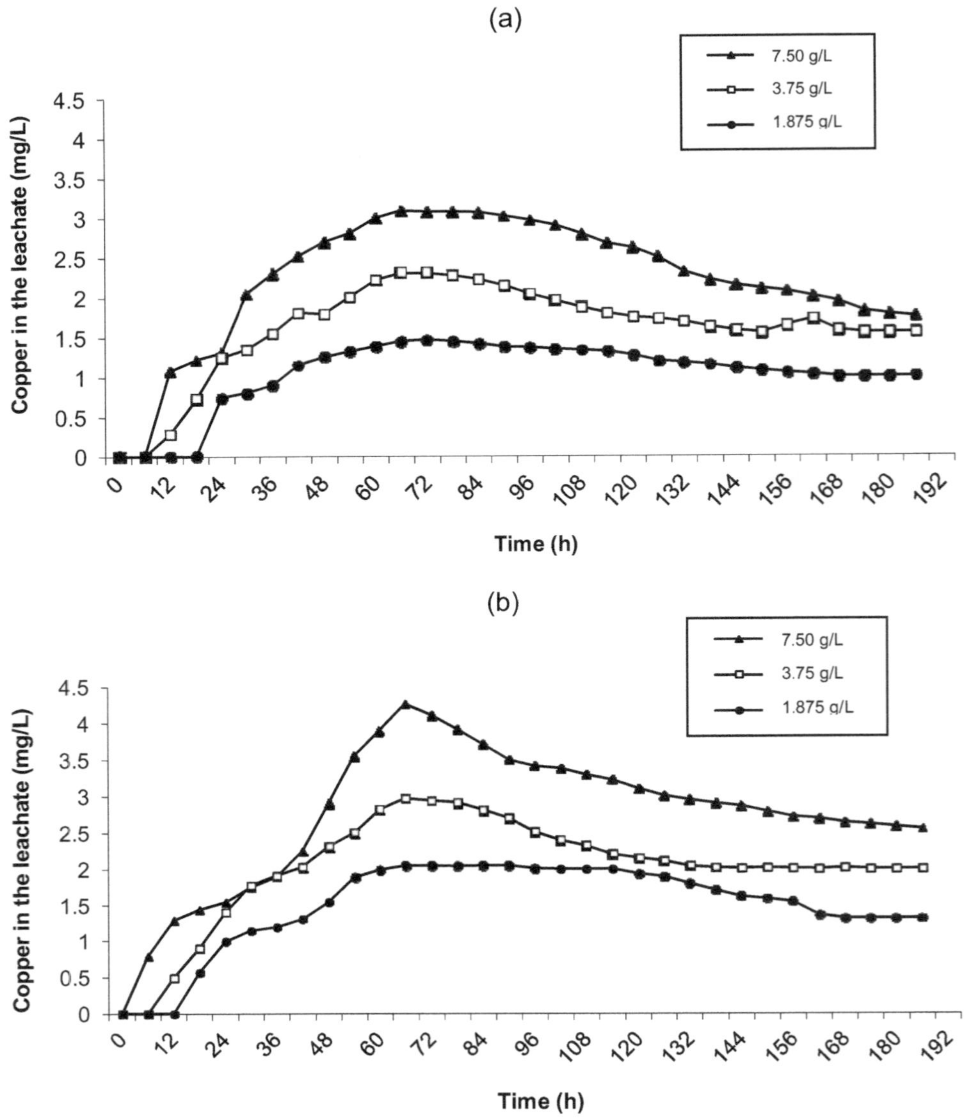

Figure 4. Leaching of copper from Kocide 101 with concentrations of 7.50, 3.75 and 1.875 g/L in (a) Egbeda soil and (b) Apomu soil after the third fungicide application.

concentrations of 1.875, 3.75 and 7.5 g/L, respectively for Apomu soil, while the concentrations of copper in the leachate for Egbeda soil were 1.27, 1.95 and 2.10 mg/L for fungicide concentrations of 1.875, 3.75 and 7.5 g/L, respectively.

According to Duncan multiple range test, these three leachate concentrations of copper as affected by the applied fungicide concentrations were significantly different from one another. These leachate concentrations reduce to 1.27, 1.95 and 2.10 mg/L for the three different fungicide concentrations, respectively when the soil in the columns was Egbeda soil. The concentrations of copper in the leachate when the applied fungicide concentrations were 3.75 and 7.50 g/L are not significantly different ($p < 0.05$). However, these two leachate concentrations of copper are significantly different from the leachate concentration when the applied fungicide concentration was 1.875 g/L. Moreover, the three fungicide concentrations considered in this work were significantly different in their effect on the concentration of copper in the leachate.

Leaching of the columns with second Fungicide application

Figure 3 shows the concentration of copper in the leachate as a function of time after the commencement of drainage for the second fungicide application. From Figure 3, the concentration of copper in the leachate from

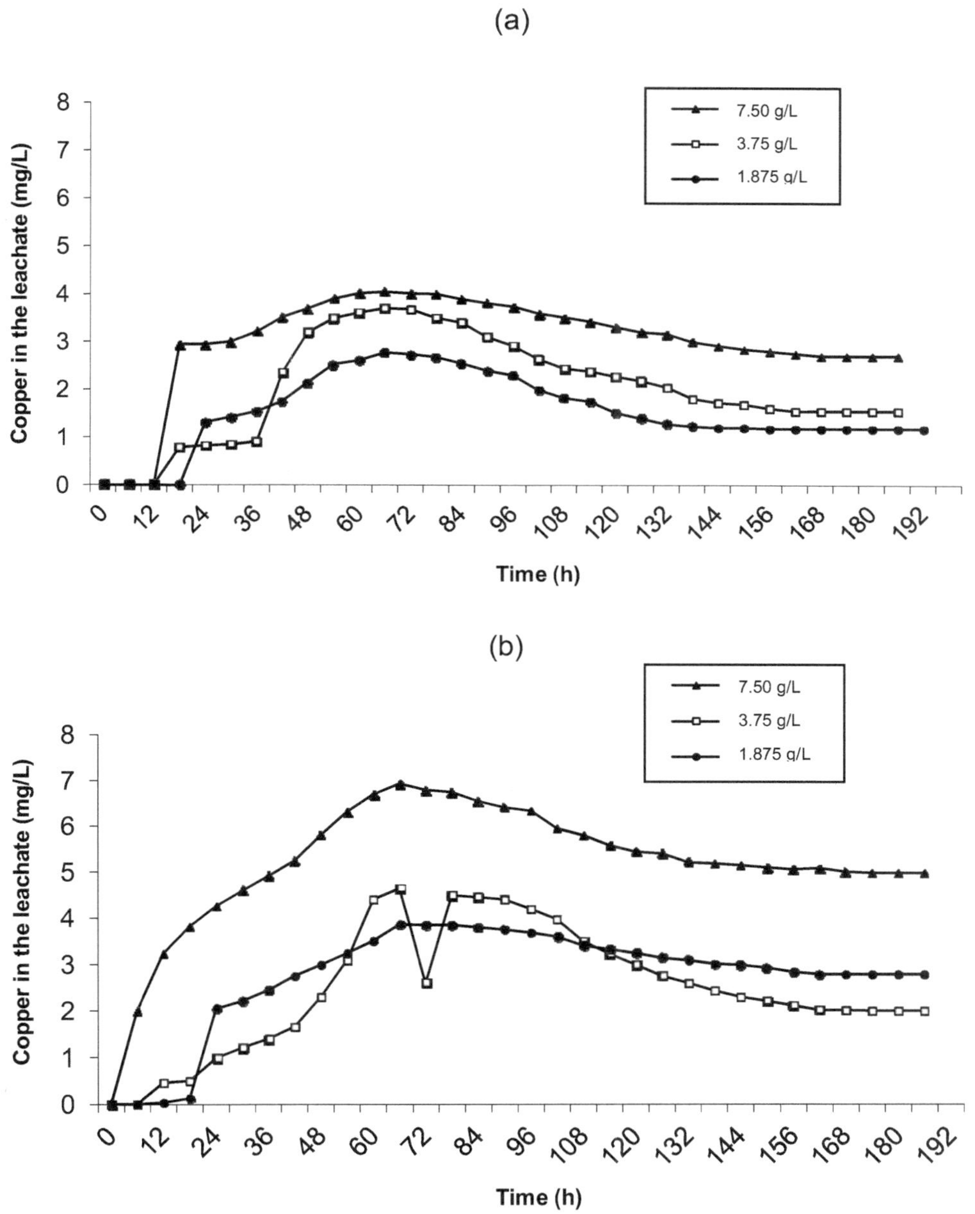

Figure 5. Leaching of copper from Kocide 101 with concentrations of 7.50, 3.75 and 1.875 g/L in (a) Egbeda soil and (b) Apomu soil after the fourth fungicide application.

Egbeda soil columns increases from 0 mg/L at time 0 h to maximum of 1.08 mg/L at 78 h after the commencement of drainage after which the concentration of copper in the leachate begins to drop. For Apomu soil, the maximum concentration of copper in the leachate was recorded at time 66 h after the commencement of drainage.

All the three factors considered that is, time after the commencement of drainage, soil and fungicide concentrations were also significant in their effects on the copper concentration of the leachate ($p < 0.01$). It is interesting to note that on the average, the concentration of copper in the leachate during the second application fell from 1.77 mg/L for the first fungicide application to 0.70 mg/L of copper in the leachate during the second application for Egbeda soil. The same trend was observed for Apomu soil where the average copper concentration of the leachate was 1.54 mg/L as against 2.11 mg/L observed for the first fungicide application. It was also observed that the amount of leachate collected during the second fungicide application and simulated rainfall was less than that which was collected during the first application. This may likely be as a result of the blockage of soil pores caused by the fungicide particles during the first application which prevents easy flow of the water and solute through the soil columns.

On the average, the concentrations of copper in the leachate during the second application were 1.42, 1.43 and 1.75 mg/L for fungicide concentrations of 1.875, 3.75 and 7.5 g/L, respectively for Apomu soil. According to Duncan multiple range test; fungicide concentrations 1.875 and 3.75 g/lt were not significantly different in their effects on the concentration of copper in the leachate. However, the two concentrations were significantly different in their effects from the fungicide concentration of 7.50 g/L. For Egbeda soil, the concentrations of copper in the leachate as affected by the fungicide concentrations of 1.875, 3.75 and 7.50 g/lt were 0.57, 0.65 and 0.89 mg/L, respectively. However, these three concentrations are significantly different ($p < 0.05$).

Leaching of the columns with third fungicide application

Figure 4 shows the concentration of copper in the leachate as a function of time after the commencement of drainage for the third fungicide application. From Figure 4, the concentration of copper in the leachate from soil columns containing Egbeda soil increases from 0 mg/L at time 0 h to maximum of 2.29 mg/L at 72 h after which the concentration begins to decline. For columns containing Apomu soil, the maximum copper concentration of 3.09 mg/L was observed at 66 h after which the concentration began to decline steadily.

All the three factors considered that is, time after the commencement of drainage, soil and fungicide concentrations were significant in their effects on the copper concentration of the leachate ($p < 0.01$). On the average, the concentration of copper in the leachate from soil columns containing Egbeda soil was 1.62 mg/L, while the concentration was 2.08 mg/L for Apomu soil. When the effect of fungicide concentrations on the concentration of copper in the leachate was investigated, it was observed that the concentrations of copper in the leachate from Egbeda soil were 1.03, 1.61 and 2.22 mg/L for fungicide concentrations of 1.875, 3.75 and 7.50 g/L, respectively, while the concentration from Apomu soil for the same three levels of fungicide concentration were 1.49, 2.01 and 2.73 mg/L. According to Duncan multiple range test, these three different leachate concentrations for both soil types are significantly different ($p < 0.05$).

Leaching of the columns with fourth Fungicide application

Figure 5 shows the concentration of copper in the leachate as a function of time after the commencement of drainage for the fourth fungicide application. From Figure 5, the concentration of copper in the leachate from soil columns containing Egbeda soil increases from 0 mg/lt at time 0 h to maximum of 3.52 mg/L at 66 h after which the concentration begins to decline. For columns containing Apomu soil, the maximum copper concentration of 5.15 mg/L was also observed at 66 h after which the concentration began to decline steadily except for column treated with fungicide concentration of 3.75 g/L. For this concentration, after the maximum copper concentration was reached at time 66 h, the concentration of copper in the leachate decreases to 2.59 mg/L at 72 h before increasing to the same value as for time 66 h after which it began to decline.

All the three factors considered that is, time after the commencement of drainage, soil and fungicide concentrations were significant in their effects on the copper concentration of the leachate ($p < 0.01$). On the average, the concentration of copper in the leachate from soil columns containing Egbeda soil was 2.19 mg/L, while the concentration was 3.46 mg/L for Apomu soil. When the effect of fungicide concentrations on the concentration of copper in the leachate was investigated, it was observed that the concentrations of copper in the leachate from columns containing Egbeda soil were 1.54, 2.03 and 3.01 mg/L for fungicide concentrations of 1.875, 3.75 and 7.5 g/L, respectively, while the concentration from Apomu soil for the same three levels of fungicide concentrations were 2.47, 2.74 and 5.18 mg/L. According to Duncan multiple range test, these three different leachate concentrations for both soil types are significantly different ($p < 0.05$).

Conclusion

The following conclusions were drawn from the study:

1) Soil type had significant effect on the concentration of the metals in the leachate.
2) Time to peak concentration of the metals in the leachate varies between 2.5 to 3.25 days.
3) Fungicide concentration had significant effect on the concentration of copper in the leachate. The three fungicide concentrations were significantly different in their effects on the copper concentration in the leachate. However, fungicide concentrations of 3.75 and1.875 g/L were not significantly different in their effects on the copper concentration in the leachate from the column containing Apomu soil for the second fungicide application. For the columns containing Egbeda soil, fungicide concentrations of 7.50 and 3.75 g/L were also not significant in their effects on the concentration of copper in the leachate for the first fungicide application.

Conflict of Interest

The author(s) have not declared any conflict of interest

REFERENCES

Arnold DJ, Briggs GG (1990). Fate of pesticides in soils predictive and practical aspect. In: Environmental fate of pesticide progress in pesticide Biochemistry and Toxiology, vol. 7. Ed. D.H. Hutson and T.R. Roberts, John Wiley and Sons, New York.

Dowdy RH, Latterrell JJ, Hinesly TD, Grossman RB, Sullivan DL (1991). Trace metal movement in an Aeric Ochraqualf following 14 years of annual sludge applications. J. Environ. Qual. 13:197-204.

Dowdy RH, Volk VV (1983). Movement of heavy metals in soils. In Chemical Mobility and Reactivity in Soil Systems, SSSA Spec. Publ. 11, ed. D.W. Nelson et al., Soil Society of America, Madison, WI, USA, pp. 227-240.

FAO (1988). FAO/UNESCO Soil Map of the World. Revised Legend with corrections. World Resources Report 60, FAO, Rome.

Giusquiani PL, Gigliotti G, Businelli D (1992). Mobility of heavy metals in urban waste - amended soils. J. Environ. Qual. 22:330-335.

Greppi M, Preti F (1999). Water quality in agriculture. In CIGR Handbook of Agric. Engineering, vol 1.H.N. Van Lier (Ed). America Society Agric. Eng. pp. 507-544.

Kordel WM, Kleim HM (1992). Determination of the fate of pesticides in outdoor lysimeter experiment. Sci. Total Environ. 123/124:421-434.

Li Z, Shuman LM (1997). Mobility of Zn, Cd and Pb in soils as affected by poultry litter extract - I. Leaching in soil columns. Environ. Pollut. 95(2):219-226.

Logan TJ (1990). Chemical degradation of soil. In : R. Lal and B. A. Stewart (eds.). Soil degradation. New - York Springer - Verlag, Adv. Soil Sci. 11:187-221.

Ojanuga AG (1975). Morphological, physical and chemical characteristics of soils of Ife and Ondo areas. Nigerian J. Sci. 9:225 – 269.

Peoples SA, Maddy KT, Kusik W, Jackson T, Cooper C, Fredrickson SA (1980). A study of samples of well water collected from selected areas in California to determine the presence of DBCP and certain other pesticide residue. Bull Environ. Contam. Toxicol. 24:611-618.

SAS Institute (1987). SAS User's Guide: Statistics, Version 5. SAS Institute, Cary, NC, USA.

Sims JT, Kline JS (1991). Chemical fractionation and plant uptake of heavy metals in soils amended with composted sewage sludge. J. Environ. Qual. 20:387-395.

Smyth AJ, Montgomery RF (1962). Soil and Water use in Central western Nigeria. Published by Government of Western Nigeria.

Soil Survey Staff (1992). Keys to Soil Taxonomy. 5th ed. SMSS technical monograph no 19. Pocahontas Press. Blacksburg, Virginia

Sposito G, Lund LJ, Chang AC (1982). Trace metal chemistry in arid-zone field soils amended with sewage sludge: I. Fractionation of Ni, Cu, Zn, Cd, and Pb in solid phases. Soil Sci. Soc. Am. J. 46:260-264.

Weaver DJ, Zalkin SF, Oshima RJ (1983). Pesticide movement to groundwater. I. survey of groundwater basin for EDB, DBCP Simazine and carbofuran State of California Dept. of Agric. P. 67.

Williams DE, Vlamis J, Pukite AH, Corey JE (1987). Metal movement in sludge - amended soils: a nine - year study. Soil Sci. 143:124-131.

10

Uptake of Pb, Zn and Cu by roots and shoots of fast growing plants grown in contaminated soil in Vietnam

Nguyen Huu Thanh[1], Tran Thi Le Ha[1], Cao Viet Ha[1], Nguyen Duc Hung[1], Phan Quoc Hung[1], Kiyoshi Kurosawa[2] and Kazuhiko Egashira[3]

[1]Faculty of Natural Resources and Environment, Hanoi University of Agriculture, Hanoi, Vietnam.
[2]Institute of Tropical Agriculture, Kyushu University, Fukuoka, 812-8581, Japan
[3]Kyushu University, Fukuoka 812-8581, Japan

The phytoremediation of soil contaminated with lead (Pb), zinc (Zn) and copper (Cu) from industrial wastewater discharge in a farming village near Hanoi City was investigated. The first growing plant species tested were the Spanish needle, water primrose, common willow herb, and water spinach. Uptake by roots and shoots per unit land area for each species was calculated and compared. The uptake of the heavy metals per unit area was higher in shoots than in roots for each plant species and was affected by the higher biomass production of the shoots. For the roots, the uptake did not differ by species, but did differ by heavy metal. For the shoots, the uptake differed by both plant species and heavy metal with the common willow herb having higher Pb and Zn uptake compared to those of the other three species. For the common willow herb, the uptake of heavy metal was in order of Pb > Zn > Cu. The results indicate that common willow herb has a superior phytoremediation capacity among the plants tested, particularly for the uptake of Pb.

Key words: Plant biomass, farming village, industrial wastewater, phytoremediation.

INTRODUCTION

In the large cities of Vietnam, surface water pollution due to the lack of wastewater treatment facilities for industrial waste is a serious problem. Some data shows high levels of heavy metals in surface water in some localities (DWRM/ESCAP/KICT (ed.), 2010). In the area around Hanoi, untreated industrial waste was discharged not only from large factories but also from small workshops including handicraft workshop. Surface water polluted by the waste was used to irrigate agricultural land, leading to heavy metal contamination of agricultural soils. Heavy metals in the soils are absorbed by crops and may adversely affect crop growth. Therefore, remediation measures to reduce soil contamination are necessary. In the area, phytoremediation, a technology of using plant to clean up pollutants in the environment is a promising method. This is due to the fact that it is an economical, energy efficient and environmental friendly method, and can be applied to large areas (Roongtanakiat, 2009).

According to Tong et al. (2004), an ideal plant for phytoremediation should be fast growing, develop a large biomass, be tolerant to and accumulate high concentrations of toxic metals in the shoot, and be easily cultivated and harvested. Much research on phytoremediation has been done using native plant species (Lorestani et al., 2011; Aziz et al., 2011; Barbafieri et al., 2011), hyperaccumulating plants (Revathi et al., 2011; Ji et al., 2011; Nazir et al., 2011), and agricultural crops (Mojiri, 2011; Poniedzialek et al., 2010; Garg and Kataria, 2009). In these studies, phytoremediation capacity was evaluated as heavy metal

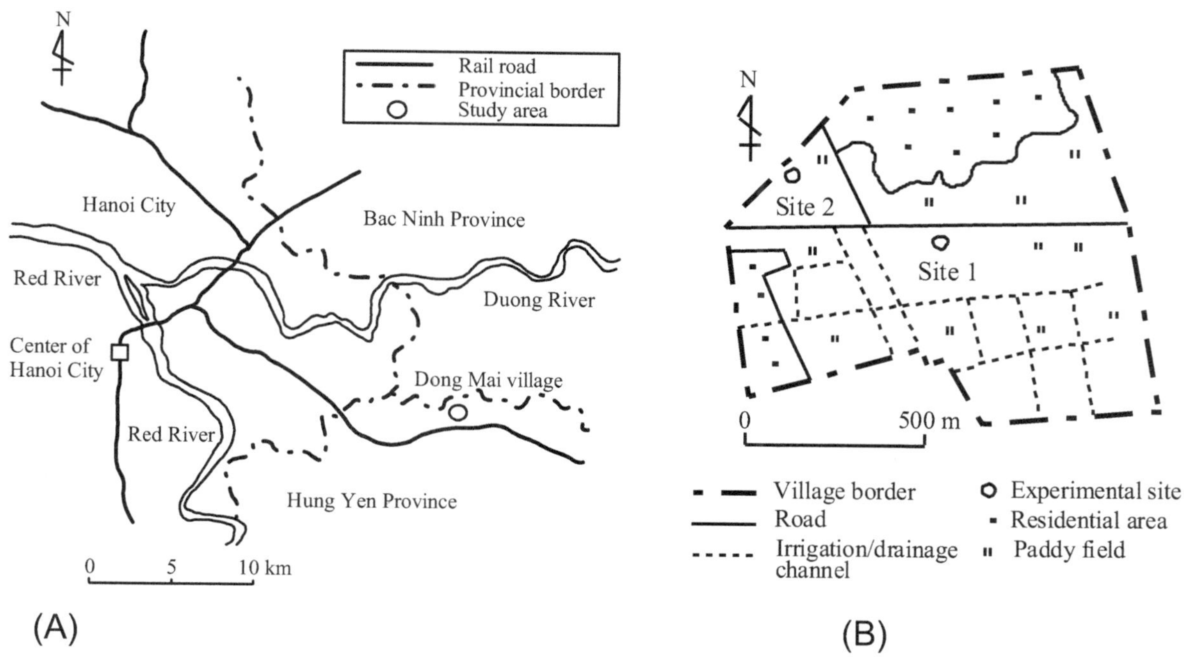

Figure 1. Study site.

uptake per unit weight of plant roots and/or shoots. However, as demonstrated by Tlustos et al. (2006) and Dai et al. (2007), phytoremediation is a function of the plant assemblage of an area, and thus the remediation capacity should be evaluated as the uptake by all plants of a single species growing per unit area of land.

Since the phytoremediation capacity of plants has not been well investigated for contaminated soils in the Hanoi area, we tested the capacity of plants of different types that are known to have fast growth in the region. By using the uptake of heavy metals by roots and shoots per unit land area, plant with high phytoremediation capacity were identified.

MATERIALS AND METHODS

Study site characteristics

Agricultural and industrial activities

The study was conducted in Dong Mai Village, Chi Dao Commune, in the Van Lam District of Hung Yen Province, Vietnam. The village is located 20 km east of the center of Hanoi City (Figure 1A). The village has a total area of 170 ha and a population of 2,300. An agricultural irrigation system has been developed, and crops are planted in the irrigated area, which occupies 98 ha of the village. Rice is planted twice a year in most of the area. Vegetables are also planted in a portion of the area every year. In addition to agricultural activity, some villagers are engaged in producing handicrafts. From 1967 to 1999, many furnaces were used to smelt Pb for handicraft production, and these were consolidated to one location in 2000. There were no systems for treating the wastes from the smelting furnaces, and waste water containing heavy metals was discharged directly to ditches. The heavy metals were transferred to agricultural land through irrigation. Although, smelting furnaces were stopped in 2006, batteries containing Pb that were used in the old furnaces were abandoned along the ditches, resulting in the continued discharge of Pb into the ditches. These batteries are still contaminating the agricultural land.

Agro-ecological setting

According to meteorological data recorded from 2001 to 2009 in Hanoi City (General Statistics Office, 2002 to 2010), the mean annual air temperature was 24.5°C with a range of monthly mean temperatures of 17.0°C in January and 29.8°C in June. Relative humidity was very high throughout the year with an annual mean of 78.6%. In the rainy season from May to October, mean rainfall was 1,470 mm and mean daily sunshine was 4.8 h, while in the dry season from November to April, mean rainfall was 244 mm and mean daily sunshine was 2.8 h. The soil types in the village area were classified as Eutric Fluvisols according to the FAO/UNESCO soil classification system (VSSS-NIAP, 1996). The village is located in a low-lying area with an elevation of 5 to 6 m above sea level that is frequently flooded in the rainy season.

Field experiments on phytoremediation of soils

Plants tested and heavy metals targeted

Four plant species were tested in the experiments: Spanish needle (*Bidens pilosa* L.), water primrose (*Ludwigia adscendens* L.), common willow herb (*Ludwigia octovalvis* L.), and water spinach (*Ipomoea aquatica* L.). All of these plants are native to the Hanoi area and share the characteristics showing fast growth, producing large amounts of biomass and reproducing prolifically. The plant names are hereafter abbreviated as follows: SN for Spanish needle, WP for water primrose, CWH for common willow herb, and WS for water spinach. Pb, Zn and Cu were targeted because high concentrations of these heavy metals were previously reported in the agricultural soils of the Van Lam District, including in the study

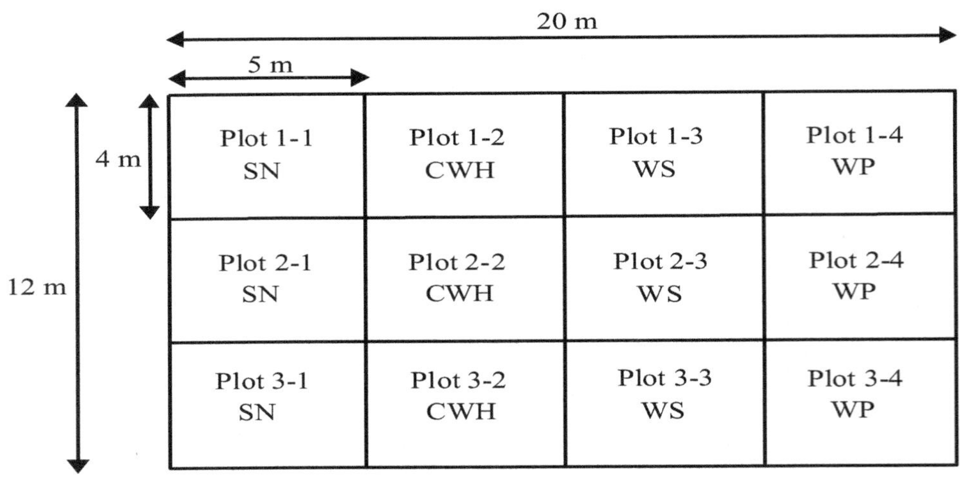

Figure 2. Arrangement of plots for testing plant species at the experimental sites

area of Dong Mai Village (Ho and Nguyen, 2003).

Experimental sites

Two experimental sites (240 m^2 divided into 12 plots) on agricultural land were designated in Dong Mai Village (Figure 1B) at 500 m (site 1) and 10 m (site 2) from the irrigation channel. At each site, each plant was planted in 3 of the 20 m^3 plots (Figure 2).

Transplantation

Plants were started in soil with low concentrations of Pb (11.5 mg/kg), Zn (38.0 mg/kg) and Cu (13.5 mg/kg) in the nursery and were transplanted to the designated plots at each experimental site on March 8, 2007. The planting density was 67 plants/m^2 for SN, WP, and WS, and 33 plants/m^2 for CWH. A total of 0.6 kg of urea fertilizer was applied to each plot 5 days after planting. The SN plots were kept moist without irrigation. Excess rain drained naturally from the plots. For WP, WS and CWH plots, the soil was kept in a wet or water-logged condition, matched to the optimal soil moisture for each plant. Except for sampling, plants were allowed to grow for 90 days after planting without disturbance, including weeding.

Plant sampling

All the plant shoots from an area of 1 m^2 at the four corners and in the center of each plot were collected by cutting all of the plants in all five sampling points of the three plots of the same species close to the ground at 30-day intervals (April 7th, May 7th, and June 6th, 2007), and biomass and heavy metals concentrations were determined. After the first and second samplings, shoots in the entire plots at each site were cut to promote regrowth of plants during the next month. The roots were sampled once on June 6th. At that time, the whole plants were taken from all five sampling points in all three plots for every plant type at both experimental sites. The reasoning for sampling over 90 days (3 months) was that shoots expand and come into contact with those of neighboring plants at 1 month. To get an average monthly growth of shoot under a constant environment, sampling was repeated three times during the 3 months. The root growth is not as fast as that of shoots in a 1 month, but it reached a high level of growth in 3 months.

Soil sampling

Soil was sampled at the two experimental sites on March 4th, 2007, immediately before the plants were transplanted. At each site, 0.5 kg of soil was taken from each of the five sampling points in each plot at a depth from 0 to 0.2 m. The samples for each plot were then combined, and a composite 0.5 kg soil sample was taken for analysis.

Analytical procedures

pH and organic matter content of soil

Soil pH was measured on a 1: 5 suspension of soil in 1 M KCl using a glass electrode. The organic matter (OM) content of soil was determined by the Walkley-Black method (Nelson and Sommers, 1996).

Concentrations of Pb, Zn and Cu in plants and soil

Plant samples (roots and shoots) were cleaned by washing with water, and fresh weight (biomass) was measured after surface drying. The samples were then oven-dried at 105°C for 10 min to neutralize enzymes, then at 70 to 80°C until the samples were completely dry. Soil samples were air-dried at room temperature for 3 to 4 days. After drying, all samples were preprocessed for heavy metal determination by finely grinding and passing the samples through a 1 mm sieve. Extraction of Pb, Zn and Cu from plant samples was done by combustion in a furnace maintained at 550°C, followed by dissolution of the ash with 3 M HCl (National Institute for Soils and Fertilizers, 1998). Soil was digested by hydrogen peroxide, hydrofluoric acid, nitric acid, and perchloric acid for Pb, Zn and Cu analyses (Amacher, 1996). Concentration of total Pb, Zn and Cu in the extracted or digested solutions was determined by atomic absorption spectrophotometry.

Data analysis

The mean difference test (t-test) was performed to assess differences between the two sites in soil heavy metal concentrations, and between the roots and shoots in biomasses productions and heavy metal uptakes. One-way analysis of

Table 1. pH, OM, and total heavy metal concentrations of agricultural soils at sites 1 and 2.

Site	pH	OM (%)	Heavy metal concentration (mg/kg)		
			Pb	Zn	Cu
1	4.67	2.84	864	204	177
2	4.32	3.52	3,352	281	180
Average	4.50	3.18	2,108	243	179
Permissible level (TCVN 7209-2002)			70	200	50

Table 2. Biomass growth of roots and shoots of the four plant species during the 3-month experimental period at sites 1 and 2 (g/m^2).

Roots/shoots	Site	SN	WP	CWH	WS
Roots	1	65	39	90	51
	2	72	30	104	47
	Average	69	35	97	49
Shoots	1	898	1,234	2,850	979
	2	930	1,181	3,338	1,017
	Average	914	1,207	3,094	998

Each value is the mean of three plots of the same species at each site.

variance (ANOVA) was used to identify significant differences in concentrations of heavy metals in soil. Two-way ANOVA was used to identify differences in heavy metal uptake by plant species and heavy metal type. Intergroup means shown to be significantly different were further analyzed by multiple comparison tests performed to assess the relative differences within plant species and/or within heavy metal types. For cases in which the interaction effect was significant between plant species and heavy metal type, the simple main-effect test was performed to determine effect of species and heavy metal type on these differences in detail. The Bonferroni method was applied for the multiple comparison and the simple main-effect tests. Microsoft Excel software was used for data analyses.

RESULTS AND DISCUSSION

Chemical properties and heavy metal concentrations of soil samples

The pH, OM, and Pb, Zn and Cu concentrations of agricultural soils at the two experimental sites are shown in Table 1. The values are the means of values from each of the 12 plots at each site. The soils at both experimental sites were strongly acidic (Table 1). The OM content indicates a medium level of soil fertility. The pH was slightly lower and the OM content somewhat higher for site 2 than for site 1. The soil Pb, Zn and Cu concentrations were all greater than the maximum permissible levels of heavy metals set by Vietnamese standards (TCVN 7209 - 2002), as shown in Table 1. According to one-way ANOVA performed on the differences between the Pb, Zn and Cu concentrations, there were no significant differences among the Pb, Zn and Cu concentrations ($F_{2,3}$=2.33, MSe=2,402,386, $P>0.05$). Further, t-test of soil heavy metal concentrations between sites 1 and 2 shows that there were no significant differences between the two sites ($t(4)=0.80$, $p>0.05$).

Biomass production of roots and shoots by plant species

Biomass production (dry matter quantity and means for three plots of the same species) of roots and shoots by plant species during the 3-month period at sites 1 and 2 are shown in Table 2. The harvested shoot mass was summed over 3 months as shoot growth for comparison with root growth, which was recorded only at the end of the 3-month period. In this table, CWH had the greatest shoot biomass per unit land area (2,850 to 3,338 g/m^2), followed by WP (1,181 to 1,234 g/m^2), and SN and WS had the least biomass (898 to 1,017 g/m^2). CWH also had the greatest root biomass per unit land area (90 to 104 g/m^2), followed by SN (65 to 72 g/m^2). WP and WS had the lowest root biomass (30 to 51 g/m^2). Shoot biomass was 31 to 39 times larger than root biomass for WP and CWH, 19 to 22 times larger for WS, and 13 to 14 times larger for SN. The paired t-test performed with data from the four species showed that shoots had significantly higher biomass than roots ($t(7)$ =4 .46, $p<0.05$).

Uptake of Pb, Zn and Cu by roots and shoots per unit weight of plant material

Table 3 shows the Pb, Zn and Cu uptake per unit weight

Table 3. Pb, Zn and Cu uptake per unit weight of roots and shoots (mg/kg) for the four plant species during the 3-month period.

Roots/shoots	Heavy metal	Site	SN	WP	CWH	WS
Roots	Pb	1	387	576	661	243
		2	951	1,061	1,839	1,663
	Zn	1	97	60	83	67
		2	119	88	134	70
	Cu	1	35	42	83	25
		2	38	67	93	42
Shoots[1)]	Pb	1	125	48	357	179
		2	244	68	419	232
	Zn	1	149	141	129	149
		2	163	145	216	154
	Cu	1	35	62	80	64
		2	51	72	74	74

SN: Spanish needle, WP: water primrose, CWH: common willow herb, WS: water spinach. Each value is the mean of three plots of the same species at each site; [1)]The mean for the sampling with 1-month interval.

of roots and shoots (mg/kg) for the four species during the 3-month period. Each value is the mean of three plots of the same species at each site. Pb, Zn and Cu uptake for roots (mg/kg) ranged from 243 to 1,839, 60 to 134 and 35 to 93, respectively, for all species. The Pb, Zn and Cu uptake for shoots (mg/kg) ranged from 48 to 419, 129 to 216 and 35 to 80, respectively for all species. The uptake of each heavy metal for each plant was generally greater at site 2 than at site 1 for both roots and shoots. According to two-way ANOVA of uptake by the roots, significant differences by species were not observed ($F_{3, 12}$=0.30, MSe=49,515, P>0.05), but the differences by heavy metal were observed ($F_{2, 12}$=11.7, MSe=1,934,736, P<0.01), with the greatest uptake for Pb. There were no significant interactions between species and heavy metal type ($F_{6, 12}$=0.23, MSe=37,419, P>0.05).

However, for the uptake by shoots, significant differences were seen for both species and heavy metal type ($F_{3, 12}$=12.9, MSe=15,883, P<0.01, and $F_{2, 12}$=35.0, MSe=43,038, P<0.01, respectively), but since the interaction was also significant ($F_{6, 12}$=8.9, MSe=10,912, P<0.01), these results should be interpreted carefully.

Paired t-test of uptake by roots and shoots of all heavy metals showed no differences in uptake by shoots and roots of each plant species (t(5)=0.85-1.26, p>0.05). Yoon et al. (2006) measured the uptake of Pb, Zn and Cu for many native plant species grown in contaminated soils. For soil Pb, Zn and Cu concentrations similar to those in our study, the maximum uptake of Zn and Cu by the roots and shoots of the plants (524 and 250 mg/kg, respectively, for Zn; 375 and 210 mg/kg, respectively, for Cu) were larger than uptake measured in our study, but the maximum Pb uptake by the roots and shoots (881 and 491 mg/kg, respectively) was either smaller or only slightly larger than the uptake we measured for CWH (Table 3).

Uptake of Pb, Zn and Cu by roots and shoots per unit land area

Table 4 shows the Pb, Zn and Cu uptakes per unit land area of roots and shoots for the four species over the 3-month period. The uptake per unit land area was calculated by the uptake per unit weight and multiplied by plant weight for the unit land area. Each value is the mean of three plots of the same species at each site. Pb, Zn and Cu uptake by the roots (mg/m2) showed ranges of 12-199, 2-15 and, 1-10, respectively, for all plant species, and uptake was the greatest in CWH among all plant species tested at each site.

Pb uptake for the shoots (mg/m^2) ranged from 61 to 1,396 and was the smallest in WP (≤83), and greatest in CWH (≥1,056). Pb uptake in the other plant species was ≤241. Zn and Cu uptake for the shoots (mg/m^2) showed ranges of 137 to 718 and 32 to 247, respectively for all plant species. Uptake was the greatest for CWH.

Paired t-tests performed on the uptake (mg/m^2) of Pb, Zn and Cu by roots and shoots showed that the uptake by the shoots was significantly higher than by roots for each plant species (t(5)=3.72 to 6.77, p<0.05), which is different from the pattern shown for uptake per unit weight of shoots/roots discussed previously. This difference is likely attributable to the larger biomass production in shoots than in roots (Table 2).

Difference in uptake per unit land area according to the plant species and heavy metal type

Two-way ANOVA of uptake by the roots showed significant differences among heavy metal types ($F_{2,12}$=7.81, MSe=8,636, P<0.01), where the uptake of Pb was higher than those of Cu and Zn and those of Cu and

Table 4. Pb, Zn and Cu uptake of roots and shoots per unit land area (mg/m^2) for the four plant species during the 3-month period.

Roots/shoots	Heavy metal	Site	SN	WP	CWH	WS
Roots	Pb	1	24	23	57	12
		2	67	33	199	78
	Zn	1	6	2	7	3
		2	8	3	15	3
	Cu	1	2	2	7	1
		2	3	2	10	2
Shoots[1]	Pb	1	115	61	1,056	180
		2	232	83	1,396	241
	Zn	1	137	175	378	148
		2	152	174	718	158
	Cu	1	32	77	238	64
		2	48	86	247	78

SN: Spanish needle, WP: water primrose, CWH: common willow herb, WS: water spinach. Each value is the mean of three plots of the same species at each site; 1The sum of values for each month.

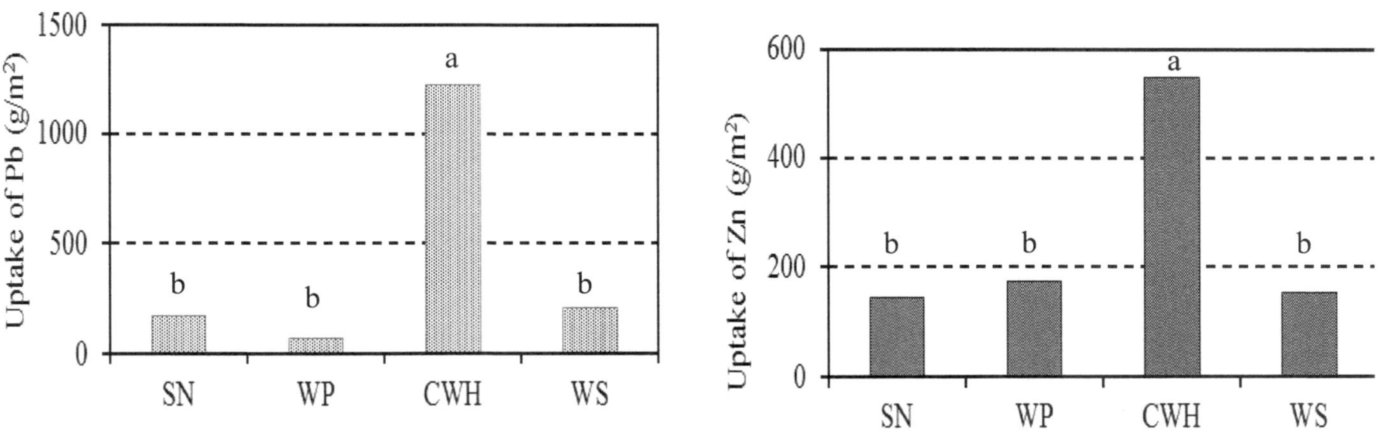

Figure 3. Uptake of Pb and Zn by the shoots for the four plant species. Each uptake value is the mean of plots of the same species at sites 1 and 2. Different letters above each column indicate values that are significantly different at the 0.05 or 0.01 levels according to simple main-effect test.

Zn were not significantly different by the multiple comparison test. The differences in uptake by the roots of plant species were not observed ($F_{3, 12}$=1.62, MSe=1,789, P>0.05), and no significant interactions between heavy metal types and plant species were observed (F6, 12=1.05, MSe= 1,156, P>0.05). In contrast, uptake by shoots showed significant differences by heavy metal type and plant species ($F_{2, 12}$=18.7, MSe=194,623, P<0.01 and F3, 12=43.3, MSe=451,249, P<0.01, respectively). A significant interaction between heavy metal type and plant species was also observed ($F_{6, 12}$=10.8, MSe=112,479, P<0.01). For the uptake of Pb and Zn by shoots, significant differences were found among the plant species by the simple main effect test (Figure 3). Uptake of Cu by shoots was not found to differ among plant species by the same test, though data not shown in Figure 3. The uptake of Pb and Zn was the greatest in CWH (Figure 3). However, uptake of Pb and Zn did not significantly differ among SN, WP and WS. Figure 4 shows the uptake of Pb, Zn and Cu by the shoots of CWH, for which a significant difference was elucidated by the simple main effect test. The uptake by shoots was in order of Pb > Zn > Cu. No differences in uptake by shoots were observed for other three plant species.

Tlustos et al. (2006) showed that maximum uptake of Pb and Zn by shoots of high biomass-producing crops (sweet clover, red clover, curled mallow, safflower and hemp) to be 120 and 104 mg/m^2, respectively. According to Dai et al. (2007), the uptake of Pb by roots plus shoots

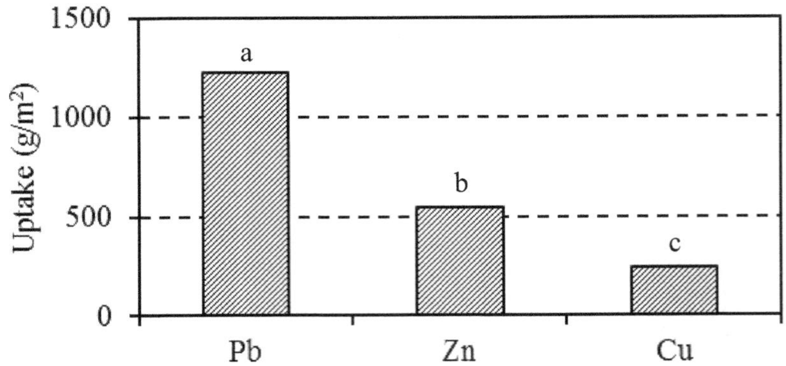

Figure 4. Uptake of Pb, Zn and Cu by the shoots in CWH. Each value is the mean of the plots of CWH at Sites 1 and 2. Different letters above the column indicate that the values are significantly different at the 0.05 or 0.01 level according to the simple main-effect test.

of maize was 465 mg/m^2. In comparison, uptake of Pb and Zn by shoots (nearly 1,200 and 500 mg/m^2, respectively) and the uptake of Pb by roots plus shoots (nearly 1,400 mg/m^2) for CWH in our study are far greater.

From these results, we can conclude that CWH has superior characteristics for the uptake of heavy metals, particularly of Pb, and can be effectively utilized for phytoremediation in this study area.

Conclusions

1. To evaluate the phytoremediation capacity, the uptake by plants per unit land area was compared. Uptake per unit land area by roots differed by heavy metal type, and uptake of Pb was the largest while uptakes of Zn and Cu were low with no difference from each other.
2. The uptake was higher in shoots than in roots for each plant species, due to the greater biomass of shoots than roots.
3. The uptake by shoots differed by both plant species and heavy metal type. Uptakes of Pb and Zn were greatest in CWH, while that of Cu did not differ among plant species. A difference in uptake among the heavy metal type was observed only in CWH, where the uptake was in order of Pb > Zn > Cu.
4. From the aforementioned higher uptake of heavy metals by the shoots of CWH, CWH has superior phytoremediation capacity, particularly, for the uptake of Pb.

REFERENCES

Amacher MC (1996). Nickel, cadmium and lead. In DL Sparks (ed.) Methods of soil analysis Part 3 Chemical methods and processes, Madison. pp. 739-768.

Aziz RA, Kamarudzaman AN, Kamaruddin NA, Salleh MN (2011). Study on phytoremediation performance and heavy metals uptake in leachate by reed beds plant. Proc. Int. Conf. Environ. Biomed. Biotechnol. 16:47-51.

Barbafieri M, Dadea C, Tassi E, Bretzel F, Fanfani L (2011). Uptake of heavy metals by native species growing in a mining area in Sardinia, Italy: discovering native flora for phytoremediation. Int. J. Phytorem. 13:985-997.

Dai Q, Yuan J, Fang W, Yang Z (2007). Differences on Pb accumulation among plant tissues of 25 varieties of maize (*Zea mays*). Frontiers Biol. China 2(3):303-308.

DWRM/ESCAP/KICT (ed.) (2010). Guideline framework for river restoration in Viet Nam. Department of Water Resources Management, Hanoi, pp. 1-32.

Garg G, Kataria SK (2009). Phytoremediation potential of *Raphanus sativus* (L.), *Brassica juncea* (L.) and *Triticum aestivum* (L.) for copper contaminated soil. Proc. 53rd Annual Meeting International Soc Syst. Sci., pp. 1-10.

General Statistics Office (2002-2010). Statistical Yearbook of Vietnam 2001-2009. Statistical Publishing House, Hanoi.

Ho TLT, Nguyen HT (2003). Heavy metal in agricultural land of Van Lam, Hung Yen. Magazine of Soil Sci. 19:167-174.

Ji P, Sun T, Song Y, Ackland ML, Liu Y (2011). Strategies for enhancing the phytoremediation of cadmium-contaminated agricultural soils by *Solanum nigrum* L. Environ. Pollut. 159:762-768.

Lorestani B, Cheraghi M, Yousefi NN (2011). Phytoremediation potential of native plants growing on a heavy metals contaminated soil of copper mine in Iran. World Academy of Science. Eng. Technol. 77:377-382.

Mojiri A (2011). The potential of corn (Zea mays) for phytoremediation of soil contaminated with cadmium and lead. J. Biol. Environ. Sci. 5(13):17-22.

National Institute for Soils and Fertilizers (1998). Handbook for analysis of soil, water, fertilizer, and plant, Publishing House of Agriculture, Hanoi.

Nazir A, Malik RN, Ajaib M, Khan N, Siddiqui MF (2011). Hyper accumulators of heavy metals of industrial areas of Islamabad and Rawalpindi Pakistan. J. Bot., 43(4): 1925-1933.

Nelson DW, Sommers LE (1996). Total carbon, organic carbon, and organic matter.. In DL Sparks (ed.) Methods of soil analysis Part 3 Chemical methods and processes, Madison. pp. 961-1010

Poniedzialek M, Sekara A, Jedrszczyk E, Ciura J (2010). Phytoremediation efficiency of crop plants in removing cadmium, lead and zinc from soil. Folia Horticulturae Ann., 22(2): 25-31.

Revathi K, Haribabu TE, Sudha PN (2011). Phytoremediation of chromium contaminated soil using Sorghum plant. Int. J. Environ. Sci. 2(2): 417-428.

Roongtanakiat N (2009). Vetiver phytoremediation for heavy metal decontamination. Technical. Bulletin no. 2009/1. Pacific Rim Vetiver Network. Office of the Royal Development Projects Board, Bangkok.

Tlustos P, Száková J, Hrubý J, Hartman I, Najmanová J, Nedělník J, Pavlíková D, Batysta M (2006). Removal of As, Cd, Pb, and Zn from

contaminated soil by high biomass producing plants. Plant Soil Environ. 52(9):413-423.
Tong YP, Kneer R, Zhu YG (2004). Vacuolar compartmentalization: a second-generation approach to engineering plants for phytoremediation. Trends Plant Sci. 9(1):7-9.
VSSS-NIAP (Vietnam Soil Science Society–National Institute of Agriculture Planning and Projection) (ed.) (1996). Soil Map of Vietnam (Scale 1:1,000,000). VSSS-NIAP, Hanoi.
Yoon J, Cao X, Zhou Q, Ma LQ (2006). Accumulation of Pb, Cu, and Zn in native plants growing on a contaminated Florida site. Sci. Total Environ. 368: 456-464.

Partial balance of N, P and K in a sole millet crop on a sandy loam soil of Northeast Nigeria

A. L. Ngala[1], H. Yakubu[1] and I. J. Tekwa[2]

[1]Department of Soil Science, University of Maiduguri, P. M. B. 1069, Maiduguri, Borno State, Nigeria.
[2]Department of Agricultural Technology, Federal Polytechnic, P. M. B. 35, Mubi, Adamawa State, Nigeria.

A study was conducted on nutrient dynamics through partial N, P and K balances during 2005 growing season. The treatments involved three levels of farmyard manure (FYM: 2.5, 5.0 and 7.5 tons/ha) combined with half (N30P15K15 kg/ha) and full (N60P30K30 kg/ha) recommended rates of NPK fertilizer and sole recommended rates of FYM (7.5 tons/ha) and NPK (N60P30K30 kg/ha) and a control. The experiment was laid out in RCB design with three replicates. Data on nutrient inputs (fertilizers and FYM) and outputs (grains and crop residues) were estimated using the budgeted method. The treatment means were statistically compared using Duncan multiple range test (DMRT). The lowest N-balance (-44.70 kg/ha/year) resulted from combination of lowest levels of FYM and NPK, and increased with increasing rates. Sole 7.5 t of FYM resulted in positive N-balance (19.00 kg/ha/year) compared to sole recommended rate of NPK (-2.40 kg/ha/year). P balance varied from as low as -1.54 kg/ha/year in control to as high as +36.16 kg/ha/year in plots which received 7.5t FYM + N60P30K30 kg/ha. All the treatments had negative K-balances ranging from -6.60 kg/ha/year in plots applied with 7.5t FYM/ha to -108.89 kg/ha/year in plots which received 2.5 t FYM/ha + N60P30K30 kg/ha, depicting high plant K-uptake relative to inputs. The grain yields varied from 789 kg/ha in control to 2440 and 2301 kg/ha in plots applied with 7.5 t FYM/ha + N60P30K30 kg/ha and 7.5t FYM/ha + N30P15K15 kg/ha, respectively. Combination of sub-optimal rates of FYM and NPK fertilizers enhanced millet yields with positive N and P balances in the soils.

Key words: Sole millet, sandy loam soil, nutrient balance, northeast Nigeria.

INTRODUCTION

The economic growth and life quality index in the semi-arid environment of northeast Nigeria largely depends on the agricultural sector, which accounts for over 25% of the gross domestic product (Nandwa, 2003). Nonetheless, the region is characterized by declining per-capita cereal production estimated at 150 to 130 kg/person in the past 35 years (Gichuru et al., 2003). Soil fertility depletion has been described as the single most important constraint to food security in the region. A larger proportion of the soils have low inherent fertility (Bationo et al., 1998). The major cause of soil fertility depletion is the imbalance caused by nutrients which are not commonly replaced when removed; resulting innegative nutrient balances (Gichuru et al., 2003). According to Sanchez et al. (1997), in the past 30 years, nutrient depletion was estimated at an average of 660 kg N/ha, 75 kg P/ha and 450 kg K/ha from about 200 million hectares of cultivated lands in 37 African counties (Nigeria inclusive). This net negative nutrient balances for African soils appear to be a contributing factor in nutrient impoverishment, reduction of agricultural productivity and the stagnation or decline in national economies. In contrast, cultivated land in temperate regions, on the average, receives nutrients in excess of those removed in crops, runoff, and erosion. During the same period of 30 years, the nearly 300 million ha of cultivated temperate-region soils had a net positive nutrient balance of at least 60 kg N/ha, 20 kg P/ha and 30 kg K/ha (Brady and Weil, 2002). While nutrient deficiencies still occur in some fields a greater majority of soils have experienced a nutrient build up (Bergstrom et al., 2005). Inherently infertile soils require external inputs of fertilizer materials for sustained fertility and productivity, especially under continuous cropping (Nyathi and Campbell (1997). The root cause of declining fertility is land degradation manifested in form of soil fertility depletion, especially under smallholder farming sector (Stoorvogel and Smaling, 1990; Smaling et al., 1997) and nutrient imba-lance on large scale farms (Nandwa and Bekunda, 1998).

Evidences and indicators of soil fertility decline in West Africa include appearances of widespread distribution of soils deficient in major macro-nutrients and micro-nutrients (ascertained from commensurate crop responses and soil analyses and local soil quality indicators), and widespread negative nutrient balances (Bationo et al., 1998). The predominantly sandy soils in northeast semi-arid Nigeria renders their management strategies less effective as a result of their low water holding capacity and low organic matter content. Soil fertility management has to focus on supplying inputs from external sources to improve organic matter content of soils and consequently its capacity as a repository for nutrients. Chemical fertilizers are the obvious sources to counter balance the nutrient depletion, but unfortunately there are few local sources of manufactured fertilizers. Because of the great distances over which imported fertilizers must be transported, their farm-delivered price is at least twice the international price (Brady and Weil, 2002). The presently high cost of chemical fertilizers coupled with unfavorable crop/fertilizer price ratios discourages their additions, especially in smallholder farms. Therefore, innovative combinations of organic and inorganic nutrient sources are to be used in order to increase nutrient inputs and to recycle the nutrients once they reach the soil. This suggests that part of the problem of nutrient deficiencies can be addressed through judicious use of organic inputs. Nyathi and Campbell (1997) found that synergism (complementary) existed when combinations of inputs to sandy soil included inorganic fertilizers, while studying the efficacy of combinations of organic and inorganic sources of nutrients. Bationo et al. (1998) also recognized the beneficial effect of continuous application of FYM along with mineral fertilizers on sandy loam soils. This research become imperative in documenting implementable information for farmers in this study area, considering the prevailing sharp nutrient decline trends evident in the region as a result of continuous cultivation without nutrient replacement. Therefore this paper is aimed at establishing optimum levels of combination of organic and inorganic nutrient sources to be applied to fertility decline.

MATERIALS AND METHODS

A field study was conducted on farmers field at Dusuman village (Latitude 11° 50.82' N and Longitude 13° 11.01' E; 320 meters above sea level) of Jere Local Government area of Borno state during 2005 rainy seasons. The aim of this study was to determine the partial nutrient balance on farmers' field. The site has an average monthly temperature of 28.5 to 32.8°C in normal years and rainfall of 525 mm with typical savanna grassland vegetation. The soil characterization was carried out prior to treatment application.

The experiment consisted of nine treatments laid out in a randomized complete block design (RCBD) and replicated three times. The treatments comprised of a control, 2.5 tones of FYM + N_{60} P_{30} K_{30} kg/ha or N_{30} P_{15} K_{15} kg/ha, 5.0 tones FYM + N_{60} P_{30} K_{30} kg/ha or N_{30} P_{15} K_{15} kg/ha, 7.5 tones FYM + N_{60} P_{30} K_{30} kg/ha or N_{30} P_{15} K_{15} kg/ha, 7.5 tones FYM and N_{60} P_{30} K_{30} kg/ha.

The land was manually cleared and harrowed with tractor driven disc. Three rows of nine plots were manually prepared using hand hoe. Each plot measured 5 × 10 m, making a total of twenty seven plots. Farmyard manure (FYM) was procured in the University of Maiduguri animal farm and applied to the plots according to the treatment levels, two weeks before planting. Chemical fertilizer in the form of compound NPK 15-15-15 was procured in the open market and applied according to the treatments. Samples of the FYM (Appendix 1) and the fertilizers were taken for nutrient content analyses. Seeds of pearl millet variety SOSAT-C-88 were dressed with Apron Star 42 WS at the rate of 10 g sachet per 4 kg seeds for protection against soil and seed borne pests and diseases and planted in the plots at a row spacing of 0.75 m apart and 0.5 m between stands in a row. Later after two weeks of planting (2 WAP) millet plants were thinned to three per stand. Plots were kept weed free manually at four weeks intervals. The plants were harvested after reaching physiological maturity stage.

Composite soil samples were collected prior to treatment application and again after harvest (post harvest samples), then crushed, sieved through 2 mm sieve and analyzed for N,P and K contents. Total N was determined using kjeldahl digestion method (Bremner and Mulvaney, 1982); available P using Bray-1 method (Olsen and Sommers, 1982) and exchangeable K using NH_4OAc method (AOAC, 1970).

Plant samples which included straw and grain at harvest were also collected and analyzed for N, P and K contents (AOAC, 1970) as well as uptake. Budgeted method (Nandwa, 2003) was used to compute the partial nutrient balances from the data analyzed. The balances were determined using the differences between the inputs (mainly from chemical fertilizer and FYM) and the outputs (from crop residues, grain and soil). All data collected were subjected to Analysis of Variance (ANOVA) using statistix 8.0 (Statistix, 2005) Analytical Software package. Differences between treatment means were compared using Duncan Multiple Range Test (DMRT) at 5% level of probability.

RESULTS AND DISCUSSION

The results of some physical and chemical properties of soil of the experimental site are presented in Table 1. The soil was low in fertility considering the low levels of organic matter (3.80 g/kg), total N (0.40 g/kg) and available P (2.92 mg/kg). In addition, the predominantly

Table 1. Physico-chemical characteristics of soil of the experimental site.

Soil characteristics	Value
pH 1:2.5 (H_2O)	6.69
pH 1:2.5 (KCl)	4.80
EC (dS/m)	0.04
Organic carbon (g/kg)	2.20
Total N (g/kg)	0.40
C : N ratio	5.50
Available P (Bray 1, mg/kg)	2.92
Exchangeable cations (ammonium acetate, cmol/kg)	
Ca	3.60
Mg	2.80
K	0.46
Na	0.05
H + Al	0.80
CEC	7.71
Base saturation (%)	89.62
Particle size distribution (%)	
Sand	76.6
Silt	6.4
Clay	17.0
Soil textural class	Sandy loam

Table 2. Nutrient balance of a millet cropping system grown under rain fed condition.

Treatment	kg/ha/year		
	N	P	K
Control	-43.41^{h*}	-1.54^{h}	-61.17^{b}
2.5t FYM + $N_{30}P_{15}K_{15}$ kg/ha	-44.70^{h}	+15.22^{f}	-86.68^{f}
2.5t FYM + $N_{60}P_{30}K_{30}$ kg/ha	-26.55^{g}	+29.86^{b}	-108.69^{h}
5.0t FYM + $N_{30}P_{15}K_{15}$ kg/ha	-8.84^{f}	+18.55^{e}	-74.66^{e}
5.0t FYM + $N_{60}P_{30}K_{30}$ kg/ha	+23.95^{b}	+31.03^{b}	-45.32^{c}
7.5t FYM + $N_{30}P_{15}K_{15}$ kg/ha	+2.43^{d}	+20.95^{d}	-53.66^{d}
7.5t FYM + $N_{60}P_{30}K_{30}$ kg/ha	+27.15^{a}	+36.16^{a}	-21.43^{b}
7.5t/ha FYM	+19.00^{c}	+7.87^{g}	-6.60^{a}
$N_{60}P_{30}K_{30}$ kg/ha	-2.40^{e}	+27.18^{c}	-96.24^{g}
SE ±	0.886	0.770	1.909

*Values followed by the same letter(s) within a column are not significantly different from each other at P (<0.05) according to Duncan Multiple Range Test (DMRT).

millet farms in order to tackle the problems of soil sandy loam soil texture in the area predisposes any amendments to reduce effectiveness because of their low water holding capacity as evidenced from the dominant sand fraction and low organic matter content (Table 1). Soil fertility management that focused on supplying inputs from external sources will improve organic matter content and consequently its capacity to supply essential plant nutrients.

Nutrient balances

The estimated N, P and K balances for the millet based cropping system are presented in Table 2. The balances were estimated using inputs from the applied fertilizers and the output by the crop (Appendix 2 and 3). Treatment effect showed variation in all the balances estimated. N-balance was most negative (-44.70 kg/ha/year) with 2.5 t FYM/ha + $N_{30}P_{15}K_{15}$kg/ha. Increasing fertilizer levels subsequently increased the N-balances. The highest effect (+27.15 kg/ha/year) was obtained with 7.5 t FYM/ha + $N_{60}P_{30}K_{30}$kg/ha which was significant ($P<0.05$) when compared with the sole 7.5 t FYM/ha. Although, this highest value ought to have been more, leaching effect of the sandy soil and microbial decomposition might have reduced the value to this level. The recommended sole FYM level (7.5 t/ha) gave positive balance (+19.00 kg/ha/year) unlike the sole $N_{60}P_{30}K_{30}$kg/ha, which gave negative balance (-2.40 kg/ha/year). The estimated results for the nutrient balance indicated that N-balances were negative with lower levels of treatment application; however, increasing the rates of NPK and FYM increased the N-balances. The negative balances with lower application rates might be attributed to crop uptakes as well as losses through leaching and soil erosion as a result of sandy nature of the soils. Rego et al. (2002) also reported similar cases with sandy soils of India. While the positive balance with the higher application rates could be attributed to the left over after satisfying microbial demand and crop uptake.

P-balance for all the fertilizer treated plots gave positive balances (Table 2), unlike the non P-treated plots. The FYM applied perhaps contained substantial amounts of Jama and Mafongoya (1999) reported that loss of P from soil will be minimal if application involved organic incorporation. Estimated K-balances revealed negative balances with all the treatments, indicating soil K mining. K accumulated in the millet straw was much more than the applied rates, and this probably accounts for the K negative balances. Olk and Cassman (1993) similarly reported a net decrease in amount of K fixed in manure plots than in control plots and this explains why the treatments with higher combination rates had higher negative K balances than the untreated ones.

Grain yields

Mean millet grain yield from different fertilizer treatments are shown in Table 3. The highest yield (2440 kg/ha) was obtained with the application of 7.5 t FYM/ha + $N_{60}P_{30}K_{30}$kg/ha and the lowest (789 kg/ha)

Table 3. Effect of FYM and NPK fertilizer on pearl millet grain yield (kg/ha).

Treatment	Mean grain yield (kg/ha)
Control	789[f*]
2.5t FYM + N_{30} P_{15} K_{15} kg/ha	1703[de]
2.5t FYM + N_{60} P_{30} K_{30} kg/ha	1930[cd]
5.0t FYM + N_{30} P_{15} K_{15} kg/ha	1978[bcd]
5.0t FYM + N_{60} P_{30} K_{30} kg/ha	2065[bc]
7.5t FYM + N_{30} P_{15} K_{15} kg/ha	2301[ab]
7.5t FYM + N_{60} P_{30} K_{30} kg/ha	2440[a]
7.5t/ha FYM	1433[e]
N_{60} P_{30} K_{30} kg/ha	1970[bcd]
SE ±	111.7

* Values followed by the same letter(s) within a column are not significantly different from each other at P (<0.05) according to Duncan Multiple Range Test (DMRT).

with the control. Application of FYM alone gave a significantly (P<0.005) lower yield compared with the application of the combined fertilizer treatments. This shows the beneficial effect of combined nutrient application. The highest yield associated with the highest application rate represents about 20% increase over non application. Application of 7.5 t FYM/ha together with sub-optimal mineral fertilizer could maintain soil fertility under continuous cropping system and consequently improve the nutrient balance as reported by Kwari et al. (1998), even though this huge amount is a constrain in this region. In addition, resource poor farmers might not be able to afford the cost of the recommended rates of mineral fertilizers. Therefore, combination of FYM and NPK fertilizers at the rates of 2.5 t FYM/ha + $N_{30}P_{15}K_{15}$ kg/ha could be beneficial as revealed by this present study. Tabo (1995) also reported similar case of achieving higher grain yield of 1420 kg/ha with the application of 2.5 t FYM + 32 kg N/ha.

Conclusion

Application of lower levels of FYM in combination with NPK fertilizer could be more economical than either sole FYM or NPK. However, the performance of millet was better with the application of higher rates of combined application of FYM and NPK (7.5 t FYM/ha + $N_{60}P_{30}K_{30}$ kg/ha). Hence, application of lower levels of farmyard manure in combination with half the recommended rates of NPK fertilizer (2.5 t FYM/ha + $N_{30}P_{15}K_{15}$kg/ha) for millet over a period of time will reduce nutrient depletion, increase nutrient balance and maintain sustainable crop production.

REFERENCES

Association of Official Analytical Chemists (AOAC) (1970). *Official methods of analysis*, Ed II. Washington, D. C.

Bationo A, Lompo F, Koala S (1998). Research on nutrient flows and balances in West Africa: State - of - the - Art. Agr. Ecosys. Environ. 71:21-38.

Bergstrom L, Bowman BT, Sims JT (2005). Definition of sustainable and unsustainable issues in nutrient management of modern agriculture. Soil Use Manage. 21:76-81.

Brady NC, Weil RR (2002). *The nature and properties of soils* (13th Ed.). Prentice-Hall, Inc. New Jersey, U.S.A. pp. 199-223.

Bremner JM, Mulvaney CS (1982). Total N. *In*: Page, A. L. (ed.). *Methods of soil analysis part 2.* Chemical and mineralogical properties (2nd ed). *Agron. Monograph.* ASA-SSSA, Madison, Wisconsin, USA 9:595-622.

Gichuru MP, Bationo A, Swift MJ (2003). Preface Note. *In*: Gichuru MP, Bationo A, Bekunda MA, Goma HC, Mafongoya PK, Mugendi DN, Murwira HM, Nandwa SM, Nyathi P, Swift MJ (Eds.). *Soil Fertility Management in Africa: A Regional Perspective.* Academic Science Publications. Nairobi, Kenya. pp. 147-186.

Jama BA, Mafongoya PL (1999). Strategies for soil fertility recapitalization in Africa: Experiences from Western Kenya and Eastern Zambia. *In*: FAO proceedings on soil fertility initiative in sub-Saharan Africa.

Kwari JD, Grema AK, Bibinu ATS (1998). Fertilizer trials for millet/legume mixtures with emphasis on nitrogen rates. *In*: Emechebe AM, Ikwelle MC, Ajayi O, Aminu-Kano M, Anaso AB (eds.). *Pearl millet in Nigerian Agriculture: Production, Processing and Research priorities.* Proceedings of the pre-season national coordination and planning meeting of the Nationally Coordination Research Programme on pearl millet. Lake Chad Research Institute, Maiduguri, Nigeria.

Nandwa SM, Bekunda MA (1998). Research on nutrient flows and balances in east and southern African: State-of-the-art. *Agr. Ecosys. Envi.* 71:5-18.

Nandwa SM (2003). Perspectives on Soil fertility in Africa. *In*: Gichuru MP, Bationo A, Bekunda MA, Goma HC, Mafongoya PK, Mugendi DN, Murwira HM, Nandwa SM, Nyathi P, Swift MJ (Eds.). *Soil Fertility Management in Africa: A Regional Perspective.* Academic Science Publications. Nairobi, Kenya. pp. 1- 50.

Nyathi P, Campbell BM (1995). Interaction effects of tree leaf, manure and inorganic fertilizer on the performance of maize in Zimbabwe. Afr. Crop Sci. J. 3(4):451-456.

Olk DC, Cassman KG (1993). Reduction of potassium fixation by organic matter in vermiculitic soils. In: Mulongoy K, Merckx R (eds.). Soil Organic Matter Dynamics and Sustainability of Tropical Agriculture. A Wiley-Sayce Co-publication, IITA/K.U. Leuven. pp. 307-315.

Olsen SR, Sommers LE (1982). Phosphorus. *In*: Page AL (ed.). *Methods of soil analysis part 2.* Chemical and microbiological properties (2nd ed). *Agron. monograph.* ASA-SSSA, Madison, Wisconsin, USA. 9:401-430.

Rego TS, Seeling B, Nageswara RV, Pardhasaradhi G, Kumar RJVDK, Myers RJK, Johansen C (2002). Nutrient balances: A guide to improving the sorghum and groundnut-based dry land cropping systems in semi-arid tropical India. *In*: Adv-Gyamfi JJ (ed.). Food security in nutrient-stressed environments: Exploiting plants' genetic capabilities. Kluwer Academic Publishers, The Netherlands. pp. 301-311.

Sanchez PA, Shepherd KD, Soule MJ, Place FM, Buresh RJ, Izac AMN, Mokwunye AU, Kwesiga FR, Ndiritu CG, Woomer PL (1997). Soil fertility replenishment in Africa: An investment in natural resource capital, *In*: Buresh RJ, Sanchez PA, Calhoun F (Eds.). *Replenishing Soil fertility in Africa.* SSSA Special publication, No.51. Madison, Wisconsin, USA.

Smaling EMA, Nandwa SM, Janssen BH (1997). Soil Fertility in Africa is at stake. *In*: Buresh RJ, Sanchez PA, Calhoun F (eds.). Replenishing Soil fertility in Africa. SSSA Special publications. SI. SSSA, Madison, Wisconsin.

Statistix (2005). Statistical Software Analytical Package, Version 8.0, Cary, USA.

Stoorvogel JJ, Smaling EMA (1990). Assessment of soil nutrient depletion in Sub-Saharan Africa: 1983-2000. Vol (I) Main report.

The Winand Staring Centre. Wageningen, The Netherlands.
Tabo R (1995). Performance of sorghum and millet varieties under varying combinations of animal manure and chemical fertilizer. *In*: *ICRISAT and collaborative programmes.* West and Central Africa. Annual Report (1995). International Crops Research Institute for Semi- Arid Tropics. pp. 53-54.

Appendix 1. Nutrient content of FYM applied to the field.

Nutrient	Value (%)
N	1.05
P	0.14
K	1.28

Appendix 2. Effect of FYM and NPK fertilizers on N, P and K uptake (kg/ha) by pearl millet leaves grown under rainfed condition, 2004 cropping season.

Treatment	N	P	K
Control	30.85[d*]	1.46[c]	8.85[d]
2.5t FYM + N_{30} P_{15} K_{15} kg/ha	93.40[bc]	3.69[bc]	21.01[bc]
2.5t FYM + N_{60} P_{30} K_{30} kg/ha	134.09[ab]	4.76[ab]	25.39[ab]
5.0t FYM + N_{30} P_{15} K_{15} kg/ha	115.55[abc]	5.03[ab]	22.55[bc]
5.0t FYM + N_{60} P_{30} K_{30} kg/ha	120.84[abc]	3.86[b]	24.04[ab]
7.5t FYM + N_{30} P_{15} K_{15} kg/ha	137.85[ab]	7.04[a]	26.40[ab]
7.5t FYM + N_{60} P_{30} K_{30} kg/ha	159.17[a]	6.46[a]	31.47[a]
7.5t/ha FYM	81.85[c]	3.26[bc]	15.21[cd]
N_{60} P_{30} K_{30} kg/ha	78.97[cd]	5.05[ab]	21.43[bc]
SE ±	16.926	0.777	2.638

* Values followed by the same letter(s) within a column are not significantly different from each other at P (<0.05) according to Duncan Multiple Range Test (DMRT).

Appendix 3. Effect of FYM and NPK fertilizer on N, P and K uptake (kg/ha) by pearl millet grain grown under rainfed condition in the field, 2004 cropping season.

Treatment	N	P	K
Control	51.99[f*]	2.41[e]	37.92[d]
2.5t FYM + N_{30} P_{15} K_{15} kg/ha	128.99[bc]	4.70[c]	84.82[abc]
2.5t FYM + N_{60} P_{30} K_{30} kg/ha	127.74[bc]	5.68[b]	105.05[a]
5.0t FYM + N_{30} P_{15} K_{15} kg/ha	108.28[d]	4.62[c]	71.25[bc]
5.0t FYM + N_{60} P_{30} K_{30} kg/ha	122.81[c]	4.82[c]	81.77[abc]
7.5t FYM + N_{30} P_{15} K_{15} kg/ha	141.35[ab]	5.70[b]	90.72[ab]
7.5t FYM + N_{60} P_{30} K_{30} kg/ha	151.62[a]	6.37[a]	103.59[a]
7.5t/ha FYM	76.08[e]	3.04[de]	59.43[cd]
N_{60} P_{30} K_{30} kg/ha	99.54[d]	3.37[d]	76.52[bc]
SE ±	4.750	0.215	8.917

* Values followed by the same letter(s) within a column are not significantly different from each other at P (<0.05) according to Duncan Multiple Range Test (DMRT).

A trial to determine the impact of soil washing on coal seam gas (CSG) dam sediments in Queensland

Lee Fergusson

Prana World Consulting, P. O. Box 1620, Oxenford, Queensland 4210, Australia.

Extracting coal seam gas (CSG) is becoming increasingly common in Australia and throughout the world. With the number of wells in Queensland alone increasing from 10 in 1995 to 4,484 by 2013 and 20,000-40,000 more expected to be drilled in the next 20 years, the production of methane from CSG is forecast to become one of Australia's largest export commodities within the next decade. There is, however, growing disquiet about the potential for CSG to negatively impact the environment because it generates a number of gaseous, liquid and solid waste streams, the composition of which are largely unknown. CSG waste streams include fugitive emissions, drilling mud, chemical fracking fluids, produced chemical fracking fluids, and produced water and brines, some of which are settled water and brines, some of which are settled into large evaporation dams to form sediments. Moreover, minimal data on the chemical and physical properties of these various waste streams resulting from rigorous environmental research have been published or made publicly available, adding to society's growing concern for CSG as an environmentally sustainable approach to extracting gas from coal. For this reason, the present study examines the chemical properties of two types of CSG dam sediment from the Bowan Basin in Queensland, and investigates the role that conventional soil washing might have in converting these low-level contaminated sediments into a "clean soil" which can be applied in on-site, beneficial reuse projects; the chemical properties of the resultant flushing solutions are also examined. The study found the majority of components in CSG dam sediments are benign, and soil washing removes or partially removes most inorganic contaminant species from CSG dam sediments; the study also found that these contaminants were mostly transferred to the flushing solution, which can be treated separately and potentially reused for livestock drinking water or irrigation water or recycled into the operating circuit.

Key words: Coal seam gas, CSG, soil washing, sediment, flushing solution.

INTRODUCTION

Coal seam gas (CSG) extraction is an industrial process developed to produce methane (CH_4) from uneconomical coal reserves. CSG uses a method called "cleat pressure drawdown" to depressurize and thereby liberate gas from underground coal formations using large quantities of pumped freshwater. According to industry predictions, by accessing this "unminable" coal CSG has the potential to increase global coal reserves by 600 billion tonnes or 70%, with India alone planning to access 350 billion tonnes of gas using non-conventional means, including

CSG (World Energy Council, 2013: 1.7). As a consequence, the scale and volume of gas production from CSG in Australia has increased exponentially over the last ten years. For example, the number of wells across Queensland has increased from 10 in 1995 to 4,484 in 2013, with an additional 20,000-40,000 wells expected to be drilled in the next 20 years (Lloyd-Smith and Senjen, 2011, citing Queensland Government forecasts); as a consequence, production of methane from CSG and other gas commodities are expected to become one of Australia's largest exports within the next decade (EnergyQuest, 2009).

Queensland's four main gas fields in the Bowan, Surat, Galilee, and Clarence-Moreton Basins together generated a combined gas production of 4 PJ in 1999, 63 PJ in 2006, and 264 PJ in 2013 (Department of Natural Resources and Mines, 2014: 1). Along with other forms of liquefied natural gas (LNG) generated by conventional processes and non-conventional processes, such as underground coal gasification (UCG), CSG will be routed via the $70 billion Curtis Island gas terminal in Gladstone and newer hubs for export to overseas (mostly Asian) markets, although recent reports suggest that some CSG wells are underperforming expectations (Chambers, 2013). Of the 4,484 CSG production wells currently generating methane in Queensland, 68 use hydraulic fracturing (or "fracking") to liberate gas from coal, but none of the 230 wells in New South Wales employs the practice. This finding is relevant to the case made against CSG, because fracking is considered a significant contributor to environmental pollution, particularly of groundwater (Lloyd-Smith and Senjen, 2011), and benzene, a known mutagen and carcinogen, has been found in the produced water of some CSG operations (Cubby, 2010).

Along with the extraction of commercial gas, CSG operations also generate a range of waste byproducts, including fugitive gas emissions (Day et al., 2103, found a CSG operation in Queensland inadvertently released between 13 tonnes and 58 tonnes of CH_4 per year, 17 tonnes per year of which could still be measured 3.0 km from the site), drilling mud, wastewater (or "produced water" in the language of the industry), and brine. For example, while it differs from well-to-well, an average CSG well will generate between 3,000 and several hundred thousand litres of wastewater per day depending on coal seam formation (CSIRO, 2012: 1); as a consequence, a total of 16.9 GL of wastewater was generated by Queensland's CSG wells in 2012 (Department of Natural Resources and Mines, 2014: 1). Problems associated with so-called "overburden fracturing", that is, when the ground surface above a CSG well subsides, have also been reported. About 3 to 5% of CSG wastewater by volume is actually solids, typically present in the liquid fraction as suspended, colloidal particles or dissolved solids; most of these solids are composed of indigenous alkanic hydrocarbons and/or "salty brines", which do not form part of the gas fraction. Thus, a minimum of approximately 480 million tonnes of brine is generated each year in Queensland; these brines are generally settled in holding ponds or evaporation dams located contiguously to clusters of wells (the footprint of dams can be up 100 ha in area). Wastewater is typically allowed to either evaporate, be recycled to livestock drinking water or irrigation water, or be used as cooling water after treatment with reverse osmosis; brines can be disposed to regulated landfill (Egan, 2012). Not a lot of information has been written about the chemical properties of wastewater generated by CSG (e.g., CSIRO, 2012); what has been written typically focuses on the potentially harmful health effects of chemicals used in fracking and their likelihood of polluting groundwater. For example, Colburn et al. (2011) revealed that of the 632 documented chemicals found in 944 fracking fluids, 353 had been assigned a Chemical Abstract Service (CAS) tracking number, 75% of which can damage the skin, eyes and respiratory and gastrointestinal systems, at least 40% can affect the brain and central nervous system, immune system and cardiovascular system, and 25% can cause cancer.

However, the systematic study of groundwater and its hydrogeology are themselves fraught with sometimes insurmountable hurdles, not the least of which is any attempt to understand the fluid dynamics and chemical behaviour of interactions occurring between 300 to 1,000 m underground. The fact that industry has not always been entirely forthcoming about the chemical components of fracking fluids or other waste streams generated by CSG only adds to society's concerns and suspicions (Gottlieb, 2013). What role environmental remediation interventions, such as soil washing and partitioning, might play in addressing organic and inorganic contaminant species in CSG wastewater and sediments, should they exist, has yet to be examined.

Washing contaminated soils and sediments has been well documented (Abumaizar and Smith, 1999; Dermont et al., 2008; Mann, 1999) and is particularly common in mine site rehabilitation (Moutsatsou et al., 2006). Soil washing is an *ex-situ* remediation process which involves mixing solids with freshwater in a low-impact agitator, thereby liberating or "partitioning" contaminants from soil. Once the washing cycle has occurred, treated solids are settled and the so-called "flushing solution" decanted for discharge or recycling, in theory leaving behind clean soil. This practice has become more widespread in commercial remediation applications in the last ten years, particularly in Europe where the greatest expertise resides, however soil washing is rarely undertaken in Australia.

Soil washing is based on the principle that an organic or inorganic contaminant will bind to the soil's fine particles, which in turn bind to coarser particles, and thus by separating the course fraction from the fine fraction using agitated fresh water, contaminants along with the using agitated fresh water, contaminants along with the finest soil fractions can be partitioned from the soil, thereby liberating contaminants from the course fraction. This process then allows the separated flushing solution to be treated and discharged or recycled in further washing cycles, leaving behind a remediated course soil fraction which can be reused as backfill or, as is intended in CSG applications, integrated into beneficial reuse programs to rehabilitate disturbed land. Some soil washing techniques also use surfactants or other chemical additives to aid in the liberation of contaminants (Chu and Chan, 2003; Torres et al., 2012), and novel agitation mechanisms, including jet reactions and ultrasonic washing (Feng et al., 2001), have been deployed.

However, to this author's knowledge, soil washing has never been applied to CSG sediments, and fundamental questions such as ease of application, sediment treatment outcomes, role of surfactants, and other basic scientific data are lacking from the literature. Moreover, even the most basic information related to the chemical and physical properties of CSG waste sediments are all but non-existent. For these reasons, the present study asked four research questions: 1) What are the chemical properties of CSG dam sediments; 2) does soil washing (with or without surfactants) reduce contaminant species (if any) in CSG dam sediments to a point where sediments can be reused as "clean soil"; 3) what are the chemical properties of flushing solutions after partitioning; and 4) do flushing solutions generated from the washing of CSG sediments meet standards for water reuse as livestock drinking water or irrigation water? Exploring methods which might allow for beneficial reuse of sediment as clean soil is aligned to the Queensland Government's priority to decommission CSG evaporation dams, although the government only identifies "creating saleable salt", injecting sediments into underground voids, or disposing it to regulated landfill in their "range of options" (2012: 5).

METHODS

Two x 33 kg samples of CSG evaporation dam sediments were collected from dam sites in the Bowen Basin, central Queensland. The two sediment samples were labelled 1 and 2, with sample 1 collected from about 500 mm under water in an operational dam, and sample 2 from a decommissioned, dewatered dam in which sediments had dried and oxidized naturally; sample 1 therefore was an un-oxidized sediment, whereas sample 2 was an oxidized and naturally weathered sediment. Visual inspection of sample 1 indicated a fine grained (1.0 to 2.0 mm), spadable, greyish sediment with brown-grey stratigraphy (possibly sand) with no obvious smell of hydrocarbons or phenols; on a scale of 0 to 5, with 5 being the most objectionable, the sample's odour was rated 0. Inspection of sample 2 also indicated a fine-grained, spadable, grayish sediment with a blue-grey stratigraphy, but this sample appeared to contain fibrous vegetable matter (possibly grass); similarly, sample 2 had no obvious smell of hydrocarbons or phenols, but a sweet odour like potatoes, which was rated between 0 to 1.

In order to answer research question 1, 1 kg subsamples of both samples 1 and 2, labelled 1A and 2A, were sent to a certified laboratory in Brisbane for analysis of the following: pH; electrical conductivity (EC); total calcium (Ca), magnesium (Mg), potassium (K), sodium (Na), and chloride (Cl); total alkalinity (the sum of bicarbonate and carbonate alkalinity); exchangeable or soluble calcium, magnesium, potassium, and sodium; cation exchange capacity (CEC is the number of exchangable cations that the sediment is capable of holding, at a given pH, which are available for exchange with the flushing solution); sodium adsorption ratio (SAR is a measure of the suitability of water for use in agricultural irrigation as determined by concentrations of solids dissolved in the water); total metal concentrations of arsenic (As), cadmium (Cd), chromium (Cr), copper (Cu), lead (Pb), mercury (Hg), nickel (Ni), and zinc (Zn); leachable metal concentrations of the same heavy metals; benzene, toluene, ethylbenzene, and xylene (BTEX); organochlorine pesticides; polycyclic aromatic hydrocarbons (PAH); total petroleum hydrocarbons (TPH); and ammonia.

The two samples were divided into two x 2 kg subsamples, labelled 1B and 1C for sample 1 and 2B and 2C for sample 2. In order to answer research question 2, these four subsamples were subjected to the following soil washing regimes. Subsample 1B was mixed with 5.0 L of fresh water without a surfactant (ratio 1:2.5 sediment:water) for 15 minutes using a slow electric mixer; the solids were separated by passing the mixture through a 100 micron geotextile membrane and retaining the flushing solution as (a); subsample solids were rewashed a second time with another 5 L of fresh water for 15 minutes using the same method; the sediment-water mixture was again dewatered using the geotextile membrane and the flushing solution decanted and retained as (b); the solid fraction of subsample 1B was retained for analysis; flushing solutions (a) and (b) were mixed together and retained for analysis as FS1B.

The second subsample 1C was washed with 4.0 L of fresh water using 40 ml (1%) of sodium lauryl sulfate ($CH_3[CH_2]_{11}OSO_3Na$) as a surfactant (ratio 1:2 sediment:water) for 15 min; the solids were separated by passing the mix through a 100 micron geotextile membrane and retaining the flushing solution as (a); subsample solids were rewashed a second time with another 4.0 L of fresh water for 15 min using the same method, but without the surfactant; the sediment-water mix was again dewatered using the geotextile membrane and the flushing solution decanted and retained as (b); the solid fraction of subsample 1C was retained for analysis; flushing solutions (a) and (b) were mixed together and retained for analysis as FS1C. Subsamples 2B and C were subjected to the same soil washing regimes as subsamples 1B and 1C, and resulted in flushing solutions FS2B and FS2C.

For the purposes of this study, the term "clean soil" referred to washed CSG dam sediments which conform to these criteria: 1) the sediments would be comparable to profiles of clean "background soils" derived from samples taken in the vicinity of the evaporation dams where samples 1 and 2 were collected (determined by data supplied to the author by the CSG dam operator); and 2) contaminants in the sediments would be at or below levels set by the Australian Standard AS-4454 for soil suitable as a "soil conditioner" (Dorahy et al., 2007).

To answer research question 3, flushing solutions which resulted from these four soil washing regimes (that is, FS1B, FS1C, FS2B and FS2C) were analysed for the following: pH; EC; total dissolved solids (TDS); total Ca, Mg, K, Na, sulfate (SO_4), chloride (Cl), and flouride (Fl); hardness; hydroxide alkalinity, total carbonate alkalinity, total bicarbonate alkalinity, and total alkalinity; and total As, Cd, Cr, Cu, Pb, Hg, Ni, and Zn. It should be noted that because saline water matrices are prone to analytical interference due to the presence of high levels of TDS (the NATA-certified laboratory used in this study applies a 2,000 mg/L cut-off limit for TDS) and individual cations and anions; for this reason, flushing solution subsamples FS1C, FS2B and FS2C required dilution prior to

Table 1. Chemical analysis of sediment sample 1 before soil washing (1A) and after two soil washing methods (1B and 1C), with percent changes between 1A and 1B and between 1A and 1C, background soil, and AS-4454.

Parameter	1A	1B	Percent change between 1A and 1B	1C	Percent change between 1A and 1C	Background soil	AS-4454
pH	9.8	10.0	—	10.0	—	8.2	5.0-7.5
EC (mS/cm)	2.1	0.46	78↓	0.2	90↓	0.84	No limit
Ca (mg/kg)	<10†	<10†	—	<10†	—	7,320	No limit
Mg (mg/kg)	<10†	<10†	—	<10†	—	2,680	No limit
K (mg/kg)	80	10	88↓	<10†	100↓	1,390	No limit
Na (mg/kg)	4,090	620	85↓	370	91↓	120	<10,000
Cl (mg/kg)	5,400	200	97↓	50	99↓	<10†	No limit
Total alkalinity as $CaCO_3$ (mg/kg)	1,320	1,180	11↓	673	50↓	230	TBD‡
Exchangeable Ca(meq/100 g)	12	10.5	13↓	8.1	34↓	22.8	No limit
Exchangeable Mg (meq/100 g)	2.6	1.5	43↓	1.0	62↓	6.2	No limit
Exchangeable K(meq/100 g)	0.8	0.3	63↓	0.1	88↓	1.0	No limit
Exchangeable Na (meq/100 g)	9.6	2.5	74↓	1.2	88↓	<0.1	No limit
CEC (meq/100 g)	25	15	40↓	10	59↓	30	No limit
SAR (meq/100 g)	189	77	60↓	50	74↓	—	No limit
As (mg/kg)	7.0	5.0	39↓	<5.0†	100↓	6.0	20
Cd (mg/kg)	<1.0†	<1.0†	—	<1.0†	—	<1.0	3.0
Cr (mg/kg)	20	14	30↓	12	40↓	20	400
Cu (mg/kg)	17	9.0	48↓	6.0	65↓	22	200
Pb (mg/kg)	11	9.0	19↓	<5.0†	100↓	19	1.0
Hg (mg/kg)	<1.0†	<1.0†	—	<0.1†	—	<1.0	60
Ni (mg/kg)	12	6.0	50↓	4.0	67↓	17	200
Zn (mg/kg)	24	10	59↓	7.0	71↓	19	250
Average percent change for K, Na and Cl	—	—	90↓	—	97↓	—	—
Average percent change for metals	—	—	41↓	—	74↓	—	—

† Below level of detection; ‡ to be determined, because AS-4454 does not have a limit on total alkalinity but requires a subjective judgment based on pH.

analysis. To answer research question 4, this study used ANZECC (2000) guidelines for livestock drinking water and irrigation water.

RESULTS

Tables 1 and 2 present the analytical findings of subsamples 1A and 2A related to research question 1. As no evidence of leachable metals, BTEX, pesticides, PAH, TPH or ammonia was found in either subsample 1A or 2A, these data have not been included in Tables 1 or 2, with all parameters below the limits of detection of 0.1 mg/L for leachable metals, 0.2 mg/kg for BTEX and ammonia (as NH_3), 0.5 µg/kg for pesticides, 0.004 µg/kg for PAH, and 3.0 mg/kg for TPH.

As shown in Table 1, subsample 1A had significantly higher pH, EC, and concentrations of sodium and chloride than background soil, although calcium, magnesium and potassium were significantly lower than background soil; most other variations between 1A and background levels follow from these concentrations of sodium and chloride, for example total alkalinity. Metal concentrations were almost identical in both subsample 1A and background soil.

Similarly, subsample 2A had a significantly higher pH and EC, and higher concentrations of sodium and chloride than background soil, although Ca, Mg and K were significantly lower than background soil, as was the case for sample 1. Metal concentrations were almost identical in both sample 2A and background soil. When AS-4454 standards are applied to subsamples 1A and 2A, these CSG sediments would not be considered acceptable for soil conditioning or on-site rehabilitation due to higher than allowable pH in both subsamples, and lead in 1A and lead and Na in 2A.

Results for soil washing regimes 1B and C are shown in Table 1 as compared with both the unwashed subsample 1A and background soil and AS-4454 guidelines. In all cases except pH, all parameters decreased when comparing 1A to 1B and 1A to 1C, with

Table 2. Chemical analysis of sediment sample 2 before soil washing (2A) and after two soil washing methods (2B and C), with percent changes between 2A and B and between 2A and C, background soil, and AS-4454.

Parameter	2A	2B	Percent change between 2A and B	2C	Percent change between 2A and C	Background soil	AS-4454
pH	10.0	10.1	—	10.1	—	8.1	5.0-7.5
EC (mS/cm)	9.4	1.4	86↓	1.7	82↓	0.17	No limit
Ca (mg/kg)	10	<10†	100↓	<10†	100↓	6,600	No limit
Mg (mg/kg)	<10†	<10†	—	<10†	—	1,840	No limit
K (mg/kg)	340	40	89↓	60	83↓	980	No limit
Na (mg/kg)	16,100	1,900	88↓	3,150	81↓	170	<10,000
Cl (mg/kg)	19,900	1,670	92↓	3,150	85↓	<10†	No limit
Total alkalinity as $CaCO_3$ (mg/kg)	6,290	1,620	75↓	2,080	66↓	345	TBD‡
Exchangeable Ca (meq/100 g)	14.7	14.6	0	14.9	2↑	24.6	No limit
Exchangeable Mg (meq/100 g)	0.8	0.7	13↓	0.8	0	5.3	No limit
Exchangeable K (meq/100 g)	0.9	0.5	45↓	0.8	12↓	0.6	No limit
Exchangeable Na (meq/100 g)	7.3	3.9	47↓	7.8	7↑	0.1	No limit
CEC (meq/100g)	24	20	17↓	24	2↑	30.7	No limit
SAR (meq/100g)	931	173	82↓	255	83↓	—	No limit
As (mg/kg)	<5.0†	<5.0†	—	<5.0†	—	<5.0†	20
Cd (mg/kg)	<1.0†	<1.0†	—	<1.0†	—	<1.0†	3.0
Cr (mg/kg)	11.1	4.0	64↓	6.0	46↓	6.0	400
Cu (mg/kg)	10	<5.0†	100↓	5.0	50↓	11	200
Pb (mg/kg)	10	<5.0†	100↓	6.0	40↓	10	1.0
Hg (mg/kg)	<1.0†	<1.0†	—	<1.0†	—	<1.0†	60
Ni (mg/kg)	6.0	<2.0†	100↓	2.0	66↓	6.0	200
Zn (mg/kg)	21	<5.0†	100↓	8.0	62↓	13	250
Average percent change for Ca, K, Na and Cl	—	—	92↓	—	87↓	—	—
Average percent change for metals	—	—	93↓	—	53↓	—	—

† Below level of detection; ‡ to be determined.

total K, Na and Cl decreasing by an average of 90% for 1B and 97% for 1C (neither sample contained Ca or Mg). Similarly, total metals decreased by an average of 41% for 1B and 74% for 1C, indicating that the addition of a surfactant in soil washing improved metal removal efficiencies from CSG dam sediments. However, lead only decreased by 19% in 1B although by 100% in 1C; while the <5.0 parts per million observed for 1C may mean that lead concentrations were below the AS-4454 guideline, with a detection limit of 5.0 mg/kg and a guideline of <1.0 mg/kg, it is not possible to determine whether lead levels in 1C actually met the clean soil criterion or not. The pHs of washed sediments from both 1B and C were the same as 1A, and do not meet the criteria for clean soil when compared to either background soil or AS-4454 guidelines.

Results for soil washing regimes 2B and C are shown in Table 2 as compared with both the unwashed sub-sample 2A and background soil and AS-4454 guideline. In all cases except pH and exchangeable Ca, NA and CEC, which increased negligibly, analytes decreased when comparing results of 2A to B and 2A to C, with total Ca, K, Na and Cl decreasing by an average of 92% for 2B and 87% for 2C (neither sample contained Mg). Similarly, total metals decreased by an average of 93% for 2B and 52% for 2C, indicating that while soil washing removed metals from sediment, the addition of a surfactant did not improve metal removal efficiencies, which was the opposite finding to sample 1. Also in reverse of sample 1, lead decreased by 100% in 2B but only 40% for 2C; for reasons noted above, it is unclear whether lead levels in 2B met the clean soil criterion or not due to limitations of detection in instrumentation. The pHs of washed sediment from both 2B and C were the same as 2A, and do not meet the criteria for clean soil when compared to either background soil or AS-4454 guidelines (Table 3).

SAR is used to determine the suitability of soil to

Table 3. Analysis of flushing solutions resulting from soil washing regimes FS1B, FS1C, FS2B and FS2C, compared to ANZECC guidelines for livestock drinking water and irrigation water.

Parameter	FS1B	FS1C	FS2B	FS2C	Livestock drinking water	Irrigation water
pH	8.9	9.1	9.1	9.2	‡	6.0-9.0
EC (mS/cm)	1.1	5.5	7.8	26	‡	0.95- 12.2*
TDS(mg/L)	754	3,610	5,070	16,900	<2,000	‡
Ca (mg/L)	<1.0†	2.0	3.0	4.0	1,000	‡
Mg (mg/L)	<1.0†	<1.0†	<1.0†	<1.0†	‡	‡
K (mg/L)	4.0	21	25	105	‡	‡
Na (mg/L)	223	1,180	1,730	7,300	‡	‡
SO_4 (mg/L)	21	23	23	72	1,000	‡
Cl (mg/L)	223	1,280	1,770	8,220	‡	<170-750**
Fl (mg/L)	<10†	<10†	<10†	13	‡	‡
Hardness (mg/L)	<1.0†	5.0	7.0	10	‡	‡
Hydroxide alkalinity (mg/L)	<1.0†	<1.0†	<1.0†	<1.0†	‡	‡
Carbonate alkalinity (mg/L)	52	210	315	998	‡	‡
Bicarbonate alkalinity (mg/L)	316	420	919	2,130	‡	‡
Total alkalinity (mg/L)	368	630	1,230	3,120	‡	‡
As (mg/L)	0.02	0.06	0.07	0.22	0.5	0.1-20#
Cd (mg/L)	0.0006	0.0005	0.0006	0.002	0.01	0.01-0.05#
Cr (mg/L)	0.39	0.28	0.25	0.86	1.0	0.1-1.0#
Cu (mg/L)	0.47	0.30	0.29	0.99	0.5	0.2-5.0#
Pb (mg/L)	0.35	0.39	0.34	1.4	0.1	2.0-5.0#
Hg (mg/L)	<0.0001†	<0.0001†	<0.0001†	<0.0001†	<0.002	<0.002#
Ni (mg/L)	0.35	0.15	0.15	0.51	1.0	0.02-2.0#
Zn (mg/L)	0.56	0.42	0.37	1.3	20	2.0-5.0#

† Below level of detection; ‡ no guideline; * EC can range from a maximum of 0.95 mS/cm for salt intolerant, sensitive species, such as apple, to 12.2 mS/cm for highly salt-tolerant species, such as Rhodes grass; an EC above 12.2 mS/cm is considered too saline for plant growth (ANZECC, 2000, p. 4.2-7); ** Cl levels range from a maximum of 170 mg/L for sensitive species, such as citrus and grape, to 750 mg/L for more tolerant species, such as cauliflower (ANZECC, 2000, p. 4.2-10); # lower concentrations in the range indicate acceptable heavy metal concentrations in "long-term" applications of irrigation water, and higher concentrations indicate acceptable heavy metal concentrations in "short-term" applications of irrigation water.

receive irrigation water by measuring the relative concentrations of Na to Ca and Na to Mg; SAR is therefore an important measure when determining the suitability of soil for revegetation, but is less important when considering general land reclamation and rehabilitation. The absence of SAR data for background soil means there is no benchmark against which to compare washed sediments, although higher concentrations of Ca, Mg and K and lower concentrations of Na and Cl suggest background soil in the vicinity of the CSG dams is probably also saline, but not sodic. Using a general SAR calculation (Na/√Ca+Mg/2, from ANZECC, 2000: 4.2-4.6), the SAR of sample 1 background soil is 1.7 meq/100 g and sample 2 background soil is 2.6 meq/100 g, indicating that subsamples 1B, 1C, 2B and C are unsuitable as revegetation media, except in the most salt-tolerant cases. This conclusion is confirmed when comparing CEC and SAR to Cl concentrations in both cases.

Analyses of the flushing solutions generated from regimes FS1B, FS1C, FS2B and FS2C are shown in Table 3 and compared with ANZECC guidelines for livestock drinking water and irrigation water. The pH of all four flushing solutions are the same, and all parameters for use of the flushing solutions as drinking water for livestock, except TDS in subsamples FS1C, FS2B and FS2C, and lead in all subsamples, have been met. While levels of lead >0.1 mg/L may be hazardous to animal health, ANZECC guidelines state that acceptable levels of lead depend on the type and age of animals drinking the water, and therefore it may be possible that lead concentrations in subsamples FS1B, FS1C and FS2B (0.35, 0.39, and 0.34 mg/L respectively) meet acceptable livestock drinking water quality, although subsample FS2C at 1.49 mg/L is unlikely to do so.

Subsample FS1B meets, and subsamples FS1C, FS2B and FS2C almost meet, the pH criteria for irrigation water, however only subsample FS1B meets the Cl criteria for Cl-tolerant species, with subsamples FS1C, FS2B and FS2C significantly higher than allowable limits, and none of the subsamples meet the criteria for EC. ANZECC guidelines for heavy metals in irrigation water provide a range of acceptable limits, depending on whether the irrigation water is to be applied over a longer or shorter timeframe, with lower limits allowed when water is applied over a longer period and the reverse when applied over a shorter period. From these criteria, it can be seen: As concentrations in FS1B, FS1C and FS2B would be acceptable for long-term application, but FS2C would only be suitable for short-term application; Cr, Cu and Ni concentrations in all flushing solution subsamples would be acceptable only for short-term application; and Cd, Pb, Hg and Zn concentrations in all subsamples would be acceptable for both short- and long-term application of flushing solutions as irrigation water.

Conclusion

The Queensland Government has recognized that CSG sediments must be managed sustainably, and where possible its highest priority is to treat and beneficially reuse sediments; untreated disposal of sediments is considered a solution of "last resort" (Department of Environment and Heritage Protection, 2012). Therefore, any attempt to treat dam sediments using soil washing and partitioning would be considered worthwhile. However, in this study, soil washing was only partially effective in converting low-level contaminated dam sediments into "clean soil" (with the role of a surfactant indeterminate), and flushing solutions did not meet the standards required for use as either livestock drinking water or irrigation water.

In answer to research question 1, this study provides evidence that the CSG dam sediments tested in this experiment contain mostly "salts" and do not contain major contaminants of concern. For example, no evidence was found for the presence of benzene, PAHs (such as benzo(a)pyrene and naphthalene), petroleum hydrocarbons, pesticides or ammonia, and heavy metals were at generally low concentrations, although the presence of lead was concerning. The primary contaminants were inorganic cationic and anionic species, such as potassium, sodium and chloride.

From this data it is reasonable to conclude that soil washing went some of the way toward removing these contaminants from CSG dam sediments, thereby potentially allowing them to be re-classified as clean, useable soil suitable for on-site land reclamation, dam decommissioning and rehabilitation programs. While not all parameters conformed to the criteria set for this experiment, the washed sediments were generally significantly lower in most contaminant species as a result of soil washing. Therefore, research question 2 was partially answered in the affirmative, with the addition of a surfactant aiding removal efficiencies in one example but not in the other.

Similarly, this study provided evidence on the main contaminants in flushing solutions generated by soil washing of CSG dam sediments to answer research question 3, but showed the four flushing solutions generated in this experiment only partially met the criteria for use as livestock drinking water or irrigation water. These findings indicate that soil washing may be a viable method to recondition CSG dam sediments for on-site beneficial reuse when organic contaminants, such as petroleum hydrocarbons and benzene, are not present in the sediment profile. However, further research on the role of different surfactants, methods of soil washing (including volumes of water and water:sediment ratios), methods of mixing and agitation, energy consumption, and other physical and chemical parameters must be carried out before concluding that soil washing provides a viable, economical or operationally useful tool to treat CSG dam sediments and thereby convert them into a soil with beneficial reuse potential.

Conflict of Interest

The author has not declared any conflict of interest.

ACKNOWLEDGEMENT

The author has received assistance in the preparation of this study from Virotec Pty Ltd in Australia, with whom the author is affiliated.

REFERENCES

Abumaizar RJ, Smith EH (1999). Heavy metal contaminants removal by soil washing. J. Haz. Mat. 70(1-2):71-86.

Australian and New Zealand Environment Conservation Council (ANZECC)(2000). Guidelines for fresh and marine water quality, volume 1. Australian and New Zealand Environment Conservation Council (ANZECC), Artarmon, NSW.

Chambers M (2013). Export fears as gas wells fall short. The Australian, November 19, 2013.

Chu W, Chan KH (2003). The mechanism of the surfactant-aided soil washing system for hydrophobic and partial hydrophobic organics. Sci. Tot. Environ. 307(1-3):83-92.

Colburn T, Kwiatkowski C, Schultz K, Bachran M (2011). Natural gas operations from a public health perspective. Hum. Eco. Risk Pers. 17:1039-1056.

Commonwealth Scientific and Industrial Research Organisation (2012). Coal seam gas: Produced water and site management. Commonwealth Scientific and Industrial Research Organisation (CSIRO), Clayton South, Victoria.

Cubby B (2010). Origin stops coal seam gas drilling after chemicals found in water. The Sydney Morning Herald, October 21, 2010.

Department of Environment and Heritage Protection (2012). Coal seam gas water management policy. Energy Resources, Department of Environment and Heritage Protection, Brisbane, December 2012.

Department of Natural Resources and Mines (2014). Queensland's coal seam gas overview. Department of Natural Resources and Mines, Report No. B/Brochures/CSG/CC13-PET001, Queensland Government, Brisbane, January 2014.

Day S, Dell'Amico M, Fry R, Javanmard H (2013). Preliminary field measurements of fugitive methane emissions from coal seam gas production in Australia. 14th International Conference on Coal Science and Technology, Pennsylvania, USA, September 29 – October 3:1445-1456.

Dermont G, Bergeron M, Mercier G, Richer-Laflèche M (2008). Soil washing for metal removal: A review of physical/chemical technologies and field applications. J. Haz. Mat. 152(1):1-31.

Dorahy CG, Pirie AD, Pengelly P, Muirhead LM, Chan KY, Jackson M, Smith A, Emery T (2007). Guidelines for using compost in land rehabilitation and catchment management. Department of Environment and Climate Change and Department of Primary Industries, New South Wales Government, Sydney.

Egan G (2012). CSG waste water trucked to Ipswich. The Queensland Times, June 13, 2012.

EnergyQuest (2009). Australia's natural gas markets: Connecting with the world. Part One: Essay, EnergyQuest, Adelaide, SA.

Feng D, Lorenzen L, Aldrich C, Maré PW (2001). *Ex-situ* diesel contaminated soil washing with mechanical methods. Min. Eng. 14(9):1093-1100.

Gottlieb B (2013). Hydraulic fracturing: How great is the risk to health? Environmental Health Policy Institute, Physicians for Social Responsibility, June 2013.

Lloyd-Smith M, Senjen R (2011). Hydraulic fracturing in coal seam gas mining: The risks to our health, communities, environment and climate. National Toxics Network, Bangalow, NSW, July 2011.

Mann MJ (1999). Full-scale and pilot-scale soil washing. J. Haz. Mat. 66(1-2):119-136.

Moutsatsou A, Gregou M, Matsas D, Protonotarios V (2006). Washing as a remediation technology applicable in soils heavily polluted by mining-metallurgical activities. Chemosphere 63(10):1632-1640.

Torres LG, Lopez RB, Beltran M (2012). Removal of As, Cd, Cu, Ni, Pb, Zn from a highly contaminated industrial soil using surfactant enhanced soil washing. Phys. Chem. Earth. Parts A/B/C 37-39:30-36.

World Energy Council (2013). Coal: Strategic insight. World Energy Council, London, UK.

13

Soil organic carbon fractions as affected by land use in the Sourthern Guinea Savanna ecosystem of Adamawa State, Nigeria

G. Y. Jamala[1] and D. O. Oke[2]

[1]College of Agriculture, Ganye, Adamawa State, Nigeria.
[2]Federal University of Technology Akure, Ondo State, Nigeria.

This study was conducted to evaluate the effects of three land use management (natural forest, crop land and natural fallow) and soil depth on soil organic carbon fractions. The study was carried out in Toungo and Jada, Adamawa State, Nigeria. Three transects were cut 100 m apart in each of the study sites and four sampling plots of 20 m × 20 m in dimension were laid in alternate positions along each transect at 50 m interval. Soil samples were collected at depth of 0-15, 15-30, 30-45 and 45-60 cm and at three points along the diagonal of each plot using a 3 cm diameter soil auger. Soil contents of mineral-associated organic carbon fraction were higher than the particulate organic carbon pool in all land use management systems. Surface soil (0-15 cm) total organic carbon (TOC) content was highest under the natural forest (1.94%) and lowest in the cropland (1.46%). From the results it was noted that, natural forest had capacity for increasing soil organic carbon to an appreciable concentration. Fallow contribution to soil organic carbon was minimal. This could be attributed to several factors such as wind erosion, grazing, bush burning and cutting of vegetation for fuel wood. The contents of total, particulate and mineral-associated soil organic carbon was significantly influenced by land use management and soil depth. All the different land use types showed highest accumulation of the various carbon fractions in the surface layer (0-15cm). This high level of organic carbon stock in the surface layer could be due to the slow of mixing of the soil.

Key words: Land use, carbon fraction, forest, mineral, particulate organic carbon.

INTRODUCTION

Soil stores two or three time more carbon than that which exists in the atmosphere as CO_2 and 2.5 to 3 times as much as that stored in plants in the terrestrial ecosystem (Post et al., 2000; Houghton and Skole, 1990). Schimel et al. (2000) pointed out that the knowledge of the spatial distribution of soil organic carbon is an important requirement for understanding the role of soils in the global carbon system.

Soil organic matter is composed of different compartments which differ from each other in biochemical composition, biological stability and carbon turnover rates (Paustian et al., 1992). Both labile, mineral associated and humified soil organic fractions may show different susceptibility to land use and management effects. In cold and/or semiarid region, the carbon stocks in labile particulate soil organic carbon fraction constitute carbon (Chan, 1997; Franzluebbers and Arshad, 1997), and is the most affected by agricultural practices. A study conducted by Bayer et al. (2006) shows that particulate soil organic carbon usually ranges between 30 to 40% of total organic carbon in most of Brazilian soils. Hot and humid environment favours microbial activity that leads to intensive decomposition and humification of labile fractions (Bayer and Bertol, 1999).

The extent to which land use management influences soil organic matter (SOM) dynamics can be best evaluated by separating SOM into fractions. These include particulate organic carbon (POC), mineral-associated organic carbon (MOC) and the humified soil organic carbon (HA, FA and Humin). The POC consists of partly decomposed plant and animal residues with a rapid turnover; it is more responsive to management factors and it is believed to make a greater contribution to nutrient cycling (Janzen et al., 1992). Great stability of mineral-associated soil organic carbon in highly weathered soils, which may be important in maintaining and restoring soil quality, determines the soil potential for acting as an atmospheric CO_2 sink in tropical and subtropical regions (Parfitt et al., 1997).

Soil organic matter is an important factor in evaluating management system quality (Doran and Parkin, 1994). In warm and wet tropical and subtropical climatic regions, rapid soil organic matter (~ 58% Carbon) decline occurs under conventional management systems involving intensive soil disturbance (Tiessen et al., 1992; Parfitt et al., 1997). The soil organic matter degradation has negative effects mainly on cation exchange capacity, nutrient availability, aggregate stability and microbial activity (Bayer and Mielniczuk, 1999).

Several studies on soil organic matter dynamics focused on its absolute amount in the savanna ecosystem. It is important also to investigate the effects of land use management and soil depth on the labile (particulate) and stable (mineral-associated) soil organic matter fractions (Scholes et al., 1997). This study therefore, examines the effects of three land use management (natural forest, cultivated land and natural fallow) and soil depth on total soil organic carbon stocks and carbon stocks in labile and stable fractions.

MATERIALS AND METHODS

The study area

Adamawa State lies between latitudes 8° N and 11° N and longitude 11.5°and 13.5°E. The State shares border with Gombe State to the North, and Borno State to the North East, while to the West it is bordered with Taraba State as well as the Republic of Cameroon to the East. Adamawa State has a total area of 39,742.12 square kilometers. The study was carried out in and around Gumti sector, the northern half of Gashaka-Gumti National Park, covering Toungo and Jada local government areas of Adamawa State (Figure 1). The area received an annual rainfall of 1000–1600 mm. Rainfall distribution is unimodal, with much of the rain falling between April and November. Day time temperatures may drop below 18°C at higher altitudes and gradually rise to 40°C. The rainy season is followed by a dry season. During this period, the area comes under the strong influence of the hammattan (November and March) a dry dusty wind blows from Sahara Desert and temperatures may be significantly cooler (GGNP, 2010; Adebayo, 1999).

Site selection and sampling procedure

Soil samples were collected from three sites under different land use management: natural forest, crop land and a natural fallow. The forest consisted of southern guinea vegetation; the crop land had been under cultivation for approximately 27 years and the fallow land had been under fallow condition for about 7 years. The natural forest is strictly restricted as a National Park and hence no crop cultivation. The crop land is extensively cultivated with crops such as Maize, Sorghum, Cowpea, Arachis *hypogeal,* Yam and livestock grazing is usually carried out on the site. Fallow land was left under natural fallow, crops cultivated before fallow includes *Arachis hypogeal,* Yam, Maize, Sorghum and Cowpea. Livestock grazing and fetching of fuel wood also takes place.

Systematic strip sampling was employed in laying out the plots. Three transects were cut 100 m apart in each of the study site and four sampling plots of 20 × 20 m in dimension were laid in alternate positions along each transect at 50 m interval. Soil core samples were collected at a depth of 0-15, 15-30, 30-45 and 45-60 cm and at three points along the diagonal of each plot using a 3 cm diameter soil auger. Samples from each plot at each depth were bulked separately and the composite soil samples taken to the laboratory and analyzed for total, particulate, mineral-associated soil organic carbon and other physico-chemical soil properties.

Physical fractionation of soils into particulate and mineral size fractions

The process of soil physical fractionation was based on the methodology developed by Cambardella and Elliott (1992). 50 g of 2 mm soil sample was transferred into a 250 ml sample bottle and 105 ml distilled water added. Soil suspension was then washed through a 53 mm sieve and the coarse fraction was separated. The soil samples above the 53 μm sieve, were considered particulate SOM while, those that pass through the sieve was mineral associated SOM. The particulate (≥53 μm) and mineral associated (<53 μm) soil organic matter fractions were dried in an oven at 75°C, and analyzed for organic carbon using the wet oxidation method.

Organic carbon determination in the various fractions

Organic carbon was determined in the bulk soil, (≥53 μm) and (<53 μm) soil fractions as total organic carbon (TOC), particulate organic carbon (POC) and mineral organic carbon (MOC) respectively, using the wet oxidation method of Walkley and Black (1934).

Statistical analyses

Data collected were subjected to analysis of variance (ANOVA) procedure for Randomized Complete Block Design (RCBD). The land use types constituted the treatments while soil depths constituted the blocks. The Statistical analysis was performed with SPSS (V. 18), mean separation was performed with Duncan's Multiple Range Test (DMRT) at $p<0.05$.

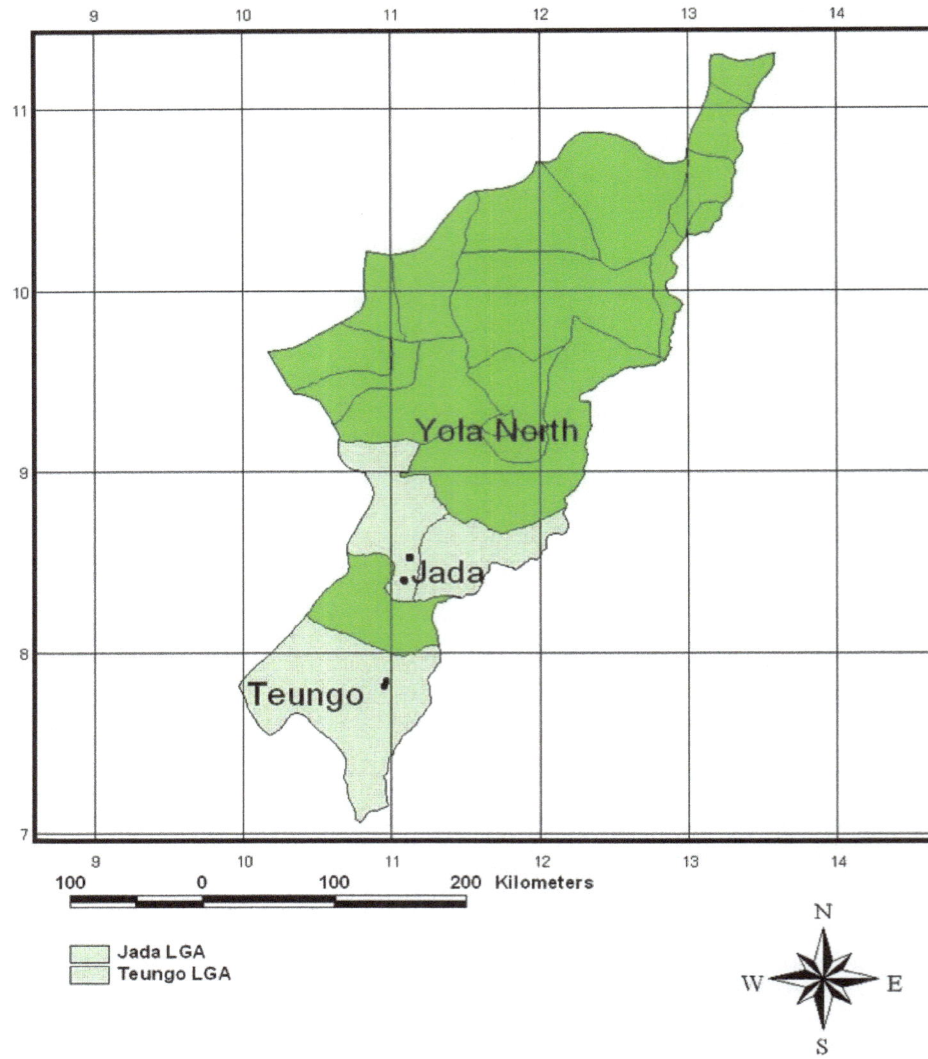

Figure 1. Map of Adamawa State, Nigeria, showing the study sites.

RESULTS AND DISCUSSION

Effect of land use management on total soil organic carbon (TSOC)

Table 1 presents the results of particle size distribution and soil texture of the various soil depths in the study sites. The soil texture in the natural forest and fallow land ranges from sandy loam to clay loam and loam to clay loam in the crop land. Compared to natural forest site, land cultivation led to a decrease in total soil organic carbon (Table 2). In the fallow land soil organic carbon status was lower than the values obtained in both natural forest and crop land. Total soil organic carbon in the natural forest was approximately 8 and 15% higher compared with crop land and fallow land respectively. This finding corroborate the report of Alexandra and Jose (2005), who reported that conversion of grassland and forest plantations to arable cropping results in the loss of 30% of the soil organic carbon originally present in the soil. This shows that continuous cultivation of these soils can accelerate depletion of the soil organic carbon content. Loss of soil organic matter can therefore, reduce soil fertility, degrade soil structure and lower water holding capacity, ultimately leading to desertification. Bationo et al. (2003) noted that one single most important biophysical constraint to food security in the West African savanna has been identified to be soil fertility degradation.

In the savanna agro ecosystem, the bush fallow remains one of the traditional methods that are used for soil fertility restoration through the buildup of soil organic matter. In this study it was noted that the contribution of seven years of fallow to the status of total soil organic carbon was minimal. This might be attributed to several factors such as livestock grazing, incidence of annual bush burning and cutting of vegetation for fuel wood which is more frequent in the study area.

Table 1. Particle size distribution and soil texture.

Site	Depth (cm)	Sand (%)	Silt (%)	Clay (%)	Texture
Natural forest	0-15	58.92	29.64	11.44	SL
	15-30	51.84	30.72	17.44	L
	30-45	39.84	18.72	41.44	CL
	45-60	34.2	19.64	46.16	CL
Crop land	0-15	39.28	42.92	17.80	L
	15-30	34.2	19.64	46.16	CL
	30-45	31.28	33.28	35.44	CL
	45-60	21.84	32.36	45.80	CL
Fallow land	0-15	71.28	19.28	9.44	SL
	15-30	65.28	15.28	19.44	SL
	30-45	44.92	17.64	37.44	L
	45-60	31.28	33.28	35.44	CL

NB: SL=Sandy Loam, L=Loam, CL=Clay Loam.

Table 2. Influence of land use management on total soil organic carbon (%).

Soil depth (cm)	Natural forest	Fallow land	Crop land
0-15	1.94^{a}	1.55^{b}	1.46^{c}
15-30	1.33^{a}	0.54^{c}	1.12^{b}
30-45	0.70a	0.54^{b}	0.55^{b}
45-60	0.70^{a}	0.34^{c}	0.54^{b}

Means with the same letter along the rows are not significantly different (Duncan's Multiple Range Test P< 0.05).

The increasing pressure on agricultural lands in the savanna is making it impossible to continue with bush fallow. Currently, fallow periods have either been shortened or totally replaced by continuous cultivation. In addition to these, most lands of the dry savanna regions are highly vulnerable to wind and water erosion resulting in soil degradation. In a 12-year fallow study, Bayer et al. (2000) found that, the increase in C and N stocks were minimal. Scherr (1999) pointed out that proper management of soil organic matter is important to food security and the protection of marginal lands.

Effect of land use on particulate and mineral-associated soil organic carbon fractions

The values obtained for mineral-associated soil organic carbon pool (the stable form) were relatively higher than the particulate soil organic carbon pool (the labile pool) (Table 3) in all land use management systems. This indicates that a high proportion of total soil organic carbon is made up of mineral-associated soil organic carbon pool in advanced humification form. The higher proportion of carbon in mineral-associated fractions than in labile fractions is probably due to climatic conditions favorable to organic matter decomposition (mainly labile fractions), as well as to physical and chemical stability of mineral-associated soil organic carbon. In tropical and subtropical soils, Parfitt et al. (1997) noted that variable charge mineral and soil organic matter interactions can promote a great soil organic matter protection against biological decomposition. Also, Bayer et al. (2001) found that the carbon pool in mineral-associated soil organic matter increased several times more than the particulate organic matter fraction and attributed it to greater physical recalcitrance of mineral-associated soil organic matter to biological decomposition.

The content of particulate organic carbon was observed to be low in almost all the land use management systems. This could be the results of increased rate of humification in these soils. Bayer et al. (2006) reported that the content of particulate soil organic carbon was lower than the mineral-associated soil organic carbon in the most Brazilian soils, this usually ranged between 30 and 40% of total soil organic carbon. According to Adejuyigbe et al. (2000), the particulate organic matter is a more labile fraction of the soil organic matter which is the most readily formed and when it decomposes, it serves as an important substrate for mineralization process in the soil. The level of this fraction therefore, could be an essential determinant of the fertility status of savanna soils. Bescansa et al. (2006) stated that,

Table 3. Influence of land use management on Particulate and Mineral-Associated Soil Organic Carbon.

Soil depth (cm)	Natural forest	Fallow land	Crop land
	Particulate soil organic carbon (%)		
0-15	0.77^{a}	0.68^{b}	0.50^{c}
15-30	0.62^{a}	0.24^{c}	0.44^{b}
30-45	0.39^{a}	0.24^{c}	0.31^{b}
45-60	0.39^{a}	0.11^{c}	0.31^{b}
	Mineral-associated soil organic carbon (%)		
0-15	1.15^{a}	0.87^{c}	0.95^{b}
15-30	0.70^{b}	0.29^{c}	0.75^{a}
30-45	0.31^{a}	0.23^{b}	0.23^{b}
45-60	0.31^{a}	0.29^{a}	0.23^{b}

Means with the same letter along the rows are not significantly different (Duncan's Multiple Range Test $P < 0.05$).

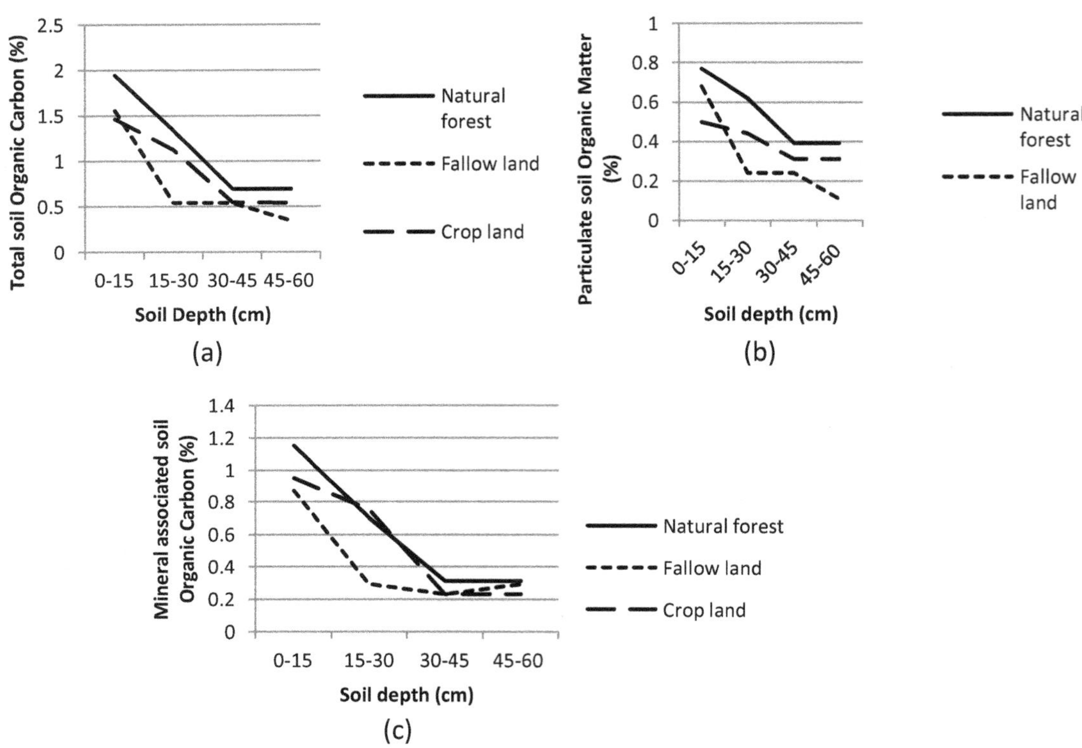

Figure 2. Changes in (a) total, (b) particulate and (c) mineral-associated organic carbon with soil depth in the various study sites.

particulate organic carbon is an accurate soil quality indicator despite soil differences, and therefore advocated that it should be considered when assessing the quality of different soil management practices.

In a study conducted by Chan (1997) and supported by Janzen et al. (1998) they pointed out that soil organic matter constituents ranged from labile compounds that mineralize during the first stage of decomposition to more resistant (mineral-associated soil organic carbon). They concluded that soil organic matter changes occur primarily in the labile fractions.

Changes in total, particulate and mineral-associated soil organic carbon with depth under the various land use management

The content of total, particulate and mineral-associated soil organic carbon was significantly influenced by soil depth (Figure 2). All the different land use types showed highest accumulation of the various carbon fractions in the surface layer (0 to 15 cm). This high level of organic carbon stock in the surface layer could be due to the slow of mixing of the soil and the litter layer formed as a result of leaf-fall from the trees.

In a 12-year experiment, Bayer et al. (2000) found that by the third year, the increase in carbon and nitrogen stocks were minimal and occurred only in the 2.5 cm top layer, however, by the 5th year, this effect had spread to the 7.5 cm depth. In the 9th and 11th year, carbon and nitrogen stocks increased through to 12.5 and 17.5 cm depths respectively. Soil organic carbon storage in deeper soil layer has been related to the development of roots systems (Pillon, 2000) and to the amount of above ground biomass addition on the soil surface (Burle et al., 2005) implying that the trees will normally improve in their respective organic carbon addition potentials, depending on the length of time and fallow period since the biomass increase with age. The organic carbon fractions were observed to decrease with depth. The top layer recorded the highest concentration of these fractions.

Conclusions

Land use management and soil depth influence contents of total, particulate and mineral-associated soil organic carbon fractions. Crop cultivation led to a decrease in total soil organic carbon, but the value was higher than the contents obtained in the fallow land. The values obtained for the contents of mineral-associated soil organic fraction were relatively higher than the particulate organic carbon pool in all land use management systems. This indicates that high proportion of total soil organic carbon composed of mineral-associated soil organic carbon. From the results it was noted that, natural forest had capacity for increasing soil organic carbon to an appreciable concentration. Fallow contribution to soil organic carbon was small. This could be attributed to several factors such as wind erosion, grazing, bush burning and cutting of vegetation for fuel wood.

The contents of total, particulate and mineral-associated soil organic carbon was significantly influenced soil depth. All the different land use types showed highest accumulation of the various carbon fractions in the surface layer (0 to 15 cm). This high level of organic carbon stock in the surface layer could be due to the slow of mixing of the soil.

REFERNCES

Adejuyigbe CO, Agbeyangi OA, Adeoye GO (2000). Soil organic matter fraction and micro arthropod population under agroforestry trees with contrasting litter quality. Proc. of the 26th Ann. Conf. Soil Sci. Soc. Nigeria. Ibadan, Oct 30th – Nov. 3rd 2000. pp. 203 – 206.

Adebayo AA (1999). Climate I &II (Sunshine, Temperature, Evaporation and Relative humidity: In Adebayo A.A, Tukur, AL (eds) Adamawa in Maps, Paraclete publishers Yola-Nigeria. pp. 15 – 30.

Alexandra B, Jose B (2005). The importance of soil organic matter. Food and Agriculture Organization of the United Nations (FAO) Soils Bulletin.

Bationo A, Mokwunye U, Vlek PLG, Koala S, Shapiro BI (2003). Soil fertility management for sustainable land use in the West African Sudano – Sahelian zone. In: MP. Giuchuru et al., (eds) Soil fertility management in Africa: A Regional perspective. Academy Science Publishers. pp. 247 – 292.

Bayer C, Mielniczuk J, Giasson E, Martin–Neto L, Pavinato A (2006). Tillage Effects on particulate and mineral – Associated organic matter in Two Tropical Brazilian soils. Soil Sci. Plant Anal. 37:389 – 400.

Bayer C, Martin-Neto L, Mielniczuk J, Pavinato A, Dieckow J (2006). Carbon sequestration in two Brazilian Cerrado soils under no-till. Soil Till. Res. 86:237– 245.

Bayer C, Martin-Neto L, Mielniezuk J, Pillon CN, Sangoi L (2001). Changes in organic matter fractions under subtropical No–Till cropping systems. Soil Sci. Soc. Amer. J. 65:1473 – 1478.

Bayer C, Mielniczuk J, Amado TJC, Martin-Neto L, Fernandes SV (2000). Organic matter storage in a sandy clay loam Acrisol affected by tillage and cropping systems in southern Brazil. Soil Till Res. 54:101–109.

Bayer C, Bertol I (1999). Características químicas de um cambissolo húmico afetadas por sistemas de preparo, com ênfase à matéria orgânica. Rev. Bras. Ci Solo., 23: 687-694.

Bayer C, Mielniczuk J (1999). Dinâmica e função da matéria orgânica. In: Santos GA, Camargo FAO (Eds.), Fundamentos da matéria orgânica do solo: ecossistemas tropicais e subtropicais. Porto Alegre: Genesis pp. 9-26.

Bescansa P, Imaz MJ, Virto I, Enrique A (2006). Particulate organic matter carbon as a soil quality indicator in semiarid Agricultural soils. Geophys. Res. Abstracts 8: 06328.

Burle ML, Mielniczuk J, Focchi S (2005). Effect of cropping systems on soil chemical characteristics, with emphasis on soil acidification. Plant Soil 190:309-316.

Cambardella CA, Elliott ET (1992). Particulate soil organic matter changes across a grassland cultivation sequence. Soil Sci.Soc.Am.J.,56:777-783.

Chan KY (1997). Consequence of changes in particulate organic carbon in vertisols under pasture and cropping. Soil Sci. Soc. Am. J. 61:1376 – 1382.

Doran JW, Parkin TB (1994). Defining and assessing soil quality. In: Doran JW, Coleman DC, Bezdicek DF, *et al.* (Eds). Defining soil quality for a sustainable environment. Madison: Soil Sci. Soc.Amer. pp. 3-21.

Franzluebbers AJ, Arshad MA (1997). Particulate organic carbon content and potential mineralization as affected by tillage and texture. Soil Sci. Soc. Amer. J. 61: 1382-1386.

GGNP (2010). A Wild World of Wonders! Taraba State, Nigeria. p. 50.

Houghton RA, Skole DL (1990). Carbon in the Earth as Transformed by Human Action (eds Turner BL et al.,), Cambridge University Press, Cambridge. p.143.

Janzen HH, Campbell CA, Izaurralde RC, Ellert BH, Juma N, McGill WB, Zentner RP (1998). Management effect on soil C Storage on the Canadian prairies. Soil Till. Res., 47:181 – 195.

Janzen HH, Campbell CA, Brandt SA (1992).Light-fraction organic matter in soils from long-term crop rotations. Soil Sci. Soc. Amer. J. 56:1799-1806,

Parfitt RL, Theng BKG, Whitton JS (1997). Effects of clay minerals and land use on organic matter pools. Geoderma 75:1-12.

Paustian K, Parton WJ, Persson J (1992). Modeling soil organic matter in organic amended and nitrogen-fertilized long-term plots. Soil Sci. Soc. Amer. J. 56: 476–488.

Pillon CN (2000). *Stocks and Quality of soil Organic Matter as Affected by No –Till Cropping System*, doctorate thesis; Federal University of Rio Grande do Sul: Porto Alegre, p. 248.

Post WM, Kwom WM (2000). Soil carbon sequestration and land use change: Process and Potential. Global change Biol. 6: 317-327.

Scherr S (1999). Soil degradation. A threat to developing – country food security by 2020. IFPRI Discussion paper 27.

Schimel D, Melillo J, Tian H, McGuire AD, Kicklighter D, Kittel T, Rosenbloom N, Running S, Thornton P, Ojima D, Parton W, Kelly R, Sykes M, Neilson R, Rizzo B (2000). Contribution of increasing CO_2 and climate to carbon storage by ecosystems of the United States. Science 287: 2004-2006.

Scholes MC, Martin R, Scholes RJ, Parsons D, Winsted E (1997). No and N_2O emissions from savanna soils following the first simulated rains of the season. Nutr. Cycl. Agroecosyst. 48:115–122.

Tiessen H, Salcedo IH, Sampaio EVSB (1992). Nutrient and soil organic

matter dynamics under shifting cultivation in semiarid northeastern Brazil. Agric Ecosyst Environ, 38: 139-151.
Walkley A, Black IA (1934). An examination of the Degtjarelf method for Determing soil organic Matter and proposed modification of the Chromic Acid Titration method. Soil Sci., 37: 29-38:51-61.

Assessing soil amendment potentials of *Mucuna pruriens* and *Crotalaria juncea* when used as fallow crops

Imoro Abukari Ziblim, Gakpo Selorm Paul and Khan Aikins Timothy

Department of Range and Wildlife Management, Faculty of Renewable Natural Resources, University for Development Studies, Nyankpala Campus. Tamale, Ghana.

This research was carried out to assess the soil amendment potentials of *Mucuna pruriens* and *Crotalaria juncea* when used as fallow crops. The research made use of randomized complete block design (RCBD) with three treatments replicated three times. The study revealed higher mean levels of percentage nitrogen, phosphorus, potassium, percentage organic carbon and soil pH after incorporation as compared to the results obtained before planting (Control). The results revealed significant difference in the levels of the above mentioned parameters before planting and after incorporation in soils under both *M. pruriens* and *C. juncea* but there were no significant differences in soils under the natural regeneration. Although there were no significant differences in the results obtained after incorporation between *M. pruriens* and *C. juncea*, *M. pruriens* recorded higher mean levels of percentage nitrogen, phosphorus, percentage organic carbon and soil pH except in potassium where the mean added level was higher in *C. juncea*. Though there was no significant difference in the soil amendment potentials of *M. pruriens* and *C. juncea*, *M. pruriens* is recommended to be used as a fallow crop since it has an added advantage of serving as a cover crop to control soil erosion.

Key words: Fallow, soil amendment, potential, *Mucuna pruriens*, *Crotalaria Juncea*.

INTRODUCTION

Fallow management and improvement is as old as agriculture itself. Fallows are still intrinsic parts of many tropical farming systems. While fallow is commonly referred to as resting period for agricultural land between two cropping cycles during which soil fertility is restored, it has more roles than just fertility restoration. Fallow functions include weed control and the interruption of pest and disease cycles. They provide cash income in times of immediate need and help to balance food supply (Styger, 1999). Physical properties such as hydraulic conductivity, bulk density, total porosity and aggregate stability all decrease as soil organic matter falls, as a result of the duration and intensity of cropping (Agboola, 1994). The degradation and irreversible destruction of soils have reached alarming proportions and every year, 5 to 7million ha of agricultural land is lost worldwide (Steiner, 1996). To maintain and improve soil fertility overtime, the incorporation of leguminous cover crops in farming systems has been advocated (Osei-Bonsu and Buckles, 1993).

Inclusion of these legumes into the farming systems has been shown to contribute nutrients to the soil through nitrogen fixation and through recycling of residues which also contribute organic matter into the soil (Anthofer, 1999).

They also play an important role in weed suppression and soil conservation. Increasing pressure on land is leading to shorter fallow periods, and soil fertility can no longer be regenerated naturally. This has led to a gradual destruction of the natural resource base, declining soil fertility and yields, and hence, reduced farm income. The integration of the leguminous cover crops into the existing farming systems to address these problems has been very successful because of the high agronomic benefits achieved from the use of these legumes (Loos et al., 2001). Soil is considered as number one growth medium in crop cultivation in the tropics. Therefore, there is the need to conserve this medium to enhance ideal plant growth and development. Different types of soils contain different plant nutrients depending on the parent material from which the soil was developed (Brady and Weil, 2008). Some soils contain high levels of plant nutrients, while others contain very low levels.

Farmers in Ghana normally leave their farmland to fallow after harvesting their crops. Most of these lands when left fallow become exposed to erosion and bushfires as a result; the expected fallowing which allows the decomposition of dead plants into the soil is not achieved. In areas where the soil is not burnt nor exposed to erosion, plants growing on the land may not be important species which can help replenish the nutrient levels of the soil. *Mucuna* and *Crotalaria* are leguminous plants which fix nitrogen into the soil. *Mucuna* and *Crotalaria* also serve as effective cover crops in controlling erosion and excessive loss of soil moisture. *Mucuna pruriens* and *Crotalaria juncea* are known to improve soil fertility but as to which one is most effective in improving the nutrient level is not known. It is for this reason that this research was conducted to ascertain soil improvement potentials of *M. pruriens* and *C. juncea*. This will help determine which of these plants will serve as an effective fallow crop in addressing problems like nutrient inadequacies and soil erosion in fallow lands to increase crop production.

MATERIALS AND METHODS

Study area

The study was carried out on the research plot of the Faculty of Renewable Natural Resources, University for Development Studies, Nyankpala Campus. Geographically, the campus lies within latitudes 9° 25'N and longitude 0° 58' W. Nyankpala is 16 km (10 miles) away from Tamale, the capital of the region with an altitude of 183 m above sea level .

The area experiences moderate annual rainfall pattern. Rains start in May and end in October. Monthly rainfall is 131.7 mm, the dry season which follows the wet season is normally between the months of November and April. During the dry season, the daily average temperature is as low as 18°C and maximum temperature reaches 44°C at the peak.

The vegetation of the area is guinea savannah (grassland) type with extensive shrubby grassland and few sparse populations of trees. The common trees found include *Azadirachta indica, Parkia biglobosa, Adansonia digitata, Tectona grandis* and *Senna siamea*. The soil type is mainly Alfisols with clayey loam and coarse-grave (SARI, 2004). The people of the area are mostly farmers growing crops like maize, rice, sorghum millet, yam, groundnut and soybeans.

Experimental design

The research made use of randomized complete block design (RCBD) with three (3) treatments namely: *Mucuna* (T_1), natural regeneration (T_2) and *Crotalaria* (T_3) replicated 3 times giving a total of 9 experimental units. Nine (9) 2 × 2 m bed was prepared with 0.5 m spacing between each plot.

Soil sampling procedure

Soil samples were collected before planting using soil auger at the depth of 0 to 20 cm from the 9 sample plots. The soil samples were collected at three different spots on each plot and were bulked to have a composite sample for each plot. The soil samples were air-dried at the laboratory and the initial determination serving as the control for the various levels of nitrogen, phosphorus, potassium, pH and organic carbon were determined. *M. pruriens* and *C. juncea* were planted using the vines and seeds, respectively. Natural regeneration of weeds was allowed to occur to serve as the second treatment (T_2). The *M. pruriens*, *C. juncea* and the natural regeneration were incorporated into the soil 10 weeks after planting and allowed to decompose for 4 weeks. The final soil samples were then collected from the 9 plots and the levels of nitrogen, phosphorus, potassium, pH and organic carbon determined again.

Laboratory analysis

The samples were air-dried, ground, sieved to pass 2 mm size screen and stored in sealed plastic for laboratory analysis. pH was determined using an electronic pH meter (Bates, 1954). Nitrogen was determined using the Kjeldahl method (Jackson, 1962; Black, 1965). The OC content of the soil was then determined using the Walkley-Black method (Walkley and Black, 1934).

Two grams of air-dried soil sample was put into a 50 ml test tube and 14 ml extraction solution was added. The mixture was shaken for a few minutes and filtered. 1 ml of standard serves 2 ml boric acid and 3 ml of mixed reagent were homogenized. The solution was allowed to stand for 1 h for the blue color to develop and the absorbance measured on spectrometer at 720 nm. Standards were also prepared and the absorbance recorded was used to calculate for the phosphorus concentration (Jackson, 1962).

Five grams of air-dried soil sample was weighed into a shaking bottle. 50 ml ammonium acetate solution (1 N, pH = 7) was added and shaken for 2 h. The solution was filtered through No. 42 Whatman filter paper. The concentration of potassium in the soil extract was determined using the flame photometer.

Statistical analysis

Data collected from the laboratory analysis were analyzed with the general statistics software (GENSTATS). Analysis of variance (ANOVA) was conducted to determine significance of treatment effects. Microsoft Office Excel was used to present results in charts with error bars showing differences in the treatments.

RESULTS

Effects of *M. pruriens, C. juncea* and natural regeneration on soil nitrogen percentage

Figure 1 shows the comparative effect of *Mucuna*, *Crotalaria* and natural regeneration on the soil nitrogen. There were significant ($p < 0.05$) increases in the levels of nitrogen added to the soil by both *Mucuna* and *Crotalaria* but there was no significant ($p > 0.05$) change in the natural regeneration.

Effects of *M. pruriens, C. juncea* and natural regeneration on available soil phosphorous (Mg/kg)

From the ANOVA test, the levels of phosphorous added to the soil by *Mucuna* and *Crotalaria* before planting and after incorporation were significant ($p < 0.05$) but no significant ($p > 0.05$) change in the natural regeneration (Figure 2).

Effects of *Mucuna pruriens, Crotalaria juncea* and natural regeneration soil potassium (Mg/kg)

Figure 3 shows substantial increase in the levels of potassium when *Mucuna* and *Crotalaria* were incorporated into the soil but only a slight increase was observed in the control.

Effects of *Mucuna pruriens, Crotalaria juncea* and natural regeneration on soil pH (H_2O)

The incorporation of *Mucuna* and *Crotalaria* biomasses into the soil resulted in a significant ($p < 0.05$) increase in the pH levels of the soil (Figure 4). Though nominally, *Mucuna* recorded slightly higher pH value after incorporation than *Crotalaria*, there was no significant ($p > 0.05$) difference between the two crops.

Effects of *Mucuna pruriens, Crotalaria juncea* and natural regeneration on soil organic carbon percentage

Figure 5 shows the comparative effect of *Mucuna*, *Crotalaria* and natural vegetation on soil organic carbon. There was significant ($p < 0.05$) difference in the levels of organic carbon added to the soil by both *Mucuna* and *Crotalaria* but there was no significant ($p > 0.05$) change in the natural regeneration.

DISCUSSION

Effects of *M. pruriens, C. juncea* and natural regeneration on soil nitrogen

The results obtained after incorporation of *M. pruriens* and *C. juncea* into the soil showed a significant ($p < 0.05$) difference between the levels of nitrogen obtained before planting and after incorporation. The increase in the mean levels of nitrogen in both *Mucuna* and *Crotalaria* could be attributed to the fact that leguminous plants are capable of fixing nitrogen through the formation of root nodules and can often provide the required quantity of nitrogen for crop production as reported by Thiessen-Martens et al. (2005). The increase in nitrogen levels could also be as a result of increasing decomposition and subsequent mineralization of biomass. However, the mean percentage nitrogen fixed by *M. pruriens* (0.13%) was higher as compared to *C. juncea* (0.09%). This could be as a result of different responses of the crops to climatic and biological processes in the study area. Cherr et al. (2006) reported that *C. juncea* retained large amounts of nitrogen and slowed decomposition due to the structural portioning of dry matter and minerals to the stem.

The results further showed that there was no significant difference in the level of nitrogen before and after incorporation of the natural regeneration. The low percentage of nitrogen in the natural regeneration soil could be as a result of the natural vegetation utilizing the already available nitrogen and not being able to replace them as they are not legumes.

Effects of *M. pruriens, C. juncea* and natural regeneration on soil phosphorous

The results obtained from the research showed that *M. pruriens* and *C. juncea* improves phosphorous availability. There was significant ($p < 0.05$) differences between the before planting and after incorporation of the plant materials of *Mucuna* and *Crotalaria*. This could be attributed to the fact that *M. pruriens and C. juncea* could access poorly available forms of phosphorous by altering the rhizosphere chemically through the exudation of compounds such as organic-acid anions (Ae et al., 1991). It was also reported by Vanlauwe et al. (2000) that in Nigeria, the legumes *M. pruriens* (velvet bean) and *C. juncea* has been found to increase the availability of phosphorus in soil when it was incorporated into the soil. Although both *Mucuna* and *Crotalaria* had a significant effects on the soil available phosphorous, the mean phosphorous added by *M. pruriens* (3.36 Mg/kg) was higher as compared to that added by *C. juncea* (2.94 Mg/kg). Also, the result obtained from the soil samples from the natural regeneration showed that there was no significant differences in the mean values of available phosphorous before and after incorporation. This could be as a result of leaching and heavy uptake of phosphorus from the soil by the natural regeneration.

Effects of *M. pruriens, C. juncea* and natural regeneration on soil potassium

The results indicate a significant ($p < 0.05$) difference in

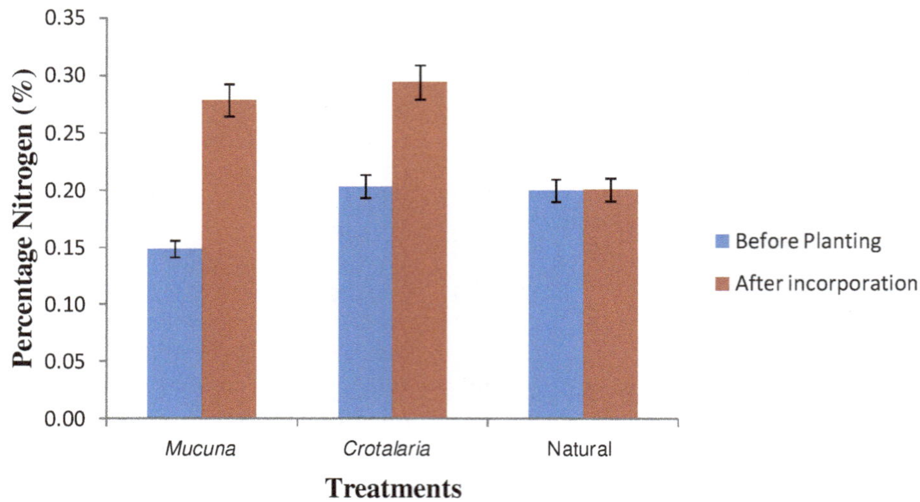

Figure 1. Effect of *M. pruriens*, *C. juncea* and natural regeneration on soil nitrogen.

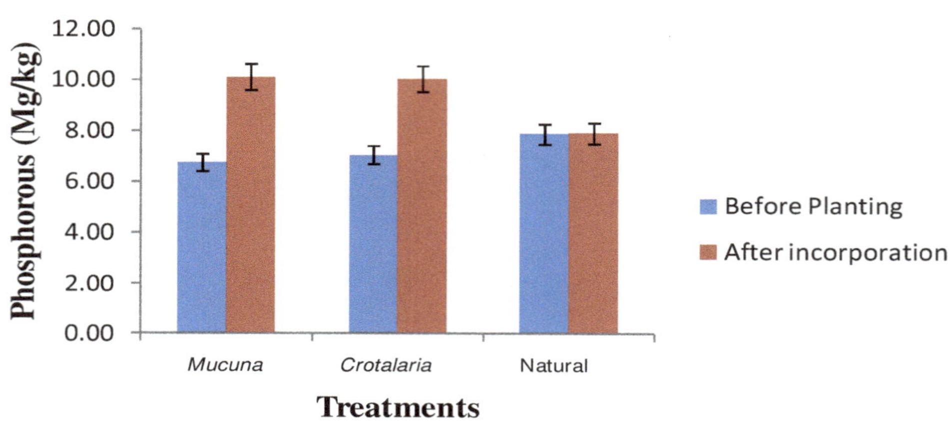

Figure 2. Effect of *M. pruriens*, *C. juncea* and natural regeneration on soil phosphorous.

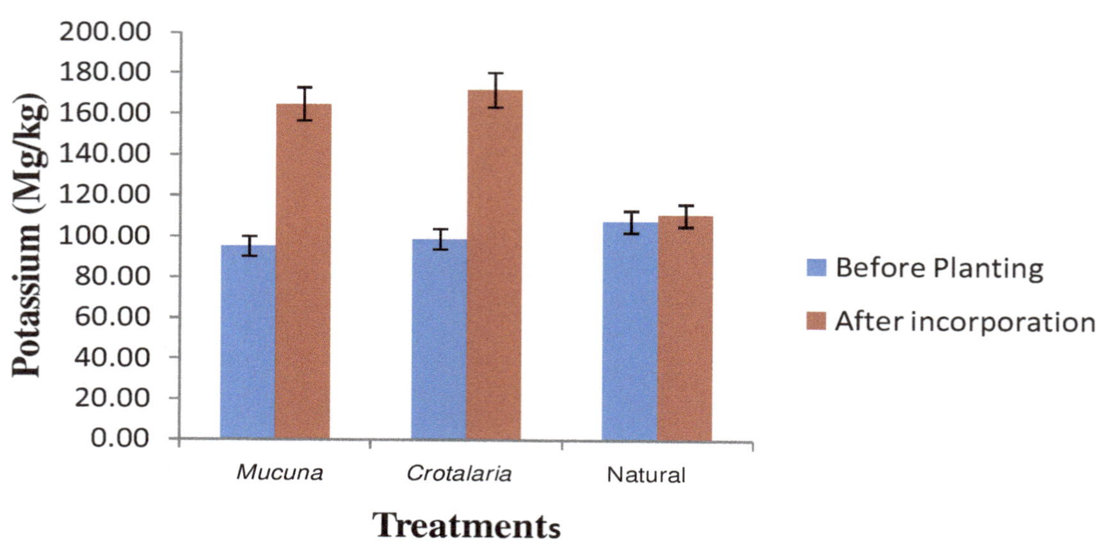

Figure 3. Effect of *M. pruriens*, *C. juncea* and natural regeneration on soil potassium.

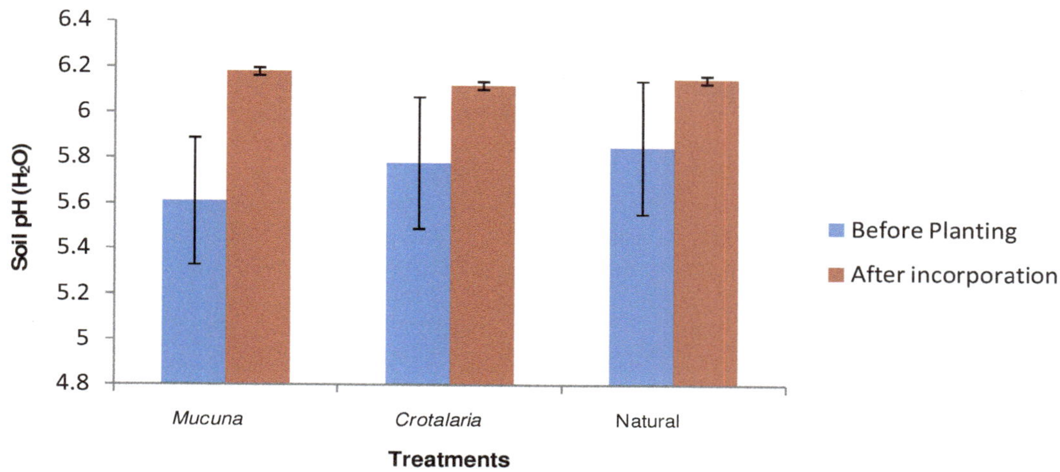

Figure 4. Effects of *M. pruriens*, *C. juncea* and natural regeneration on soil pH (H_2O).

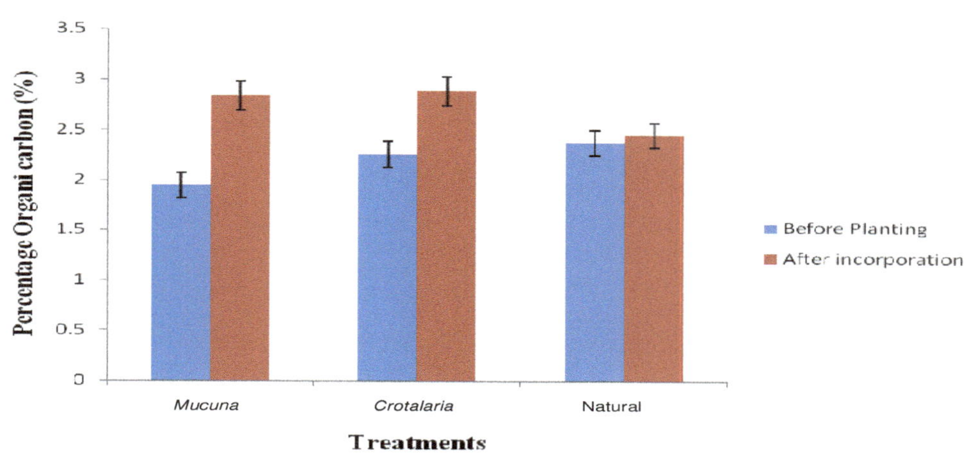

Figure 5. Effects of *M. pruriens*, *C. juncea* and natural regeneration on soil organic carbon.

the levels of potassium between *M. pruriens* and *C. juncea* but showed no significant difference in the natural regeneration before planting and after incorporation of the plant materials in the soil. This confirms the findings of Feichtinger et al. (2004) that, the use of *M. pruriens* and *C. juncea* can conserve the moisture, builds up organic matter and improves the properties of terrestrial soil and activity of microbes supporting the mineralization rate and release of potassium in greater proportion in the soil. Although the incorporation of *Mucuna* and *Crotalaria* led to increase in potassium levels, *C. juncea* (89.71 Mg/kg) added a higher amount of available potassium as compared to *M. pruriens* (69.51 Mg/kg). This could be attributed to the rooting system of the two plants where *Crotalaria* has deeper roots which could take up more residual potassium than that of *Mucuna*. The insignificant change in potassium levels in the natural regeneration may be due to lack of cover which promotes leaching of the available potassium. It could also be attributed to the fact that the natural vegetation takes up these available K without replenishing them.

Effects of *M. pruriens, C. juncea* and natural regeneration on soil pH

The result showed a significant ($p < 0.05$) difference between the soils samples under *M. pruriens, C. juncea* and the natural regeneration. The soils from the legume incorporation showed a higher mean pH levels with *M. pruriens* also indicating a higher mean pH level (6.18) as compared to *C. juncea* (6.12). The higher pH value of soil under both *M. pruriens* and *C. juncea* could be due to the plant biomass incorporated into the soil resulting in the conversion of organic element to inorganic form which increases soil pH (Weischt and Claviecles, 1993). Shoko and Tagwira (2005) also noted that legumes such as *Crotalaria* and *Mucuna* have the potential to improve soil pH and the availability of organic matter exchangeable bases. The pH largely controls plant nutrient availability and microbial reaction. This is evident in the study carried out when the amount of nitrogen, phosphorus, and potassium was high. This is in line with the findings of Hussein (1997), who reported that most of the essential plant nutrients are readily available in between pH 5.2 and 8.8. However, the low pH of the soils under the control may be due to the fact that the organic matter was not able to decompose as a result of changes in temperature with time.

Effects of *M. pruriens, C. juncea* and natural regeneration on soil organic carbon

The result revealed significant ($p < 0.05$) difference in levels of organic carbon in *M. pruriens* and *C. juncea* between before planting and after incorporation of their plant materials but no significant change was observed in the natural vegetation. The increase in the levels of organic carbon in the *Mucuna* and *Crotalaria* treated soils could be as a result of the plant residues added to the soil and hence are the primary source of organic matter which can release organic carbon into the soil. The increase in soil organic carbon by these legumes could also be as a result of *C. juncea* and *M. pruriens* in that they develop an organic layer above the mineral soil and this layer generally improves physical (soil aeration, water retention, resistance to erodability etc.) and biological properties (build-up of soil micro-organisms, nutrients etc.) which enhances the productivity of the soil and thereby increasing organic carbon levels of the soil. This is in line with Juo et al. (1995) who reported that the contribution of root biomass from legumes such as *C. juncea* and *M. pruriens* to the nutrient and organic matter of the soil is believed to be important for soil fertility maintenance and carbon sequestration, as the below-ground biomass forms a substantial proportion of the total biomass in an ecosystem. The low level or insignificant change in the organic carbon in the natural regeneration could be as a result of low amount of organic materials on the natural regeneration plot before incorporation.

Conclusion

The results obtained from the research showed an appreciable or significant effect of *M. pruriens* and *C. juncea* on the soil as it led to an increase in the chemical parameters studied in terms of the levels of nitrogen, phosphorous, potassium, organic carbon and soil pH after incorporating the organic materials into the soil. The natural regeneration plot did not show any significant change in the levels of the various parameters assessed. Apart from the natural regeneration plot where there was no significant change in the results obtained, *M. pruriens* and *C. juncea* showed significant change in the levels of nitrogen, phosphorous, potassium, pH and organic carbon after incorporation of the plant materials. *M. pruriens* added higher levels of nitrogen, phosphorous, organic carbon and pH to the soils as compared to *C. juncea* which showed a higher level only in soil potassium *M. pruriens*. This could mean that nominally *M. pruriens* contributes more to soil nutrient improvement when used as fallow crops than *C. juncea* but both crops could be used as fallow crops.

REFERENCES

Ae N, Arihara J, Okada K (1991). Phosphorus uptake mechanisms of pigeon pea grown in alfisols and vertisols. In Phosphorus Nutrition of Grain Legumes in the Semi-Arid Tropics. (Eds C. Johansen K, Lee K.Sahrawat KL) (ICRISAT Publishing, India). pp. 91-98.

Agboola AA (1994). A recipe for continuous arable crop production in the forest zone of western Nigeria, *In*: Alternatives to Slash-and-Burn Agriculture: Proceedings of 15th International Soil Science Congress, Acapulco, Mexico Sanchez PA, Van Houten, H, Eds., International Centre for Research in Agroforestry and International Society of Soil Sciences, Nairobi, Albany, NY: Delmar Thomson Learning. P. 106.

Anthofer J (1999). Experimentation and experiences of farmers with *Mucuna* cover crop system in Ghana. Proceedings of the 2nd CIEPCA workshop, October 1999, Cotonou. ASHS Press, Arlington, VA. Appl. Soil Ecol. 27:197-210.

Bates RG (1954). Electronic pH determination. John Willey and Sons, Inc. New York. pp. 87-92.

Black CA (Ed) (1965). Methods of Soil Analysis. Agron. no. 9 part 2. Am. Soc. Agron. Madison WI. pp. 1379-1396.

Brady N, Weil R (2008). The Nature and Properties of Soils. 14th ed. Prentice Hall, Upper Saddle River, NJ. P. 975. ISBN 13-978-0-13-227938-3.

Cherr CM, Scholberg JMS, McSorley R (2006). Green manure as nitrogen source for sweet corn in a warm-temperate environment. Agron. J. 98:1173-1180.

Feichtinger F, Erhart E, Hartl W (2004). Net nitrogen (N)-mineralization related to soil organic matter (SOM) pools. Plant Soil Environ. 50:273-276.

Hussein J (1997). Introduction to soil management. Module 2. University of Zimbabwe Press. Harare. pp.67-89.

Jackson ML (1962). Soil Chemical Analysis. Prentice Hall, New York. P. 498.

Juo ASR, Dabiri A, Franzluebbers K (1995). Acidification of a kaolinitic Alfisol under continuous cropping and nitrogen fertilization in West Africa. Plant Soil 171:245-253.

Loos H, Zschekel W, Schiller S, Anthofer J (2001). Integration of *Mucuna* improved fallow systems into cropping systems of the Brong Ahafo Region. Paper presented at international conference organized by the Soil Science Society of Ghana, 26 February-3 March, Tamale.

Osei-Bonsu P, Buckles D (1993). Controlling weed and improving soil fertility through the use of cover crops: experience with *Mucuna* sp. In Benin and Ghana. West Afr. Farm. Syst. Res. Bull. 4:2-7.

Savannah Agricultural Research Institute (SARI) (2004). Annual report for the year 2004, Nyankpala. P. 22.

Shoko MD, Tagwira F (2005). Assessment of the potential of vegetable and grain soybeans as break crops in sugarcane production systems in Zimbabwe. Proc. Afr. Crop Sci. Soc. 7:59-65.

Steiner KG (1996). Causes of Soil degradation and approaches to sustainable soil management. GTZ Weikersheim: Margraf.

Styger E (1999). Indigenous fruit trees of Madagascar. Potential components of agroforestry systems to improve human nutrition and restore biological diversity. Agroforestry Syst. 46:289-310.

Thiessen-Martens JR, Entz MH, Hoeppner JW (2005). Legume cover crops with winter cereals in southern Manitoba: Fertilizer replacement values for oat. Can. J. Plant Sci. 85(3):645-648.

Vanlauwe B, Nwoke OC, Diels J, Sanginga N, Carsky RJ, Deckers J, Merckx R (2000). Utilization of rock phosphate by crops on a representative toposequence in the Northern Guinea savanna zone of Nigeria: response by *Mucuna pruriens*, Lablab purpureus and maize. Soil Bio. Biochem. 32:2063-2077.

Walkley A, Black IA (1934). An examination of the Degtyareff method for determining soil organic matter and a proposed modification of that chronic acid titration matter. Soil Sci. 37:29-38.

Weischt W, Claviecles CN (1993). The persisting ecological constraints of tropical agriculture. The Longman group UK. Limited. pp. 104-110.

The effects of soil properties and vegetation cover on the sedimentation of forest roads

Aidin Parsakhoo[1] , Majid Lotfalian[2] and Hamid Jalilvand[2]

[1]Department of Forestry, Faculty of Forest Science, Gorgan Agricultural Sciences and Natural Resources University, Golestan Province, Gorgan City, Iran.
[2]Department of Forestry, Faculty of Natural Resources, Sari Agricultural Sciences and Natural Resources University, Mazandaran Province, Sari City, Iran.

This study was conducted to investigate the interaction of soil properties and vegetation cover on soil loss from forest road prism. Rainfall simulations were carried out on road surface, fillslope and cutslope. Runoff and sediment samples were collected every 4 min and then total soil loss was measured. Samples from top soil were randomly collected for analysis and grouping soil into A, B and C. Results showed that on cutslope, the highest soil loss was detected for soil group B, where sediment concentration in runoff was 21.83 g L^{-1} and vegetation cover was 0 to 30%. On fillslope, the highest soil loss was detected for soil group C, where sediment concentration in runoff was 18.07 g L^{-1} and vegetation cover was 10 to 40%. On road surface, the highest soil loss was detected for soil group A, where sediment in runoff was 8.99 g L^{-1} and vegetation cover was 2 to 5%.

Key words: Forest road, sediment, soil groups, rainfall simulation, principal component analysis (PCA).

INTRODUCTION

Globally, soil erosion is one of the most important environmental problems which threaten soil and water resources (Cerdà, 2007). An Iran, due to the arid climate and the human land use is affected by soil erosion and land degradation. Soil properties is one of the main parameters that affect runoff and soil loss processes. Water erosion of forest soil is naturally very low and can be neglected, but after clear cutting for road construction, the bare soil is exposed to rainfall and other erosive agents (Croke et al., 2001; Foltz et al., 2009). Road erosion is also found during the construction of the roads and the embankments and soil development only took place at a low rate. So, under this situation the inherent sensitivity of forest soil to erosion is appeared especially on steep slopes (Jordán and Martínez-Zavala, 2008). If a forest road is planned and constructed without considering region erodibility, the engineers would be faced to high cost to solve road sedimentation problem. Road is the main type of transportation system in Hyrcanian forests of Iran (Parsakhoo et al., 2009). Most of these roads traversed through hilly and mountainous areas. These mountainous roads experience numerous hazards such as water erosion and landslides which cause disruption, injuries and losses to life and economy

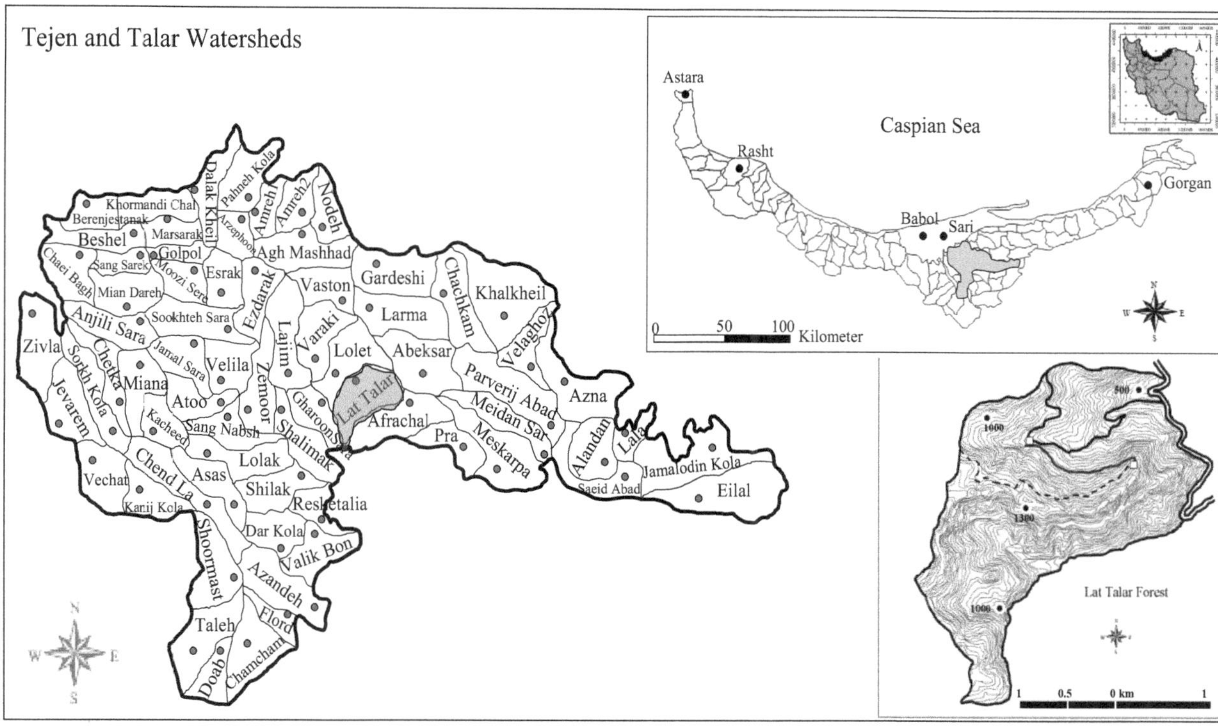

Figure 1. Geographical position of the study area.

(Mohammadkhan et al., 2011).

Kavian et al. (2010) reported that soil initial moisture, percentage of soil organic matter, bulk density and sand percent are most effective factors in runoff generation. Also, the results illustrated that percentage of soil organic matter, soil initial moisture and silt percent affect on soil erosion. Najafian et al. (2010) showed that both life form and vegetation cover significantly influenced the amount of runoff and sedimentation. Greatest runoff was found for forbs while grasses had lowest amount of runoff. Also bare ground and 100% cover showed the greatest and lowest amount of runoff, respectively. The concentration of sediment for shrubs was significantly greater than grasses and in 100% cover it was significantly lower than 50% and bare ground. Scientific understanding of factors influencing soil loss from forest road can effectively be used by managers in the design and sustainable management of road networks (Pappas et al., 2008; Olive and Marion 2009). Road prisms, including cutslopes, road surfaces, and fillslopes, can be important contributors of sediment to streams in forested watersheds (Bold et al., 2009; Morgan, 2005).

Interaction between soil properties and vegetation cover is one of the main factors responsible for soil loss (Fu et al., 2009). Monitoring of soil loss using rainfall simulator and runoff plots was cost-effective and provided valuable information about soil erosion caused by road construction. Thus, this study was conducted to (i) grouping soil properties for different parts of forest roads and (ii) investigate the effects of the interaction of some soil properties and vegetation cover on runoff and soil loss from road prism and forest lands located in Hyrcanian zone of Iran.

MATERIALS AND METHODS

Description of the study area

Lat Talar forest within the watershed number of 71 in Hyrcanian zone of Iran was selected as the study area. The research site has a total area of 3801 ha. The mean annual temperature of the study area is 15°C. Minimum altitude is about 300 m and the maximum 1650 m. The mean slope was 50%. The location of the research is in latitude 53° 9′ 40″ to 53° 13 ′55″ E and longitude 36° 12′ 55″ to 36° 15′ 45″ N. The region has very moist to mid moist and cold climate with a mean annual precipitation of 800 mm. The bedrock is typically marl, marl lime and limestone with a soil texture of loam and clay loam. Forest stands are dominated by *Fagus orientalis* Lipsky, *Carpinus betulus* and herbaceous species of *Carex silvatica*, *Buxus hyrcanus, Berachypodium silvaticum*, *Ruscus hyrcanu*s, *Phyllitis scolopendrium, Rubus hyrcanus* L. and *Polypodium auidinum.* Forest roads in the studied area are used by truck, motorcycle, Landrover and nissan. The traffic density is 5 vehicles per day. *Philonotis marchica* (Hedw.) Brid. and *Rubus hyrcanus* L. are dominant species on cutslopes and fillslopes of road (Figure 1).

Data collection

Totally, 38 rainfall simulations were randomly done all over the area

Figure 2. Rainfall simulator and runoff collection by the gauge.

for road surface (10 replications), fillslope (13 replications) and cutslope (15 replications). A portable single nozzle rainfall simulator simulated rain with drop size of 3 mm for duration of 20 min at intensity of 32.4 mm h^{-1} (Figure 2). Water with a mean temperature of 23°C was rained using a Schlick r86510 nozzle mounted 2 m above the ground onto a squared area of 0.48 m^2 that is bordered by a steel structure on cut slope and road surface. Runoff and sediment samples were collected every 4 min by gauge and then total soil loss was measured. The clear water was separated from deposited sediment. Sediment was air-dried for a week in temperature of 30°C. Besides, it was oven-dried at 105°C for at least 2 h and then weighted using digital balance. 25 samples of the top soil (0 to 20 cm deep) were randomly collected from cutslope, fillslope and road surface. Soil texture was determined by the Bouyoucos hydrometer method. Lime percentage (T.N.V or $CaCO_3$) was measured using the NaOH titration method. Soil organic carbon was determined using the Walkley–Black technique. Soil bulk density was measured using a relation of soil dry weight to volume of sampling cylinder (484 cm^3) (Koc et al., 2008). Runoff coefficient is calculated as Equation 1 (Figure 2).

$$RC = \frac{R_h}{P_h} \times 100 \quad (1)$$

Where, RC is runoff coefficient in %, R_h is runoff height in mm and P_h is rainfall height in mm.

Statistical analysis

In grouping process of plots, each table of soil data sets was analyzed using principal component analysis (PCA) in PC-ORD software. Then, data were statistically analyzed using GLM procedure in SAS program. Student Newman Keuls (SNK) multiple comparison test at probability level of 5% was used to compare means among groups and diagram designed by Excel software.

RESULTS AND DISCUSSION

Effects of vegetation cover and soil groups on soil loss from cutslope

Figure 3 shows the PCA correlation circle for the soil variables on cutslope. Soil properties on cutslope were classified into three groups based on PCA. The purpose of this analysis is to define groups of items based on their similarities. The properties of each group have been recorded in Table 1.

The runoff and soil loss in soil groups B and C was significantly more than that of soil group A (Table 2). A soil with a high percentage of silt and clay particles has a greater erodibility than a sandy soil under the same

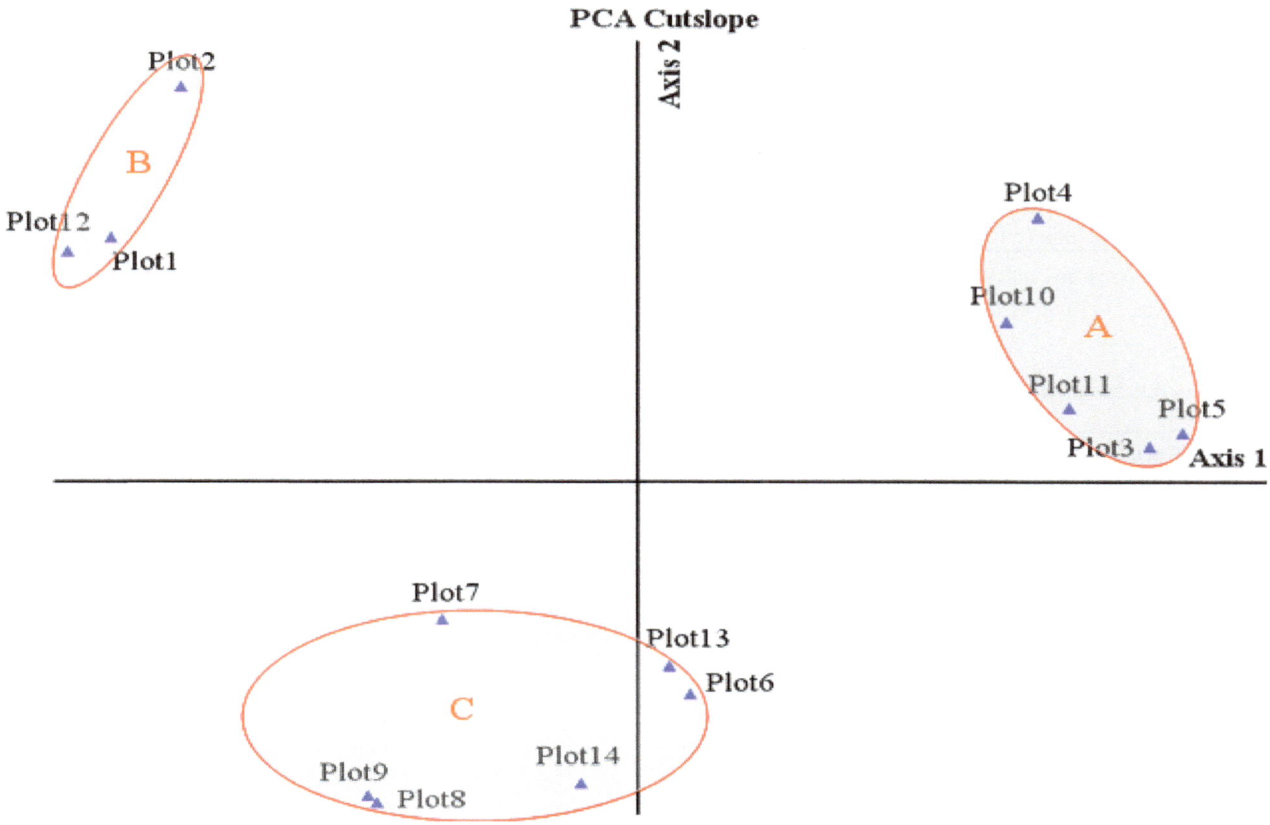

Figure 3. PCA analysis of soil on cutslope.

Table 1. Characteristics of the soil groups in cutslope.

Soil groups	Bulk density (g cm^{-3})	Moisture (%)	Organic matter (%)	$CaCO_3$ (%)	Clay (%)	Silt (%)	Sand (%)
A	1.2±0.1	10.7±0.9	1.96±0.1	36.5±5.6	12.3±2.3	13.1±1.2	74.6±1.1
B	0.9±0.1	25.3±0.2	4.64±0.3	18.4±0.1	23.3±4.5	27.4±2.5	49.2±5.7
C	1.3±0.1	15.9±3.7	0.82±0.1	23.2±1.8	24.7±4.0	25.3±3.4	49.9±5.6

Table 2. Effect of soil groups on runoff and sedimentation from cutslope.

Soil groups / Variables	A	B	C	ANOVA, *p*
Runoff coefficient (%)	22.22^{B}	41.99^{A}	48.05^{A}	0.0122
Time to runoff (s)	240.67^{A}	169.80^{B}	182.67^{B}	0.0500
Sediment concentration (g L^{-1})	11.27^{A}	19.32^{A}	10.39^{A}	0.5083
Total soil loss (g m^{-2} h^{-1})	81.70^{A}	252.70^{A}	152.80^{A}	0.5690

In a row, means with the same letter are not significantly different based on Student-Newman-Keuls test; Alpha=0.05.

conditions. Although all soils are potentially susceptible to water erosion, silts, silt loams and loams are most at risk because there is no adhesion forces among silt particles, (Ziegler et al., 2001).

The low-vegetated road cutslope increase their runoff coefficient. The soil loss was lower than 90 g m^{-2} h^{-1} on

Table 3. Effect of vegetation cover on runoff and sedimentation from cutslope.

Vegetation cover / Variables	0-30%	30-45%	ANOVA, *p*
Runoff coefficient (%)	42.69^{A}	24.31^{B}	0.0420
Time to runoff (s)	180.13^{B}	233.33^{A}	0.0405
Sediment concentration (g L^{-1})	16.29^{A}	10.85^{A}	0.3858
Total soil loss (g m^{-2} h^{-1})	216.09^{A}	80.58^{B}	0.0325

In a row, means with the same letter are not significantly different based on Student-Newman-Keuls test, Alpha=0.05.

Table 4. Interaction of soil and vegetation cover on runoff and sedimentation from cutslope.

Soil groups	A		B	C		ANOVA, *p*
Plant cover	0-30%	30-45%	0-30%	0-30%	30-45%	
Runoff coefficient (%)	20.8	22.5	48.0	44.1	33.4	0.348
Time to runoff (s)	162.0	256.4	182.7	182.7	118.0	0.004
Sediment concentration (g L^{-1})	11.8	11.2	21.8	10.4	9.3	0.433
Total soil loss (g m^{-2} h^{-1})	84.9	78.4	296.3	152.8	81.0	0.370

the high vegetated road cutslope and greater than 200 g m^{-2} h^{-1} on the low-vegetated ones in study area (Table 3). Vegetation provides a protective layer or buffer between the atmosphere and the soil by means of its canopy, roots, and litter components (Mohammad and Adam, 2010). Many studies have emphasized the importance of vegetation cover on soil loss. On cutslope, the highest soil loss was detected for soil group B, where sediment concentration in runoff was 21.83 g L^{-1} and vegetation cover was 0 to 30% (Table 4). There was no slope class of 30 to 45% in soil group B.

Effects of vegetation cover and soil groups on soil loss from fillslope

Figure 4 and Table 5 show the PCA correlation circle for the soil variables and the properties of each group on fillslope, respectively. Minimum runoff coefficient and sediment concentration as well as soil loss was observed in soil group B (P<0.05), this might be due to the high content of $CaCO_3$ and sand (Table 6). Soil texture determines the rate at which water drains through a saturated soil; water moves freely through sandy soils than it does through clayey soils (Clinton and Vose, 2003). In humid Mediterranean mountainous area it was detected that the sand content and clay content of the soil have significant correlations with runoff rates on the three parts of the roads, affecting in a negative or positive manner respectively. The organic matter content contributed to reduce the runoff rate on the road surface and on the fillslope. Soil loss was negatively correlated to the sand content and organic matter content, but positively correlated to clay content (Jordán-López et al., 2009).

Shixiong et al. (2006) indicated that vegetation cover could significantly reduce sediment yield from unpaved roads. The above-ground components of the vegetation, such as leaves and stems, partially absorb the energy of the erosive agents of water and wind, so that less is directed at the soil, whilst the below-ground components, comprising the rooting system, contribute to the mechanical strength of the soil (Morgan and Rickson, 1995). In current study runoff and soil loss from fillslope plots with higher percentage of vegetation cover was lower than that of plots with lower vegetation cover, but this difference was not significant which could be due to the effects of soil properties, different plant forms or different amount of plant residues (Table 7). On fillslope, the highest soil loss was detected for soil group C, where sediment concentration in runoff was 18.07 g L^{-1} and vegetation cover was 10 to 40% (Table 8).

Effects of vegetation cover and soil groups on soil loss from road surface

Figure 5 and Table 9 show the PCA correlation circle for the soil variables and the properties of each group on road surface, respectively. Minimum runoff coefficient and sediment concentration as well as soil loss was observed in soil group C (P<0.05), this might be due to high content of $CaCO_3$, organic matter and sand (Table 10). Calcium carbonate of soil is an effective factor on conjunction of clay particles (Feiznia et al., 2005). There was no significant difference between soil loss from plots with vegetation cover of 2-5 and 5-8% (Table 11).

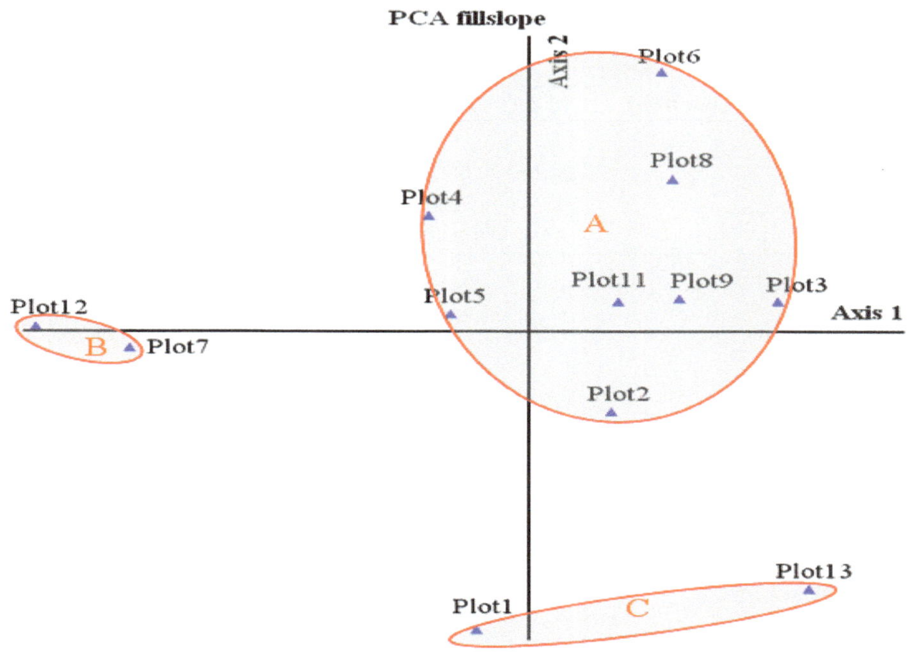

Figure 4. PCA analysis of soil on fillslope.

Table 5. Characteristics of the soil groups in fillslope.

Soil groups	Bulk density ($g\ cm^{-3}$)	Moisture (%)	Organic matter (%)	$CaCO_3$ (%)	Clay (%)	Silt (%)	Sand (%)
A	1.4±0.1	19.1±1.3	1.6±0.5	33.8±4.5	30.4±2.8	23.5±4.5	46.2±3.7
B	1.4±0.1	18.9±3.3	1.4±0.1	36.3±0.0	16.7±0.0	16.2±0.0	67.0±0.0
C	1.3±0.1	17.7±0.8	1.7±0.0	30.0±4.3	26.7±4.2	23.9±5.6	42.4±5.8

Table 6. Effect of soil groups on runoff and sedimentation from fillslope.

Soil groups / Variables	A	C	B	ANOVA, *p*
Runoff coefficient (%)	9.62^{B}	16.08^{A}	9.81^{B}	0.0426
Time to runoff (s)	619.38^{A}	272.50^{B}	754.00^{A}	<0.0001
Sediment concentration ($g\ L^{-1}$)	10.05^{A}	13.84^{A}	2.34^{B}	0.0388
Total soil loss ($g\ m^{-2}\ h^{-1}$)	47.43^{B}	77.30^{A}	17.24^{C}	0.0024

In a row, means with the same letter are not significantly different based on Student-Newman-Keuls test; Alpha=0.05.

Table 7. Effect of vegetation cover on runoff and sedimentation from fillslope.

Vegetation cover / Variables	10-40%	40-70%	ANOVA, *p*
Runoff coefficient (%)	11.00^{A}	10.18^{A}	0.3104
Time to runoff (s)	555.63^{B}	640.75^{A}	<0.0001
Sediment concentration ($g\ l^{-1}$)	10.26^{A}	7.68^{A}	0.9650
Total soil loss ($g\ m^{-2}\ h^{-1}$)	48.05^{A}	47.04^{A}	0.5516

In a row, means with the same letter are not significantly different based on Student-Newman-Keuls Test, Alpha=0.05.

Table 8. Interaction of soil and vegetation cover on runoff and sedimentation from fillslope.

Soil groups / Plant cover (%)	A		C		B		ANOVA, *p*
	10-40	40-70	10-40	40-70	10-40	40-70	
Runoff coefficient (%)	9.2	10.7	18.3	13.9	18.6	0.96	0.0667
Time to runoff (s)	642.5	550.0	445.0	100.0	490.0	1018	<0.0001
Sediment concentration(g L^{-1})	11.5	5.6	18.1	9.6	3.3	1.37	0.4314
Total soil loss (g m^{-2} h^{-1})	51.5	35.2	119.6	35.0	32.3	2.14	0.0874

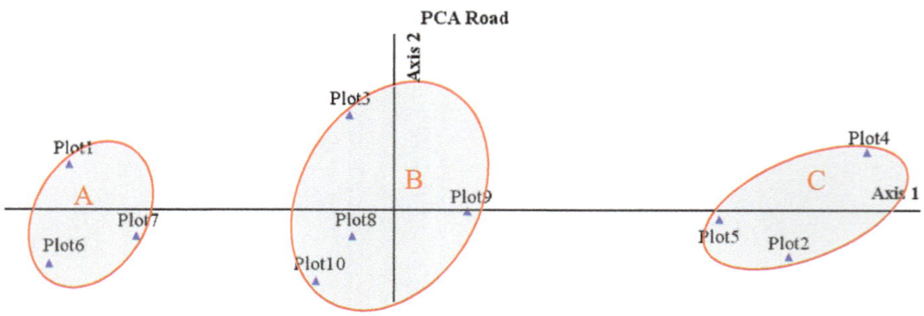

Figure 5. PCA analysis of soil on road surface.

Table 9. Characteristics of the soil groups in road surface.

Soil groups	Moisture (%)	Organic matter (%)	$CaCO_3$ (%)	Clay (%)	Silt (%)	Sand (%)
A	7.3±0.6	0.8±0.1	36.0±1.4	23.8±1.7	21.9±3.8	54.3±4.0
B	6.2±0.5	0.9±0.0	41.9±1.7	19.5±1.5	16.9±2.6	63.5±2.9
C	5.0±0.0	1.0±0.1	48.9±3.8	14.1±2.3	11.0±2.9	74.8±3.5

Table 10. Effect of soil groups on runoff and sedimentation from road surface.

Soil groups / Variables	A	B	C	ANOVA, *p*
Runoff coefficient (%)	75.04^{A}	74.41^{A}	36.68^{B}	<0.0001
Time to runoff (s)	53.67^{B}	58.50^{B}	104.00^{A}	<0.0001
Sediment concentration (g L^{-1})	7.34^{A}	7.26^{A}	6.98^{A}	0.7589
Total soil loss (g m^{-2} h^{-1})	168.55^{A}	171.46^{A}	85.70^{B}	0.0012

In a row, means with the same letter are not significantly different based on Student-Newman-Keuls Test, Alpha=0.05.

On road surface, the highest soil loss was detected for soil group A, where sediment concentration in runoff was 8.99 g L^{-1} and vegetation cover was 2 to 5% (Table 12).

Conclusions

Soil properties and vegetation cover are two effective factors on sediment yield from different parts of forest road. In this study it was concluded that the runoff and soil loss was more in soils which had high content of silt and clay. The maximum sedimentation and soil loss from cutslope was observed in soil group B due to high moisture and low bulk density. In fillslope, minimum sedimentation and soil loss was detected in soil group B, because high content of sand and $CaCO_3$ in this group. This finding was also observed for road surface and forest ground in soil group C and A, respectively. It is possible to repair cut and fillslopes with revegetation methods. Moreover, soil properties can be improved

Table 11. Effect of vegetation cover on runoff and sedimentation of road surface.

Vegetation cover / Variables	2-5%	5-8%	ANOVA, *p*
Runoff coefficient (%)	69.15[A]	59.36[B]	0.0214
Time to runoff (s)	48.00[B]	85.83[A]	<0.0001
Sediment concentration (g L^{-1})	6.86[A]	7.43[A]	0.6514
Total soil loss (g m^{-2} h^{-1})	148.64[A]	142.34[A]	0.9371

In a row, means with the same letter are not significantly different based on Student-Newman-Keuls test, Alpha=0.05.

Table 12. Interaction of soil and vegetation cover on runoff and sedimentation from road.

Soil groups	A		B		C		ANOVA, *p*
Plant cover (%)	2-5	5-8	2-5	5-8	2-5	5-8	
Runoff coefficient (%)	70.9	77.1	78.9	72.9	43.6	33.2	0.8116
Time to runoff (s)	61.0	50.0	34.0	66.7	58.0	127.0	<0.0001
Sediment concentration (g l^{-1})	8.9	6.5	5.9	7.7	8.3	6.3	0.1607
Total soil loss (g m^{-2} h^{-1})	191.4	157.1	149.7	178.7	130.6	63.3	0.0344

through mulching, hydro-seeding and etc.

REFERENCES

Bold KC, Wood F, Edwards PJ, Williard KW, Schoonover JE (2009). Using photographic image analysis to assess ground cover: a case study of forest road cutbanks. Environ. Monitor. Assess. 1:15-21.

Croke J, Hairsine P, Fogarty P (2001). Soil recovery from track construction and harvesting changes in surface infiltration, erosion and delivery rates with time. For. Ecol. Manage. 143:1-3. http://dx.doi.org/10.1016/S0378-1127(00)00500-4

Cerdà A (2007). Soil water erosion on road embankments in eastern Spain. Sci. Total. Environ. 378:151-155. http://dx.doi.org/10.1016/j.scitotenv.2007.01.041 PMid:17382993

Clinton BD, Vose JM (2003). Differences in surface water quality draining four road surface types in the southern appalachians. Southern J. Appl. For. 27(2):100-106.

Feiznia S, Khadjeh M, Ghauomian J (2005). The study of the effect of physical, chemical and climate factors on surface erosion sediment yield of loess. Pajouhesh. Sazandegi. 66:14-24.

Fu B, Newham LTH, Field JB (2009). Modelling erosion and sediment delivery from unsealed roads in southeast Australia. Math. Comput. Simul. 79:2679-2688.http://dx.doi.org/10.1016/j.matcom.2008.10.001

Foltz RB, Copeland NS, Elliot WJ (2009). Reopening abandoned forest roads in northern Idaho, USA: Quantification of runoff, sediment concentration, infiltration, and interrill erosion parameters. J. Environ. Manage. 90:2542-2550.http://dx.doi.org/10.1016/j.jenvman.2009.01.014 PMid:19264387

Jordán-López A, Martínez-Zavala L, Bellinfante N (2009). Impact of different parts of unpaved forest roads on runoff and sediment yield in a Mediterranean area. Sci. Total. Environ. 407:937-944. http://dx.doi.org/10.1016/j.scitotenv.2008.09.047 PMid:18992920

Jordán A, Martínez-Zavala L (2008). Soil loss and runoff rates on unpaved forest roads in southern Spain after simulated rainfall. For. Ecol. Manage. 255:913-919.http://dx.doi.org/10.1016/j.foreco.2007.10.002

Koc B, Eren I, Ertekin FK (2008). Modelling bulk density, porosity and shrinkage of quince during drying: The effect of drying method. J. Food Eng. 85: 340-349.http://dx.doi.org/10.1016/j.jfoodeng.2007.07.030

Kavian A, Azmodeh A, Soleimani K, Vahabzadeh G (2010). Effect of soil properties on runoff and soil erosion in forest lands. J. Rang. Water. Manage. 63(1):89-104.

Morgan RP (2005). Soil erosion and conservation. ISBN 1-4051-1781-8. Blackwell Publishing Company, P. 304.

Morgan RPC, Rickson RJ (1995). Slope stabilization and erosion control: A bioengineering approach. Published by E & FN Spon, an imprint of Chapman & Hall, 2-6 Boundary Row, London SE1 8HN, UK. 293 pp. ISBN 0 419 15630 5

Mohammad AG, Adam MA (2010). The impact of vegetative cover type on runoff and soil erosion under different land uses. Catena 81:97-103.http://dx.doi.org/10.1016/j.catena.2010.01.008

Mohammadkhan S, Ahmadi H, Jafari M (2011). Relationship between soil erosion, slope, parent material, and distance to road (Case study: Latian Watershed, Iran). Arab. J. Geosci. 4:331-338. http://dx.doi.org/10.1007/s12517-010-0197-z

Najafian L, Kavian A, Ghorbani J, Tamartash R (2010). Effect of soil properties on runoff and soil erosion. Rangeland. 4:334-347.

Olive ND, Marion JL (2009). The influence of use-related, environmental, and managerial factors on soil loss from recreational trails. J. Environ. Manage. 90:1483-1493.http://dx.doi.org/10.1016/j.jenvman.2008.10.004 PMid:19062152

Parsakhoo A, Hosseini SA, Lotfalian M, Jalilvand H (2009). Soil loss and displacement by heavy equipment in forest roads. Int. J. Sed. Res. 24:227-235.http://dx.doi.org/10.1016/S1001-6279(09)60029-9

Pappas EA, Smith DR, Huang C, Shuster WD, Bonta JV (2008). Impervious surface impacts to runoff and sediment discharge under laboratory rainfall simulation. Catena 72:146-152.http://dx.doi.org/10.1016/j.catena.2007.05.001

Shixiong Cao C, Chen L, Gao W, Chen Y, Yan M (2006). Impact of planting grass on terrene roads to avoid soil erosion. Landscape Urban. Plan. 78:205-216.http://dx.doi.org/10.1016/j.landurbplan.2005.07.007

Ziegler AD, Sutherland RA, Giambelluca TW (2001). Interstorm surface preparation and sediment detachment by vehicle traffic on unpaved mountain roads. Earth Surf. Process. Landf. 26:235-250.http://dx.doi.org/10.1002/1096-9837(200103)26:3<235::AID-ESP171>3.3.CO;2-K http://dx.doi.org/10.1002/1096-9837(200103)26:3<235::AID-ESP171>3.0.CO;2-T

Influence of depths and soil pH on forms of magnesium in soils of four parent materials (*Rhodic paleudults, Rhodic tropudalfs, Oxic tropudalfs* and *Aquic tropossamment)*

Okpamen S. U.[1] , Ilori E. G.[2], Agho I.[3,4], Nkechika A.[1], Maidoh F. U.[1] and Okonjo P. N.[4]

[1]Chemistry Division, NIFOR, P.M.B 1030, Benin City, Edo State, Nigeria.
[2]Edo State Institute of Technology and Management, Usen, Edo State, Nigeria.
[3]Biochemistry Division, NIFOR, P.M.B 1030, Benin City, Edo State, Nigeria.
[4]Analytical Laboratory Division, NIFOR, P.M.B 1030, Benin City, Edo State, Nigeria.

The role of depths and soil pH in the content and occurrence of the forms of magnesium in soils of the four parent materials in Edo and Delta States (Coastal plain sand, Shale, Basement complex and Alluvium*)* was investigated. Soils samples were collected from profiles that developed on these parent materials, with dimensions 1.2 x 1.2 x 1.5 m and demarcated according to depths ranging from 0 to 150 cm. Reagents were applied under regulated conditions and durations ranging from 10 min to 1 h. The results showed that different forms of magnesium were extracted from varying depths in each profile. In the alluvium the non available Mg content in the study ranged from 7.79 to 24.34 c mol kg^{-1} and was highest in the top soils of the profile examined,. The exchange Mg was between 0.16 to 0.24 c mol (+) kg^{-1} decreasing also across depths, in the coastal plain sand, the non available Mg ranged from 17.56 to 19.97 c mol kg^{-1} with greater amount realized from the bottom depths. Exchange Mg pool ranged from 0.12 to 0.52 c mol kg^{-1} with higher records on surface depths of 0-3, 3-15 and 15-45 cm. In the Basement complex the non available Mg ranged from 12.26 to 18.97 c mol kg^{-1} with greater value recorded from 45 to 62 cm depths. While exchangeable Mg ranged from 0.22 to 0.46 c mol kg^{-1} and 3.22 to 6.25 c mol kg^{-1} respectively with both forms distributed oppositely across the depth. The structural Mg ranged from 6.28 to 8.84 c molkg increasing down the depths. The water sol Mg in the four parent materials were distributed in similar pattern in three out of the four parent materials, it ranged from 0.16 to 0.52 c mol kg^{-1} and 0.09 to 0.19 c mol kg^{-1} in coastal plain sand and alluvium. In Shale it ranged from 0.21 to 0.58 c mol kg^{-1}, while in the basement complex 0.25 to 0.40 c mol kg^{-1} with values decreasing downs the depths across the four parent materials. The soil ph across the four parent materials ranged from 4.41 to 5.63 decreased from surface to the bottom.

Key words: Soil pH, horizon, depths, profile, available magnesium.

INTRODUCTION

Depths and pH in soil profiles are important factors to be considered in plant nutrient availability and quantification study because plant nutrients are located in different depths or horizons in the soils (Buzzle, 2011). Depth in this context refers to the vertical distance across the soil profile from the topmost surface to the last point dug at the bottom that is accessible (Henson, 2007; Buzzle, 2011). Soil pH is the soil reaction in the area examined and refers to the sum total effects of the hydrogen and associate acid ions concentration in the soil (Manjula, 2009). SQMN (2011) emphasized that depth and soil pH have important influence on nutrient availability status of most soils in the tropics stating that weathering is a major factor that determines the forms and level of these nutrients in the soil. Plant roots are the major physical and biological instruments that are used for absorption of nutrients from the soil and the rate of plant nutrient adsorption is influenced by depths (Mitchall et al., 2005). Depth influence the adsorption of nutrients by plant roots as most plants have their functional roots and root hairs located in the top soil region of the profiles. Usually plants have vast arrays of root network scattered in this region, while perennial crop has its' growth and development among other factors tailored to the ability of the crops to utilized nutrients at all depths across the profile. Tennakoon (2003) reported that arable crops such as maize, millet, vegetables and peas utilizes and sourced their nutrients from the top soils of profiles but added that most palms would however source for their nutrients in diverse depths or horizons far as 45 to 200 cm. Depths are influenced by the geological and climatic characteristics of the soil micro environment which exposes the lithographical constituents within the soil structure in a micro scale. The parent materials and the geological structure of a profile is a direct function of the component materials that make up the physico-chemical constituents of the soil (Brady and Weil, 1999). These however are mainly of plant and animal materials with added geo-mineral constituents that are bio-chemically synthesized into both macro and micro nutrients which become plant nutrients in soils. Kamaljit (2005) stated that, plant nutrients are utilized at different depths where they occur, and are taken from vast array of reserves of nutrient in these horizons with biophysical mechanism by plants. Soils developed from various parent materials in the states are differentiated and classified into several soil types and series. Such soils are characterized into several land capability and suitability status which can influence the cropping and farming systems done in any area. Tennakoon (2003) and Fayemi and Lombin (1975) stated that the magnesium forms and content of Nigerian soils varied with the magnesium content of different parent materials and the climatic locations. Okpamen (2011) however added that the content and forms of magnesium in a soil is dependent on the status of the primary magnesium bearing mineral found in such soils, adding that it is subjected to vast environmental forces such as weathering and other physical factors. Weathering occurs at different areas in the states but the intensity of occurrence is dependent on the prevailing bio-physico chemical factors present in the area, and it has been established that weathering occurs higher in the top soils than in lower depths (Brady and Weil, 1999).

MATERIALS AND METHODS

Study areas, site description and sample collection

The influences of depths and soil pH on forms of magnesium in soils developed on four parent materials in Edo and Delta states (Coastal plain sand, Shale, Basement complex and Alluvium) were investigated. Soil samples were collected from four locations representing four parent materials. The locations were NIFOR developed on coastal plain sand, Uhunmora developed on Shale; Sobe developed on Basement complex all in Edo state while the forth location is in Koko in Delta state developed on alluvium parent material all in two ecological zones of Edo and Delta states of Nigeria. Samples were collected from depths demarcated, and a physical assessment of profile pits dogged on fields was made. Then, these soil samples were stored in labeled polythene bags and transported to the laboratory for physico-chemical analysis namely; Soil pH (H_2O), exchangeable acidity, exchangeable bases, organic matter.

Particle size analysis and effective cation exchange capacity (ecec)

Characterization of soil of the four parent materials (coastal plain sand, shale, alluvium, and basement complex)

The sites of these parent materials chosen were NIFOR, an Agricultural Research Institute in Edo State about 24 km from the state capital Benin City. It is located on latitude 06°33'N and longitude 05°37'E and on altitude 149.4 m. NIFOR soil is classified into four soil series namely; Alagba, Kulfo, Ahieria and Orlu (Ogunkunle, 1983) which is Typic/Rhodic paleudults (FAO/UNESCO, 1994) the pits were dug on Orlu/Alagba series of the institute. Samples were collected from pits dug with dimensions 120 x 120 x 150 cm and vegetation is rainforest with mean annual temperature of 31°C and a bimodal rainfall regime. The second site is at Uhunmora; Cocoa Research Institute of Nigeria substation in Edo State, the soil is described as Rhodic tropudualf (FAO/UNESCO, 1994). Site was on old Cocoa plot about one kilometer from the administrative block of the institute. The pit was 120 x 120 x 150 cm dimensions and soil samples were collected from depths by similar process described above. This site is located in North East central end of Edo State; temperature is same as NIFOR, though rainfall is within 1575 to 1725 mm across the year not as high as in NIFOR. Koko is the third site; it is in Delta state, Southern part of former Bendel State soil samples were collected from the profile pits dug on an old cassava farm with 120 x 120 x 150 cm dimension, located 500 m north of Total Bitumen plant Plc. The area is with high rainfall > 2500 mm per annum an average temperature of 25°C and fresh water swamp vegetation, soils are hydromorphic and humus with high water table. The profile was studied, and described while samples collected were bagged for laboratory analysis. The forth site was at Sobe farm settlement plot, where 120 x 120 x 150 cm area was laid out and dug for sampling. The soils in the site are on a gentle sloping surface with stones and pebbles as common features, it occurs in the derived savanna zone under forest soil, derived from metamorphic and igneous rocks from inselbergs of granite, gneiss, Schist and quartzite (Fagbami,1985) of the old Bendel State now Edo and Delta States. Its rainfall is about 1450 to 1650 mm $year^{-1}$ the least in the study with average bimodal temperature of 31°C.

Laboratory analysis of the physico-chemical properties

Soil samples were air dried and passed through a 2 mm sieve, while some samples were ground to pass through 0.5 mm sieve and stored for determination of organic matter and total nitrogen. Analyses of samples were carried out using the known conventional procedures outlined as follow:

1. Soil pH (H_2O): This was determined in 1 : 1 soil – water suspension. 20 g sample of air dried soil was passed through a plastic 2 mm sieve and weighed into 100 ml beaker. 20 ml distilled water was added and stirred thoroughly and allowed to equilibrate in 30 min. A glass electrode pH meter was used to read the pH of the suspension after standardization with pH buffer 7.0 and 4.0 as outlined by Schofield and Taylor (1955).
2. Exchangeable acidity: The exchangeable acidity was extracted with IN KCl and determined by titration with 0.05M NaOH using phenolphthalein as the indicator (Jackson, 1964).
3. Exchangeable bases and organic matter: the exchangeable cations, Na, K, Ca and Mg were extracted with 1 N Ammonium Acetate solution (NH_4OAC) buffered at pH 7.0. Exchangeable Na, K and Ca content of the extracts were determined on flame photometer, while exchangeable Mg was read on Atomic Absorption Spectrophotometer (Jackson, 1964). Organic Matter Content was determined by wet dichromate-acid oxidation as outlined by Walkley and Black (1934) and modified by Piper (1942).
4. Particle size analysis and effective cation exchange capacity (ECEC): Particle size analysis was carried out using hydrometer method outlined by Bouyoucos (1951) as modified by Day (1965). ECEC was determined by summation method. (Hearse, 1971) (i) Exchangeable acidity (EA) (H^+ and Al^{3+}) (ii) Total exchangeable bases (TEB).

Determination of forms of magnesium in the four parent materials

Magnesium forms in the soils were fractionated in the following ways

1. Water soluble magnesium content procedure: This form of magnesium was determined with deionized water in 1:10 soil – water ratio. 5 g of the soil samples were weighed into 250 ml plastic bottles, 50 ml of deionized water was added to each sample in the bottles and shaken on a mechanical shaker for 30 min and filtered with whattman No.42 filter paper into a 100 ml volumetric flask. The magnesium in the extract was determined with an Atomic Absorption Spectrophotometer (AAS).
2. Exchangeable magnesium content: the residue from the water extract was treated with 1N NH_4OAC solution buffered at pH 7.0 at soil : solution ratio of 1 : 10 for 30 min. The suspension was then filtered and magnesium in the filtrate was determined using AAS.
3. Difficultly exchangeable magnesium pool; 1N HNO_3 extractable magnesium: the residue from (ii) above in the centrifuge tube was treated with 50 ml of 1N HNO_3 acid. The suspension was boiled for 10 min in a digester and the content allowed cooling for 20 min before centrifuging. They were then decanted and the extracts read with the AAS.
4. 1N HCl extractable magnesium: To the residue from the 1N HNO_3 acid extract was added 50 ml of 1N HCl and stirred with a glass rod to re - suspend it and the suspension placed on a digester to boil for 1 h, before cooling. The suspension was then centrifuged, to cool for 20 min before filtration was carried out on suspension. The magnesium content of the extract was determined on an AAS.
5. Fixed magnesium reserve pool: This form of magnesium was obtained by computation using the difference between the 1 N HCl extractable Mg values and the exchangeable Mg in the samples. The difference represents the fixed Mg reserve pool in the samples. This was in accordance with the procedure of (Flinck, 1962).
6. Structural magnesium content: This form of magnesium was also obtained by computation. The result obtained from the difference between the total Mg content and the 1N HCl extractable Mg content was tagged structural Mg or residual magnesium.
7. Total magnesium reserve pool: total magnesium in the sample was determined by the sodium carbonate fusion procedure as describe by Jackson (1958). 0.2 g of air dried soil was ground to pass through 100 µ sieve and weighed into platinum crucible in duplicate. 5 g of anhydrous Na_2CO_3 was added to the weighed sample and the mixture was thoroughly mixed. An additional 1 g of Na_2C0_3 was weighed and spread evenly on top of the mixtures. And was covered at an angle and ignited using a mekker burner for about 10 min. The cover was lifted with a pair of tongs to let in air through a slit when frothing sound had ceased. Heating was continued until the soil carbonate mixture melted. The crucible was gently rotated to allow the melt spread evenly on the sides of the crucible. The cover was then removed and heating discontinued. The crucible was then put into a 400 ml beaker, distilled water added to cover the crucible and the beaker heated cautiously to dissolve the melt. The platinum crucible was lifted and the sides washed into the beaker with a wash bottle. 5 ml of 6 N HCl was added to the beaker and warmed over a low flame to avoid too vigorous effervescence with loss of material. When all the melt had dissolved, the solution containing gelatinous silica was filtered into a 250 ml volumetric flask. The residue was washed several times with hot distilled water and the washings were added to the volumetric flask. Magnesium in the filtrate was determined by atomic absorption spectroscopy (Okpamen et al., 2011).

Categorization of forms of magnesium

1. Available magnesium: The available magnesium comprise of the water soluble magnesium and the exchangeable magnesium (H_2O Sol. Mg + 1N NH_4OAC Mg) in the study (Chevalier, 1971; Fayemi and Lombin, 1975).
2. Non available magnesium: The non available magnesium consists of all the other forms extracted from the samples less the available magnesium. This comprises the acid magnesium, fixed Mg Reserve and structural Mg. (Hearse, 1971)

Statistical analysis of the data

The statistical method as described Alika (1997) and an SPSS package were employed for data obtained from the laboratory analysis. Simple linear and multiple correlations were carried out on factors that had influences on the magnesium species studied. The level of probability used in tests of significance was 1 % and 5 % respectively.

RESULTS AND DISCUSSION

The content and forms of Magnesium extracted was examined relative to profiles depths and soil pH in this study, it is to ascertain the level of influences both parameters had on the presence and contents of forms of magnesium in the four parent materials. Depths and soil pH were correlated with water soluble Mg, exchangeable Mg and non available Mg and results showed that soil pH in the various profiles and parent materials were low and they fell within acid range. The values recorded fluctuated with depths for two of the parent materials examined except in the basement complex and Shale parent materials. In the Basement complex, a soil pH of 5.6 was recorded at the surface of 0 to 15 cm and declined to 5.2 at a depth of 30 cm and remains constant. While the

Table 1. Correlation of Depth and Soil pH with Forms of Magnesium.

		Depth	Soil ph	Water Mg	Exch. Mg	Non avail. Mg
Pearson correlation	Depth	1.000	-0.489	-0.912	-0.346	0.015
	Soil pH	-0.489	1.000	0.235	0.693	-0.550
	Water Mg	-0.912	0.235	1.000	0.391	-0.077
	Exch. Mg	-0.346	0.693	0.391	1.000	-0.930
	Non. avail. Mg	0.015	-0.550	-.077	-0.930	1.000
Sig. (1-tailed)	Depth	-	0.202	0.015	0.284	0.491
	Soil ph	0.202	-	0.352	0.097	0.168
	Water Mg	0.015	0.352	-	0.258	0.451
	Exch. Mg	0 .284	0.097	0.258	-	0.011
	Non. avail. Mg	0.491	0.168	0.451	0.011	-

Water mg, Water soluble magnesium; Exch. Mg, exchangeable magnesium; non avail Mg; non available magnesium.

various forms of magnesium varied across depths in different pattern in all the parent materials, water soluble Mg and exchangeable Mg pool which represents the available magnesium had values decreasing from topmost depths (0 to 3 cm) to bottom (90 to 150 cm) in all the profiles examined (Table 1). The non available Mg forms had values varying increasingly with depths from the surface to the bottom of the profiles. The soil formed on shale (CRIN) location had a pH of 5.32 on surface, 4.41 in mid profile and 4.46 at the bottom. The amount of magnesium recorded were; water soluble Mg; 0.58 to 0.42 cmolkg^{-1} (0 to 75 cm) and 0.32 to 0.21 cmolkg^{-1} (75 to 150 cm), exchangeable Mg values ranged from 1.03 to 0.83 cmolkg^{-1} (0 -150 cm) (Table 2). While the non available Mg form ranged from 7.79 to 24.34 cmolkg^{-1} (between 0 - 150 cm) in the four parent materials studied. In coastal plain sand parent material the pH of the soil were 5.34 on surface 5.16 in mid profile and 5.32 at depth while the various forms of Mg ranged from water soluble Mg; 0.16 to 0.52 cmolkg^{-1} (0 to 150 cm) increasing from top to bottom. Exchangeable Mg ranged from 0.42 to 0.12 cmol kg^{-1} (0 to 150 cm) while the non available Mg ranged from 17.54 to 19.97 cmolkg^{-1} this results indicates an increase to a depth of 150 cm. In alluvium the pH ranged from 5.15 on surface to 4.57 at depth while the different forms of magnesium were; water soluble Mg having 0.18 to 0.09 cmolkg^{-1} between 0-150 cm, exchangeable Mg ranged from 0.24 to 0.16 cmolkg^{-1} (0 - 150 cm) while the non available Mg in this parent material ranged from 7.79 to 9.86 cmolkg^{-1} (0 - 150 cm) decreasing from top to bottom, this represents the least values extracted for this form of magnesium in the study. In Basement complex, the soil pH ranged from 5.63 to 5.2 but the various forms of magnesium extracted ranged from water soluble Mg having 0.25 to 0.40 cmolkg^{-1} (0 to 150 cm) Exchangeable Mg pool ranged from 0.22 to 0.46 cmolkg^{-1} (0 to 62 cm) and the non available Mg pool ranged from 12.28 to 18.59 cmolkg^{-1}. The pattern of distribution of the various forms of magnesium in the respective parent materials can further be categorized into two fashions; the water soluble Mg and exchangeable Mg that constitute the available magnesium, tended towards the surface depths of most of the profiles. The reasons attributed for this finding is the prevalence organic matter in the surface depths of the profiles that are open to bio-physical and chemical weathering agents in the environments. These actions cause the breakdown of the complexes in minerals and organic materials which exists at different stages which available magnesium constitute a part (Table 1).

This finding is in line with work done by Tennakoon (2003) and Kamaljit (2005) who studied the status of major elements in soils of Sri lanka and those of grassland in Central Queensland Austrilia and reported high amount of K, Mg, and Ca in higher depths. Secondly, the non available magnesium which constitute the acid, fixed Mg reserve, structural and total Mg pool are distributed in similar pattern as the exchangeable Mg. They tend to increase down the profiles; higher values were recorded for these forms of magnesium at depths almost in all the four parent materials. This finding is corroborated also by the earlier reports of Fayemi and Lombin (1975) that had worked on status of magnesium in soils of south western Nigeria and reported that these forms are usually higher in depths than the available forms. Kamaljit (2005) concluded that the variability in amount of nutrients may have been as a result of extent of weathering across depths and other biophysical and chemical activities which had varied in such areas. From Table 1, there was significant negative correlation between depths and the distribution of water soluble Mg in the study (r = -0.912**) and between non available magnesium and exchangeable magnesium distribution (r= -0.930**). There was also a fair positive correlation between soil pH and exchangeable Mg (r = 0.693) while most other parameters examined fall short of a significant statistical positive and negative correlations which would have imparted more on this investigation or work however

Table 2. Distribution of depths and soil pH with forms of magnesium in the study.

Depth (cm)	Soil pH	Forms of magnesium			Parent materials
		H_2O Sol. Mg	Exch. Mg	Non avail. Mg	
0-3	5.34	0.16	0.42	17.56	Coastal plain sand
3-15	5.43	0.27	0.51	15.94	
15-45	5.32	0.31	0.52	16.30	
45-75	5.16	0.30	0.48	19.84	
75-90	4.93	0.48	0.20	19.97	
90-150	5.32	0.52	0.12	19.17	
0-3	5.32	0.58	1.03	19.72	Shale
3-15	4.91	0.49	0.64	24.31	
15-45	4.41	0.51	0.66	24.13	
45-75	4.41	0.42	0.75	21.62	
75-90	4.98	0.32	0.76	21.78	
90-150	4.46	0.21	0.83	20.40	
0-3	5.15	0.18	0.24	9.86	Alluvium
3-15	5.09	0.12	0.23	9.82	
15-45	4.99	0.19	0.20	0.18	
45-75	4.65	0.22	9.67	8.07	
75-90	4.87	0.12	0.17	7.79	
90-150	4.57	0.09	0.16	7.79	
0-3	5.63	0.40	0.46	12.28	Basement complex
3-15	5.48	0.31	0.37	17.64	
15-45	5.58	0.20	0.36	18.59	
45-75	5.40	0.25	0.22	19.73	
75-90	na	na	na	na	
90-150	na	na	na	na	

na, Not available; cm, centimeters H_2O sol, water soluble; Exch Mg, exchangeable magnesium; Non Avail Mg, non available magnesium.

the correlation of the non exchangeable Mg with soil pH was only slightly negative correlated (r = -550, Table 1). The results above implicated the following; that a strong linearity exists between those soils properties that had significant correlation with each other such as stated above in this study. This findings is in line with earlier reports given by Fayemi and Lombin (1975) working on soils of south western Nigeria who reported a significant relationship between soil ph and exchangeable magnesium and depth respectively.

Conclusion

In summary, the pattern of distribution of the various forms of magnesium in the respective parent materials have been categorized into two; the water soluble Mg and exchangeable Mg that constitute the available magnesium which appears to be concentrated in the surface depths and reasons attributed for this, was the prevalence of organic matter in the surface depths of the profiles where many bio-physical and chemical weathering activities in the environments seem to be occurring. The other forms of magnesium tagged the non available magnesium also responded to depths by continuous downward increased in amount. An indication that these forms of magnesium are more in the bottom depths than in the surface, this suggests slower weathering actions to be occurring at depths. Brady and Weil (1999) stated that the non available forms of most cations in their parent forms are rendered to their primary and secondary forms by processes of breakdown sponsored by actions of weathering. The relevant of this work is hinged on the fact that the two classes of magnesium defined from the various forms in this work are depths and soil ph related to a great extent.

REFERENCES

Alika JE (1997). Statistics and Research Methods. 1st. Edition Ambik Press Benin City, Nigeria, P. 157.

Bouyoucos CJ (1951). Recalibration of the Hydrometer method for

making the mechanical analysis of Soils. Agron. J. 43:434-438.
Brady NC, Weil R (1999), Nature and properties of Soils 12th Edition, pp. 1-108.
Buzzle, (2011). Different Types of Soil, Soil Sciences (38). http; //twit er.com/ BuzzleDot.com
Chevalier M (1971) Principia Aspect of Potassium and Magnesium Nutrition of plant Potassium/magnesium fertilizer. Manuring subject 16:53.
Day PR (1965) Particle Fractionation and particle size analysis in Methods of Soil Analysis. C.A Black (ed). Madison, Wisconsin.
FAO/UNESCO (1994). Soil map of the world 1:500,000 Vol. 1, legend sheet and memoirs Paris P. 55
Fagbami (1985). Soils of Bendel state. Federal department of agriculture, final report Dr.Fagbami report Department of Agronomy, University of Ibadan. August 1985 pp.1-77.
Fayemi AA, Lombin G (1975). Evaluation of various soil test for available - magnesium in soil of Western Nigeria. Soil Sci. Soc. 122:59-64.
Flinck A (1962). K-alium status of some Sudan clay soils. Plants Soil 16:292-331.
Hearse PR (1971). A Textbook of Soil Chemical Analysis, John Murray, London. pp. 402-407.
Henson IE. (2007). Modeling oil palm yield based on source and Sink. Oil palm Bull. 54;27-51
Jackson ML, Bear [1964]. chemical composition of the soil., Prentice-Hall Inc., Englewood Cliffs, New Jersey, p. 498.
Kamaljit K, Sangha. AC, Rajask K. Jolata B, David JM (2005). Impact of Clearing on Soil Ph and availability of Nutrient in grazing System of Central Queensland, Australia. Plant Sciences Unit, Primary Industries Research Centre University Rock Hampton Qld 4702, Australia.
Manjula (2009). MG4 Soil and Plant Nutrition; Nutrient Management, Soil Test and Diagnostic Service Laboratory, Gardener Core Manual. Published by University of Missouri Extension guidelines 1993-2013.
Mitchall RL, James K, Soga K (2005). Fundamentals of soil behaviour. 3rd Edition John Wily and son, Inc. P. 129.
Ogunkunle O (1983). Updating and classification of Acid sand soil with particular Reference to soils of NIFOR main station, J. Nig. Inst. Oil. Palm Res. 6(23)234-255.
Okpamen SU, Uzu FO, Aghimien AE (2011) Fractionation of Magnesium in Soils of Rhodic paleudults, J. Phys. Sci, Afri. Res. Development Network ARD/297/2011.
Schofield RK, Taylor AU (1955). The measurement of soil pH". Soil Sci. Soc. Am. Proc. 49:164-167.
SQMN (2011) Soil Quality; Management Nutrient. Soil Quality for Environmental health,> http;//soil quality .org/practices/nutrient management.html.
Tennakoon A (2003). Soil Fertility Status and Fertilizer Recommendation For Coconut in Sri lanka. Soil and Plant Nutrition Division Coconut Research Institute Lunuwila Sri lanka, P. 13.
Walk!ey A, Black IA (1934). An examination of the Degijareff method for determining soil organic matter and a proposed modification of chronic acid titration method. Soil sci. soc. Am. Proc. 37:29-38.
Febvre-Chevalier C (1971). Constitution ultrastructurale de Globigerina bulloides d'Orbigny 1826 (Rhizopoda-Foraminifera): Protist., 3, 3:311-324.

Evaluation of the effect of *Ficus thonningii* (blume) on soil physicochemical properties in Ahferom district of Tigray, Ethiopia

Daniel Hagos Berhe[1], Agena Anjulo[2], Abdu Abdelkadir[2] and Sue Edwards[3]

[1]Axum Agricultural Research Centre, P. O. Box 230, Axum, Ethiopia.
[2]Wondo Genet College of Forestry and Natural Resources, P. O. Box 128, Shashemene, Ethiopia.
[3]Institute for Sustainable Development (ISD), P. O. Box 171, Code 1110, Addis Ababa, Ethiopia.

A study to evaluate the effect of *Ficus thonningii* (Blume) on soil physicochemical properties was conducted in Ahferom district of Tigray, Ethiopia. For the soil physico-chemical property study, two factors (distance from the tree trunk and soil depth from the ground level) arranged in randomized complete block design (RCBD) with six replications was involved. The distance factor had three levels viz. at half of the canopy radius under the tree, canopy edge (radius of the canopy) and at three times canopy radius away from the trunk outside the canopy. The depth factor had two levels viz. surface (0 - 15 cm) and subsurface (15 - 30 cm) soil layers. Data were collected on soil physicochemical properties viz. soil texture, bulk density, moisture content, soil N, soil P and soil K, %OC, pH and electrical conductivity (EC). The collected data were subjected to ANOVA using the general linear model of SAS. Results of soil physicochemical properties revealed that except for soil texture, the studied soil physicochemical properties (soil bulk density, moisture content, soil N, soil P, soil K, %OC, pH, EC) were enhanced under *F. thonningii* canopy zone as compared to outside canopy zone. Hence, retention and planting of *F. thonningii* is recommended. The N and P stocks of soils under and away the canopy of *F. thonningii* are above the recommended range for optimum growth of most crops while K stock is rated low for optimal crop growth. Due to the deficient levels of soil K, sustainable crop production in the district requires supplementary application of K fertilizers.

Key words: Soil texture, Nitrogen, Phosphorus, Potassium, carbon stock, bulk density.

INTRODUCTION

Ethiopia faces high rates of deforestation (Kindeya, 2004). Vegetative cover has dwindled and soil fertility and productivity have declined. In some parts of northern Ethiopia, the land has been completely degraded and is no longer capable of supporting human population. Badege and Abdu (2003) reported that 94% of the Ethiopian population relies on wood based and biomass fuel for household energy. Scarcity of firewood has become acute in many parts of the country causing a continuous rise in prices, thus increasing the economic burden on the household budget. As a result, animal dung and crop residues are increasingly being used for household fuel rather than being added to the soil to improve soil fertility, thus further exacerbating the problems of environmental degradation. To tackle the multifaceted challenges of Ethiopian agriculture, there are different options in which agroforestry is one (Neufeldt et al., 2009). However, agroforestry should not be taken as universal remedy for all land use problems, but it may be considered as a potential alternative to some of the wasteful land use practices that exist in the country (Badege and Abdu, 2003).

Kindeya (2004) argued that proposed agroforestry options should start with the species to which farmers are most accustomed to. These would be the native multipurpose trees that have evolved under smallholder environmental and socioeconomic conditions. Hence, strategies to improve smallholder crop production should begin by investigating the merits and demerits of existing indigenous farming systems as knowledge of the contribution of trees to soil fertility is vital to encouraging tree planting by individual farmers (Jiregna et al., 2005). Studies by Kindeya (2004) and Teklay and Malmer (2004) have also revealed that the existence of indigenous tree species with great untapped potential for growing with agricultural crops.

Against this background, to gain immense benefits like fuel wood, food, shade, fodder, medicine, ecological services including soil fertility and microclimate amelioration from trees, farmers in Ahferom district of Tigray Region plant different indigenous and exotic trees. Among the indigenous species grown and managed by farmers in the district is the drought tolerant tree, *Ficus tonningii*; a multipurpose indigenous tree that easily propagates and provides medicine, wood, cash and year round fodder for animals.

Though *F. thonninigii is* widely grown in Ahferom and used as fodder in times of feed shortage in the district, it is still among the indigenous tree species least considered in agroforestry development practices in most parts of the country. Its fodder potential and nutritive value, its interaction with other crops in the farmland and its effects on soil physicochemical properties have not been scientifically quantified and documented.

Therefore, based on the foregoing information, this initiative was taken to investigate the effects of *F. thonningii* on some physicochemical properties of the soil under and away from its canopy at different soil layers in Ahferom district of Tigray, Ethiopia.

MATERIALS AND METHODS

Description of the study site

The study was conducted in central zone of Tigray Regional State, Ethiopia, in *Ahferom* district. The district has a geographical location of 14° 07' to 14°39' N and 38°57' to 39° 18' E. It is situated at 960 km north of the capital Addis Ababa and 181 km North West of *Mekelle* town, the regional capital. The district is characterized by rugged topography with altitude ranging from 1400 to 3200 m (BoARD, 2010). The rainfall distribution of the study area is unimodal, mostly falling within the months of June to September, exhibiting high temporal and spatial variability. The annual rainfall ranges from 450 to 650 mm, and with average annual temperature range of 25 to 34°C. The dominant soil types in the district vary in ranges of clay, loam, sandy loam and silty loam (BoARD, 2010).

Experimental details

There were two factors involved in the experiment, viz. distance from the tree trunk and depth from the ground level. The distance factor had three different treatment levels; at half of the canopy radius under the tree, canopy edge (radius of the canopy) and at three times canopy radius away from the trunk outside the canopy as control following the procedure by Jiregna et al. (2005) and Pandey et al. (2000). And the depth factor had two levels: Surface soil layer (at 0 to 15 cm) depth representing the top soil and the other lower layer (at 15 to 30 cm) depth to represent the subsurface soil layer. The design employed had a 3 * 2 factorial arrangement of treatments in randomized complete block design (RCBD) replicated six times, totalling 3 * 2 * 6 = 36 experimental units or samples.

Description of the studied trees for soil physicochemical property

Six isolated individual *F. thonningii* trees, outside the influence of other trees, were randomly selected for physicochemical attribute study. The trees were sampled from a locality within the same agroecology. Each individual tree was considered as a replicate. Each tree's canopy area was determined and divided into radial transects. Besides, tree height, canopy cover as area of tree influence and trunk diameter measurements were done on the six mature *F. thonningii* trees that were used for soil physicochemical property attribute study. The studied *F. thonningii* trees had mean trunk diameter of 42.67 cm and mean height of 11.25 m. The canopy diameter cover ranged from 8.00 to 10.75 m with mean canopy diameter cover of 9.28 m. The potential area of influence of each *F. thonningii* tree ranged from 50.24 to 90.72 m^2 with mean area of influence of 68.41 m^2.

Soil sampling

Four transects extending from the base of the tree trunk were laid out in East, West, North and South directions. From the four transects, soil samples were taken at three distances (at half of the canopy radius under the tree, canopy edge and at three times canopy radius away from the trunk outside the canopy) from the tree base and from two different depths (0 to 15 and 15 to 30 cm) of each respective distances with six replications. Overall there were 3 × 2 × 6 × 4 = 144 soil sub samples; 72 for the surface soil and 72 for the subsurface.

Samples within the same radial distances and same depths which represents sub samples, were pooled or bulked to achieve a composite sample; in total 18 composite sample out of the 72 sub samples from the upper 0 to 15 cm and 18 composite samples out of 72 sub samples from the lower 15 to 30 cm subsoil layer were collected, that is, four samples within same radial distance and same depth were pooled to form a single composite sample. And to avoid contamination between the layers while taking the soil samples, a 40 cm × 40 cm pit was dug and then the soil samples were taken by scratching the wall of the soil profile for the respective depth. First, a soil sub sample was taken for the 15 to 30 cm then for the 0 to 15 cm layer. By bulking the four sub samples and quartering a composite sample of 1 kg was prepared for the chemical and physical analysis purpose of soil properties: But for bulk density and moisture measurement a separate soil sample with core sampler was collected.

Table 1. Soil physical fertility parameters as influenced by radial distance and depth in Ahferom district.

Soil fertility parameter	Soil depth (cm)	Radial distance			Overall
		Mid canopy	Canopy edge	Canopy gap	
% Sand	0 - 15	19.00±2.48	21.83±3.02	24.00±2.29	21.61±1.50[a]
	15 - 30	21.67±1.76	22.67±2.28	24.33±2.11	22.89±1.15[a]
	Overall	20.33±1.50[A]	22.25±1.81[A]	24.17±1.47[A]	
% Silt	0 - 15	36.33±4.34	36.00±50	34.17±5.35	35.50±2.68[a]
	15 - 30	35.67±4.82	35.67±4.52	34.17±3.95	35.17±2.42[a]
	Overall	36.00±3.09[A]	35.83±3.21[A]	34.17±3.17[A]	
% Clay	0 - 15	44.67±5.19	42.17±5.76	41.83±6.62	42.89±3.20[a]
	15 - 30	42.67±5.53	41.67±5.70	41.5±5.25	41.94±2.98[a]
	Overall	43.67±3.62[A]	41.92±3.87[A]	41.67±4.03[A]	
BD (g cm^{-3})	0 - 15	1.12±0.02	1.25±0.024	1.38±0.03	1.25±0.03[b]
	15 - 30	1.18±0.02	1.34±0.02	1.44±0.03	1.32±0.03[a]
	Overall	1.15±0.02[C]	1.29±0.02[B]	1.41±0.02[A]	
% MC	0 - 15	22.80±0.78	19.99±0.76	17.95±0.52	20.25±0.61[a]
	15 - 30	21.22±0.75	19.11±0.66	15.91±0.40	18.75±0.63[b]
	Overall	22.01±0.57[A]	19.55±0.50[B]	16.93±0.44[C]	

BD = Bulk density; MC = moisture content. Means along the same column (soil depth) and rows (radial distance) with different superscripts are significantly different (P<0.05); Values are Mean ± SEM.

Soil analysis and laboratory methods

Before chemical analysis, the collected samples were first air dried, then ground and sieved to separate the <2 mm fraction at Shire Soil Laboratory. Then the soil samples were delivered to National Soil Testing Laboratory Addis Ababa and were analyzed for organic carbon content by wet oxidation method of Walkley and Black (Schnitzer 1982); total nitrogen by the Kjeldahl method (Bremner and Mulvaney, 1982). Available P was determined by Olsen method (Olsen et al., 1954); Available K by neutral ammonium acetate extraction (Merwin and Peach, 1951); soil pH and EC was determined in a 1:2.5 soil to water suspension (Jackson, 1973); texture was determined by the Hydrometer (Bouyoucos); soil moisture content by oven drying at 105°C and bulk density by weighing oven dried (105°C) soil samples with known volume (Brady and Weil, 2002).

After chemical analysis, soil BD, soil depths and concentration[1] of soil nutrients (Carbon, Phosphorus and Potassium) were used to calculate the soil nutrients from concentration basis to an amount or nutrient stock basis (soil C, soil P and soil K stocks) according to the procedure followed by Enideg (2008).

Statistical analysis

The data collected were subjected to two way analysis of variance (ANOVA) using the General Linear Model (GLM) procedures of SAS (SAS, 2002). Comparison of treatment means was performed using Fisher's Least Significant Difference test at $P < 0.05$ probability level. Unless otherwise stated, any mention of significance is based on the 5% level. Besides, Pearson correlation analyses were performed to reveal the relationships between organic carbon, soil depth and distance from the tree with other soil physicochemical properties.

RESULTS AND DISCUSSION

Soil physical properties

Soil texture

Soil texture, as factor that might affect soil physicochemical and to some extent soil biological properties, was not significantly affected (Table 1) both laterally as function of distance from tree trunk and vertically by the presence of *F. thonningii* trees. The non significant differences in the mean proportions of sand, silt and clay fractions between the soils under the canopies of *F. thonninigii* and in the open farmland suggests that the soils, are texturally similar, being clay for the surface and clay loam for the subsurface and having been derived from the same parent material, under the same climate, and similar topography and vegetation cover. Hence, any observed difference between the soil under the tree canopies and in the open farmland is most likely due to the effects of the trees on the soil, rather than to mineralogical or textural differences between the soils. And it is discussed as it is induced by the presence of *F. thonningii* trees in the landscape.

[1] Nutrient concentration in this study refers to the amount of nutrients in %, ppm, mg per g, mg per kg etc.

Soil bulk density (BD)

Comparisons of soil BD showed that there was a highly significant ($P<0.0001$) variation of BD with increasing distance from tree and a significant ($P=0.0002$) difference with soil depth (Table 1). However, the interaction effect of distance from tree trunk and soil depth was not significant ($P>0.05$). The surface and subsurface soil under canopy were 14.31 and 12.37% lower in density than the respective depth of soil outside the canopy. The surface soils under canopy were 5.7% lower than the subsurface soil under canopy and the surface soil outside canopy were 3.83% lower than the subsurface soil outside the canopy.

BD findings for this study are in concert with the findings of other scholars (Aweto and Dikinya, 2003; Tadesse et al., 2000; Pandey et al., 2000). For instance, results of Tadesse et al. (2000) on BD under and outside *Millettia* trees revealed that both the BD of the surface soils (0.61 g cm^{-3}) and the subsurface soils (0.76 g cm^{-3}) under the trees were lower than the bulk density of the surface soils (0.69 g cm^{-3}) and the subsurface soils (0.80 g cm^{-3}) in the open areas. Pandey et al. (2000) also reported lower BD under the canopy of *A. nilotica* in a traditional agroforestry system in central India, while Aweto and Dikinya (2003) reported lower BD under the canopies of *P. africanum* and *C. apiculatum* as compared to open rangeland from Botswana. Moreover, Jiregna et al. (2005) also reported lower BD levels under *Cordia africana* and *Croton macrostachyus* canopies as compared to open farm while Enideg (2008) reported though not significantly higher bulk density outside the canopy of *F. thonningii* as compared to the canopy zone in Ethiopia.

Lower soil BD under the tree species' canopies is presumably due to the effect of litter addition to the soil. This has resulted from organic matter build up in the soil under the canopies relative to levels in soil outside the canopies. Also, the higher concentration of tree roots near the base of the trees may have had the effect of loosening the soil, thereby reducing soil BD. Furthermore, the soil outside the tree canopies dries out more, being exposed to direct solar radiation. This not only accelerates thermally induced soil organic matter decomposition, but results in the shrinking of organic matter and clay colloids, thereby making the soil more compact.

Soil moisture content (MC)

The analysis of variance for soil MC revealed that soil MC was highly significantly ($P<0.0001$) affected by distance from the tree trunk and significantly ($P=0.0002$) affected by soil depth (Table 1). However, the interaction effects of soil depth and distance from the tree trunk was not statistically significant ($P>0.05$). The surface and subsurface soils under canopy were 19.19 and 26.00% higher than the surface and subsurface soils outside the canopy zone. Besides, the surface soil under canopy zone were 6.12% higher than the subsurface soil under canopy zone while the surface soil outside the canopy zone were 12.87% higher than the subsurface soil outside the canopy zone.

Results of the present study on percent soil MC corroborate with the findings of Tadesse (2000). Results of percent soil MC under and outside *Millettia* trees found by the authors revealed that both the MC of the surface soils (19.60%) and the subsurface soils (10.00%) under the trees were higher than the MC of the surface soils (15.90%) and the subsurface soils (8.90%) in the open areas. Besides, Abebe (2006) reported a significant variation in MC that varies with tree species and with geographical location of tree species from the Highland of Harargie in Ethiopia.

The relatively higher soil MC under *F. thonningii* canopy as compared to soil MC beyond canopy and higher surface soil as compared to subsurface soils might be due to variation in Soil organic matter (SOM). SOM makes the soil to retain water by increasing its surface area and improving the soil structure. It might be also due to the shading effect of the tree. Generally, the soil outside the tree canopies might dry out more, being exposed to direct solar radiation whereas the shade provided by the trees of *F. thonningii* would have enhanced the MC under their canopy.

Soil chemical properties

Soil pH

The analysis of variance revealed a statistically significant difference in soil pH with distance ($P<0.0001$) from tree trunk and soil depth ($P=0.0063$) (Table 2). However, the interaction effect of soil depth and distance from tree trunk interaction for soil pH was not significant ($P=0.8318$). Soil pH showed an increasing trend with increasing distance from tree trunk and increasing soil depth. The mean soil pH values of the surface and subsurface soil beyond the canopy zone were 4.92 and 4.79% higher than the respective soil depths under canopy zone. Moreover, the mean soil pH of the subsurface soil of the under canopy and beyond canopy were 1.52 and 1.40% higher than the surface soil of their immediate soil depths. Generally, lower mean soil pH values were reported for soils under the tree canopy as compared to soils outside canopy and for the surface soil as compared to their immediate subsurface soils. The generally lower soil pH values under canopy of *F. thonningii* as compared to canopy gap might be due to several mechanisms that release H^+ ions, such as soil base cation uptake (or depletion) by the tree, decomposition of organic matter to organic acids and

Table 2. Soil chemical fertility parameters as influenced by radial distance and depth in Ahferom district.

Soil fertility parameter	Depth (cm)	Radial Distance			Overall
		Mid canopy	Canopy edge	Canopy gap	
pH	0 - 15	6.6±0.04	6.80±0.07	7.05±0.07	6.82±0.06^{b}
	15 - 30	6.7±0.04	6.93±0.06	7.12±0.05	6.92±0.05^{a}
	Overall	6.65±0.03^{C}	6.87±0.05^{B}	7.08±0.04^{A}	
EC (dSm^{-1})	0 - 15	0.03±0.00	0.05±0.00	0.07±0.01	0.05±0.01^{b}
	15 - 30	0.04±0.00	0.06±0.00	0.10±0.01	0.06±0.01^{a}
	Overall	0.04±0.00^{C}	0.05±0.00^{B}	0.08±.01^{A}	
%N	0 - 15	0.22±0.02	0.17±0.01	0.14±0.01	0.17±0.01^{a}
	15 - 30	0.18±0.01	0.15±0.01	0.13±0.01	0.15±0.01^{b}
	Overall	0.20±0.01^{A}	0.16±0.01^{B}	0.13±0.01^{C}	
%OC	0 - 15	1.84±0.09	1.29±0.04	1.02±0.05	1.38±0.09^{a}
	15 - 30	1.41±0.03	1.10±0.05	0.93±0.06	1.15±0.05^{b}
	Overall	1.62±0.08^{A}	1.20±0.04^{B}	0.97±0.04^{C}	
C/N	0 - 15	8.85±0.47	7.73±0.20	7.30±0.17	7.96±0.23^{a}
	15 - 30	7.92±0.29	7.33±0.10	7.43±0.19	7.56±0.13^{b}
	Overall	8.38±0.30^{A}	7.53±0.12^{B}	7.37±0.12^{B}	
P (ppm)	0 - 15	21.10±0.91	20.17±0.79	15.34±0.70	18.87±0.75^{a}
	15 - 30	20.39±0.73	15.55±0.67	14.23±0.70	16.72±0.75^{b}
	Overall	20.75±0.57^{A}	17.86±0.85^{B}	14.79±0.50^{C}	
K (ppm)	0 - 15	51.5±1.71	49.83±1.51	33±2.71	44.78±2.31^{a}
	15 - 30	50.67±1.67	34±2.67	31.5±2.74	38.72±2.45^{b}
	Overall	51.08±1.16^{A}	41.91±2.80^{B}	32.25±1.85^{C}	

Means along the same column (soil depth) and rows (radial distance) with different superscripts are significantly different ($P<0.05$); Values are Mean ± SEM.

CO_2, root respiration and nitrification. Rhodes (1997) suggested that increased accumulation of aboveground biomass and associated cation uptake by the tree component of agroforestry systems as possibly one of the causes for decreased pH in soils.

According to FAI (1977), soils having pH value in the range 6.50 to 8.70 are considered normal, that do not require treatment, and are optimum for most crops. The mean soil pH of the soil under canopy of *F. thonningii* ranged from 6.6 to 6.93 while the soil pH of the soil beyond canopy ranged from 7.05 to 7.12. As the soil pH under canopy and outside canopy of *F. thonningii* are within the normal recommended range, the soils can support most agricultural crops.

Findings of this research are in agreement with Pandey et al. (2000) and Hailemariam et al. (2010). Pandey et al. (2000) observed soil pH value that differed among canopy positions ($P<0.0001$) and soil depths ($P<0.0001$), and soil pH, averaged across all the depths, under mid canopy was 7.71% lower compared to that under canopy gap, while Hailemariam et al. (2010) found, lower pH value under canopy (7.96) than beyond canopy (8.22) showing a decrease of 3.10% under *Balanites aegyptiaca* at Goblel and Korbebite sites from Northern Ethiopia. From a study of single tree influence on soil properties in forest ecosystem, Rhodes (1997) noticed a lower pH that ranges from 4.90 to 6.10 and 5.10 to 6.80 under mid canopy and canopy edge respectively to 5.50 to 6.90 beyond canopy under *Pseudotsuga menziesii*, *Pinus ponderosa, Libocedrus decurrens* which is also in line with the present study findings.

Contrary to the present study, other researchers (Tadesse et al., 2000; Jiregna et al., 2005) reported higher pH values for both surface and subsurface soils under tree canopy as compared to outside canopy zone. Tadesse et al. (2000) reported, though not significant, higher pH value under canopy of *Millettia* as compared to open fields. Moreover, Jiregna et al. (2005) from their study on trees on farms and their contribution to soil fertility parameters in Badessa, Eastern Ethiopia reported that the presence of isolated *C. africana* and *C. macrostachyus* trees on farms in Badessa area had no significant influence on pH.

Soil electrical conductivity (EC)

The analysis of variance for soil EC revealed that soil EC were significantly affected by distance ($P<0.0001$) from tree trunk and soil depth ($P=0.0010$) (Table 2). Whereas EC were not significantly ($P=0.1144$) affected by the interaction of distance from tree trunk and soil depth. The surface soil at canopy gap were 1.78 times higher than the surface soil at canopy zone while the subsurface soil at canopy gap were 1.88 times higher than the subsurface soil under canopy zone. The subsurface soil under canopy was 1.20 times higher than immediate surface soil whereas the subsurface soil under canopy gap was 1.27 times higher than the immediate surface soil. The generally lower soil EC under tree canopies as compared to soils outside canopy might be due to the increased accumulation of aboveground biomass and associated cation uptake by the *F. thonningii* tree.

FAI (1977) suggested that soils with EC value of below 0.80 dS m^{-1} are considered normal and suitable for all crop types. The EC soils under canopy of *F. thonningii* ranged from 0.03 to 0.06 dS m^{-1} while that of beyond canopy ranged from 0.07 to 0.10 dS m^{-1}. Hence, the soils under and outside the canopy of *F. thonningii* are suitable for most crops.

The findings of the current study seem to be in agreement with Bhojvaiw et al. (1996) that conducted a research on reclaiming sodic soils for wheat production by *Prosopis juliflora* and found increased soil EC under *P. juliflora* as compared to that of the non sodic farm soil in India.

Soil organic carbon (SOC) and total soil carbon stock (SCS)

SOC was significantly affected by both the main effects ($P<0.0001$) of distance from tree trunk and soil depth and their interactions ($P=0.0006$) (Table 2). SOC decreased with increasing distance from tree trunk and with increasing soil depth. The surface SOC of canopy zone was 1.54 times higher than the respective soil subsurface outside the canopy zone while the SOC of subsurface soil were 1.35 times higher than the subsurface soil of outside canopy zone. Similarly, SOC of the surface soil under canopy was 25% higher than the immediate subsurface soil. Besides, SOC of the subsurface soil under canopy was 9.14% more than the immediate subsurface soil. The high SOC and carbon stock under canopy as compared to soil beyond canopy might be due to the organic matter inputs from fine root degeneration and litter fall. The SOC of the soil under canopy of *F. thonningii* of the present study ranged from 1.10 to 1.84% under canopy zone and from 0.93 to 1.02% under canopy gap. Hence, the soils under canopy zone and the surface soil outside canopy zone can be categorized under high SOC range whereas the subsurface soil outside canopy is in the medium range of SOC value according to Bhandari and Tripathi (1979 cited in Agena, 2009).

For the same species, Enideg (2008) reported SOC of surface and subsurface soils under canopy zone higher by 46.69 and 23.01%, respectively as compared to the surface and subsurface soil beyond the canopy. Many authors also reported a higher level of SOC concentration under the canopy of trees as compared to beyond the canopy and a higher level of SOC concentration of surface soils compared to their immediate subsurface soil for other species which is in concert with the present investigation. For instance, Abebe (1998) from his evaluation of the contribution of scattered *C. africana* trees to soil properties in cropland and rangeland ecosystems in Western Oromia, Ethiopia found soil organic carbon concentration of the top soil under canopy of *C. africana* in both the rangeland and cropland ecosystems that were respectively 35 and 43% higher than their immediate subsurface soil. Besides, he also found greater by 16, 17 and 6% SOC under canopy at a distance of 0.5, 2 and 4 m as compared to 15 m distance in the open area beyond canopy. Similarly, from their study on *Millettia ferruginea* impacts on soil fertility Tadesse et al. (2000) observed significantly higher SOC in the surface and subsurface soil beneath canopy as compared to the open field outside canopy of *Millettia*.

Soil carbon stock[2] (SCS) of the surface soil at the mid canopy and canopy edge were 46.51 and 14.66% higher than the surface SCS beyond canopy while the subsurface SCS of the mid canopy and canopy edge were 29.00 and 10.46% higher than their respective subsurface SCS beyond canopy (Figure 1).

SCS of surface (0 to 15 cm) and subsurface (15 to 30 cm) soil layers were done separately and the data were later pooled to represent SCS (Mg ha^{-1}) of the 0 to 30 cm soil layer. The results thus obtained are given in Figure 2. The SCS of the present study ranged from 22063 to 30812 kg ha^{-1} for the soils under tree canopy and from 19974 to 21030 Kg ha^{-1} for the soils outside tree canopy. The results of the pooled SCS revealed that the SCS of the 0 to 30 cm under mid canopy and canopy edge respectively were 38 and 12.63% higher than the SCS of the 0 to 30 cm under canopy gap. Comparing the SCS of the present study (19974 to 30812 kg ha^{-1}) with those reported by Enideg (2008) for the same species, the SCS of the present study was higher. Enideg (2008) reported a SCS that ranged from around 13000 to 26000 kg ha^{-1}, this variation in SCS might be the soil depth (15 cm of this against 10 cm of Enideg) considered.

Soil nitrogen (N)

Soil N concentration also showed significant variation

[2] Nutrient stock refers to the amount of nutrients on hectare basis

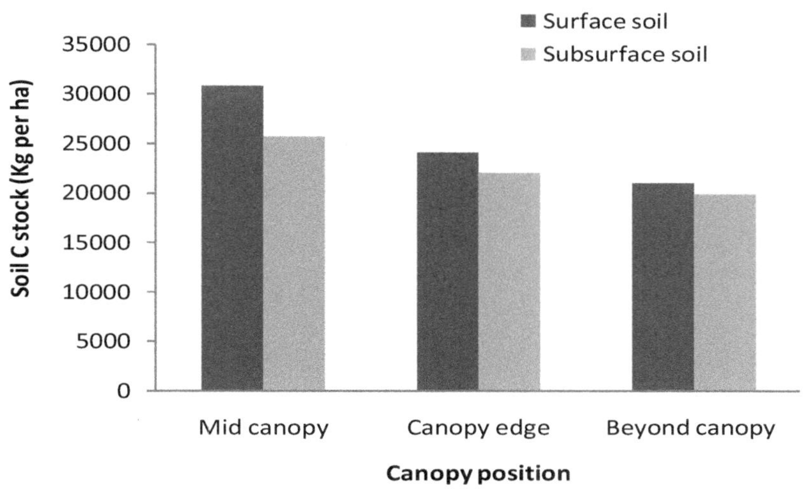

Figure 1. Soil carbon stock as influenced by canopy positions and soil depths.

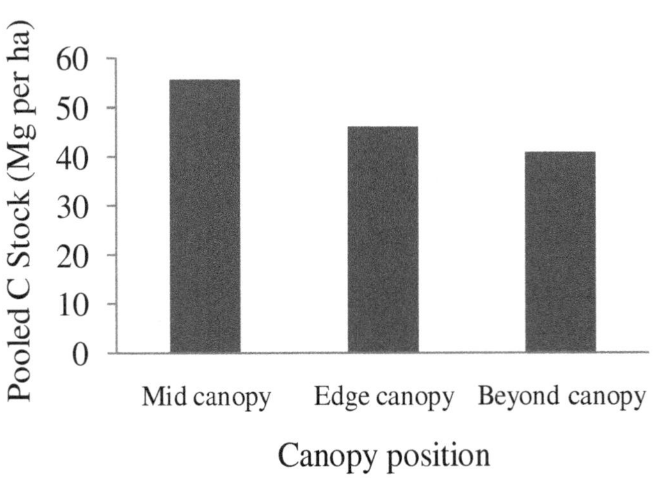

Figure 2. Pooled soil C stock at the 0 to 30 cm layer as influenced by canopy positions.

between soil depths and among the different distances from tree trunk (P<0.0001 and P<0.0001, respectively) (Table 2). However, soil N was not significantly (P=0.2336) affected by the interaction of soil depth and distance from tree trunk. Soil N concentration showed decreasing trend with increasing distance from tree and with increasing soil depth. The surface soil under canopy was 36.91% more in soil N concentration than the respective soil outside tree canopy. Likewise, the subsurface soil N concentration under tree canopy was 32% higher than the soil outside canopy zone. Moreover, the surface soil under canopy was 16.16% higher than the immediate subsurface soil under canopy while the surface soil outside canopy zone was 12% higher than its immediate subsurface soil. The higher total soil N accumulation under canopy zone as compared to outside canopy zone of the present study might be attributed to high accumulation of organic matter under tree canopies. Reports of the present study agree with the findings of Enideg (2008) and Jiregna et al. (2005) who found higher level of soil total N under canopy as compared to the open field. The former author reported an increase of in average total nitrogen under *F. thonningii* canopy by 85% in the surface soil and by 63% in the subsurface soil depths as compared to soils in the open pasture. Yet, the latter authors reported total soil N of surface and subsurface soils higher under tree canopies by 22 to 26% for *C. africana* and 12 to 17% *C. macrostachyus* than the corresponding soils away from the tree canopies. Besides, Tadesse et al. (2000) found significantly higher total soil N of both surface and subsurface soil under the canopy of *Millettia* as compared to open area outside the canopy zone of *Millettia*.

Carbon to nitrogen ratio (C/N)

The concentration of soil organic carbon to soil nitrogen (C/N) ratio differed significantly for all the distances (P=0.0002) from tree trunk and soil depths (P=0.0375) but it did not differ for the interaction (P=0.0761) of distance from tree trunk and soil depths (Table 2). The soil C/N ratio decreased with increasing distance from tree trunk and increasing soil depth. The surface and subsurface soil C/N under canopy were 13.58 and 2.58% higher than the surface and subsurface soil C/N outside canopy, respectively. Besides, the surface soil C/N under canopy and outside canopy were 8.74% higher and 1.83% lower than their immediate subsurface soils. The higher C/N ratio under canopy zone as compared to outside canopy zone of the present study might be attributed to high accumulations of organic matter (OM) under tree canopies through litter fall and root degeneration.

Table 3. Correlation of selected soil physical and chemical properties with organic carbon, distance from the trees and soil depth in Ahferom district.

Property	R		
	Organic carbon	Distance	Depth
Moisture content	0.82**	- 0.78**	- 0.28^{ns}
Bulk density	- 0.85**	0.86**	0.27^{ns}
pH	- 0.65**	0.80**	0.22^{ns}
Electrical conductivity	- 0.89**	0.79**	0.27^{ns}
Nitrogen	0.91**	- 0.70**	- 0.30^{ns}
Organic carbon	-	- 0.81**	- 0.36*
C:N ratio	0.56**	- 0.52**	- 0.25^{ns}
Phosphorus	0.60**	- 0.74**	- 0.33^{ns}
Potassium	0.62**	- 0.75**	- 0.30^{ns}

*, ** and ^{ns}=significant at P<0.05, P<0.01 and non-significant at P<0.05, respectively. R=correlation coefficient.

Results of the present study are not in agreement with findings of different authors (Enideg, 2008; Tadesse et al., 2000). For instance, Enideg (2008) reported a C/N ratio that increases with increasing distance from *F. thonninigii* and with increasing soil depth. Yet, Tadesse et al. (2000) observed lower C/N ratio under *Millettia* tree than in the open areas for both the surface and the subsurface soils, despite irregularities at the canopy edges.

Soil phosphorus (P)

Soil P concentration was affected by the main effects of distance (P<0.0001) from tree trunk and soil depth (P<0.0001) and their interaction (P<0.0001) effect (Table 2). Soil P concentration decreased with increasing distance from tree trunk and increasing soil depth. The mean soil P concentration of the surface and subsurface soil under canopy were 34.56 and 26.24% higher than the respective surface and subsurface soil outside the canopy zone. Besides, the soil P concentration of surface soil under and outside tree canopy were 14.85 and 7.75% higher than their immediate subsurface soils. The higher soil P accumulation under canopy zone as compared to outside canopy zone of the present study could be due to high accumulation of OM under tree canopies.

In support to the present study, Tadesse et al. (2000) observed available soil P concentration in the surface soils that were significantly higher under the trees than in the open fields; and the surface soil values were higher than the subsurface. Whereas Enideg (2008) reported even though the average P content under canopy was 12 and 5% higher than the open pasture in the surface and subsurface soil depths respectively for the same species; there was no significant difference in P content between the soils under canopy and open pasture, contradicting the present investigation.

Soil potassium (K)

Mean soil K concentration showed a highly significant variation with increasing distance (P<0.0001) from the base of the tree trunk and soil depth (P<0.0001) and their interaction (P<0.0001) effects (Table 2). The surface soil K concentration under canopy zone was 60.00% higher than the surface soil K concentration outside canopy zone while the subsurface soil K concentration under canopy was 38.80% the subsurface soil outside canopy zone. The soil surface K concentration under canopy zone and outside canopy zone were 19.69 and 3.83% higher than the respective soil K concentration outside canopy zone. The higher soil K accumulation under canopy zone as compared to outside canopy zone of the present study could be due to high accumulation of OM under tree canopies. In support of this study Tadesse et al. (2000) reported surface soils that were all significantly higher in soil K concentration under the *Millettia* trees than in the open fields; and the surface soil K concentration values that were higher than the subsurface ones. Besides, Abebe (2006) reported that soil K that were highly significantly (P<0.01) affected by all forms of the main effects (tree species, distance from the tree, soil depth and location) and all forms of their interaction effects from Harrargie of Ethiopia for *Acacia albida, C. africana* and *C. macrostachyus* while Enideg (2008) reported similar results under canopy of *F. thonningii* from Gondar, Ethiopia.

Pearson correlation analysis of soil physicochemical properties

Simple correlation analyses were carried out for selected soil physicochemical properties with soil organic carbon, soil depth and distance away from the tree trunk (Table 3). The correlation coefficients revealed that percent SOC was positively and highly significantly correlated with

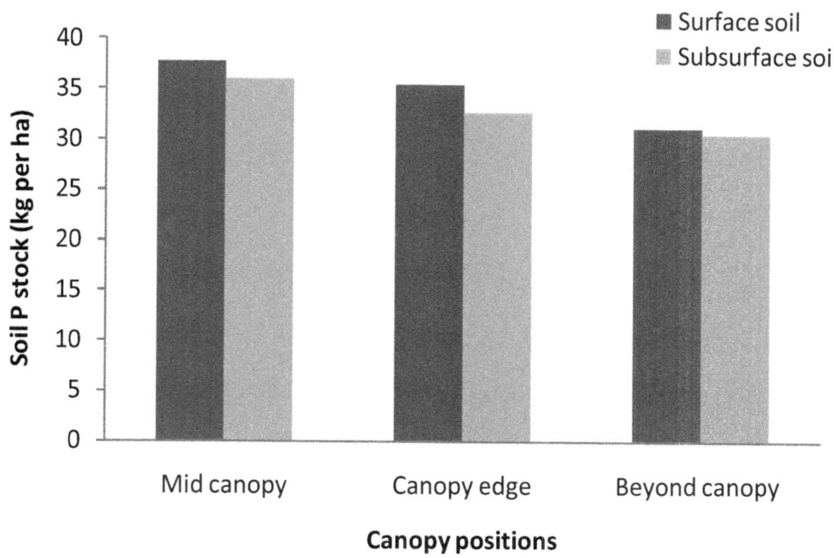

Figure 3. Soil P stock as influenced by canopy position and soil depth.

most (percent MC, C/N and soil P and soil K concentration) of the selected soil physicochemical properties except with soil BD, soil pH and EC which it correlated negatively, yet highly significantly.

The analysis of simple linear correlation coefficients revealed that there was a positive and highly significant correlation between distance from tree trunk and soil BD, soil pH and soil EC while there was a negative but highly significant correlation between distance from tree and most of the selected soil physicochemical properties viz. percent soil MC, soil N, soil OC, C to N ratio, soil P and soil K concentrations. This highly significant correlation revealed that the SOC greatly influenced important soil chemical properties. Hence, maintaining SOC of the soil may improve the important soil physicochemical properties.

Except for percent SOC, that significantly and negatively correlated with depth, the simple linear correlation coefficients revealed that all the selected soil physicochemical properties had a non-significant correlation with soil depth. Soil BD, soil pH and soil EC correlated positively with depth. Nevertheless, the rest of soil physicochemical properties, viz. MC, SOC, C/N, soil N, soil P and K soil concentration, correlated negatively with soil depth.

From his study on the importance of *F. thonningii* in soil fertility improvement in Gondar Zuria of Ethiopia, Enideg (2008) observed that soil N that had strong positive correlation ($P<0.01$) with SOC while P showed significant correlation with SOC ($P=0.01$) and soil N ($P=0.05$). From his correlation results the author revealed that P in these soils is predominantly found in organic forms not an inorganic chemical binding. Similar results were also reported by Abebe (2006) from Harargie of Ethiopia under *A. albida, C. africana* and *C. macrostachyus.*

Soil stock of macronutrients (kg/ha)

Soil phosphorus stock (Kg/ha)

Soil P stock of the soils varied from 31.19 to 37.78 Kg/ha for the surface soil while it ranged from 30.58 to 36.05 kg/ha for the subsurface (Figure 3). Soil P stock of soils under canopy ranged from 32.76 to 37.78 kg/ha while it ranged from 30.58 to 31.19 kg/ha for soils beyond canopy. The P stock of the surface soil under mid canopy and canopy edge were 21.13 and 13.63% higher than the surface soil beyond canopy. Besides, the subsurface soil under mid canopy and edge of canopy were 17.89 and 7.13%, respectively higher than the subsurface soil under canopy gap. The generally higher soil P stock under canopy as compared to way canopy could be due to higher OM accumulation under canopy.

FAI (1977) regarded soils having P stock of higher than 22.40 kg per ha to be high in P to support crop growth. Hence, the soils under study have soil P level which is above the high level required for normal crop growth.

Soil potassium stock (kg/ha)

Soil K stock of the surface soil differed from 31.19 to 37.78 kg/ha, while the soil K stock of the subsurface varied from 30.58 to 36.05 kg/ha (Figure 4). K stocks of the soils under canopy zone of the tree ranged from 73.07 to 93.20 kg/ha while it ranged from 67.86 to 70.86 kg/ha for soils outside canopy. The K stock of the surface soil under mid canopy and canopy edge were respectively 31.53 and 22.38% higher than the surface soil beyond canopy. Similarly, the subsurface soil under mid canopy and edge of canopy were 32.90 and 8.54% higher than the subsurface soil under canopy gap. Soils

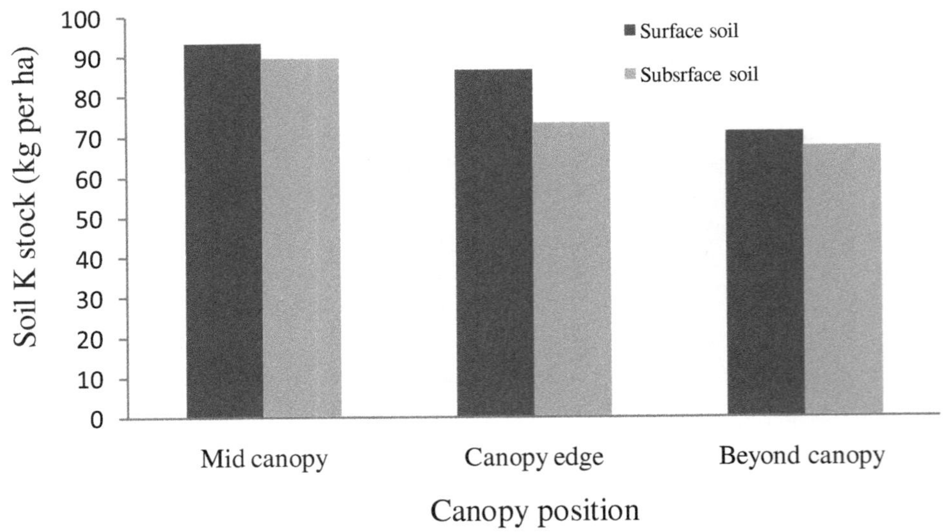

Figure 4. Soil K stock as influenced by canopy position and soil depth.

under canopy had higher level of soil K stock as compared to those outside canopy and surface soils had as well, higher level of soil K stock as compared to subsurface soils of their immediate layer. This higher level of soil K stock under canopy as compared to outside canopy could be due to the high accumulation of OM under canopy.

However, higher levels of soil K stock were reported under canopy as compared to outside canopy; all soils under study (both under and outside canopy and surface and subsurface soils) were rated as low (less than 137 kg per ha) in soil K according to FAI (1977). This trend was also evident from soils under *Tithonia diversifolia;* a plant which is rich in foliar N, P and K (Jama et al., 2000).

The soil P of the present study, however, is lower than those reported by Enideg (2008) for the same species from Gondar Zuria of Ethiopia. The author recorded soil P that ranged 864.19 to 1017.16 kg/ha for soil under canopy and from 792.85 to 847.44 kg/ha. This high variation in soil P stock could be due to the fact that he considered total P, while available P was considered in this study.

CONCLUSION AND RECOMMENDATION

F. thonningii, selected as useful multipurpose tree by farmers in the landscape, positively and significantly influenced soil nutrients to be higher via protection against leaching, translocation of nutrients from deeper to the surface layer and accumulation of litter which created a temporary nutrient pool in the surface soils under the canopies. The current study revealed that presence of *F. thonningii* trees in the land use have enhanced soil physico-chemical properties by increasing soil moisture content, %N, P, K, %OC and C/N ratio both vertically and horizontally while by lowering soil bulk density, pH and EC. However, its presence had no effect on soil texture (sand, silt and clay fractions) as texture is more related to parent material than tree influence. Hence, retention of *F. thonningii* in the farmland and grazing fields as well as its planting in degraded areas within its agro-ecological zone should be widely considered. Since N and P stock of soil under and away *F. thonningii* showed sufficient level and even above the recommended range for optimum growth of most agricultural crops, its inclusion into agricultural system can save fertilizer input cost for applying the respective nutrients. On the other hand, since K stock is rated low for optimal crop growth, sustainable crop production in the district with integration of *F. thonningii* requires application of supplementary K fertilizers.

ACKNOWLEDGEMENTS

We gratefully acknowledge the financial assistance from Institute for sustainable development (ISD) and thank Axum Agricultural Research Centre staff members for their help in data collection.

REFERENCES

Abebe N (2006). Soil Fertility Status under Indigenous Tree Canopies on Farmland in Selected Areas of Eastern and Western Haragehe Zone. M.Sc. Thesis. Haramaya University, Haramaya.

Abebe Y (1998). Evaluation of the Contribution of Scattered *Cordia africana* Trees to Soil Properties in Cropland and Rangeland Ecosystems in Western Oromia, Ethiopia. M.Sc. Thesis, ISSN 1402 - 201X (1998:36) SLU, Sweden.

Agena A (2009). Component Interactions and Their Influence on the Production of Apple Based Agroforestry System in Wet Temperate Zone of Himachal Himalayas. Doctoral Thesis. Dr. Yashwant Singh Parmar University of Horticulture and Forestry, Nauni, Solan-173-230 (H.P.), India.

Aweto AO, Dikinya O (2003). The Beneficial Effects of Two Tree Species on Soil Properties in a Semi-Arid Savanna Rangeland in Botswana. Land Contam. Reclam. 11(3):339-334.

Badege B, Abdu A (2003). Agroforestry and Community Forestry for Rehabilitation of Degraded Watersheds in the Ethiopian Highlands. The Second EAF International Symposium on Contemporary Development Issues in Ethiopia: Addis Ababa, Ethiopia (July 11-13, 2003).

Bhojvaiw EP, Timmer VR, Singh G (1996). Reclaiming Sodic Soils for Wheat Production by *Prosopis juliflora* (Swartz) Dc Afforestation In India. Agroforest. Syst. 34:139-150.

BoARD (2010). Ahferom District Bureau of Agriculture and Rural Development 2010 Annual Reports, Ahferom District, Central Tigray, Inticho.

Brady NC, Weil RR (2002). Nature and Properties of Soils. 13th ed. New York, USA. P. 2.

Bremner JM, Mulvaney CS (1982). Nitrogen-Total. In: Page AL, Miller RH, Keeney DR (eds). 1982. Methods of Soil Analysis. Part 2. Chemical and Microbiological Properties. Agronomy Monograph No. 9, 2nd eds. Am. Soc. Agron. Madison, WI. pp. 595-622.

Enideg D (2008). Importance of *Ficus thonningii* Blume in Soil Fertility improvement and Animal Nutrition in Gondar Zuria, Ethiopia. M.Sc. Thesis, University of Natural Resources and Applied Life Science, Vienna.

FAI (1977). Hand Book of Fertilizer Usage. Fertilizer Association of India, New Delhi. India.

Hailemariam K, Kindeya G, Yamoah C (2010). *Balanites aegyptiaca*, a Potential Tree for Parkland Agroforestry Systems with Sorghum in Northern Ethiopia. J. Soil Sci. Environ. Manag. 1(6):107-114.

Jackson ML (1973). Soil Chemical Analysis. Prentice Hall of India Pvt. Ltd., New Delhi. India.

Jama B, Palm CA, Buresh RJ, Niang A, Gachengo, C, Nziguheba G, Amadalo B (2000). *Tithonia diversifolia* as a Green Manure for Soil Fertility Improvement in Western Kenya: A Review. Agroforest. Syst. 49:201-221.

Jiregna G, Rozanov A, Legesse N (2005). Trees on Farms and Their Contribution to Soil Fertility Parameters in Badessa, Eastern Ethiopia. Biol. Fertil. Soils 42:66-71.

Kindeya G (2004). Dry Land Agroforestry Strategy for Ethiopia. Paper Presented at the Drylands Agroforestry Workshop 1st-3rd September 2004. ICRAF Headquarters, Nairobi-Kenya.

Merwin HD, Peech M (1951). Exchangeability of soil potassium in sand, silt and clay fraction as influenced by the nature of complementary exchangeable cations. Soil Sci. Soc. Am. 15:125-128.

Neufeldt H, Wilkes A, Zomer RJ, Xu J, Nang'ole E, Munster C, Place F (2009). Trees on Farms: Tackling the Triple Challenges of Mitigation, Adaptation and Food Security. World Agroforestry Centre Policy Brief 07. World Agroforestry Centre, Nairobi, Kenya. P. 2.

Olsen SR, Cole CV, Watanabe FS, Dean LA (1954). Estimation of Available Phosphorus in Soil by Extraction with NaHCO3; U.S. Department of Agriculture Circular, U.S. Government Printing Office, Washington, D.C. P. 939.

Pandey CB, Singh AK, Sharma DK (2000). Soil Properties under *Acacia nilotica* Trees in a Traditional Agroforestry System in Central India. Agroforest. Syst. 49:53-61.

Rhodes CC (1997). Single-Tree Influences on Soil Properties in Agroforestry: Lessons From Natural Forest and Savannah Ecosystems. Agroforest. Syst. 35:71-94.

SAS (2002). User's Guide: Statistics Version 9.00. SAS Institute, Inc., Cary, North Carolina. USA.

Schnitzer M (1982). Total Carbon, Organic Matter and Carbon. In: Page AL, Miller RH, Keeney DR (eds) Methods of soil analysis. Part 2, 2nd edn. Am. Soc. Agron. Madison, WI. Agronomy Monograph 9:539-577.

Tadesse H, Legesse N, Olsson M (2000). *Millettia ferruginea* from Southern Ethiopia: Impacts on Soil Fertility and Growth of Maize. Agroforest. Syst. 48:9-24.

Teklay T, Malmer A (2004). Decomposition of Leaves from Two Indigenous Trees of Contrasting Qualities under Shaded-Coffee and Agricultural Land-Uses During the Dry Season at Wondo Genet, Ethiopia. Soil Biol. Biochem. 36:777-786.

Farmers' perception of soil fertility problems and their attitudes towards integrated soil fertility management for coffee in Northern Tanzania

Godsteven P. Maro[1] , Jerome P. Mrema[2], Balthazar M. Msanya[2] and James M. Teri[1]

[1]TaCRI, Lyamungu P. O. Box 3004, Moshi, Tanzania.
[2]Department of Soil Science, Faculty of Agriculture, Sokoine University of Agriculture, P. O, Box 3008, Morogoro, Tanzania.

A study was conducted in Hai and Lushoto districts, Northern Tanzania to establish the farmers' perception of soil fertility problems and their attitudes towards integrated soil fertility management (ISFM) for coffee, thereby identifying the appropriate intervention strategies. The study was based on a structured questionnaire involving 126 respondents. Both farmers' awareness of the problem and their attitudes were highly significant (at p<0.01). Age, household size, and adoption of improved coffee varieties, affected farmers' awareness significantly (p<0.05). As for farmers' attitudes, six of the eight predictors were significant (p<0.05). Age, household size, adoption of new varieties and total farm income were highly significant (p<0.01). Age, total land under coffee and total off-farm income negatively affected farmers' attitudes. As farmers get older, they tend to refrain from innovation. Larger farms are likely to exert more pressure on the available organic resources. With multiple farms, distant farms are likely to receive less attention. When off-farm income was considered, multiple ventures compete for the farmers' time, resources and attention. For the two districts, ISFM interventions will make a better impact to younger and more energetic farmers with sufficient lands for commercial coffee production and to farmers who depend largely on this resource for their livelihood.

Key words: Soil fertility, farmers' perception, integrated soil fertility management (ISFM), coffee, Tanzania.

INTRODUCTION

Coffee is one of the major export crops in Tanzania contributing to 23% of the agricultural GDP (Anonymous, 2007). It contributes directly to the livelihoods of over 420,000 farm families and indirectly to over 2 million people employed in the coffee value-chain industry (Carr et al., 2003). Arabica coffee contributes 65% of the Tanzanian total coffee export. The Tanzanian coffee, especially the washed Arabica is one of the best in the world ranked among the rare category of "Colombian Milds" used to blend other inferior coffees.

Coffee is also grown in many countries in East and Central Africa. Other important coffee producers are Ethiopia, Uganda, Kenya, Rwanda and Burundi. According to statistics from International Coffee Organization (ICO, 2011), total production for the six countries was 10.6, 11.4 and 12.9 million bags for 2008, 2009 and 2010, respectively. Tanzania's share was 11.14, 6.2 and 7.08%, while Kenya's share was 5.08,

5.51 and 6.56%. Ethiopia and Uganda together commanded over 70% of the share for all the three years. The Tanzanian average smallholder coffee productivity per hectare ranges between 250 and 300 kg of parchment which is very low compared to the potential yield of over 1000 kg per tree (Baffes, 2003; Hella et al., 2005). In Kenya, coffee yields were reported to have fallen from 892 $kg.ha^{-1}$ in 1980 to 284 $kg.ha^{-1}$ in 2006, much lower than average yields for Arabica coffee worldwide of 698 kg/ha and yields of 1160 kg/ha in Rwanda and 995 kg/ha in Ethiopia.

Soil fertility degradation is one of the major problems facing coffee productivity in Tanzania. It is defined by Stocking and Murnaghan (2000) as the loss of soil physical and nutritional qualities. It has been an issue of concern throughout the Sub-Saharan Africa (SSA), and cuts across many different soils and crops (Okalebo et al., 2007). In Tanzania, the problem covers all coffee growing zones and all types of coffee growers (Envirocare, 2004). Reports from Kenya indicate that decline in coffee yields were caused by farmers' reluctance to invest in fertilizers (Condliffe et al., 2008) which translates to poor soil fertility.

Integrated soil fertility management (ISFM) has been cited by many authors, including Okalebo et al. (2007), Gumbo (2006) and Raab (2002), as the key approach in raising productivity levels in agricultural systems while maintaining the natural resource base. It is described by Vanlauwe and Zingore (2011) as a set of soil fertility management practices that necessarily include the use of fertilizer, organic inputs, and improved germplasm combined with the knowledge on how to adapt these practices to local conditions, aiming at maximizing agronomic use efficiency of the applied nutrients and improving crop productivity. Because of the pressing need for global food security, many articles have been published which relate ISFM to the production of annual food crops like maize (Ikerra et al., 2007; Kimani et al., 2007), and rice (Kaizzi et al., 2007), giving lesser attention to perennial crops like coffee. It is no longer wondering then that the role of ISFM for coffee in Tanzania and the socio-economic perception of it have not been studied to any significant detail.

The coffee producing zone of Northern Tanzania comprises four regions, namely Arusha, Kilimanjaro, Manyara and Tanga (a total of 12 districts). Coffee production is both historical and traditional, especially in Kilimanjaro region which was the first to grow coffee as a commercial crop (Maro et al., 2010). Annual coffee production trend for the zone indicates a decline over the years. A number of constraints have been suggested as the cause of this decline. Currently, as reflected during the coffee stakeholders' forum (Tanzania Coffee Board [TCB], 2009), soil fertility degradation has emerged as the most limiting factor. This is however, a very generic perception which needs to be studied in detail by targeting specific locations and farming communities.

The current study was therefore conducted in Hai and Lushoto districts to establish the magnitude of soil fertility problem as perceived by farmers in the two districts, and to establish the baseline farmers' attitudes towards ISFM, thereby identifying the appropriate intervention strategies.

METHODOLOGY

A structured questionnaire was administered to farmers in Hai and Lushoto districts to solicit the farmers' opinion on soil fertility and coffee productivity. The two districts were selected as representative of coffee growing areas of Northern Tanzania, and also representative of soils with contrasting geological backgrounds; originating from volcanic and gneissic parent material, respectively. The coffee areas in the districts were categorized by altitudinal zones: low (900-1100), medium (1100-1400) and high (>1400) m above mean sea level and respondents were randomly selected on basis of having at least 50 coffee trees. A total of 60 respondents were interviewed in Lushoto and 66 in Hai, making a total of 126 respondents. Generic questions included personal details (gender, age, level of education, position in the household, household size and sources of coffee management information) and farm details (size, number of trees and varieties). Additionally, respondents were requested to give an account of their knowledge of soil problems, source of ISFM knowledge if any, experience in industrial fertilizer use with coffee and negative effects if any, usage of organics (manure, coffee processing by-products, mulches, green manure plants), major and subsidiary income sources and income ranges last season.

The data were processed and analyzed by using the Statistical Package for Social Sciences (SPSS version 16) (SPSS Inc, 2007). The analysis involved computations of mean and frequency, together with two linear regressions: one on farmers' appreciation of soil fertility problem and the other on farmers' attitude towards ISFM.

Defining the variables

The degree of appreciation of soil fertility deterioration as a problem (aP) was described as a mean of two ratings, one qualifying the farmers' knowledge of their soils (0, 1 and 2 for no, slight and basic knowledge, respectively) and the other qualifying farmers' understanding of soil related problems (0 = no idea, 1 = could identify other problems, 2 = could identify crop-related problems and 3 = was able to identify nutritional disorders). The ratings were categorized as 0, 0.5, 1.0, 1.5 and 2.0 for unaware, slightly aware, moderately aware, sufficiently aware and fully aware, respectively. The assumption was that, as noted by D'Emden et al. (2005), awareness of a problem is a motivator in devising (or adopting) problem-solving techniques.

Attitude towards ISFM (α) was described as a mean of eight ratings including the two stated above (R_{soil} and R_{prob}) and six others. R_{ind}, R_f and R_b are dummy variables qualifying whether a farmer uses (1) or does not use (0) industrial fertilizers, farmyard manures or coffee by-products, respectively. R_{fp} and R_{bp} at the scale of 0, 1, 2 and 3 are the ratings qualifying farmers who do not process because they do not use farmyard manure or pulp, those who use the organics raw without any processing, those who just heap the material to stabilize in the open, and those who compost the material in a pit. R_{train} is a rating that qualifies whether and how many times last year a farmer received training on ISFM (an aggregate of four topics – soils, ISFM, identification of nutritional problems and making of organic composts): 0 = no training, 1 = trained once, 2 = trained twice and 3 = trained more than twice. The resultant ratings varied between 0 and 2, and were clustered at maximum values in terms of readiness to adopt ISFM interventions

Table 1. Description of clustered ratings.

Cluster	Maximum value	Description
0	0	Minimum likelihood of adoption
0.1-0.5	0.5	Will need a lot of time and conviction to adopt
0.6-1.0	1.0	Will need some time and conviction to adopt
1.1-1.5	1.5	Will need little time and conviction to adopt
1.5-2.0	2.0	Will adopt readily.

Table 2. A comparison of the selected predictors per district.

Predictor	Unit	Means	95% C.I		Notes
			Lower	Upper	
A	Years	60.83	58.37	63.29	Coffee is a crop for old people
ED	Rating	1.23	1.09	1.37	Majority primary, fewer ordinary
HS	Rating	2.37	2.21	2.54	2 to 8 persons per household
FEX	Years	30.08	27.3	32.86	People with immense coffee exp.
LS	Ha	0.8	0.68	0.92	Typical smallholders
CV	0=no, 1=yes	0.33	0.24	0.41	Adoption of 24-41%
FI	Rating	9.13	8.39	9.86	600,000 to 900,000 TZS
NFI	rating	1.83	1.05	2.6	Maximum of 200,000 TZS

as shown in Table 1.

Descriptive statistics

The two variables aP and α were exposed to descriptive statistics following the models of Nkamleu (2007) and Zhou et al. (2008) which involved physical counts and percentage frequency, and were compared per district.

Regression modelling

The defined variables aP and α were separately exposed to a linear regression model as functions of demographic predictors (age and level of education of the household head, the size of the household, farm and non-farm income) as defined by Doss (2003) and farm related predictors (such as land size and types of coffee trees). Both models used the same predictors as shown in the example below which represents aP

$$aP = b_0 + b_1A + b_2ED + b_3HS + b_4FEX + b_5LS + b_6CV + b_7FI + b_8NFI + \ell$$

Where: b_0 represent the constant, b_1A = coefficient related to age, b_2ED = coefficient related to level of education, b_3HS = coefficient related to household size, b_4FEX = coefficient related to coffee farming experience in years, b_5LS = coefficient related to total coffee land size, b_6CV = coefficient related to coffee varieties (whether improved varieties are adopted), b_7FI = coefficient related to farm income last year, b_8NFI = coefficient related to non-farm income last year, e = random error of prediction.

Each of the eight predictors were then assessed in terms of the significance level at which it influences the farmers' awareness of soil fertility decline as a problem on one hand, and the farmers' readiness to adopt ISFM interventions on the other.

RESULTS AND DISCUSSION

The significance of predictors per district

The eight selected predictors were compared per district (t-test) and were all highly significant ($p < 0.01$). Means and their 95% confidence intervals are shown in Table 2. Average age of respondents was around 60 years, implying that coffee is still held by old people. This observation was in line with Morris and Venkatesh (2000), Mateos-Planas (2003) and Tiamiyu et al. (2009). Education level was mainly primary, with fewer cases of post-primary education. Majority of households have 2-8 persons, which is average for many Tanzanian households (ILFS, 2001; Kamuzora, 2001). With the mean coffee farming experience of 30 years, it implies that most of the coffee farmers in the study districts have immense experience in their business, and their perception of soil problems and best ways to manage soil fertility should be considered in devising appropriate ISFM packages (Douthwaite et al., 2002).

Land size of mean 0.8 ha (CI 0.68-0.92) implies that the people we are dealing with are truly smallholders who are resource-poor, and therefore, the ISFM packages should have that in mind. An average of 33% of the respondents have adopted the new improved varieties released by TaCRI. This implies that there is still an uphill task for TaCRI and other coffee stakeholders to promote these varieties among farmers.

The distribution of farm and off-farm incomes in

Table 3. A summary of farm and off-farm incomes in 2009/2010.

Category	Farm income (%)	Off-farm income (%)
None	0	74.6
<0.3 m	14.4	11.0
0.3 – 2.0 m	76.0	9.6
>2.0 m	9.6	4.8

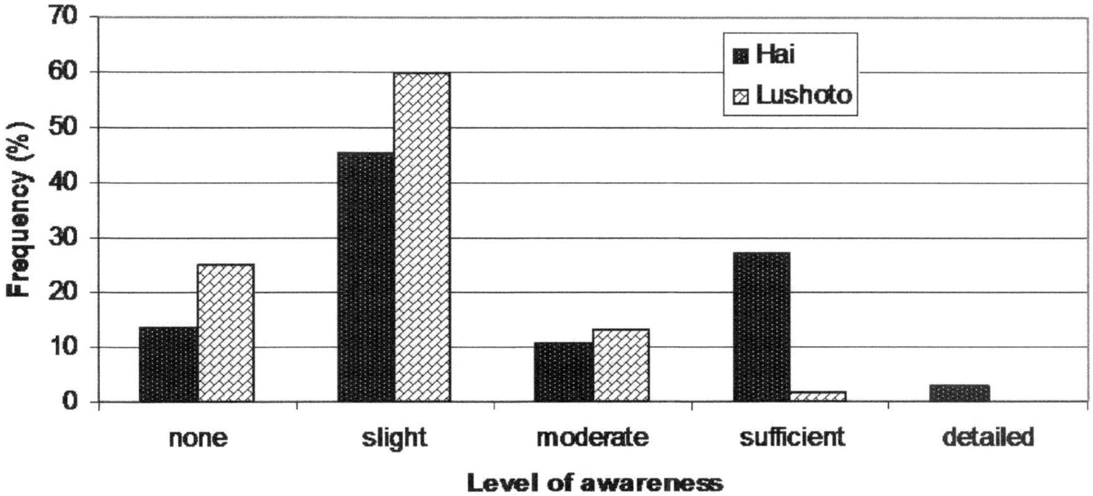

Figure 1. Distribution of awareness of soil fertility decline as a problem.

2009/2010 is given in Table 3. Farm income appears to be fairly normally distributed with the majority ranging between 0.3 and 2 m Tanzania Shillings (equivalent to US$ 190-1250 at the current exchange rate of TZS 1,600 per US$). With off-farm income, 74.6% of the respondents reported to have none, thus depending entirely on the farm for their livelihood. Those who have subsidiary off-farm incomes (25.4%) may portray variable pictures as regards farm attention. For some, it may be a deterrent factor, keeping the farmer busy with the off-farm ventures at the expense of the farm. For elite farmers however, a subsidiary off-farm income can act as a buffer against fluctuating coffee prices, and/or a stimulant in adopting good agricultural practices (Karki and Bauer, 2004).

The distribution of variables per district

The frequency of farmers' awareness of soil fertility degradation as a problem is shown in Figure 1. The majority of respondents from Lushoto are either unaware (25%) or slightly aware (60%). On the other hand, 9% had sufficient awareness and 0% fully aware. In Hai, the unaware and slightly aware groups were 13.6% and 45.4%, respectively, while 3.0% are fully aware, 10.6% moderately aware and 27.3% sufficiently aware. The results appear to correlate well with the respondents' levels of education, whereby 6.67% in Lushoto and 25.53% in Hai reached post-primary education.

As for attitude towards ISFM (Figure 2), the distribution of respondents in Hai was fairly normal, with a peak at 50% for moderate attitude group, tailing at very low (1.5%) and very high (7.6%). The Lushoto distribution was rather irregular with only one interesting feature that the percentages that have moderate and high attitudes are equal at 40% each, therefore constituting the bulk of the sample. The percentage of respondents with moderate to very high attitudes was 84% for Hai and 92% for Lushoto, implying that ISFM intervention will have greater impact in the latter.

Analysis of regression models

A summary of the regression models for problem appreciation and attitude towards ISFM is given in Table 4.

Problem appreciation

The regression model for problem appreciation (aP) was

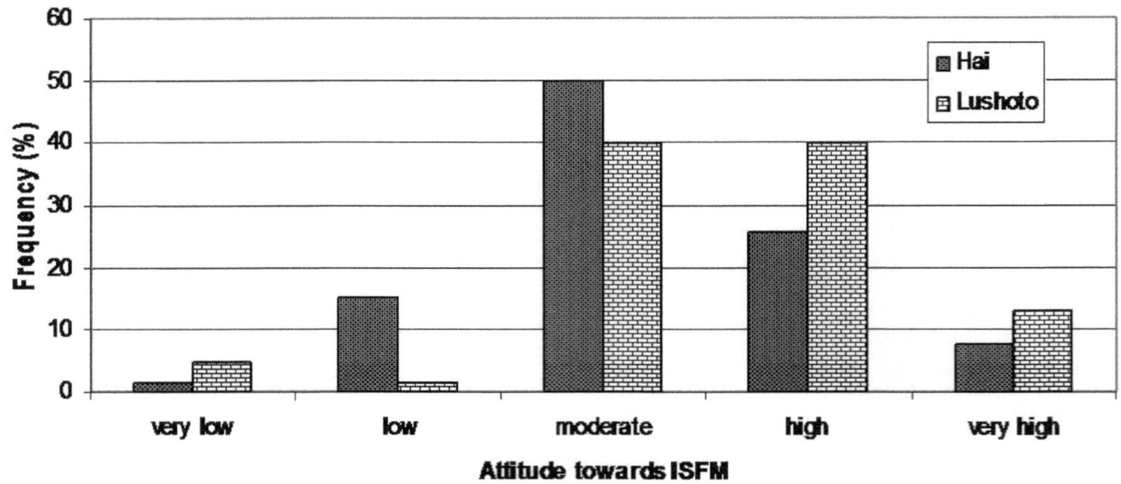

Figure 2. Distribution of attitudes of farmers towards soil fertility management.

Table 4. Model summaries for problem appreciation and attitude towards ISFM.

Predictor	Problem appreciation			Attitude towards ISFM		
	β	t	Sign	β	t	Sign
Age	-0.163	-1.597	0.113	-0.350	-3.103	0.002
Level of education	0.041	0.447	0.656	0.113	1.319	0.190
Household size	0.251	2.761	0.007	0.235	2.785	0.006
Years growing coffee	0.079	0.763	0.447	0.288	2.530	0.013
Coffee land size	0.165	1.743	0.084	-0.185	-2.083	0.039
New varieties adoption	0.228	2.553	0.012	0.422	5.022	0.000
Farm income last year	0.087	0.956	0.341	0.227	2.659	0.009
Off-farm income last year	-0.110	-1.159	0.249	-0.145	-1.654	0.101
(Constant)		1.747	0.083		3.953	0.000

highly significant (at $p<0.01$) even though there was a rather poor correlation (Adjusted R^2 of 0.133) among the parameters entered. Only household size and adoption of improved coffee varieties were highly significant ($p<0.01$), while age was significant at $p<0.05$. Age was seen to negatively affect the farmers' awareness of soil fertility problem as older people tend to become more passive about what happens in their farms (Truong and Yamada, 2002). The rest did not show any statistical significance; including level of education. The relationship between household size and problem appreciation is not very clear. However, if family members are trained in diagnosing unusual characteristics in the field, the bigger the hosehold size, the more likely it is for problems to be identified.

During the survey in Lushoto, 75% of respondents who had slight to sufficient awareness about soil fertility degradation also had considerable information about soil fertility management. Similar observations had been noted at Makueni District, Eastern Kenya by Kimiti et al. (2007). Their indigenous technical knowledge (ITK) showed that "mishai" trees (*Albizzia maranguensis)* contribute in restoration/maintenance of soil fertility. Other ITKs learnt during the survey include the "tugutu" bush (*Adhatoda engleriana* Lindau, family Acanthaceae) which is also medicinal (Moshi et al., 2005). It has been tested with other crops and found to have high nutrient release potential. A formulation for making liquid fertilizer from their leaf extract was described. This opens an avenue for further research on the nutrient content of the"tugutu" leaves and ways in which this, where present, can be integrated in the local ISFM packages for coffee.

Attitude towards ISFM

The regression model was also highly significant (at $p<0.01$). Of the 8 parameters used in predicting α (attitude towards ISFM), 4 were highly significant (Age, household size, adoption of new varieties and total farm income) and 2 were significant at $p<0.05$ (land size and coffee farming experience). These observations are partly in line with those of Jamala et al. (2011). Level of education showed positive but insignificant influence on farmers' attitudes. The significance of education level in affecting adoption was reported by Barungi and Maonga (2011), Tiamiyu et al. (2009), Ono (2006) and Ani et al. (2004), which does not appear to be true in the study areas.

Age, total land under coffee and total off-farm income had negative B, β and t values. Age showed to negatively influence the capacity and willingness to adopt new approaches including ISFM. This is in line with the observations by Nzomoi et al. (2007). The fact that total coffee land showed negative relationship with attitude towards ISFM (contrary to the observation by Karki and Bauer, 2004) can only be explained in two scenarios: larger farms exerting more pressure on the limited amounts of available organic sources of nutrients like FYM; and farmers having multiple farms, some a distance away from their households, thus precluding efforts to use organic sources in those distant farms (Vanlauwe and Giller, 2006; Nkamleu, 2007). Off-farm income showed negative influence on farmers' attitudes, observations that are in line with those of Adolwa et al. (2010). If this source of income contributes substantially to the total family income, the farmers' attention gets skewed from coffee towards the other ventures.

Conclusions

The results from this study led to the following conclusions:

The perception of soil fertility degradation as a problem in the study areas is influenced by several household and farm variables. Among the eight predictors, only the age of household head, the size of the household and adoption of new improved varieties showed to be responsible for variation in perception, with the former having a negative impact. More farmers in Hai are aware of soil fertility problem than in Lushoto.

Attitudes towards ISFM showed to be highly influenced by age, household size, adoption of new varieties and total farm income; and moderately influenced by total land under coffee and number of years spent by the household head in coffee business. Again here, age showed a negative relationship to attitude towards ISFM, implying that older people are usually skeptical in adopting new approaches. The percentage of respondents with moderate to very high attitudes was higher for Lushoto than Hai.

In the two districts, ISFM interventions will make a better impact to younger and more energetic farmers with enough land for commercial coffee production and who depend largely on this source for their livelihood. These are the ones who can easily adopt improved varieties and good agricultural practices, including ISFM practices like mulching, composting of farmyard manure, coffee pulp and other field residues. Long-term plans should be to encourage younger people to take up the coffee farming business, build the capacity to monitor the soil fertility regularly and give quick, site-specific recommendations. Also, promotion of the improved coffee varieties among farmers should continue.

Future directions

This is the first in a series of studies aimed at developing an effective and spatial ISFM decision support system for coffee in Northern Tanzania. It has effectively opened up our knowledge of soil fertility problems as perceived by farmers. The next steps will be to explore the extent of the problem through soil fertility surveys, and then decide on the right ISFM packages that will make impact in the study areas. The findings will be useful for TaCRI in planning for ISFM intervention in the Northern Zone.

REFERENCES

Adolwa IS, Esilaba AO, Okoth PO, Mulwa MR (2010). Factors influencing uptake of integrated soil fertility management knowledge among smallholder farmers in Western Kenya. Proc. KARI Biennial Scientific Conference. P. 9.

Ani AO, Ogunnika O, Ifah SS (2004). Relationship between socio-economic characteristics of rural women farmers and their adoption of farm technologies in Southern Ebonyi State, Nigeria. Int. J. Agric. Biol. 6(5):802-805.

Anonymous (2007). The National Economic Situation. Ministry of Planning, Economy and Empowerment. United Republic of Tanzania (URT) Topcom Printers, Dar es Salaam, Tanzania. P. 217.

Baffes J (2003). Tanzania's Coffee Sector: Constraints and Challenges in a Global Environment.http://www.aec.msu.edu/fs2/inputuseworkshop/Baffes_Tanzania_Coffee.pdf. Site visited on 30th June 2007.

Barungi M, Maonga BB (2011). Adoption of soil management technologies by smallholder farmers in Central and Southern Malawi. J. Sustain. Dev. Afr. 13(3): 28-38.

Carr MKV, Stephens W, Van der Vossen HAM, Nyanga A (2003). Tanzania Coffee Research Institute Strategic Action Plan 2003-2008: Contributing towards a profitable and sustainable coffee industry in Tanzania. Report to TaCRI, ICPS, Cranfield University, Silsoe, UK. P. 161.http://www.dev.tacri.org/uploads/media/TaCRI_Strategic_Action_Plan_2003-2008_01.pdf visited 21 February 2012

Condliffe K, Kabuchi W, Love C, Ruparell R (2008). Kenya coffee: A cluster analysis. Harvard Business School. P. 34. http://www.isc.hbs.edu/pdf/Student_Projects/Kenya_Coffee_2008.pdf visited 22 February, 2012.

D'Emden FH, Llewellyn RS, Burton MP (2005). Adoption of conservation tillage in Australian cropping regions: An application of duration analysis. Elsevier Sci. Direct, Technol. Forecasting Soc. Change. 73:630-647.

Doss CR (2003). Understanding Farm Level Technology Adoption: Lessons Learned from CIMMYT's Micro Surveys in Eastern Africa. CIMMYT Economics Working Paper 03-07. Mexico, D.F.: CIMMYT.

Douthwaite B, Manyong VM, Keatinge JDH, Chianu J (2002). The adoption of alley farming and mucuna: Lessons for research, development and extension. Kluwer Academic Publishers, Agroforest. Syst. 56:193-202.

Envirocare (2004). A study on the importance of coffee industry in Kilimanjaro region. A NORAD Project, Moshi, Tanzania. P. 33.

Gumbo D (2006). Integrated soil fertility management. Technical Brief,

Practical Action Southern Africa, Harare, Zimbabwe. P. 5.
Hella JP, Mdoe NS, Lugole JS (2005). Coffee baseline report for Tanzania Coffee Research Institute. Bureau for Agricultural Consultancy and Advisory Service, Sokoine University of Agriculture, Morogoro, Tanzania. P. 40.
ICO (International Coffee Organization) (2011). Coffee export statistics by country of origin. http://www.ico.org/historical/2000+/PDF/EXPORTS.pdf (visited 30 January, 2012)
Ikerra ST, Semu E, Mrema JP (2007). Combining Tithonia diversifolia and Minjingu Phosphate Rock for improvement of P availability and maize grain yields on a Chromic Acrisol in Morogoro, Tanzania. In: A. Batiano et al (eds). Advances in ISFM in Sub-Saharan Africa: Challenges and Opportunities. Springer. pp. 333-344
ILFS (Integrated Labour Force Survey) (2001). Household analysis in Tanzania. Chapter 10: 123-128. http://www.tanzania.go.tz/ilfs%5CChapter%2010.doc visited 21 February, 2012.
Jamala GY, Shehu HE, Garba AT (2011). Evaluation of factors influencing farmers' adoption of irrigated rice production in Fadama soil of North Eastern Nigeria. J. Dev. Agric. Econ. 3(2):75-79.
Kaizzi CK, Ssali H, Nansamba A, Vlek PLG (2007). The potential benefits of Azolla, velvet bean and N fertilizers in rice production under contrasting systems in Eastern Uganda. In: A. Batiano et al (eds). Advances in ISFM in Sub-Saharan Africa: Challenges and Opportunities. Springer. pp. 423-433.
Kamuzora CL (2001). Poverty and family size patterns: Comparison across African countries. Research Report No. 01.3, REPOA: 44 pp. http://www.repoa.or.tz/documents/01.3_Poverty_and_Family_Kamuzora_fin_17-04.pdf visited 21 February, 2012
Karki LB, Bauer S (2004). Technology adoption and household food security. Analyzing factors determining technology adoption and impact of project intervention: A case of smallholder peasants in Nepal. Proceedings of Deutscher Tropentag Workshop, 5-7 October, 2004, Humboldt University, Berlin. pp. 1-8.
Kimani SK, Esilaba AO, Odera MM, Kimenye L, Vanlauwe B, Bationo A (2007). Effects of organic and mineral sources of nutrients on maize yields in three districts of Central Kenya. In: A. Batiano et al (eds). *Advances in ISFM in Sub-Saharan Africa:* Challenges and Opportunities. Springer. pp. 353-357.
Kimiti JM, Esilaba AO, Vanlauwe B, Batiano A (2007). Participatory diagnosis in the Eastern Drylands of Kenya: Are farmers aware of their soil fertility status? In: A. Batiano et al (eds). Advances in ISFM in Sub-Saharan Africa: Challenges and Opportunities. Springer. pp. 961-967.
Maro GP, Kitalyi A, Nyabenge M, Teri JM (2010). Assessing the impact of land degradation on coffee sustainability in Kilimanjaro region, Tanzania. *In:* Proceedings of the 23rd ASIC Conference, 3 -8 October, 2010, Nusa Dua, Bali, Indonesia. pp. 607-614.
Mateos-Planas X (2003). Technology adoption with finite horizons. Elsevier Science-Direct. J. Econ. Dyn. Control 28:2129-2154
Morris MG, Venkatesh V (2000). Age differences in technology adoption decisions: Implications for a changing work force. Personnel Psychology. 53(2):375-403.
Moshi MJ, Kagashe GAB, Mbwambo ZH (2005). Plants used to treat epilepsy by Tanzanian traditional healers. J. Ethnopharmacol. 97:327-336.
Nkamleu GB (2007). Modelling farmers' decisions in ISFM in Sub-Saharan Africa: A multinomial logit analysis in Cameroon. In: Batiano, A. et al (eds). Advances in ISFM in Sub-Saharan Africa: Challenges and Opportunities. TSBF-CIAT, Nairobi, Kenya. pp. 891-903.
Nzomoi JN, Byaruhanga JK, Maritim HK, Omboto PI (2007). Determinants of technology adoption in the production of horticultural export produce in Kenya. J. Bus. Manage. 1(5):129-135.
Okalebo JR, Othieno CO, Woomer PL, Karanja NK, Semoka JMR, Bekunda MA, Mugendi DN, Muasya RM, Bationo A, Mukhwana EJ (2007). Available technologies to replenish soil fertility in East Africa. In: Batiano, A. et al (eds). Advances in ISFM in Sub-Saharan Africa: Challenges and Opportunities. TSBF-CIAT, Nairobi, Kenya. pp. 45-62.
Ono Y (2006). Technology adoption in a community of heterogeneous education level: Who are your good neighbours? Econ. Bull. 15(8):1-11.
Raab RT (2002). Fundamentals of Integrated Soil Fertility Management. IFDC Training materials for the "Training Program on Integrated Soil Fertility Management (ISFM) in the Tropics", Lome, Togo. P. 10.
Stocking M, Murnaghan N (2000). Land degradation: Guidelines for field assessment. Overseas Development Group, University of East Anglia, Norwich, UK. P. 120.
Tanzania Coffee Board (TCB) (2009). Way forward in the Tanzanian coffee sector. Proceedings of the first Coffee Stakeholders' Conference, Arusha, Tanzania. P. 39.
Tiamiyu SA, Akintola JO, Rahji MAY (2009). Technology adoption and productivity difference among growers of new rice for Africa in savanna zone of Nigeria. Tropicultura. 27(4):193-197.
Truong TNC, Yamada R (2002). Factors affecting farmers' adoption of technologies in farming system: A case study in O'Mon district, Can Tho province, Mekong Delta. Omorice 10:94-100.
Vanlauwe B, Giller KE (2006). Popular myths around soil fertility management in Sub-Saharan Africa. Agric. Ecosyst. Environ. 116:34-46.
Vanlauwe B, Zingore S (2011). Integrated soil fertility management: An operational definition and consequences for implementation and dissemination. Better Crops. 95(3):4-7.
Zhou S, Herzfeld T, Glauben T, Zhang Y, Hu B (2008). Factors affecting Chinese farmers' decisions to adopt a water saving technology. Can. J. Agric. Econ. 56:51-61.

The effects of tillage systems on soil physical properties and water conservation in a sandy loam soil in Eastern Kenya

J. M. Miriti[1] , G. Kironchi[2], A. O. Esilaba[3], C. K. K. Gachene [2], L. K. Heng[4] and D. M. Mwangi[2]

[1]Kenya Agricultural Research Institute (KARI), ARC Muguga South Centre, P.O. 30148-00100 Nairobi, Kenya.
[2]Faculty of Agriculture, University of Nairobi, P.O. Box 30197-00100 Nairobi, Kenya.
[3]Kenya Agricultural Research Institute (KARI) Headquarters, P.O. Box 57811-00200 Nairobi, Kenya.
[4]International Atomic Energy Agency (IAEA), Soil and Water Management and Crop Nutrition Section, Wagrammer Strasse 5, P.O. Box 100, 1400 Vienna, Austria.

Soil water conservation through tillage is widely accepted as one of the appropriate ways of addressing soil moisture constraints in rainfed agriculture. A field experiment was conducted for 6 seasons (2007 to 2009) to evaluate the effects of 3 tillage practices namely ox-plough (OP), subsoiling-ripping (SR) and tied-ridge (TR) on soil physical and hydro-physical properties of a sandy loam soil. Soil surface roughness, bulk density (BD), total porosity, crust strength, saturated hydraulic conductivity, steady infiltration rates and soil water storage were evaluated. Results showed that, mean inter-row BD (1.51 Mg m^{-3}) and crust strength (3.05 MPa) with SR tillage were 7 and 15% greater ($P \leq 0.05$) than with OP, respectively. Inter-row infiltration rates for OP (7.9 cm h^{-1}) were more than two fold greater than for SR (3.6 cm h^{-1}) and TR (3.3 cm h^{-1}). Soil surface roughness with TR was significantly greater than with OP and SR tillage systems. Soil water storage for TR (222 mm) exceeded ($P \leq 0.05$) that for OP and SR by 30 mm (16%) and 32 mm (17%), respectively. This study concluded that, ripping soil along planting line that are prone to surface crusting without disturbing the soil between crop rows was not effective as a conservation tillage method.

Key words: Tillage, soil physical properties, soil water conservation, sandy loam soil, infiltration rate.

INTRODUCTION

Agricultural production in the arid and semi-arid areas of Kenya is highly dependent on rainfall because water for irrigation is scarce or farmers cannot afford the technology. In order to have successful rainfed crop production in these areas, rain water conservation is essential (Barron et al., 2003). The success of on-farm soil water conservation however depends upon many soil factors such as soil bulk density (BD), porosity, soil surface sealing and crusting, surface roughness, hardpans, hydraulic conductivity, and infiltration rates as they determine the hydrological properties of soil (Strudley et al., 2008). Soil compaction causes low porosity, reduced infiltration, increased penetration resistance and limited root growth. Soil surface roughness (configuration) influences wind and water erosion by decreasing soil detachment and transport caused by erosion (Cogo et al., 1983). Soil surface roughness can however change considerably with rain, wind and soil cultivation events (Guzha, 2004). Crusting is a soil surface characteristic that provides information

Table 1. Seasonal rainfall during the study period.

Season	Rainfall (mm)	Rainy days
Long rains of 2007	44	7
Long rains of 2008	125	18
Long rains of 2009	54	10
Short rains of 2007	203	30
Short rains of 2008	222	18
Short rains of 2009	200	29

on soil strength or penetration resistance. Higher crust strength has been reported to reduce soil water storage due to the reduction of crust conductance and infiltration rates (Gicheru et al., 2004; Shivonje et al., 2005).

Large areas of semi-arid areas of Kenya are covered by sandy soils which are prone to crusting and compaction leading to low rain water use efficiency (Biamah et al., 1993). It has been suggested that, tillage can improve the physical and hydro-physical properties of the soils and consequently increase rainwater harvesting and crop yields (Gachene and Kimaru, 2003, Rockstrom et al., 2009, Strudley et al., 2008). Ox-plough (OP) tillage using mouldboard plough followed by hand weeding with a hoe is a common tillage method used by farmers in semi arid areas of Kenya. This tillage system has been reported to encourage crust and hardpan formation (Biamah et al., 1993; Gitau 2004). In such soils, deep tillage in form of subsoiling-ripping (SR) has been recommended to improve soil hydraulic conditions (Rockström and Lars-Ove, 1999; Soltanabadi et al., 2008). It has also been reported that, tillage practices that increase soil surface roughness such as tied ridging and ripping can increase soil water storage and availability to crop because they are able to capture rainfall and increase the time for infiltration to take place resulting in better crop yields (Guzha, 2004). The challenge faced by farmers when using tied ridging in Kenya is the high labour required because this tillage system is not mechanized. According to Rockström et al. (2009) and Stevens (2000), SR is an attractive tillage system that farmers can use to harvest and conserve rainwater because less time is required in the tillage operations. Temesgen (2007) however noted that subsoiling a sandy loam soil along crop rows with minimum soil disturbance between rows resulted in reduced yields when compared to ploughing because of increased runoff.

Soils of Kenya are diverse and there is need to determine the appropriateness of different water conservation tillage systems. Currently, there is very little information available on appropriateness of subsoiling and ripping tillage technologies on the crust prone soils widely cultivated by farmers in the dry areas. Previous studies on ripping in Kenya reported by Rockström et al. (2009) was focused on crop yields and did not study its effects on soil properties and soil moisture conservation. The objective of this study was therefore, to determine the comparative effectiveness of SR, tied-ridge (TR) and OP tillage systems in improving soil hydraulic properties and soil water conditions in sandy loam soils in semi-arid eastern Kenya.

MATERIALS AND METHODS

Site characteristics and treatments

The field experiment was conducted at Kampi ya Mawe (1°50'S, 37°40'E, elevation 1155 m) in the experimental field of Kenya Agricultural Research Institute. Average annual rainfall is 600 mm with two peaks in April/May and in October/December (Jaetzold et al., 2006). The first and second rainy seasons were referred to as long rainy season and short rainy season, respectively. The field experiment was conducted for 6 rainy seasons: three long and three short rainy seasons between 2007 and 2009. Seasonal rainfall received at the site during the study period is given in Table 1. The soil is classified as chromic Luvisol (FAO-UNESCO, 1994) formed on undifferentiated basement system rock (gneisses). The soil surface 0 to 14 cm layer texture is sandy loam (80% sand, 7% silt, and 13% clay) with low organic carbon (0.61%) content (Table 2). The land slope at the study site was 0 to 2%.

The treatments consisted of 3 tillage practices: TR, SR and OP. SR tillage (about 24 cm depth) was done along the planting line (90 cm interval) with an ox driven soil subsoiler to break the hardpan. The subsoiler and ripper used were fabricated at the University of Nairobi in the Department of Environmental and Biosystems Engineering (Figure 1).

Tied ridges in TR tillage were made manually as this tillage method is not yet mechanized. Cross ties (small ridges joining big ridges) were made at 1 m intervals to prevent water movement along the furrow and create small check dams for rainwater harvesting. In each plot, there were 35 check dams. For the OP tillage treatment, the entire plot was ploughed (about 14 cm depth) using an oxen drawn mouldboard plough as practiced by farmers in the area. In each tillage method, 3 cropping systems namely sole maize (3.7 plants m^{-2}), sole cowpea (5.5 plants m^{-2}) and maize/cowpea intercrop (planted in the same row but in alternate hill) were compared; in a randomized split-plot design with tillage practices as the main plots and cropping systems as the sub-plots. When comparing the tillage systems, OP tillage treatment was used as control while sole maize was the control when comparing cropping systems. No crop rotation was done during the study. Each treatment was replicated 3 times. A ridge was constructed around the experimental block to prevent water run-on.

Data collection and analysis

At different stages of crop cycle, undisturbed soil cores were taken between and within crop rows for determination of soil saturated hydraulic conductivity, bulk porosity density and porosity changes. This was achieved by driving stainless steel cylinder (100 cm^3, approximately 4.0 cm in length, diameter 5.6 cm) in the soil using a hammer. The samples were then carefully trimmed both sides to the size of the cylinder and then taken to the laboratory for soil physical properties analysis. The saturated hydraulic conductivity of the soil was determined using the constant head method described by Klute and Dirksen (1982). After this determination, the soil cores were put in an oven and dried at 105°C to a constant weight. BD of the soil was calculated as the mass of dry soil divided by the core ring volume. Total soil porosity was calculated from BD as:

$$\text{Porosity (\%)} = \left(1 - \frac{BD}{PD}\right) \times 100 \tag{1}$$

Table 2. Soil characteristics at the Kampi ya Mawe experimental site.

Soil horizon	Depth (cm)	Sand (%)	Clay (%)	Silt (%)	BD (Mg m^{-3})	pH	Organic Carbon (%)
Ap	0-14	80	13	7	1.48	6.04	0.61
AB	14-27	63	29	8	1.54	5.32	0.50
Bu_1	27-50	48	39	13	1.44	5.78	0.57
Bu_2	50-83	47	37	16	1.33	5.28	0.58
BC	83-94	57	23	20	1.57	5.47	0.70
C	90-106+	61	32	7	1.59	5.32	0.72

Figure 1. The ripper and subsoiler used in the study.

Where BD is the soil BD (Mg m^{-3}) and PD is average particle density (2.65 Mg m^{-3}).

Soil surface roughness was measured immediately after the tillage operations and within the crop season. Readings were taken from 3 randomly selected positions in each plot. A pin-meter similar to that described by Moreno et al. (2008) was used to measure surface roughness (Figure 2). The instrument consisted of a graduated flat metal board on which a 130 × 90 cm frame with 20 needles that slide up and down was fitted. The spacing from one pin to the other was 20 cm. Each needle was fitted with a small foot to prevent it from penetrating the soil.

During the height measurement, the whole device was placed horizontally and the needles were then allowed to slide down until their feet freely touched the soil surface and then locked in position. The height of each of the needles above the top of the frame was recorded after taking the readings on the graduated board. The needles were then pulled up, locked and the instrument moved to the next measuring position. Measurements were replicated in 3 positions per plot and the following the formula was used to calculate surface roughness:

$$\text{Surface roughness (\%)} = LOG\left(STDEV\right)\times 100 \quad (2)$$

Where LOG is the logarithm, STDEV is the standard deviation of the pin height measurements. Penetration resistance was measured at the soil surface using a hand-held cone penetrometer (Stiboka penetrograph, Eijkelkamp Agrisearch Equipment, Netherlands). The penetration resistance was measured within and between crop rows at 13, 26, 56, and 101 days after sowing (DAS) during short rains cropping season of 2009. Ten soil crust strength measurements were taken at randomly selected positions in sole maize plot. The penetrometer springs and cone sizes were adjusted depending on the soil strength. The penetration resistance which represents the crust strength was calculated as:

$$CR = I \times \left(\frac{C_s}{AC}\right) \quad (3)$$

Where CR is the resistance (N cm^{-2}), I is the impression on the scale (cm), Cs is the spring constant (N cm^{-1}), AC is the area of the cone (cm^{2}).

Field infiltration rates were measured using a disc permeameter as described by Perroux and White (1988). Measurements were taken at planting, mid season, and at harvest. A 20 cm diameter steel ring was gently forced, approximately 5 mm into the soil surface, taking care not to disturb the soil and after organic material at soil surface was removed. A spirit level was placed over the ring to ensure the ring was horizontal. The clear polycarbonate disc was placed on the ring and adjusted until a 5 mm gap between the soil surface and the bottom of the disc was achieved. The valve was opened to allow water to flow from the side tube filling the gap between the soil surface and the disc. As water entered the soil, the time taken for a 5 cm depth of water drop in the graduated big tube was recorded. Water in the big tube was refilled immediately after

Figure 2. The pin meter used in the study

80 cm of water entered the soil. Infiltration test was then continued until the time taken by 5 cm of water in big tube to enter the soil was constant. Infiltration rate and cumulative infiltrations were then calculated from the recorded data. Soil water in the profile (0 to 110 cm) was monitored using a neutron probe (CPN 503-DR Campbell Pacific Nuclear, Martinez, CA). The measurements were done weekly during the crop cycle at 15, 30 45, 60 80, and 100 cm soil depths from aluminum access tubes installed in the plot along the middle crop row.

The data was subjected to analysis of variance (ANOVA) using the Genstat (2007) software package. Significant treatment differences were obtained using Fisher's Protected LSD test. Effects were considered significant in all statistical analysis for P values ≤ 0.05.

RESULTS

Soil bulk density (BD) and total porosity

Soil surface (0 to 10 cm) BD and total porosity data measured between the crop rows (inter-row) is presented in Table 3. BD and porosity were significantly affected by the tillage methods. The BD after sowing was greatest in SR (1.47 Mg m^{-3}), followed by TR (1.41 Mg m^{-3}) and lowest in OP tillage (1.34 Mg m^{-3}) which indicated that, OP significantly ($P \leq 0.05$) reduced inter-row BD by 5% (0.13 Mg m^{-3}) compared to untilled soil in SR plots. Initial total porosity was greatest for OP (49%) and least for SR (44%) with significant ($P \leq 0.05$) differences among the tillage systems.

There was a temporal increase in inter-row BD and a decrease of soil porosity as the season progressed. The greatest increase in BD 10% (from 1.34 Mg m^{-3} to 1.48 Mg m^{-3}) was recorded in the OP plots while BD changes in SR and TR plots were 6% (from 1.47 Mg m^{-3} to 1.56 Mg m^{-3}) and 5% (from 1.41 Mg m^{-3} to 1.48 Mg m^{-3}) respectively, by the end of season. Total porosity also declined in similar proportions. BD and total porosity within rows was not significantly different among the tillage methods.

Saturated hydraulic conductivity and infiltration

Soil surface (0 to 10 cm) saturated hydraulic conductivity (Ks) measured between crop rows (inter-row) is presented in Table 3. The Ks in TR (0.81 cm h^{-1}) was significantly ($P \leq 0.05$) lower than in OP (6.24 cm h^{-1}) and SR (4.33 cm h^{-1}) plots after sowing. The Ks data shows that, though there was a significant tillage differences in Ks at sowing, the Ks at 30 DAS and at harvest was smaller and not significantly different among tillage systems. The overall reduction in Ks ranged from 5.61 to 0.38 cm h^{-1} and the greatest decline was in OP followed by SR and least in TR plots. The Ks values within crop rows were not significantly different ($P \leq 0.05$) among tillage treatments during the crop cycle but they double the Ks observed in the inter-row.

Soil surface (0 to 5 cm) steady infiltration rates measured using a disc permeameter is given in Table 4. The highest inter-row infiltration rate was observed in the OP tillage (7.9 cm h^{-1}) and was two fold greater than the infiltration rate observed in SR and TR tillage systems. SR resulted in the least infiltration rates (27.4 cm h^{-1}) within crop rows compared to OP but the difference was not significant. The infiltration within the rows in the ridges was not measured because water was being lost through the sides of the ridge during the process of measurement. When averaged across OP and SR tillage systems, the infiltration rates within crop rows was 32.2 cm h^{-1}, significantly greater than the inter-row

Table 3. Effect of tillage treatments on inter-row row soil surface (0 to 10 cm) BD, porosity and saturated hydraulic conductivity at sowing, 30 DAS and at harvest during the short rainy season of 2009.

Tillage	Bulk density (Mg m^{-3})				Porosity (%)				Hydraulic conductivity (cm h^{-1})			
	Sowing	30 DAS	Harvest	Mean	Sowing	30 DAS	Harvest	Mean	Sowing	30 DAS	Harvest	Mean
Ox-plough	1.34	1.39	1.48	1.40	49.43	47.55	44.15	47.04	6.24	1.37	0.63	2.75
Subsoiling-ripping	1.47	1.49	1.56	1.51	44.53	43.77	41.13	43.14	4.33	1.07	0.97	2.12
Tied-ridge	1.41	1.45	1.48	1.45	46.79	45.28	44.15	45.41	0.81	0.80	0.43	0.68
Mean	1.41	1.44	1.51		46.92	45.53	43.14		3.79	1.08	0.68	
LSD (0.05)	0.02†	0.01‡	NS§		1.32†	1.59‡	NS§		1.51†	0.84‡	1.67§	

† = for comparing tillage means within a soil property, ‡ = for comparing date means within a soil property, § = for comparing the interaction, NS = not significant.

infiltrations rates (5.75 cm h^{-1}).

Soil crust strength

Soil crust strength is an important soil physical characteristic that provides information on ability to allow water movement into the soil. The temporal variations in soil crust strength with the tillage treatments during the long rain season of 2009 are shown in Figure 3. The inter-row crust strength responded to tillage and the crust strength for OP (0.15 MPa) was 213 and 166% significantly lower than for SR (0.47 MPa) and TR (0.40 MPa) respectively, 13 DAS. The crust strength increased with time in all tillage systems. When averaged across the tillage systems, crust strength at the end of the season was 4.8 MPa compared to crust strength (0.34 MPa) soon after tillage. OP maintained the least crust strength throughout the entire crop season. The low crust strength measured on 13 and 26 DAS in the TR plots was because of the loose soil deposited at the ridge basin during weeding and ridge reconstruction. In general, the crust strength recorded along crop rows was higher by a factor of 6 compared to the inter-row crust strength. Tillage did not have a significant effect on the crust strength along crop rows.

Soil surface roughness

Soil surface roughness is an unevenness of the soil surface. Temporal variations in soil surface roughness with the tillage treatments during are shown in Figure 4. Surface roughness was significantly different among the tillage methods and the effect of tillage was largest soon after cultivation and the effect decreased as the season progressed. The soil surface roughness was consistently highest in TR (64 to 79%) tillage and lowest in OP (6 to 50%). Surface roughness soon after tillage was 79, 58, and 50 for TR, SR and OP, respectively. At the time of harvesting, surface roughness had declined to 64, 20 and 6% in TR, SR and OP plots respectively, indicating that the greatest decline was with OP, followed by SR and then TR tillage.

Soil water storage

Seasonal rainfall at the site ranged from 44 to 222 mm per season during the study period (Table 1). The short rains of 2008 was the wettest season (222 mm), followed by short rainy season of 2007 (203 mm), short rains season of 2009 (200 mm), long rains season of 2008 (125 mm), long rains season of 2009 (54 mm) and finally the long rains season of 2007 (44 mm). There were crop failures during the long rains seasons of 2007 and 2009 failed due to low rainfall and hence the 2 seasons are not discussed in this paper.

Soil profile (0 to 110 cm) water content averaged over 4 seasons (long rainy season of 2008 and short rainy seasons of 2007, 2008, and 2009) responded significantly to tillage, cropping systems, and tillage × cropping systems interactions (Figure 5). When averaged across seasons and cropping systems, the highest soil water storage was observed under TR (222 mm) and exceeded ($P \leq 0.05$) that for OP and SR by 30 mm (16%) and 32 mm (17%), respectively. There was no significant soil water content difference between SR and OP tillage systems. The soil water content under cowpea sole crop and maize/cowpea intercrop cropping systems were similar but significantly less than the soil water content under sole maize cropping system, when averaged across tillage systems over the 4

Table 4. Effect of tillage treatments on steady infiltration rates between the crop rows (inter-row) and within rows during the short rains of 2009.

Tillage	Inter-row	Within row
Ox-plough	7.9	37.0
Subsoiling-ripping	3.6	27.4
Tied-ridge	3.3	-
Mean	5	32.2
LSD (0.05)	1.7	NS

NS = not significant

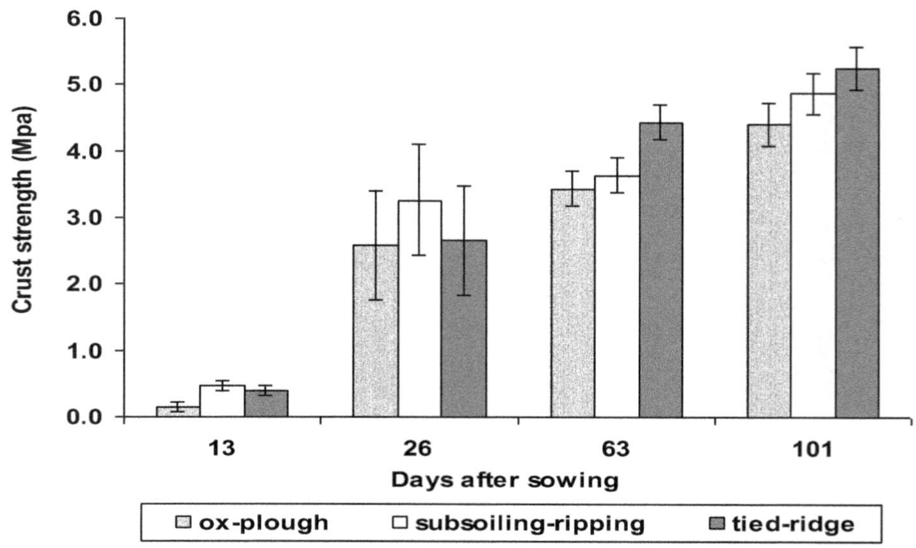

Figure 3. Effect of tillage on inter-row soil crust strength during short rains of 2009 season. Bars=$LSD_{0.05}$.

seasons. The tillage × cropping systems interactions indicated that, maximum soil water content (225 mm) was recorded in sole maize grown under TR and minimum soil water content (179 mm) in sole bean grown under SR tillage system.

DISCUSSION

Soil bulk density and total porosity

The low BD and high porosity values observed in OP plots is attributed to soil disturbance during ploughing. The high initial BD in SR compared to OP is because the compacted inter-rows which were not tilled during land preparation. Though the inter-row BD after sowing was different ($P \leq 0.05$) among tillage systems, it was statistically the same at the end of the season in all tillage treatments. Similar observations were reported by Šimanskaitė (2008). The temporal increase in BD during the crop season could be due to a combination of rain drop impact and soil compaction due to repeated human traffic during weeding, data collection and other crop management activities. The exposure of the hardpan during the construction of ridges caused the observed high inter-row BD in TR plots. BD within rows was not different among tillage methods but TR maintained the lowest BD than OP and SR because of the relatively lose soil on the ridge.

Soil crust strength

Ploughing caused a breakdown of soil surface sealing resulting in weak crust strength in OP, and by not ploughing the inter-rows in SR tillage resulted in soil surface sealing and strong crust. These results are in agreement with Gicheru et al. (2004) who observed that bare minimum tillage where the weeds were controlled by herbicides had a stronger crust strength at the soil surface

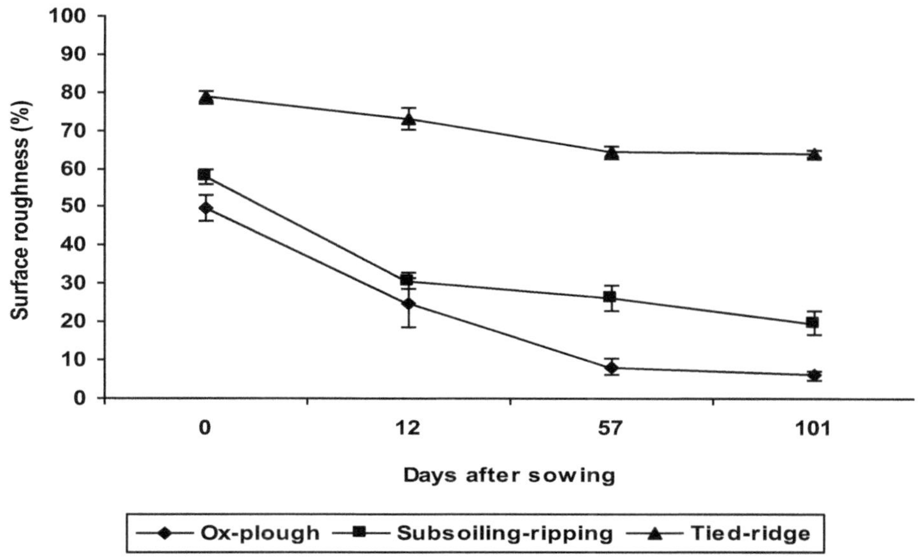

Figure 4. Soil surface roughness decline as affected by tillage during short rains of 2009 season. Bars=$LSD_{0.05}$

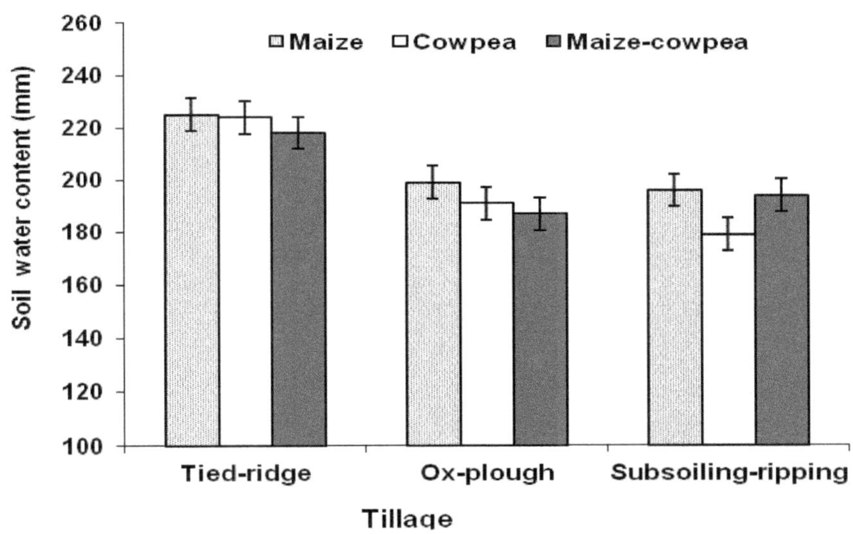

Figure 5. Soil profile (0 to 110 cm) water storage as affected by tillage and cropping systems (averaged for 4 seasons). Error bars =$LSD_{0.05}$.

(0 to 5 cm) than in treatments where the crust were disturbed by hand hoe weeding. Crust strengths at the base of the ridge basins were higher than under OP because of the exposure of hard subsoil during the making of the ridges. The crust strength between crops rows was always higher than within crop rows due to compaction through trampling. The general increase in crust strength with time within crop rows is attributable to natural formation of crust under raindrop impact since there was minimal human interference.

Saturated hydraulic conductivity and steady infiltration rate

Previous studies show that high BD, strong soil crust and low total porosity have a negative influence on Ks and infiltration rates (Bhattacharyya et al., 2006, Gicheru et al., 2004). In our study, the low inter-row Ks and infiltration rates observed in SR and TR plots is a reflection of the high BD and crust strength, and low total porosity in the inter-row spaces. The temporal decline in inter-row Ks during the crop cycle is consistent with the findings of Mwendera and Feyen (1993). They attributed the Ks decline to the increase in soil compaction and total porosity decline. The insignificant changes in Ks observed in TR are due to the exposed hardpan.

In this study, we observed that SR did not result in higher in-row infiltration rates when compared to the OP tillage system. There are two possible explanations to this unexpected result. One of them is that, there was downward movement of fine soil particles into the soil profile during the rains leading to the clogging and sealing of the macro pores in the ripped furrows. Clogging of macro-pores causes a discontinuity of water transmitting pores (Kooistra et al., 1984; Römkens et al., 1990). Low infiltration rates in SR tillage could also have been caused by slow lateral water movement in SR treatments because downward and outward of water during infiltration depend on the capacity of soil to transmit water both vertically and laterally (Michael, 1978). Ploughing the entire plot in the OP treatments, resulted in increased total porosity in the whole plot which probably encouraged faster water lateral movement in the plots while lateral water movement in SR tillage was restricted by the low total porosity in the unploughed sections between crop rows. These results are supported by the Guzha (2004) who reported that tilled soil gave significantly higher infiltration rates than untilled soil. The infiltration rates within crop rows were greater than the inter-row infiltrations rates due to the observed stronger crust strength and lower total porosity in the inter-row spaces.

Soil surface roughness

The high surface roughness observed in the TR is due to the raised ridges and basins created during tie ridging; while the surface roughness in SR was greater than in OP tillage because of the ripped deep furrows. The rapid reduction in surface roughness particularly in OP is a reflection of unstable soil which looses the clods formed after tillage due to raindrop impact, erosion and human interference during data collection and weeding. Lal (1997) reported that tilled sandy soils are susceptible to soil erosion after ploughing as soil aggregates are broken down leading to loss of soil structure. The clay content of the study soil at the plough layer was less than 13% implying that the soil had unstable structure. TR maintained the greatest surface roughness than the other tillage systems because the ridges were reconstructed every time weeding was done. The high surface roughness observed in TR would ensure that substantial rainfall is trapped and conserved in the soil profile.

Soil water storage

The variations in tillage effects on soil hydro-physical properties (BD, porosity, crust strength, infiltrations, and surface roughness) led to the observed differences in soil water storage among the tillage systems. The highest soil water content recorded in TR treatments is due to increased depression water storage induced by the high surface roughness in the tillage system. According to Hansen et al. (1999), depression storage of tilled soil increases with increasing surface roughness. This was apparent in the field soon after the rains when we observed higher depression storage (water pools) in TR than in the other tillage treatments. The capacity for TR to capture and store rainfall would ensure more water will be available to crop in TR tillage than the other tillage systems (Guzha, 2004). SR tended to produce less soil water content than OP because of the low steady infiltration rates associated with the high BD and soil crust strength and low total porosity conditions developed between the crop rows under SR tillage system.

Temesgen (2007) observed that, runoff in treatments with subsoiling and furrowing along planting rows (without disturbing the soil in between) exceeded runoff in ploughed treatments which he ascribed to the high BD (low porosity) in the un-ploughed inter-row spaces. Soil water retention under SR tillage systems would probably have been improved by mulching as it protects soil structure from destruction by the raindrop impact thus, maintaining good soil porosity for infiltration (Gicheru et al., 2004). The mulching option was however not feasible at the study site because the crop residue is mainly for livestock feed.

Soil water content in cowpea sole crop was always less ($P \leq 0.05$) than maize sole crop treatments, suggesting that cowpea utilized more soil moisture than maize. The observation that cowpea sole crop and maize/cowpea intercrop treatments tended to give similar soil water content implies that both treatments had similar water consumption.

Conclusion

The results confirmed that sandy loam soils are prone to crust formation and compaction which lead to unfavorable soil hydro-physical properties. Ox-plough tillage resulted in significant improvement in soil surface hydraulic conditions. Subsoiling-ripping tillage system conserved lesser soil water compared to OP tillage system (control) because of the unfavorable soil surface hydraulic conditions (high BD and soil crust strength, and low total porosity) that developed between the crop rows under the SR tillage. The significantly higher soil water conserved under TR tillage was attributed mainly to high soil surface roughness which enhanced depression water storage. Based on these results, the conclusion is that the subsoiling and ripping tillage method used in this study was inferior to the conventional OP tillage method in conserving soil moisture in the sandy loam soils. The results presented in this study, therefore, clearly demonstrated that maintenance of good soil porosity, infiltration and surface roughness is critical when selecting tillage systems for increased soil water conservation.

ACKNOWLEDGEMENTS

This work was supported by Kenya Agricultural Productivity Programme (KAPP), Kenya Arid and Semi-arid Lands Programme (KASAL) and the International Atomic Energy Agency (IAEA). Special thanks are due to Dr. Patrick Gicheru and Isaya Sijali of KARI-Kabete Centre for providing the penetrometer used in the study and their technical support. The contribution of Peter Wakaba, Elijah Nyangena and Bonface Muliru in data collection is highly appreciated.

REFERENCES

Barron J, Rockström J, Gichuki F, Hatibu N (2003). Dry spell analysis and maize yields for two semi-arid locations in east Africa. Agric. For. Met. 117:23-37.

Bhattacharyya R, Praksh V, Kundu S, Gupta HS (2006). Effect of tillage and crop rotations on pore size distribution and soil hydraulic conductivity in sandy clay loam soil of the Indian Himalayas. Soil Till. Res. 86:129-140.

Biamah E, Gichuki FN, Kaumbutho PG (1993). Tillage methods and conservation methods in East Africa. Soil Till. Res. 27:105-123.

Cogo NP, Moldenhauer WC, Foster GR (1983). Effect of crop residue, tillage-induced roughness and runoff velocity on size distribution of eroded soil aggregates. Soil Sci. Soc. Am. J. 47(5):1005-1008.

FAO-UNESCO (1994). FAO-UNESCO. Soil map of the world 1:5,000,000 vol. vi. Africa UNESCO Paris.

Gachene CKK, Kimaru G (2003). Soil fertility and land productivity. Technical Handbook P. 30. RELMA/Sida Publication, Nairobi Kenya.

GenStat (2007). GenStat Release 9.1, Lawes Agricultural Trust (Rothamsted Experimental Station). GenStat Ninth Edition.

Gicheru PT, Gachene C, Mbuvi J, Mare E (2004). Effects of soil management practices and tillage systems on soil surface soil water conservation and crust formation on a sandy loam in semi-arid Kenya. Soil Till. Res. 75:173-184.

Gitau AN (2004). Mechanical behaviour of hardsetting Luvisol soil. PhD *Thesis,* University of Nairobi, Kenya.

Guzha AC (2004). Effects of tillage on soil microrelief, surface depression storage and soil water storage. Soil Till. Res. 76:105–114.

Hansen B, SchjØnning P, Sibbesen E (1999). Roughness indices for estimation of depression water storage capacity of cultivated soil surfaces. Soil Till. Res. 53:103-111.

Jaetzold R, Schmidt H, Hornetz B, Shisanya C (2006). Farm management handbook of Kenya.Natural conditions and farm information. Vol II, part C1- East Kenya. Ministry of Agriculture, Kenya. 2nd Edition.

Klute A, Dirksen C (1982). Hydraulic conductivity and diffusivity. In Klute, A. (Ed.), Methods of Soil Analysis. American Society of Agronomy. *Agron.* Monogr. 9. ASA, SSSA, Madison, WI. USA.

Kooistra MJ, Bouna J, Boersma OH, Jager A (1984). Physical and morphological characterization of undisturbed and disturbed ploughpans in a sandy loam soil. Soil Till. Res. 4:405-417.

Lal R (1997). Long-term tillage and maize monoculture effects on a tropical Alfisol in western Nigeria. I. Crop yield and soil physical properties. Soil Till. Res. 42:145-160.

Michael AM (1978). Irrigation, theory and practice. Vikas Publishing House PVT Ltd. New Delhi, India. P. 801.

Moreno RG, Diaz Alvarez MC, Tarquis AM, Barrington S, Requejo SA (2008). Tillage and soil type effects on soil surface roughness at semiarid climatic conditions. Soil Till. Res. 98:35-44.

Mwendera EJ, Feyen J (1993). Predicting tillage effects on infiltration. Soil Sci. 155:229-235.

Perroux KM, White I (1988). Designs for disc infiltrometers. Soil Sci. Soc. Am. J. 52:1205-1215.

Rockström J, Kaumbutho P, Mawalley J, Nzabi AW, Temesgen M, Barron J, Mutua J, Damagaard-Larsen S (2009). Conservation farming strategies in East and and Southern Africa: yields and rain water productivity from on-farm action research. Soil Till. Res. 103:23-32.

Rockström J, Lars-Ove J (1999). Conservation tillage systems for dryland farming: on-farm research and extension experiences. E. Afr. Agric. For. J. 65(2):101-114.

Römkens MJM, Prasad SN, Parlange JY (1990). Surface sealing development in relation to rainstorm intensity. Catena 17:1-11.

Shivonje FM, Okwach GE, Kironchi G (2005). Soil crust formation as affected by slope gradient on clay soils of semi-arid Machakos District, Kenya. In: Mugendi DN, Kironchi G, Gicheru PT, Gachene CKK, Macharia PN, Mburu M, Mureithi JG, Maina F, (eds), Capacity building for land resource Management to meet the challenges of food security in Africa. 21st Annual Conference of Soil Science Society of East Africa (SSSEA), 1st -5th December, 2003, Eldoret, Kenya, pp. 123-136

Šimanskaitė D (2008). The impact of soil tillage minimization on sandy light loam soil. Agron. Res. 6:349–357

Soltanabadi DH, Miranzadeh M, Karimi M, Varnamkhasti MG, Hemmat A (2008). Effect of subsoiling in condition of strip tillage on soil physical properties and sunflower yield. J. Agric. Tech. 4(2):11-19.

Stevens PA (2000). Ripping with draught animals: IMAG-DLO experiences in Zambia. *In*: Biamah, E.K., Rockstrom J. and Okwach G.E., (eds.). Conservation tillage for dryland farming. technical options and experiences in Eastern and southern Africa. Nairobi: Regional Land Management Unit (RELMA), Swedish International Development Cooperation Agency (Sida), 2000. (RELMA Workshop Report Series 3):42-53.

Strudley MW, Green TR, Ascough JC (2008). Tillage effects on soil hydraulic properties in space and time: State of the science. Soil Till. Res. 99:4–48

Temesgen M (2007). Conservation tillage systems and water productivity implications for smallholder farmers in semi-arid Ethiopia. PhD *thesis*, Delft University of Technology Delft, Netherlands.

Properties and management of gravelly soils developed on ferruginous cuirass in Mali

Drissa DIALLO

Unité des Sciences du Sol, DER STA IPR/IFRA de Katibougou BP6 Koulikoro, Mali.

In Mali, different types of gravelly soils on Cuirass were identified in association with a native vegetation including many species of tree, shrub and herbaceous. All these soils are developed by degradation of ferruginous cuirass, a material largely observed in West Africa Sudanian and Sudano-Guinean zones. Gravelly soils of Mali have a variable depth (< 10 to 95 cm depth) and gavel content (15 to 60%) and are ranged in three textural classes (silty, sandy loam and clayey loam). Their marginal status for agriculture are always justified by pedologists but they are used both for cropping and breeding (when annual rainfall ≥ 800 mm) or only for breeding (when annual rainfall < 800 mm). In manual cultivation system, many farmers prefer gravelly soils where weed control is easy. In mechanized agriculture, the gravelly soils on cuirass are used when deep soils of lowlands are not available. Agriculture use of cuirass lands favours deforestation, increase runoff, soil erosion and inundation risk at watershed scale. So, better management systems are needed and research programs must identify techniques for harvesting runoff from cuirass table and its stockage for complementary irrigation. However, the field experimentations must identify ways to improve fodder production and conservation.

Key words: Ferruginous cuirass, gravelly soils, land use, landscape degradation, West Africa contrasted climate zones.

INTRODUCTION

In the Sudanese zones of West Africa, there are large areas where the ferruginous Cuirass is present, covered or not with a shallow or moderate depth gravelly material. Studies of this landscape had permitted production of a relatively abundant literature (Maignien, 1958; Tricart and Cailleux, 1974; Michel, 1973; Leprun, 1977, 1979; Boeglin, 1990; Bantsimba, 2001). The genesis of Cuirass materials, their mineralogical and geochemical characteristics, as well as their influence on the current dynamic of reliefs have been studied with special attention in regional geology and geomorphologic research programs, and mineral exploration. Moreover, the description and evaluation of soils located on cuirassed reliefs were made during pedological surveys (Anonyme, 2001). These soils have generally been recognized marginally for agriculture and native vegetation development (Anonyme, 2001; PIRT, 1983). Gravely soils on Cuirass deserve to be reviewed in order to develop new strategies for land management in the current context of general degradation and reduced

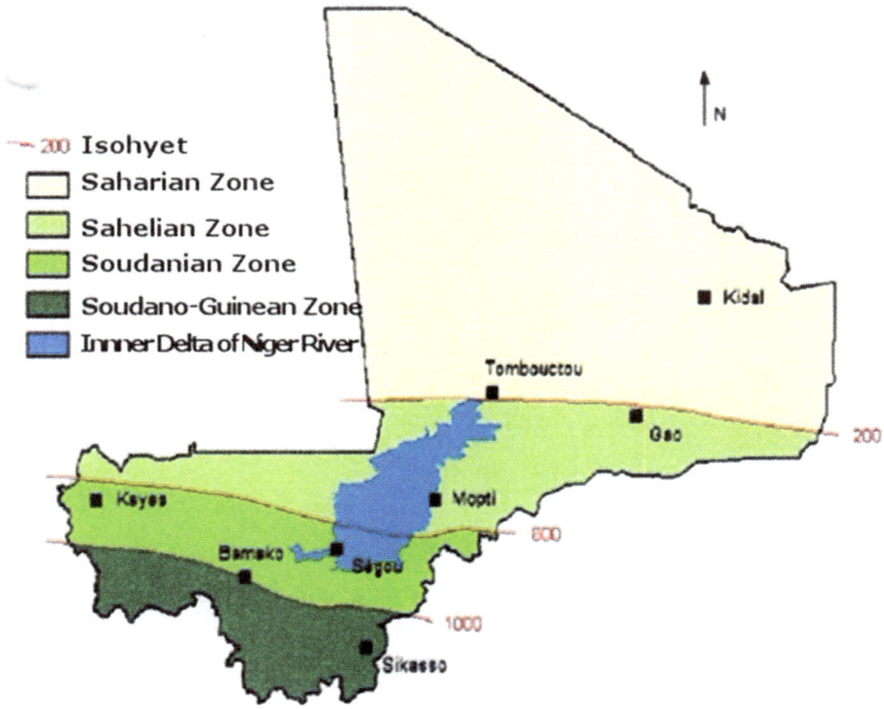

Figure 1. Agroecological map of Mali according Labo SEP, quoted by Coulibaly (2003).

Figure 2. Gravelly soil profile observed in Bagoe watershed, Sikasso region (Photo I: Kassogué).

productivity of Sudano-Sahelian ecosystems. The present paper aims to underline the pedological properties of shallow and moderate depth gravelly soils), their capability and current uses and a discussion about research needs for new strategies to manage this landscape.

STUDY AREA

Localisation of the study area

Gravelly soils developed on ferruginous Cuirass in Mali are mainly surveyed in Sudanian and Sudano-Guinean zones (Figures 1 and 2), which are largely represented in West Africa.

Climate context

The current climate of Sudanian and Sudano-Guinean zones where cuirass materials are observed in West Africa, is characterized by a high variability of parameters as annual rainfall: more than 1300 mm (in south) to 600 - 550 mm (in north). Some current climatic characteristics measured at Sikasso station, located at 5°6'W and 11°3'N are shown in Tables 1 and 2 according to Malian Meteorological Service (Sangaré, 2011). In the Sudano-Gunean zone, maximum temperature is over 30° all the year. Air humidity is very low from January to March (43 to 50% for maximum value and 13 to 17% for minimum value), in relation with Harmattan (a dry wind) presence. When Monsoon come in the Sudanian zones (around April and May) air humidity increases and get its most high values (more than 90% for maximum value and around 60% for minimum value) during the period July to October. The mean annual rainfall, for the period 1980 to 2008 was 1105.4±16 mm; only in four months (June to

Table 1. Means of temperature and air humidity at Sikasso Station from 1980 to 2009 (Sangaré, 2011).

Parameter		J	F	M	A	M	J	J	A	S	O	N	D
Temperature (°C)	Min	18	21	24	26	25	23	22	22	22	22	20	17
	Max	33	35	37	37	35	33	30	30	31	33	34	33
Air humidity (%)	Min	15	13	17	29	40	52	60	64	59	46	25	17
	Max	48	43	50	70	81	89	94	95	95	91	79	61

Table 2. Means of rainfall and evaporation at Sikasso Station from 1980 to 2009 (Sangaré, 2011).

Parameter	J	F	M	A	M	J	J	A	S	O	N	D
Rainfall (mm)	1.3±0.0	5.0±1.6	9.6±12.4	48.4±8.2	99.1±39.6	155.1±17.3	221.2±53.1	285.1±37.7	198.0±19.0	76.2±101.2	6.6±3.5	0.0±0.0
Evaporation Piche (mm)	269.5±33	282.7±34	303.6±48	218.8±37	164.6±34	103.4±18	70.9±12	54.8±8	59.4±8	96.1±24	175.6±27	232.8±29

Table 3. Geochemistry and mineralogy of Cuirass and gravel in Kangaba sector (Mali) according to DNGM, ORSTOM, CNRS (1986).

Element	Cuirass (%)	Gravels (%)
SiO_2	33.04	58.76
Al_2O_3	19.02	17.01
Fe_2O_3	43.9	18.74
TiO_2	0.89	1.34
MgO	0.2	0.23
K_2O	0.37	0. 38
MnO	0.03	0.02
Quartz	17.5	38.74
Gibbsite	11.98	0.02
Kaolinite	1.74	43.04
Goethite	45.27	7.67
Hematite	14.67	11.84

September), rainfall was superior to evaporation.

Paleoclimatic considerations are usually mentioned when talking about cuirass materials currently observed in West Africa and other tropical regions. Cuirass materials and reliefs are the result of a long evolution, more than 100 Ma (Tardy, 1991), influenced by alternating humid and arid climates in the past, during geological times. These considerations allowed scientists to consider Cuirass materials as an important paleoclimatic marker in the tropics (Gunnell, 1996). Most Cuirass land in West Africa are actually located in very contrasted climate conditions (wet from May-June to October and severely dry the rest of the year).

Geological context

Cuirass reliefs and the associated soils in West Africa are formed from different rock types, granite, shale, sandstone, greenstone (Boeglin, 1990; Bantsimba, 2001). According to their mineralogical characteristics, different types of cuirass are distinguished: bauxite or aluminous cuirass, alumino-ferruginous cuirass and ferruginous cuirass (Boeglin, 1990). Ferruginous cuirass, on which are located shallow and moderate depth gravelly soils, which is subject of the present paper, is characterized by the preponderance of goethite, Fe_2O_3 and SiO_2 (Table 3). The associate gravel (≥ 2 mm) is rich in goethite, SiO_2 and kaolinite. Geomorphologic studies showed the important role of cuirass in the organization of actual relief in the Sudanese zones of West Africa, where different categories of surface covered by cuirass are observed (Michel, 1973). Among these surfaces, there are glacis which are quaternary reliefs characterized by the presence of ferruginous cuirass and gravelly soils (Michel, 1973).

The formation process of cuirassed materials has been described by many researchers (for example, Tricart and

Cailleux, 1974; NcFarlane, 1976; Tardy, 1991). It is widely known today that cuirassed materials (or laterite) are the result of enrichment of superficial geological material in iron or alumina, or a mixture of both, and their induration. It should be noted, that the enrichment in iron, for example, can be the result of one of the two following process:

i) A relative accumulation by departure of other elements;
ii) An absolute accumulation as result of migration in the cuirass development site.

About current evolution of old ferruginous cuirass materials (mainly formed during the old and middle Quaternary), ubiquitous in West Africa, Leprun (1977) explained the process running on a toposequence in Thion (Burkina Faso), located on biotite and amphibole migmatite. This evolution is characterized by:

i) Old soil profile internal degradation by geochemical weathering;
ii) Surface degradation by mechanical process.

So, the cuirass is attacked in its mass and basics by hydrodynamic and geochemical mechanisms which magnitude increases with the degree of slope; the result is lithosols transformation in gravelly soils. At the landscape scale, relief formation process will be strongly influenced by current destruction of old cuirassed material and will differ, depending on the climate zone, the anterior cuirassed relief and the original rock.

METHODOLOGY

This article is based on a synthesis of accumulated data on Cuirassed reliefs and soils identified in Mali during studies for rural development program. These data were supplemented by detail soil descriptions, indigenous knowledge and agricultural surveys.

Extracting data from Malian Atlas of land resources

The inventory project of Mali land resources led from 1975 to 1981 provided important data on the major soils of the country and the state of land use and management. Some data from these studies are synthesized in this work.

Soil profile description and soil sampling

Some supplementary soil profiles (about 50) have been described in Malian Sudanian and Sudano-Guinean zones. The main criteria for describing soils *in situ* were the tickness of the profile, horizons colour, texture and porosity, the behavior of roots and the traces left by water saturation above hardpan (cuirass). Soil samples were taken in representative soil profiles.

Laboratory analysis

These analyses mainly concern the determination of particle size distribution with the international method (Baize, 1984), measurements of pH by electrometric method in the ration 1/2.5 and carbon by Anne method (Baize, 1984).

Agricultural practices surveying

Surveys were conducted with 51 farmers distributed in 17 villages. The questions focused on indigeneous knowledge of soil resources, land use practices, actual agricultural techniques application to gravelly soils and crops grown on the gravelly soils.

CHARACTERISTICS AND ALTERNATIVE MANAGEMENT OPTIONS OF GRAVELLY SOILS ON CUIRASS

Characteristics of gravelly soils on Cuirass

Typology and morphologic characteristics

Soil surveys in Mali led to the distinction of seven soil-vegetation units under the name of Cuirass land (*Terrains cuirassés*, TC), noted TC1 to TC7 (PIRT, 1983). These units are spatially distributed in annual rainfall context range from 250 to over 1100 mm/yr. Only soil profiles of units TC2 (formed on alluvial material) and TC7 (formed in colluvial and often alluvial material) are not gravelly, except at the lower layers of the profile, the contact between alluvium or colluviums and Cuirass. Soils of the other units (Table 4), subjects of the present article, are mainly gravelly, derived from the degradation of ferruginous cuirass. They occupy different topographic positions (upland and slope of different gradients) and are different types of soil, according to PIRT (1983): Cambic Cuirorthids, Cuirustalfs aridic, Typic Cuirustalf, mollic Cuirorthents.

Morphological descriptions of these soils are depicted as follows:

i) The thickness of the soil profile (always less than 100 cm above ferruginous cuirass); profile depth is usually variable in each soil type and from a soil type to another (Table 4);
ii) Ferruginous gravels on the surface, but also in the soil profile;
iii) Ferruginous blocks on the surface;
iv) the hue of topsoil (0 to 10 cm) colour is variable according to Revised Standard Soil colour Charts (10YR in flat position and 7.5YR on slope); it is generally 7.5YR or 5YR under surface.

Texture, pH and organic matter status

Soils content in gravel is variable (15 to 60%) and three textural classes are distinguished: silty, sandy loam, clayey loam. All these soils have an acidic pH (5.1 to 6.0), and organic matter content generally close to 2%.

Table 4. Some characteristics of gravelly soils on ferruginous cuirass in Mali.

Characteristic		Soil vegetation units				
		TC1	TC3	TC4	TC5	TC6
Annual rainfall (mm) of reference station		250 – 550	550 -800	800-1150	> 800	>1100
Typical soil		Cambic Cuirorthids	Cuirustalf aridic	Typic Cuirustalf	Mollic Cuirorthents	Typic Cuirustalfs
Soil depth (cm)		20 to 70	25 to 85	60	< 10 to 35	< 10 to 35
Presence of ferruginous block and gravel		+	+	+	+	+
Vegetation species	Total number of species observed with high frequency	21	29	46	34	45
	3 main woody species	(1) (2) (3)	(1) (3) (4)	(4) (5) (6)	(4) (6) (7)	(8) (9) (10)
	3 main herbaceous species	(a) (b) (c)	(d) (e) (f)	(f) (g) (h)	(f) (i) (g)	(g) (j) (k)

(1) Pterocarpus lucens (2) Boscia senegalensis (3) Combretum micranthum (4) *Combretum glutinosum*; (5) *Bombax costatum* (6) *Pterocarpus erinaceus* (7) *Lanea microcarpa* (8) *Isoberlinia doka* (9) *Detarium microcarpum*, (10) *Anona senegalensis*; a)*Schoenefeldia gracilis* (b) *Aristia SPP* (c) *Cenchrus biflorus* (d) *Loudetia togoensis* (e) *Diheteropogon ragerupii* (f) *Andropogon pseudapricus* (g) *Andropogon gayanus* (h) *Pennisetum pedicellatum* (i) *Ctenium SPP* (j) *Cochlospermum glutinosum*.; (+) means Yes. NB: Soil typology refers to Malian Land Resources Atlas (PIRT, 1983).

Soil-Water relationship

Gravelly soils on Cuirass have low water storage capacity when they are located on slope (4 to 6%). The average annual runoff coefficient is high enough: 41.2% in Djitiko watershed, according to measurements on experimental plot (Diallo et al., 2004). Due to runoff process, water volume lost on a surface bearing shallow gravelly soil (<40 cm) can reach 504,000 to 307,500 m^3/ ha/yr in Sudanian zone, depending on the isohyets and land use (Diallo and Roose, 2008).

Associated native vegetation

Vegetation installed on gravelly soils includes trees and shrubs in the woody stratum, and mainly grasses in the herbaceous layer. Many plant species are observed (Table 4). This biodiversity is mainly a function of the rainfall range. Only in the case of TC5, where the thickness of the soil is very low (usually less than 30 cm), soil material has more impact, than rainfall (greater than 800 mm/yr), on native vegetation. Woody vegetation of this unit occurs generally as groves scattered in abundant herbaceous stratum (PIRT, 1983). A site of unit TC5 is shown in Figure 3.

Alternative management options of soils

The characteristics of gravelly soils on cuirass, especially thin profile and high content in gravel were always arguments for soil scientists who drew attention to the marginal status of these soils. All these constraints are well known by farmers who practice mechanization. The current uses of these soils (Table 5) are linked to climatic factors agricultural techniques evolution and land availability in each agricultural village.

Climatic factors and land use

In situation where annual rainfall is 800 mm and plus, gravelly soils are used both for cropping and breeding. In this climate context TC4, TC5 and TC6 are observed as indicated in Table 5. The units TC4 and TC6 have the same assignments: sorghum, millet, maize, peanut and often cotton. The unit TC5 which occupies a flat topography, favorable to stagnation of rainwater, is managed as a garden and assigned to a wide range of crops (sorghum, maize, peanut, yam, sweet potatoes and various vegetable crops) in a general context of female and manual agriculture. In this case, crops are generally installed on boards or mounds (around 30 cm high), made to increase the volume of soil exploited by the roots of crops and also to mitigate the negative effects of any excess standing water (Figure 4).

When annual rainfall is less than 800 mm, gravelly soils on cuirass are exclusively used for breeding (units TC1 and TC3).

Agricultural techniques evolution and land use

Agricultural techniques evolution has an impact on gravelly soils use. In manual cultivation system, many farmers prefer gravelly soils where weed control, is relatively easy. In rural areas where mechanization with animal traction is widespread, the peasants are not interested in gravelly shallow soils and are driven to the plains and lowlands with deep soils, where they can easily practice conventional tillage (Figure 5). So, according to photographs interpretation, in Bélékoni watershed (120 km^2) for example, gravelly soils on

Figure 3. Groves scattered in herbaceous stratum in TC5 sites observed during the rainy season in Bagoé watershed, Sikasso region (a) I. Kassogué (b) T. Diallo.

Table 5. Current use of gravelly soils on Cuirass.

Rainfall and land use	Soil vegetation units				
	TC1	TC3	TC4	TC5	TC6
Annual rainfall (mm) of the reference station	250 – 550	550 - 800	800 - 1150	> 800	>1100
Breeding	(+)	(+)	(+)	(+)	(+)
So, Mi, Ma, Pe	(-)	(-)	(+)	(+)	(+)
Ri	(-)	(-)	(-)	(+)	(-)
Co	(-)	(-)	(+)	(-)	(+)

So = sorghum; Mi= millet; Ma = maize; Pe= peanut; Ri = rice; Co = cotton. (+) means Yes and (–) means No.

cuirass represented 62% of the cultivated area before 1960. Today this percentage is very low (inferior to 20%). With the development of mechanized agriculture, many village communities look gravelly soils as women landscapes. Indeed, these soils are occupied by many small plots, manually owned by women.

Environmental impacts of agricultural use of gravelly soils on cuirass

Problems arising from the cultivation of gravelly soils depend on topographic position, land use and agriculture techniques. Concerning soils of TC5, the problems are not perceptible enough. But after heavy rains, runoff and fine particles lost by erosion are possible. Cultivation of gravelly soils located on slope (TC4 and TC6), especially with the mechanization, can quickly lead to the disappearance of the dense tree cover. That is a major problem, because the presence of cuirass (not far of surface) and gravel does not allow recovery after cultural abandonment. Moreover, the slope (> 4%) is favorable to high runoff in Sudanian zone of West Africa and erosion of fine particles (clay and organic matter); it is a mass of sterile gravel that will be left on site. Many farmers give evidence of this rapid depletion of fine particles of gravelly soils.

DISCUSSION

Gravelly soils fundamental properties

Ferruginous gravels and blocks presence can be explained by the current degradation process affecting cuirass materials (parental materials of these soils), as described by Leprun (1977). The dominance of red in soil colour is due to the presence of iron oxides and the acidity seems a consequence of the large amounts of silica in the parental material.

The low agricultural quality of gravelly soils on cuirass is mainly explained by the low thickness of soil profile, its

(a) (b)

Figure 4. Mounds confectioned on TC5 for cropping in Bagoé watershed, Sikasso région (a) peanut plantation (photo T. Diallo); (b) sorghum and earth peas (photo I. Kassogué).

Figure 5. Conventional tillage on a deep soil in Bagoé watershed, Sikasso region (photo H.Sangaré).

high content in ferruginous gravel and the low water storage capacity.

Current land use and sensibility to degradation

In Mali, the widespread use of gravelly soils for breeding is justified by the existence of high forage availability. The cultivation of these soils in some situations is explained by low availability of land resources and strong pressure on the better one (deep soils). In a climate context characterized by an annual rainfall inferior to 800 mm, it is very likely that gravelly soils are cultivated in general, because growing of crops is highly uncertain (low rainfall and low water storage capacity of soils are serious constraints).

The high sensibility of gravel soils to degradation can be explained by:

i) the thickness of the soil over hardpan which is favourable to runoff;
ii) the low levels of stable macro-aggregates measured in these soils (Diallo et al., 2004). These authors reported that stable macro-aggregates rate is variable from 147 to 225 g/kg of gravelly soil against 560 to 630 g/kg of brown vertic soil located below.

Concerning soils of unit TC5, the physical degradation is not perceptible enough according to the flat topography; it is also not conductive to runoff.

RESEARCH NEEDS FOR A BETTER UTILIZATION OF CUIRASS ENVIRONMENT

Better utilization of ferruginous cuirass environment, occupying large areas in the Sudanian and Sahelian zones of Mali and elsewhere in West Africa, is necessary in the current context:

i) Generalization of land and ecosystems degradation;
ii) Necessity to improve food production for a growing population.

So, studies must be thorough in order to better enhance their impluvium function and their potential fodder.

Research needs for improving impluvium function of cuirass table

Today it is necessary to develop research in two directions:

i) Development of runoff harvesting from cuirass table and its stockage techniques on the landscape in upstream of rivers;
ii) Development of micro irrigation systems with the stocked water on the landscape.

Research needs for improving potential fodder of gravelly soils

Improvement of the potential fodder requires field experimentations in order to:

i) identify better herbaceous species according to their biomass productivity and their agrostological quality;
ii) identify better production and conservation techniques of fodder.

Conclusion

Agricultural use of cuirass lands faces real physical constraints (mainly thin material, heavy duty gravel and low water storage capacity) and has been identified by pedologists since long time. The techniques developed by some farmers (construction of boards and mounds) appear to be a significant alternative to improve soil depth and profitability of rainwater. However, pastoral potential of these lands appear significant. Globally, current management practices for cropping on gravelly soils are poor and contribute to acceleration of degradation process (particularly runoff) affecting all the toposequence. Better management systems have to be developed for better use of gravelly soils on cuirass and the concerned landscape in upstream of rivers. In this order, development of runoff harvesting from cuirass table and its stockage for complementary irrigation can be experimented in the local context. In other words, field experimentations are necessary to improve biomass (from herbaceous species) production and conservation for livestock use.

REFERENCES

Anonyme (2001). Les aptitudes agricoles et pastorales des sols dans les pays du CILSS. www.case.ibimet.cnr.it/SVS/data/publications/LesSolsDesPaysDuCILSS.pdf: P. 173.

Baize D (1984). Guide des analyses courantes en pédologie, INRA, Paris, P. 172.

Bantsimba CJH (2001). Contribution à l'étude des couvertures latéritiques du Sénégal oriental par imageries électrique et radar. Thèse de doctorat de l'Université Cheikh Anta Diop de Dakar P. 124.

Boeglin JL (1990). Evolutions minéralogique et géochimique des cuirasses ferrugineuses de la région de Gaoua (Burkina Faso). Thèse de doctorat de l'Université Louis Pasteur de Strasbourg P. 188.

Coulibaly A (2003). Profil fourrager Mali. Publication FAO, P. 25 www.fao.org/ag/agp/AGPC/doc/Counprof/Mali/malifr.htm. Accessed May 29, 2014.

Diallo D, Roose E (2008). Gestion de l'eau et des sols sur toposéquences cuirassées en Afrique occidentale : limites des méthodes traditionnelles et perspectives. In Roose E ; Albergel J., De Noni G., Sabir M., Laouina A : Efficacité de la GCES en milieu aride, AUF, EAC, et IRD, éditeurs, Paris pp. 92-96.

Diallo D, Barthès B, Orange D, Roose E (2004). Comparaison entre stabilité des agrégats ou des mottes et risques de ruissellement et d'érosion en nappe mesurés sur parcelles en zone soudanienne du Mali. Sécheresse, 15(1):57-64.

DNGM ORSTOM, CNRS (1986). Cuirasses latéritiques et minéralisation aurifères de la région de Kangaba au Mali: cartographie, pétrographie, minéralogie, géochimie et télédétection. Rapport de première phase. Strasbourg-Bamako annexes P. 131.

Gunnell Y (1996). Géographie comparative des héritages cuirassés sur les terres cristallines de l'Inde du Sud et d'Afrique de l'Ouest. Ann.Géo. 591:451-479.

Leprun JC (1977). Géochimie de la surface et formes du relief. IV. La dégradation des cuirasses ferrugineuses. Etude et importance du phénomène pédologique en Afrique de l'Ouest. Sci. Géol. Bull. Strasbourg, 30(4):265-273.

Leprun JC (1979). Les cuirasses ferrugineuses des pays cristallins de l'Afrique occidentale sèche. Génèse.Transformation. Dégradation. Sci. Géol. Mém. 58:224.

Maignien M (1958). Le cuirassement des sols en Afrique tropicale de l'Ouest, Sols Afr. 4(4):5-41.

Michel P (1973). Etude géomorphologique des bassins des fleuves Sénégal et Gambie, tome2. Editions ORSTOM, Paris. P. 715.

NcFarlane MJ (1976). Laterite and landscape. Academic Press, P. 151.

PIRT (1983). Les Ressources Terrestres du Mali (Atlas, Rapport Technique, Annexes). New York: Kenner Printing Company.

Sangaré H (2011). Usage, occupation des terres et érosion Hydrique dans le bassin versant de la Bagoé: cas du sous-bassin versant agricole de Sinani (sous préfecture de N'Kourala). Mémoire de DEA, Université de Bamako.

Tardy Y (1991) Mineralogical composition and geographical distribution of African and Brasilian periatlantic laterites. The influence of continental and tropical paleoclimats during the past 150 millions years and implication of Indian and Austria. J. Afr. Earth Sci. 12(1/2):283-295.

Tricart J, Cailleux A (1974). Le modelé des régions chaudes, 2è édition. SEDES, Paris. P. 345.

Integrating scientific and farmers' perception towards evaluation of rain-fed agricultural technologies for sorghum and cowpea productivity in Central Kenya

Njeru P. N. M.[1], Mugwe J.[2], Maina I.[3], Mucheru-Muna M.[2], Mugendi D.[4], Lekasi, J. K.[1], Kimani S. K.[1], Miriti J.[1], Esilaba A. O.[3] and Muriithi F.[3]

[1]KARI Muguga South, P.O Box 30148-00100, Nairobi.
[2]Kenyatta University, P.O Box 43844, Nairobi.
[3]KARI Headquarters, P.O Box 57811-002004, Nairobi.
[4]Embu University College-Nairobi University, P.O Box 6- 60100, Embu.

Soil fertility degradation remains the major biophysical cause of declining per capita crop production on smallholder farms in Central Kenya highlands. A study was conducted to compare farmers' perception and biophysical data on selected water harvesting and integrated soil fertility management technologies on sorghum (*Sorghum bicolor* (L.) Moench) and cowpea (*Vigna unguiculata* L.) production in Central highlands of Kenya. Three hundred and seventy one smallholder farmers were invited to evaluate thirty six plots laid out in Partially Balanced Incomplete Block Design (PBIBD) replicated three times. The treatment which was ranked best overall rated as 'good' by the farmers was farmers practice with a mean score of (2.78) and yielding (3.5 t/ha) under sorghum alone plus external soil amendment of 40 kg P /ha+20 kg N /ha. This was closely followed by tied ridges and contour furrows overall rated as 'good' by the farmers under sorghum alone plus external soil amendment of 40 kg P /ha+20 kg N /ha+manure 2.5 t/ha and 40 kg P /ha+40 kg N /ha+manure 5 t/ha both with a mean score of (2.7) and yielding (3.0 t/ha) and (2.9 t/ha) respectively. Generally, all experiment controls were overall scored as 'poor' yielding as low as 0.3 t/ha to 0.6 t/ha. Therefore, integration minimal addition of organic and inorganic inputs on highly valued traditional crops with adequate rainfall under normal farmers practice in semi arid lands could be considered as an alternative option contribution to food security in central highland of Kenya.

Key words: Food security, water harvesting, integrated soil fertility management, Central Kenya.

INTRODUCTION

Smallholder farms in Central highlands of Kenya are characterized by unreliable rainfall distribution and declining soil fertility that are unsuitable for sustainable rain-fed agriculture in semi-arid lands (SALs) (Miriti, 2011). Approximately 83% of Kenya's land surface is classified as ASALs, that is characterized by low and erratic rainfall (100-900 mm per annum) which is not suitable for sustainable rain-fed agriculture. Agricultural production is affected by the high variability of rainfall onset, distribution and frequent droughts which usually occur during the growing season, often resulting in depressed yields and persistent crop failures (Keating et al., 1992; Miriti et al. 2012). Agricultural intensification requires the maintenance of soil physical, chemical and biological quality. The loss of nutrients through plant nutrient mining, removal of crop residues, erosion, leaching or volatilization, and the deterioration of soil physical properties can independently or interactively result in yield reduction (Bielders et al., 2002). Appropriate technologies need to be employed on the basis of their ability to maintain soil fertility (mulching; e.g., Batiano et al., 2004), to conserve the soil against soil losses by wind erosion (mulching or ridging), and to improve soil physical conditions in the topsoil. This suggests that only low-input technologies may be currently suitable and need to be adopted through a known crop intensification technologies that could be enhanced in Arid and semi-arid lands (ASALs) of Kenya. Therefore more research is needed to find out comparison of scientific research as compared to farmers perceptions towards these technologies.

Soil fertility degradation is a major biophysical cause of declining per capita crop production and food security on smallholder farms in Central Kenya (Bationo et al., 2004; Kimani et al., 2007). The soil fertility decline is as a result of a combination of processes such as high rates of soil erosion, nutrient leaching, removal of crop residues, continuous cultivation of the land without adequate fertilization and fallowing (Njeru et al., 2011; Okalebo et al., 2006). The average annual loss in soils nutrients of 42 kg Nitrogen (N), 3 kg Phosphorus (P) and 29 kg Potassium (K) ha^{-1} in Kenya is among the greatest in Africa (Smaling et al., 1997). The rising cost of inputs has led to many smallholder farmers reducing or abandoning the use of chemical fertilizer altogether in Central Kenya (Gachimbi, 2002). The Kenyan ASALs are also experiencing low crop production due to a combination of biophysical problems such as factors such as low rainfall, surface sealing, unavailability of high quality manure, declining soil fertility due to continuous cultivation and crust formation that reduces soil water availability to crops (Gicheru, 2002; Gitau, 2004). Several recent studies have yielded little evidence on occurrence of dry spells to increase the frequency of rain water use efficiency in ASALs of Africa (Stroosnijder, 2009). This has been contributed by mixed crop-livestock systems being currently projected to see reduction in crop production throughout most East Africa regions due to climate change by 2050 (Thornton et al., 2010). Therefore, food security situation is expected to continue deteriorating and could worsen in future if water harvesting and integrated soil fertility technologies are not taken up quickly particularly in Kirinyaga West District of Central Kenya. Traditional crops such as sorghum and cowpea are considered as the crop for the future which can contribute to food security (Fongod et al., 2012).

Improving agricultural productivity is crucial for resolving food crises, enhancing food security and accelerating pro-poor growth. Most food security research and development programmes tend to focus high and medium potential on promoting technologies for a limited number of major crops such as maize and beans in high potential areas, and neglected the high valued traditional crops which are drought tolerant and provides local nutrition in the vulnerable areas. Yet, sorghum and cowpea are locally important for food and household nutrition, and provide income opportunities for the most vulnerable people and women in particular. These premium crops have potential to diversify the farming systems, adapt to spread risks and are more resilient to climatic variations and climate change. This study assessed the farmer's evaluation on selected water harvesting and integrated soil fertility management technologies on sorghum and cowpea productivity in Ndia West division of Kirinyaga West District in Central highland of Kenya.

MATERIALS AND METHODS

Figure 1 shows Ndia West division of Kirinyaga West District indicating the location of the study site in central Kenya. The figure also shows the level of poverty level in this areas and this indicates the need for employing water harvesting and integrated soil fertility management technologies in the area.

Site description

The study was conducted in Ndia West division of Kirinyaga West District which represents an area of declining potential occupying total area of 1437 km^2. The District lies between latitudes 0°1° and 1°40° south and longitudes 37° and 38° east at an altitude of 1,480 to 6,800 m above the sea level. The total population of the district is 509,157 individuals out of which 30.2% are considered food poor. The population density is 309 people per km. The district has about 97,970 farm families working in the agricultural sector occupying about 96,938 farm holdings with an average farm size of 1.25 ha per family (Government of Kenya, 2007). There are four major Agro-ecological zones (AEZs) in the District (LH1, UM1, UM2 and UM3), with maize-beans, horticulture, French beans, dairy, coffee and banana production being the major crops. It receives a mean rainfall between 900-2,700 mm per annum and has temperatures of between 14 and 30°C. The soils in this district are volcanic which are known as andisols favorable for maize crop production. In addition, the district has two rain seasons- long and short rains between March-June and July-December respectively. The question is how to ensure that this area that previously was high potential does not continue to deteriorate into a low potential area (Jaetzold et al., 2007).

Soil physical characteristics of Kirinyaga West District

The results of the three types of soil texture sand (13%), silt (78%) and clay (10%) has indicated that the soil types is silty loams in Kirinyaga West District.

Experimental design

The treatments were arranged in a factorial structure each

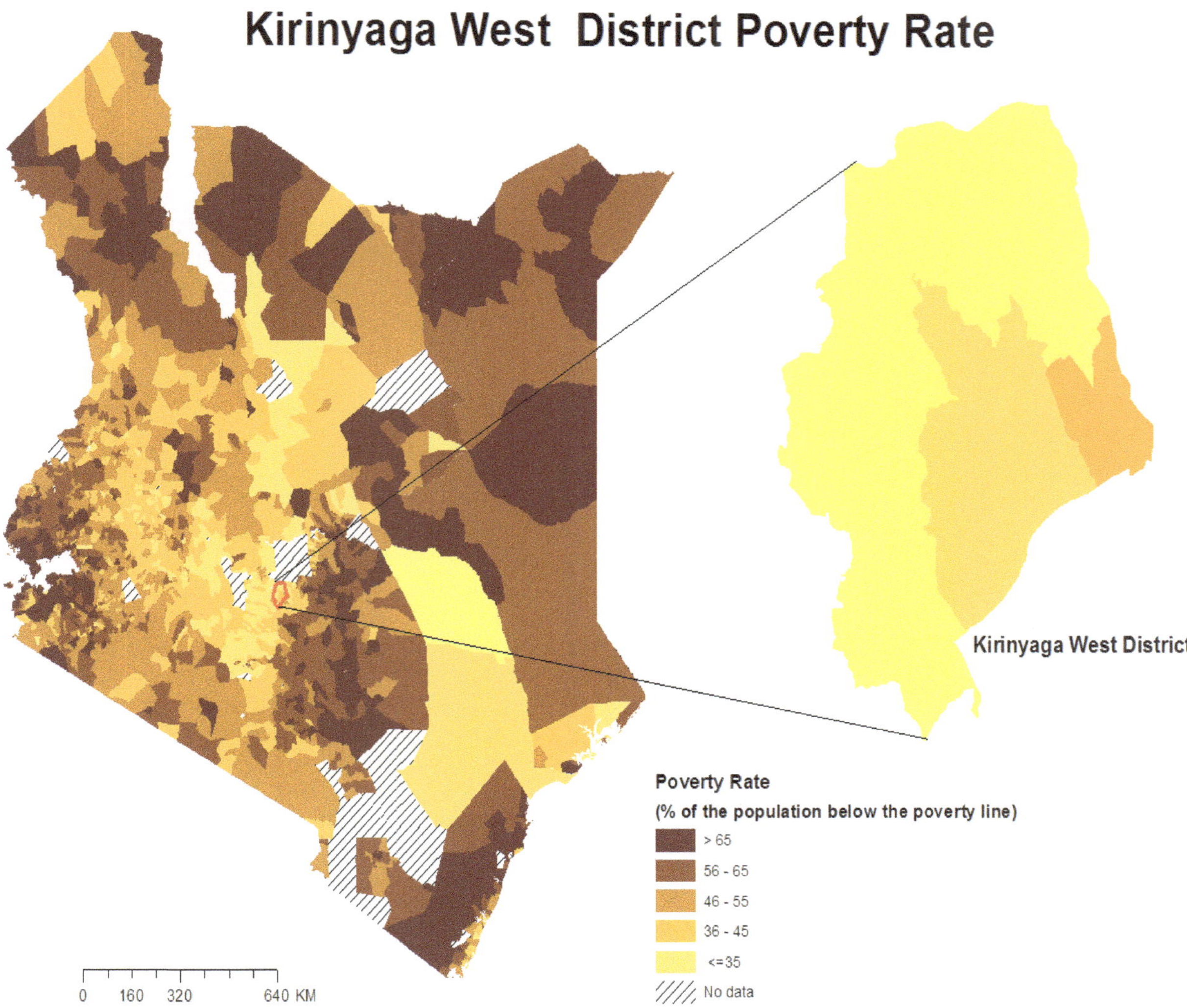

Figure 1. Map of Kenya showing the study area in Kirinyaga West District of Central Kenya.

treatment being a combination of one of the 3 levels of water harvesting techniques, cropping systems were 2 levels and the soil fertility amendment options were 6 levels thus giving a total of 36 treatments.

Water harvesting techniques (3 levels) includes:

1. Tied ridges;
2. Contour furrows, and
3. Conventional tillage/farmers practice.

Cropping systems (2 levels) includes:

1. Sole sorghum (Gadam)
2. Sorghum and cowpea (M66) intercrop;

Soil fertility amendment options (6 levels) includes:

1. Control
2. 40 kg P /ha+40 kg N /ha
3. 40 kg P /ha+20 kg N /ha
4. 40 kg P /ha+40 kg N /ha+Manure 5 t/ha
5. 40 kg P /ha+20 kg N /ha+Manure 2.5 t/ha
6. Manure 5 t/ha

They were laid out in a Partially Balanced Incomplete Block Design (PBIBD) with six incomplete blocks per replicate each containing six treatments, replicated 3 times making a total of 108 plots. Treatments were assigned to blocks randomly. The plot size was 6 m × 4 m. The dry land sorghum (Gadam) and cowpea (M66) varieties were used as the test crops. Then at the end of the short rain 2011 season, smallholder farmers were invited for a field day to evaluate each plot by scoring in a scale of good, fair and poor according to their own observation on crop performance and this was compared with scientific data collected on crop productivity. They were all given equal opportunity to evaluate 108 plots in the field experiment. They were also asked the kind of water harvesting and soil fertility management they used in their farms.

Data analysis

Social data was coded and analyzed by use of SPSS version 17. Data was analyzed by use of descriptive analysis where frequencies of scores for each treatment were computed. Dependency tests were also conducted to find out if there was a relationship between gender and the treatment score. The difference between treatment scores and gender was declared

Table 1. Distribution (%) of water harvesting technologies practiced in Ndia West Division.

Types of water harvesting	(%)
Tied ridges	34.7
Contour furrow	28.2
Convectional tillage	34.7
Others	2.4
Total	**100**

N=371.

significant at P ≤ 0.05. The biophysical data on crop yield was analyzed using statistical Analysis of Variance (ANOVA) using SAS version 8. Differences between treatment effects were declared significant at P ≤ 0.05.

RESULTS

Water harvesting technologies available in Ndia West division

Table 1 shows water harvesting technologies which were in use on-farm in Ndia West Division, Kirinyaga West District. The table indicate that tied ridges and convectional tillage were the widely used means of water harvesting technologies where (34.7%) of respondents used them on their farms while (2.4%) of respondents did not use any of the above technologies.

Farmer's evaluation on treatment performance

In Ndia division, the farmers' criteria for distinguishing plots was on a scale of good, fair and poor that included crop yield and performance. Table 2 underscore the value of taking into consideration the visual and morphological crop characteristics used by farmers as a key criterion for scientific crop evaluation and development during short rains 2011. The table show that treatment under farmers practice with sorghum alone plus soil amendment of 40 kg P /ha+20 kg N /ha attracted the highest preference of farmers who rated it as 'good' with a mean score of 2.78 and was ranked number one out of 36 treatments. This was followed by tied ridges and contour furrows under sorghum cropping system plus the same soil amendment of 40 kg P /ha+20 kg N /ha+manure 2.5 t/ha both with a mean score of 2.7. Also contour furrow under sorghum cropping system with optimal application of fertilizers and manure was also ranked under 'good' category. The results show that all the technologies ranked 'good' included combination of fertilizers and manure, or stand alone application.

The results further indicate that the overall rating of top 8 treatments was rated in the 'good' category by the smallholder farmers. This was contrary to the experiment controls which were rated as 'poor' treatment. The results further indicated that the experiment controls with tied ridges were rated in the 'fair' category.

Treatment score by gender

The results indicated that there was no significant difference ($p \geq 0.05$) regarding scoring by gender in all the 36 treatments of experiment which were ranked in the scale of good, fair and poor. However, there was a highly significant difference ($p < 0.001$) on rating of treatments by smallholder farmers in Ndia Division, Kirinyaga West District.

Field experiment results

The result in Table 3 underscore the scientific crop evaluation from the field experiment during short rains 2011. The results show the effect of various water harvesting and integrated soil fertility management options for sorghum and cowpea productivity on biomass and grain yield. the table shows three types of water harvesting, two cropping system and six fertility amendment levels but only fertility levels that differed significantly from one another ($p=0.0001$) in terms of sorghum grain yield. The three levels of water harvesting and the two cropping systems did not differ significantly in terms of grain yield among themselves ($p=0.8413$) and ($p=0.7168$) respectively. The total dry matter amount varied significantly among levels of cropping system and fertilizer application ($p= 0.0216$ and 0.0001) respectively. However the total dry matter amount did not vary significantly across water harvesting methods ($p=0.5743$). The sorghum biomass were significantly different among cropping system ($p=0.0020$) while water harvesting and fertility levels did not differ significantly ($p=0.3930$ and 0.0698).

Combination effect

The results further indicated that sorghum without manure application did not differ significantly in yield production with plots that did not receive fertilizer application. However, plots that received fertilizer and no manure gave slightly higher sorghum yield as compared to plots that received manure and no fertilizer (Table 2). The highest sorghum yield (3.5 t/ha) was recorded from farmer practice in sole sorghum cropping system with external nutrient replenishment of 40 kg P /ha+20 kg N /ha, followed by 3.0 t/ ha under tied ridges, sole sorghum cropping system with combination of half rate of fertilizer and manure. In the third place were three treatments (2.9 t/ha) under tied ridges and contour furrow, sole cropping system in both water harvesting and one intercrop under contour furrows with all under maximum nutrient

Table 2. Farmers' rating on water harvesting, cropping systems and ISFM technologies in Ndia division.

Water harvesting	Cropping systems	Soil fertility management regimes	Mean score	Mean rank	Overall rating
Farmers practice	Sole crop	40 kg P/ha+20 kg N/ha	2.78	2893.87	Good
Tied ridges	Sole crop	40 kg P/ha+20 kg N/ha+Manure 2.5 t/ha	2.7	2763.26	Good
Tied ridges	Sole crop	40 kg P/ha+40 kg N/ha+Manure 5 t/ha	2.7	2767.22	Good
Contour furrows	Sole crop	40 kg P/ha+20 kg N/ha+Manure 2.5 t/ha	2.68	2753.38	Good
Contour furrows	Intercrop	40 kg P/ha+20 kg N/ha+Manure 2.5 t/ha	2.63	2667.23	Good
Farmers practice	Sole crop	Manure 5t/ha	2.62	2641.8	Good
Tied ridges	Intercrop	40 kg P/ha+20 kg N/ha+Manure 2.5t/ha	2.6	2590.86	Good
Tied ridges	Sole crop	40 kg P/ha+40 kg N/ha	2.53	2480.37	Good
Farmers practice	Sole crop	40 kg P/ha+20 kg N/ha+Manure 2.5t/ha	2.49	2443.79	Fair
Tied ridges	Intercrop	40 kg P/ha+40 kg N/ha+Manure 5t/ha	2.49	2462.85	Fair
Contour furrows	Sole crop	40 kg P/ha+40 kg N/ha	2.48	2403.52	Fair
Farmers practice	Sole crop	40 kg P/ha+40 kg N/ha	2.46	2384.7	Fair
Tied ridges	Sole crop	40 kg P/ha+20 kg N/ha	2.45	2397.72	Fair
Tied ridges	Intercrop	40 kg P/ha+40 kg N/ha	2.44	2310.8	Fair
Contour furrows	Intercrop	40 kg P/ha+20 kg N/ha	2.43	2368.56	Fair
Tied ridges	Sole crop	Manure5t/ha	2.43	2338.03	Fair
Farmers practice	Intercrop	40 kg P/ha+40 kg N/ha	2.39	2280.98	Fair
Farmers practice	Intercrop	40 kg P/ha+20 kg N/ha	2.37	2243.88	Fair r
Farmers practice	Intercrop	40 kg P/ha+40 kg N/ha+Manure 5 t/ha	2.37	2256.27	Fair
Contour furrows	Sole crop	Manure 5t/ha	2.34	2208.08	Fair
Contour furrows	Intercrop	40 kg P/ha+40 kg N/ha+Manure 5 t/ha	2.33	2180.3	Fair
Contour furrows	Sole crop	40 kg P/ha+20 kg N/ha	2.31	2155.78	Fair
Farmers practice	Intercrop	Manure 5t/ha	2.31	2139.53	Fair
Contour furrows	Sole crop	40 kg P/ha+40 kg N/ha+Manure 5 t/ha	2.29	2123.15	Fair
Contour furrows	Intercrop	40 kg P/ha+40 kg N/ha	2.28	2107.65	Fair
Tied ridges	Intercrop	40 kg P/ha+20 kg N/ha	2.27	2098.91	Fair
Farmers practice	Intercrop	40 kg P/ha+20 kg N/ha+Manure 2.5 t/ha	2.03	1740.38	Fair
Contour furrows	Intercrop	Manure 5t/ha	2.02	1687.24	Fair
Tied ridges	Intercrop	Manure 5t/ha	2	1686.05	Fair
Farmers practice	Sole crop	40 kg P/ha+40 kg N/ha+Manure 5t/ha	1.76	1385.2	Fair
Tied ridges	Sole crop	Control	1.74	1412.16	Fair
Tied ridges	Intercrop	Control	1.56	1160.19	Fair
Contour furrows	Sole crop	Control	1.47	1053.38	Poor
Contour furrows	Intercrop	Control	1.32	865.29	Poor
Farmers practice	Sole crop	Control	1.18	677.2	Poor
Farmers practice	Intercrop	Control	1.08	534.38	Poor

N=371; Test statistics: Kruskal-H test; Chi-square =1237.5; d.f =35; p=0.000.

replenishment of 40 kg P /ha+20 kg N /ha . However the four treatments yield did not differ significantly from one another ($p<0.05$). The lowest sorghum yield (< 2.0 t/ha) was observed in treatments regarded as 'control' with neither fertilizer nor manure regardless of other intervention (water harvesting method or cropping system). Total dry matter was more in plots with farmer practice, sole cropping and maximum (40 kg N/ha) N fertilizer application but no manure (7.7 t/ha), followed by (7.1 t/ha) from plots under tied ridges, sole cropping with half rate (20 kg N/ha) fertilizer and (2.5t/ha) manure. Others that did well in total dry matter production were treatments under farmer practice, sole and intercrops with half rate of (20 kg N/ha) N fertilizers and (2.5 t/ha) manure (6.9 t/ha). All these top producers did not differs significantly from one another ($p<0.05$).

DISCUSSION

Farmer's evaluation on treatment performance

The results (Table 1) indicated that 34.7% of the respondents both used tied ridges and convectional

Table 3. The effects of water harvesting, cropping system and soil fertility regimes on sorghum yields in Ndia division.

Water harvesting method	Cropping system	Soil fertility management regimes	Total dry matter (t/ha)	Biomass+husks (t/ha)	Grain yield (t/ha)
Farmer practice	Sole crop	40 kg P/ha+20 kg N/ha	7.0	3.5	3.5
Tied ridges	Sole crop	40 kg P/ha+20 kg N/ha+manure 2.5 t/ha	7.1	4.1	3.0
Contour furrows	Sole crop	40 kg P/ha+40 kg N/ha	6.4	3.5	2.9
Tied ridges	Sole crop	40 kg P/ha+40 kg N/ha	6.3	3.4	2.9
Contour furrows	Intercrop	40 kg P/ha+40 kg N/ha	6.0	3.1	2.9
Farmer practice	Sole crop	Manure 5 t/ha	6.6	3.8	2.8
Farmer practice	Sole crop	40 kg P/ha+40 kg N/ha	7.8	5.1	2.7
Contour furrows	Sole crop	40 kg P/ha+20 kg N/ha+manure 2.5t/ha	6.1	3.4	2.7
Farmer practice	Intercrop	40 kg P/ha+40 kg N/ha	4.9	2.2	2.7
Tied ridges	Intercrop	40 kg P/ha+20 kg N/ha+manure 2.5t/ha	5.8	3.2	2.6
Tied ridges	Intercrop	40 kg P/ha+40 kg N/ha+manure 5t/ha	5.6	3.0	2.6
Contour furrows	Intercrop	40 kg P/ha+20 kg N/ha	4.9	2.3	2.6
Farmer practice	Intercrop	Manure 5 t/ha	6.2	3.7	2.5
Contour furrows	Sole crop	40 kg P/ha+40 kg N/ha+manure 5t/ha	5.9	3.4	2.5
Tied ridges	Sole crop	Manure 5 t/ha	6.7	4.3	2.4
Contour furrows	Intercrop	40 kg P/ha+20 kg N/ha+manure 2.5t/ha	5.7	3.3	2.4
Tied ridges	Intercrop	40 kg P/ha+20 kg N/ha	5.7	3.3	2.4
Contour furrows	Sole crop	Manure 5 t/ha	4.9	2.5	2.4
Farmer practice	Intercrop	40 kg P/ha+20 kg N/ha+manure 2.5t/ha	6.0	3.7	2.3
Contour furrows	Intercrop	40 kg P/ha+40 kg N/ha+manure 5t/ha	7.9	5.6	2.3
Contour furrows	Intercrop	Manure 5 t/ha	5.4	3.1	2.3
Farmer practice	Intercrop	40 kg P/ha+20 kg N/ha	6.2	4.0	2.2
Farmer practice	Sole crop	40 kg P/ha+20 kg N/ha+manure 2.5t/ha	6.2	4.1	2.1
Tied ridges	Sole crop	40 kg P/ha+20 kg N/ha	5.8	3.7	2.1
Tied ridges	Intercrop	40 kg P/ha+40 kg N/ha	5.2	3.1	2.1
Farmer practice	Sole crop	40 kg P/ha+40 kg N/ha+manure 5t/ha	5.1	3.0	2.1
Tied ridges	Intercrop	Manure 5 t/ha	5.0	2.9	2.1
Contour furrows	Sole crop	40 kg P/ha+20 kg N/ha	6.0	4.0	2.0
Tied ridges	Sole crop	40 kg P/ha+40 kg N/ha+manure 5t/ha	5.9	3.9	2.0
Farmer practice	Intercrop	40 kg P/ha+40 kg N/ha+manure 5t/ha	5.7	3.7	2.0
Tied ridges	Sole crop	Control	1.7	1.1	0.6
Tied ridges	Intercrop	Control	1.5	0.9	0.6
Contour furrows	Sole crop	Control	1.9	1.4	0.6
Contour furrows	Intercrop	Control	2.4	1.9	0.5
Farmer practice	Sole crop	Control	1.4	1.0	0.4
Farmer practice	Intercrop	Control	1.0	0.7	0.3
Means			5.6	3.3	2.3
CV			19.0	25.3	22.9
LSD			1.96	1.54	0.98

tillage as their main land preparation methods and soil fertility management on their on-farm. This was followed by contour furrows 28.2%. This could be as a result of lack of knowledge on water harvesting methods they use when growing sweet potatoes and tobacco in the area. The consistently high preference (Table 2) by farmers on overall rating as 'good' on farmers practice, tied ridges and contour furrow under sorghum alone with organic and inorganic inputs at half dose application of Nitrogen was an indication that farms in Kirinyaga West district require minimal nutrient replenishment. Soil fertility degradation is one of the major problems facing crop productivity in central Kenya. It is defined by Stocking and Murnaghan (2000) as the loss of soil physical and nutritional qualities. It has been an issue of concern throughout central Kenya, and cuts across many different soils and crops (Okalebo et al., 2007). Integrated soil fertility management (ISFM) has been cited by many authors, including Okalebo et al. (2007), Gumbo (2006) and Raab (2002), as the key approach in raising productivity levels in agricultural systems while maintaining the natural resource base. It is described by Vanlauwe and Zingore (2011) as a set of soil fertility management practices that necessarily include the use of fertilizer, organic inputs, and improved germplasm combined with the knowledge on how to adapt these practices to local conditions, aiming at maximizing agronomic use efficiency of the applied nutrients and improving crop productivity in this region. Because of the pressing need for global food security, many articles have been published which relate ISFM to the production of annual food crops like maize (Ikerra et al., 2007; Kimani et al., 2007), and rice (Kaizzi et al., 2007), giving lesser attention to perennial crops like coffee. It is no longer wondering then that the role of ISFM for sorghum and cowpea in central Kenya and the socio-economic perception of it have not been studied to any significant detail. Studies by Mugendi et al. (2010) and Gachimbi (2002) have also reported that farms in central Kenya require nutrient replenishment every season from manures, fertilizers and also from return of crop residue in their farms. It has also been reported by Njeru et al. (2010) and Mairura et al. (2007) that soil fertility can be accessed through visual observation of crop performance and yield. Therefore, soil fertility degradation has emerged as the most limiting factor in all the cropping systems of central Kenya. This is however, a very generic perception which needs to be studied in detail by targeting specific locations and farming communities in the region.

The results (Tables 2 and 3) further indicate that water harvesting technologies played a major role in moisture conservation where most of the technologies were highly ranked by the farmers. Soil desiccation is either caused by low and unreliable rainfall, poor water harvesting techniques and unsustainable farming practices in central Kenya. Low and declining soil fertility arises from continuous cultivation where levels of soil replenishment, by whatever means, are too low to mitigate the process of soil mining, whereby the soil fertility is not replaced by new inputs together with appropriate water harvesting structures on-farm (Shisanya et al., 2009). Kimani et al. (2004) has also reported that no matter how effectively other constraints are remedied, per-capita food production in central will continue to decrease unless soil fertility depletion is addressed. These results are in agreement with those reported by Miriti et al. (2012) found that farmer perception can be fairly compared with scientific data especially when soil fertility is closely related to the soil's water holding capacity.

The results (Table 2 and 3) from the farmers responded and biophysical data shows that the best four treatments of tied ridges, contour furrow and farmers practice under intercrop of sorghum and cowpea were generally ranked fairly and they were dominated by sole cropping system with soil amendment of 40 kg P/ha+20 kg N/ha+Manure 2.5t/ha. Studies have indicated that most African soils are inherently low in organic carbon (<20 to 30 mg kg^{-1}) due to; low root growth of crops and natural vegetation, continuous cultivation of crops and rapid turnover rates of organic materials with high soil temperature and microfauna (Bationo and Waswa., 2011). Furthermore, there are indications that up to 0.69 tons loss of carbon ha^{-1} $year^{-1}$ in the soil surface layers are common in Africa even with high levels of organic inputs (Nandwa, 2003). For a long-term productivity of agro-ecosystems and protection of the environment, it is necessary to develop and implement management strategies that maintain the quality of soil, through integrated soil fertility management options (Godsteven et al., 2013).

Eghball and Power (2005) has also reported similar results that a general increase in organic carbon after application of manure and mineral fertilizer over time can improve crop productivity. These results further indicate that a combination of both organic and inorganic fertilizers is better at improving soil fertility compared to their sole applications. These results are in agreement with those reported by Kathuku et al. (2011) that there was an increase in yield and soil nutrient availability in soil that was added mineral N fertilizer combined with manure when compared to their sole applications. The manure only treatment showed a reduction in crop yield in the first season and this could have resulted from the reaction of P with organics in manure to form more stable compounds making it unavailable for plant uptake (Waldrip et al., 2012). These findings are in line with Yuste et al. (2007) who reports that the higher the carbon inputs added to the soil the higher the soil respiration. Therefore, in order to increase crop productivity, and reduce production risks, better use of available rainfall and improved nutrient use efficiency is required in central Highlands of Kenya.

The results (Tables 1 and 3) have indicated that all experiment intercrops were basically ranked fairly by the farmers. This could be as a result of nutrient competition since cowpeas are heavy nutrient miners as they are associated with interspecific competition in mixed stands. The same results have been reported by Katsaruware et al. (2009) that crop yield reduction can be experienced in intercrops where they are associated with interspecific competition in mixed stands and the absence of interspecific competition in the monocrops. The results further indicate that probably intercropping sorghum with cowpea depressed sorghum yields which influenced farmers to rank them in fair category. This outcome for sorghum (Table 2 and 3) could be in line with reports for maize from Kenya (Nadar, 1984) and in Tanzania (Jensen et al., 2003) where maize grain yields reduction of 46-57 and 9% occurred when maize was intercropped with cowpea due to the competition for moisture between the two crops. Alternatively due to slow mineralization of manure which could graduate to good category with given number of seasons (Lekasi et al., 2003). The results by Miriti (2011) reported that cowpea was a nutrient competitor for maize production in semi-arid areas of eastern Kenya. The results (Tables 2 and 3) had a very clear trend on farmer's perception and crop yield that for all those treatments regarded as 'controls' were poorly ranked by the farmers and had lower crop yields regardless of water harvesting and cropping system. This is in line with cultivation in the same piece of land for a continuous cultivation results to nutrient depletion and requires nutrient replenishment (Mugwe et al., 2009; Miriti et al., 2003). Land degradation as a result of various biophysical factors contributing to reduction in agricultural productivity has contributed to farmers being discouraged from adoption of these water conservation structures due to labour shortage and land tenure uncertainty (Demelash and Stahr, 2010).

Conclusion

The results reported in the study demonstrate that smallholder farmers' knowledge can provides a consistent treatment evaluation. There was clear evidence from the study that no difference was noted in terms of scoring of treatments by gender in Ndia Division, Kirinyaga West District. Therefore, both genders could be used by agricultural extension services and researchers for evaluation of other related scientific work in this study area. Kirinyaga West district is characterized by low and erratic rainfall and generally fragile ecosystems which are not suitable for sustainable rain-fed agriculture. The results have indicated that there are needed to incorporate various water harvesting technologies and minimum soil nutrient supplements during seasons with low rainfall distribution regardless of water harvesting technology employed in semi-arid areas. This also suggests that only low-input technologies are currently suitable and need to be adopted through a known crop intensification technologies that could be enhanced in these areas of central highland of Kenya. The results have also demonstrated a very clear message to smallholder farmer, extension services and other stakeholders that there is need for water harvesting technologies and nutrient replenishment on-farm every season for increased sorghum and cowpea productivity in central highland of Kenya.

ACKNOWLEDGEMENT

This work was supported by International Development Research Centre (IDRC - Grant Number 105790-003) and Arid and Semi-arid Lands-Agricultural Productivity Research Project (ASAL-APRP). We would like to acknowledge the Director KARI, Centre Director KARI-Muguga and Kenyatta University supervisors for their facilitation. Authors would also like to acknowledge the contributions by staff from KARI-Muguga south, KARI-headquarters, KARI-Embu and KARI-Njoro for their assistance. We are grateful to all the area agricultural extension officers and farmers who participate in this work.

REFERENCES

Bationo A, Waswa BS (2011). New challenges and opportunities for integrated soil fertility management in Africa. In Bationo, A, Waswa,B, Okeyo, JM, Maina F, Kihara J (eds). Innovation as key to the green revolution in Africa. Exploring the scientific facts, Springer science+Business media. p. 1.

Bationo A, Kimetu J, Ikeera S, Kimani S, Mugendi D, Odendo M, Silver M, Swift MJ Sanging N (2004). The African Network for soil biology and fertility: New challenges and opportunities. In: Managing Nutrient Cycles to Sustain Soil Fertility in Sub-Saharan Africa. Bationo, A. (Ed.). Academy Science Publishers and Tropical Soil Biology and Fertility Institute of CIAT, Nairobi, Kenya. pp 1-23.

Bielders CL, Michels K, Bationo A (2002). On-farm evaluation of ridging and residue management options in a Sahelian millet cowpea intercrop. Soil quality changes. Soil Use Manage. 18:216-222.

Demelash M, Stahr K (2010). Assessment of Integrated soil and water conservation measures on key soil properties in south Gonder, North-Western Highlands of Ethiopia. Soil Sci. Environ. Manage. 1:164-176.

Fongod AGN, Mih AM, Nkwatoh TN (2012). Morphological and agronomic characterization of different accessions of sweet potatoes (*Ipomoea batatas*) in Cameroon. Int. Res. J. Agric. Sci. Soil Sci. 2:254-245.

Gachimbi LN (2002). Technical report of soil survey and sampling results: Embu-Mbeere Districts, Kenya. LUCID working paper series number: KARI-kabete, Nairobi, Kenya. p. 9.

Gicheru PT (2002). The effect of different soil management practices on crust formation, soil moisture conservation and crop growth in Machang'a area, Mbeere District, Kenya. Ph.D Thesis. University of Nairobi, Kenya.

Gitau AN (2004). Mechanical behaviour of hardsetting Luvisal soil. Ph.D Thesis, University of Nairobi, Kenya.

Godsteven P. Maro1, Jerome P. Mrema, Balthazar M. Msanya, James MT (2013). Farmers' perception of soil fertility problems and their attitudes towards integrated soil fertility management for coffee in Northern Tanzania. J. Soil Sci. Environ. Manage. 4(5):93-99

Government of Kenya (2007). Government of Kenya. Ministry of planning and national development. The Kenya Vision 2030; A competitative and Prosperous Kenya. Nairobi, Kenya.

Gumbo D (2006). Integrated soil fertility management. Technical Brief, Practical Action Southern Africa, Harare, Zimbabwe. p. 5.

Ikerra ST, Semu E, Mrema JP (2007). Combining Tithonia diversifolia and Minjingu Phosphate Rock for improvement of P availability and maize grain yields on a Chromic Acrisol in Morogoro, Tanzania. In: A. Batiano et al (eds). Advances in ISFM in Sub-Saharan Africa: Challenges and Opportunities. Springer pp. 333-344

Jaetzold R, Schmidt H, Hornet ZB, Shisanya C.\A (2007). Farm Management Handbook of Kenya. Natural Conditions and Farm Information. 2nd Edition. Vol.11/ C. Central Province. Ministry of Agriculture/GTZ, Nairobi, Kenya.

Jensen JR, Bernhard RH, Hansen S, McDonagh J, MØberg JP, Nielsen N.E, Nordbo, E (2003). Productivity in maize based cropping systems under various soil-water-nutrient management strategies in semi-arid Alfisol environment in East Africa. Agric. Water Manage. 59:217-237.

Kaizzi CK, Ssali H, Nansamba A, Vlek PLG (2007). The potential benefits of Azolla, velvet bean and N fertilizers in rice production under contrasting systems in Eastern Uganda. In: A. Batiano et al (eds). Advances in ISFM in Sub-Saharan Africa: Challenges and Opportunities. Springer pp. 423-433.

Katsaruware RD1, Manyanhaire IO (2009). Maize-cowpea intercropping and weed Agronomy and Soil Science: Madison, WI, USA. Suppression in leaf stripped and detasselled maize in zimbabwe. EJEAFChe. 8(11):1218-1226.

Keating BA, Wafula BM, Watiki JM (1992). Exploring strategies for increased productivity. The case for maize in semi-arid eastern Kenya. Proceeding of symposium held in Nairobi, Kenya, 10-11 December 1990. pp 90-99. Goanna Pry Ltd, Canberra, Austria.

Kimani SK, Esilaba AO, Odera MM, Kimenye L, Vanlauwe B, Bationo A (2007). Effects of organic and mineral sources of nutrients on maize yields in three districts of Central Kenya. In: A. Batiano et al (eds). *Advances in ISFM in Sub-Saharan Africa:* Challenges and Opportunities. Springer pp. 353-357.

Kimani SK, Macharia JM, Gachengo C, Palm CA, Delve RJ (2004). Maize production in the central highlands of Kenya using cattle manures combined with modest amounts of mineral fertilizer. Uganda J. Agric. Sci. 9:480-490.

Lekasi JK, Tanner JC, Kimani SK, Harris PJC (2003). Cattle manure quality in Maragwa district, Central Kenya: effect of management practices and development of simple method of assessment. Agric. Ecosyst. Environ. 94:289-298.

Mairura FS, Mugendi DN, Mwanje JI, Ramisch JJ, Mbugua PK, Chianu JN (2007). Integrating scientific and farmers' evaluation of soil quality indicators in Central Kenya. Geoderma 139:134–143.

Miriti, JM (2011). Effect of conservation tillage on water and nitrogen use efficiency in maize-cowpea cropping systems in semi-arid eastern Kenya. Ph.D Thesis. University of Nairobi, Kenya.

Miriti JM, Kironchi G, Esilaba AO, Heng LK, Gachene CKK, Mwangi DM (2012). Yield and water use efficiencies of maize and cowpea as affected by tillage and cropping systems in semi-arid Eastern Kenya. Agric. Water Manage. 115:148-155.

Miriti JM, Thomas DB, Gachene CCK (2003). Soil chemical properties, sunflower growth and yields as affected by tillage, compost and mulch application. E. Afr. Agric. For. J. 68(3):159-163.

Mugendi E, Gitonga N, Cheruiyot R, Maingi J (2010). Biological Nitrogen Fixation by Promiscous Sobean (*Glycine max* L. Merril) in the central Highlands of Kenya: Response to Soil Amendends. World J. Agric. Sci. 6:381-387.

Mugwe J, Mugendi D, Mucheru-Muna M, Merckx R, Chianu J, Vanlauwe B (2009). Determinants of the decision to adopt integrated soil fertility management practices by smallholder farmers in the central highlands of Kenya. Expl. Agric. 45:61-75.

Nadar HM (1984). Intercropping and intercrop component interaction under varying rainfall conditions in eastern Kenya. II: Maize/cowpea intercrop. E. Afr. Agric. For. J. 44:176-181.

Nandwa SM (2001). Soil organic carbon (SOC) management for sustainable productivity of cropping and agroforestry systems in Eastern and Southern Africa. Nutr. Cycl. Agroecosyst. 61:143-158.

Njeru PNM, Maina I, Miruka M, Amboga JS, Gitari J, Rono B, Mahasi M, Murithi F (2011). Soil fertility characterization of smallholder farms under group and individual management in Central Kenya. Proceedings on held at Maputo, Mozambique on 10th to 13th Octomber. Afri. Crop Sci. Conf. Prod. 10:1-4.

Njeru PNM, Hjalten J, Kimani SK (2010). A book on the effect of push-pull management and different fertilization strategies on maize crop yield in central Kenya. Lambert Acadamic publishers, German, SSBN 978-3-8433-6376-1.

Okalebo JR, Othieno CO, Woomer PL, Karanja NK, Semoka JRM., Bekunda MA, Mugendi DN, Muasya RM, Bationo A, Mukhwana EJ (2006). Available technologies to replenish soil fertility in East Africa. Nutr. Cycl. Agroecosyst. 76:153–170.

Okalebo JR, Othieno CO, Woomer PL, Karanja NK, Semoka JMR, Bekunda MA, Mugendi DN, Muasya RM, Bationo A, Mukhwana EJ (2007). Available technologies to replenish soil fertility in East Africa. In: Batiano, A. et al (eds). Advances in ISFM in Sub-Saharan Africa: Challenges and Opportunities. TSBF-CIAT, Nairobi, Kenya. pp. 45-62.

Raab RT (2002). Fundamentals of Integrated Soil Fertility Management. IFDC Training materials for the "Training Program on Integrated Soil Fertility Management (ISFM) in the Tropics", Lome, Togo. p. 10.

Smaling EMA, Nandwa SM Jansen BH (1997). Soil fertity is at stake. In. Replenishing Soil Fertility in Africa. Special publications No. 51. Amer. Soci. Agro. Soil Sci Madison, WI, USA.

Stocking M, Murnaghan N (2000). Land degradation: Guidelines for field assessment. Overseas Development Group, University of East Anglia, Norwich, UK. p. 120.

Stroosnijder L (2009). Modifying land management in order to improve efficiency of water use in the African highlands. Soil Till. Res. 103:247–256.

Thornton PK, Jones PG, Alagarswamy G, Andresen J, Herrero M (2010). Adopting to climate change: Agricultural system and household impact in East Africa. Agric. Syst. 103:73-82.

Vanlauwe B, Zingore S (2011). Integrated soil fertility management: An operational definition and consequences for implementation and dissemination. Better Crops 95(3):4-7.

Waldrip HM, He Z, Griffin TS (2012). Effects of organic dairy manure on soil phosphatase activity, available soil phosphorus, and growth of sorghum-sudangrass. Soil Sci. 177(11).

Yuste J, Baldocchi DD, Gershenson A, Goldstein A, Misson L,Wong S (2007). Microbial soil respiration and its dependency on carbon inputs, soil temperature and moisture. Global Change Biol. 13:2018–2035.

Simulation of sediment transport to *Sawah* rice fields by applying the water erosion prediction project (WEPP) model to a watershed in Ghana

E. T. Atakora[1] , N. Kyei-Baffour[1], E. Ofori[1] and B. O. Antwi[2]

[1]Department of Agricultural Engineering, Kwame Nkrumah University of Science and Technology, Kumasi, Ghana.
[2]Soil and Water Management Division, CSIR- Soil Research Institute of Ghana, Academy Post Office, Kumasi, Ghana.

Inland valleys are being used under the Sawah technology for rice production to reduce rice imports to Ghana. Sawah technology is assumed to benefit from geological fertilization. However, there is no quantitative information on runoff and sediment flows in the agricultural watershed of Ghana. This study was carried out at Biemso in the southern part of the country. The aim was to estimate runoff and sediment transport using the water erosion prediction project (WEPP) model (version 2006.500), from hillslope to the valley bottom where rice is cultivated using the Sawah technology. A digital elevation model (DEM) was created from ground survey and used to select the various plots (hillslopes) and to select slope input parameters. Four plots (hillslopes) were selected for the model simulation. Data on local daily values of rainfall and on minimum and maximum temperatures were used to set a CLIGEN model station file to determine climate input parameters for the model. Rainfall characteristics (erosivity and distribution) were analysed. Soil erodibility was also determined. Soil and crop management input parameters required by the model were identified and or estimated from field measurements and secondary sources. The model was run for two management scenarios: Fallow and continuous maize systems. The results of the simulation showed that 2.9 to 3.9 and 6.8 to 10.2 t/ha/year of sediments were eroded from upper catchment to valley bottom under fallow system and maize, respectively. The range of values for runoff produced under fallow was 17.4 to 40 mm whereas that under maize system is 158.7 to 233.62 mm. The study has shown that land use system in the study area has a great influence on geological fertilization. In addition, the valley bottom where rice is produced under the Sawah system is enriched with organic matter from upslope.

Key words: Geological fertilization, sediment, *sawah*, inland valleys, runoff, water erosion prediction project (WEPP) (2006.500) model.

INTRODUCTION

Soil erosion is a dynamic process whereby sediments are detached, transported, and accumulated at a place different from where they are originated, mostly in low-lying areas or in water bodies downstream. Sediment transport in a catchment can have both negative and positive impacts.

The negative impacts include the removal of nutrient-rich topsoils in upland areas and subsequent reduction of agricultural productivity in those areas, reduction of irrigation conveyance capacities in canals and reservoir storage volumes, reduction in irrigation water quality increasing water turbidity, changes in river channels and subsequent increase in flood vulnerability of the floodplain farmlands and residential areas. Sediment transport, however, is not always a negative environmental process. Whenever soil erosion occurs, there may be downstream benefits such as deposition of rich sediments for promotion of agricultural activities. One system that benefits from such rich sediment eroded from uplands is the Sawah technology.

The Sawah technology involves bunding, puddling and leveling of rice fields, provision of irrigation and drainage facilities. The benefit of growing rice under the Sawah system is assumed to be the flow of nutrients and sediment from the upper catchment (Wakatsuki et al., 2005). This in principle is supposed to increase rice yields. However, while making a case for the *sawah* eco-technology as a means of coping with the widespread land degradation problem and the attendant decline in soil productivity in West Africa, Obalum et al. (2012) recently decried the lack of data on the extent of geological fertilization of the lowlands in the region as one of the areas needing research.

Soil erosion has been of much concern especially to countries like Ghana whose economies depend on agriculture. Investigations carried out by the Soil Research Institute of Ghana in 1989 revealed that 29.5% of the country is subject to slight to moderate interrill and rill erosion, 43.3% to severe interrill and rill and gully erosion, and 23% to very severe interrill and rill and gully erosion (Quansah et al., 1989). Several studies conducted in the country indicates that erosion by water has contributed immensely to the loss of rich soil nutrients from upland soils to low-lying areas (Asiamah et al., 2000; Quansah, 2001; Quansah et al., 2002; Akyea, 2009; Amegashie, 2009). In Ashanti Region for example, reports by Quansah et al. (2000) and Adama (2003) indicate that eroded sediments which are transported to low-lying areas contains higher concentrations of soil organic matter and plant nutrients than the soil from which it was eroded.

Estimating soil erosion rates and sediment transport to low-lying areas has been challenging since they are affected by multiple factors. As a result, many erosion and sediment yield models have been developed to estimate runoff and sediment flows in a catchment. However, many of such models fail to consider the depositional portions of a landscape profile. Again, the practical uses of such models are limited due to uncertainties in their input requirement (Bogardi et al., 1985).

The Water Erosion Prediction Project (WEPP) model (Flanagan et al., 2001) is one well-validated erosion prediction tool that is considered as an important analytical tool for global studies (Favis-Mortlock et al., 1996; Bhuyan et al., 2002). The WEPP, a continuous physically-based simulation model, can estimate erosion and soil loss in "Sawah" site since it has the capability of estimating the spatial and temporal distribution of soil loss for an entire hillslope or for each point on a hillslope. Again, it has been demonstrated to produce reasonable results in Uganda (Biteete-Tukahirwa, 1995), Kenya (Ampofo et al., 2002; Saenyi, 2002) and Tanzania (Chitrakar, 2002). Therefore, the WEPP model was considered appropriate for this study.

The main objective of this study was to use the WEPP model to estimate runoff and soil losses from hillslopes to the valley bottom where rice is grown under the "Sawah" technology.

The WEPP model

WEPP is a process-oriented model which is based on modern hydrological and erosion science to calculate runoff and erosion on a daily basis. It is based on fundamentals of infiltration, surface runoff, plant growth, residue decomposition, hydraulics, tillage, management, soil consolidation and erosion mechanics (Flanagan et al., 2001).

The WEPP model accommodates many processes related to soil erosion, including rill and interrill erosion, infiltration, percolation, sediment transport and deposition, surface runoff, residue and canopy effects, tillage effects, and evapotranspiration. Therefore, it can provide reasonably accurate and quick estimates for soil erosion (Renschler and Lee, 2005). The most notable advantages of the model include the capability of estimating spatial and temporal distributions of net soil loss and extrapolation to a broad range of conditions, which may not be practical or economical to test in the field (Flanagan and Nearing, 1995; Flanagan et al., 2001).

The hillslope component of the WEPP erosion model requires a minimum of four input data files to run. These are: climate, slope, soil and crop/management files. The climate file includes daily values for precipitation, temperatures, solar radiation, and wind information. A stand-alone programme called CLIGEN is used to generate either continuous simulation climate files or single storm climate files. The slope file requires information on slope orientation, slope length, and slope steepness at points down the profile. Many types of non-uniformities on a hillslope can be simulated through the use of strips or Overland Flow Elements (OFE's). Each OFE on a hillslope is a region of homogeneous soils, cropping, and management. The soil file contains information on the number of soil layers, depth of soil layer from the soil surface and soil properties, such as interrill and rill erodibilities, critical shear, percentages of sand, clay, and organic matter, cation exchange capacity, hydraulic conductivity and gravel concentration. The management file contains information needed to define initial conditions, tillage practices, plant growth parameters, residue management, and crop management.

The WEPP hillslope profile erosion model uses a steady-state sediment continuity equation to calculate erosion on a rill-interrill basis as:

$$\frac{\partial G}{\partial x} = D_i + D_r \qquad (1)$$

where x represents distance downslope (m); G is sediment load ($kgs^{-1}m^{-1}$); D_i is interrill sediment delivery to the rill ($kgs^{-1}m^{-2}$), and D_r is rill erosion rate ($kgs^{-1}m^{-2}$).

The rill erosion rate is calculated using the equation:

$$D_r = D_c + \left(1 - \frac{G}{T_c}\right) \qquad (2)$$

where D_c is detachment capacity by rill flow ($kgs^{-1}m^{-2}$), and T_c is sediment transport capacity in rill ($kgs^{-1}m^{-1}$). The D_c is calculated as:

$$D_c = K_r + (\tau - \tau_c) \qquad (3)$$

where K_r is a rill erodibility parameter (sm^{-1}), τ is flow shear stress acting on soil particles (Pa), τ_c is the rill detachment threshold parameter (critical shear stress of the soil) (Pa).

Finally, the Interrill erosion rate is calculated using the equation:

$$D_i = K_{adj} I_e \sigma_{ir} SDR_r \left(\frac{R_s}{w}\right) \qquad (4)$$

where K_{adj} is the adjusted interrill erodibility (kgs m^{-4}); I_e is the effective rainfall intensity (ms^{-1}); σ_{ir} is the interrill runoff rate (ms^{-1}); SDR_r is the sediment delivery ratio in (%); R_s is the rill spacing (m), and w is the width of the rills (m).

MATERIALS AND METHODS

Description of study site

This study was conducted at a 'Sawah' technology experimental site located at Biemso in the Ashanti Region of Ghana. This study area is located in the Ahafo Ano South District of Ashanti Region (Figure 1). Its geographic location is within N 06°, 52' 53.2" and W 001°, 50' 47.3". The mean annual precipitation in the region is 1301 mm (averaged for the period from 1974 to 2004). The rainfall pattern in the area is bimodal (with two peaks). The geology of this study site consists of rock of the Lower Birimian formation comprising *phyllites*, *greywackes*, *schists* and *gneiss, with* the soils classified under the Akumadan–Bekwai/Oda Complex association in the local soil surveys (Adu, 1992). Agriculture is the main land use in the area. The valley is mostly used for rice cultivation under the system called "Sawah". The most important crop commonly grown in the adjoining uplands (hillslopes) is maize (*Zea mays*). Maize cultivation covers about 50% of the land in this study area.

Climatic data

Climatic data for the period 1974 to 2004 were collected from the Ghana Meteorological Agency. The mean monthly precipitation, mean monthly maximum and minimum temperatures for the period were calculated and used as monthly climatic parameters. For the simulation year, 2004, daily values of precipitation and maximum and minimum temperatures were used. Also, rainfall erosivity and rain distribution were estimated using the Modified Fournier Index (MFI) method (Arnoldus, 1980) and Precipitation Concentration Index (PCI) method (Oliver, 1980) respectively as follows:

$$MFI = \sum p^2 / P \qquad (5)$$

$$PCI = 100 \frac{\sum p^2}{P^2} \qquad (6)$$

where p is monthly precipitation and P is annual precipitation.

Climate dataset required by the model was built by setting up a CLIGEN par file. To build a CLIGEN input parameter file for the study area, the existing international precipitation and temperature data from Accra, and in GDS file format was downloaded from: http://hydrolab.arsusda.gov/nicks/nicks.htm. This was further used with local data collected from Ghana Meteorological Agency to search for a similar US station and also to set up the CLIGEN par file from the local data. The values were then used to generate the needed WEPP climate file using CLIGEN 4.3 to set up a CLIGEN par file for the study area.

Crop/management

Based on secondary data analysis and field visits the major crop which is maize, was identified at the hillslope. Two management scenarios were considered for model simulation: fallow land and continuous maize cropping. Data were collected on plant growth and harvest parameters, temperature and radiation parameters, canopy, leaf area index (LAI) and root parameters, residue parameters, tillage parameters and initial plant conditions. Initial plant conditions and plant growth parameters were determined using field measurement and personal communication with technicians from the Soil Research Institute. However, harvest parameters, temperature and radiation parameters, canopy, leaf area index (LAI) and root parameters, residue parameters and tillage parameters were obtained from the WEPP User's Manual (Flanagan et al., 2001).

Slope

The slope data of the catchment were collected through ground survey. Data collection was done using a dumpy level, a leveling staff and a handheld GPS. The data obtained from ground survey were converted to x, y and z coordinates using Microsoft Excel. The file was then converted into a comma separated value (csv) file and saved. The csv data file was then imported manually into an ArcMap with the aim of converting the data into an ArcGIS shape files. These were then used to create a digital elevation model (DEM). From the DEM, topographic attributes such as slope gradient, aspect, hill shade land use map, flow direction and flow accumulation were determined. Four plots (hillslopes) were extracted through overlay operations.

Soil

Based on the DEM created from ground survey, a soil survey was carried out. Two types of soil samples – disturbed and undisturbed – were collected from each plot. Samples were taken from both upslope and foot slope positions of each plot.

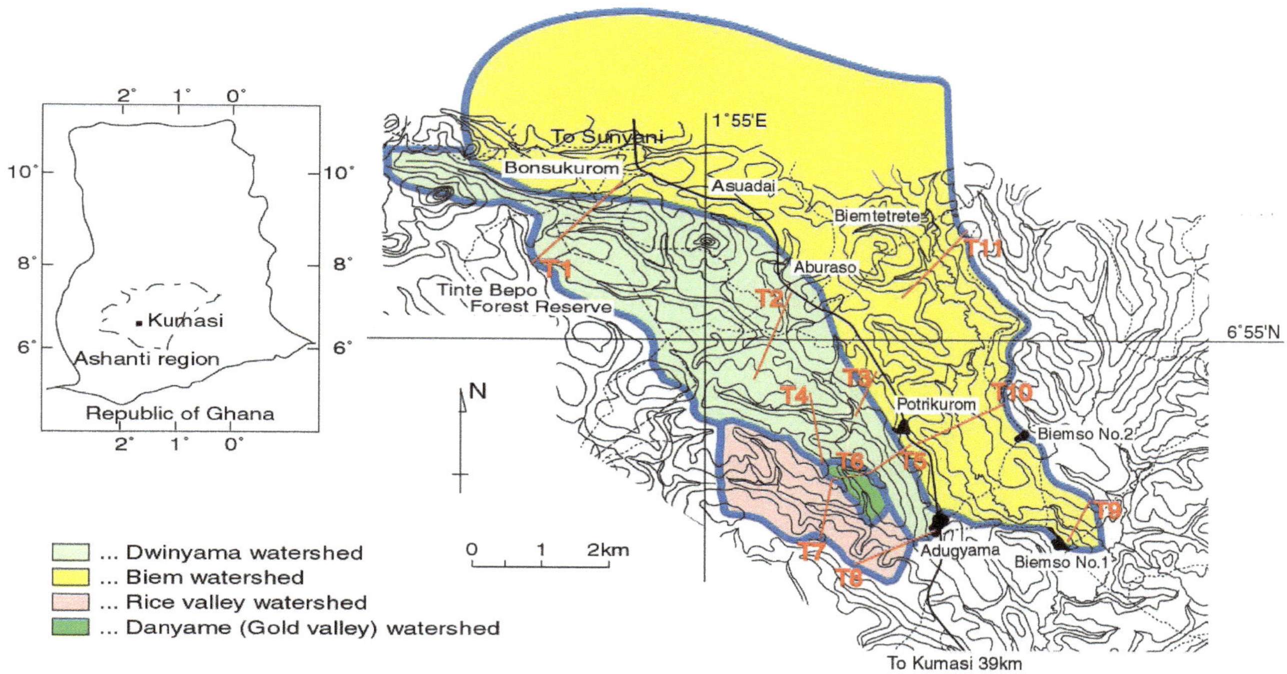

Figure 1. Location of this study area and sampling sites at Biemso in Ghana.

In the laboratory, soil bulk density was determined by the metal core sampler method (Blake and Hartge, 1986). The particle size distribution was determined by the method proposed by Van Reeuwijk (2002). The pH of the soil samples was potentiometrically measured in the supernatant suspension of a 1: 2.5 soil: liquid (pH-H_2O) as described by Peech (1965). The Walkley-Black procedure as described by Nelson and Sommers (1996) was used to determine organic matter concentration in the soil. The cation exchange capacity (CEC) was obtained by a method described by Black (1965). Infiltration rates were measured with minidisk infiltrometer and hydraulic conductivity was determined using the method proposed by Zhang (1997) for dry soils.

In addition to soil parameters which were determined in the laboratory, the model requires data on soil erodibility, critical shear stress and soil albedo. These parameters were estimated using formulae documented by Flanagan and Livingston (1995) as follows:

For cropland soils containing 30% or more sand, the equations are:

$$K_i = 2728000 - 192100 \times VFS \tag{7}$$

$$K_r = 0.00197 + 0.00030\, VFS + 0.03863 e^{-1.84\, ORGMAT} \tag{8}$$

$$\tau_c = 2.67 + 0.065 CLAY - 0.058\, VFS \tag{9}$$

where; K_i = interrill erodibility parameter (kg s m^{-4}); K_r = rill erodibility parameter (sm^{-1}); τ_c = critical shear parameter (Nm^{-2}); VFS = percent very fine sand (%) $\leq 40\%$ (if > 40%, use 40%); ORGMAT = organic matter in the soil surface (%) > 0.35% (if < 0.35%, use 0.35%); CLAY = percent clay (%) $\leq$ 40% (if > 40%, use 40%).

For cropland soils containing less than 30% sand, the equations are:

$$K_i = 6054000 - 55130\, CLAY \tag{10}$$

$$K_r = 0.0069 + 0.134 e^{-0.20 CLAY} \tag{11}$$

$$\tau_c = 3.5 \tag{12}$$

Where in Equations 10 and 11, CLAY ≥ 10% (if < 10%, use 10%)

$$Albedo = \frac{0.6}{e^{0.4 \times OM}} \tag{13}$$

WEPP 2006.500 was used in continuous-simulation mode to predict runoff and sediment yield for ten years. Each plot was considered as an overland flow element (OFE) and had its own set of input files.

RESULTS AND DISCUSSION

Topographical nature of the catchment

As shown in Figure 2, the valley occupies the northern part of the catchment. Among the four plots selected forthis study, Plot 3 had the highest slope followed by Plot 4. Also, Plot 1 which had the lowest slope had the largest area (Table 1).

Rainfall erosivity and rain distribution

Figure 3 illustrates the Modified Fournier Index (MFI) for the individual years. The MFI values ranged from 118.43 mm (in 1983) to 229.88 mm (in 2002) with a mean value of 169.08 mm (Figure 4). Only 2 years, 1983 and 1996 had MFI values slightly less than 120 mm. The remaining

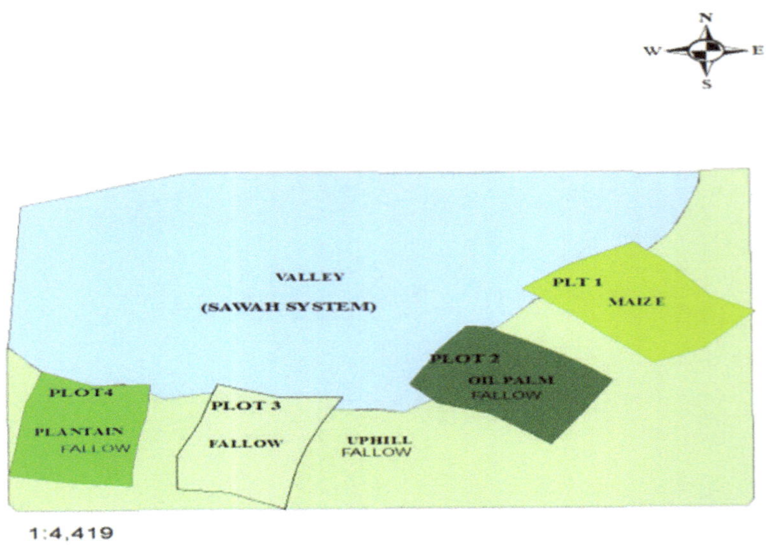

Figure 2. Schematic description of Biemso catchment.

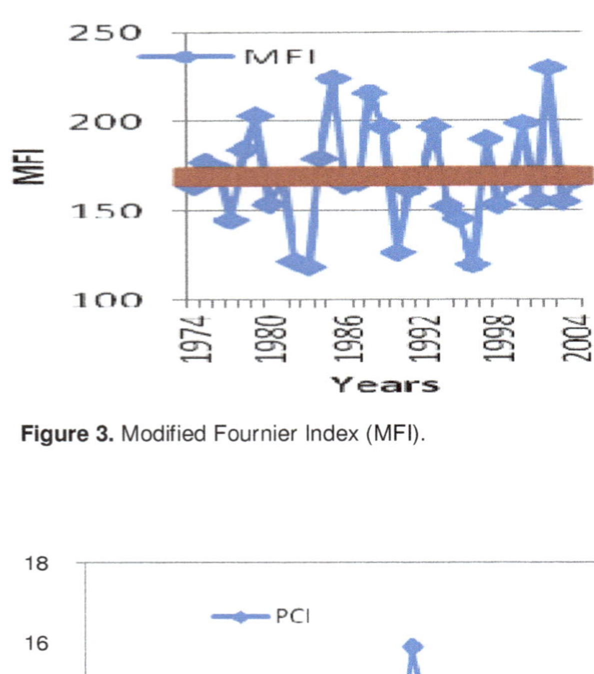

Figure 3. Modified Fournier Index (MFI).

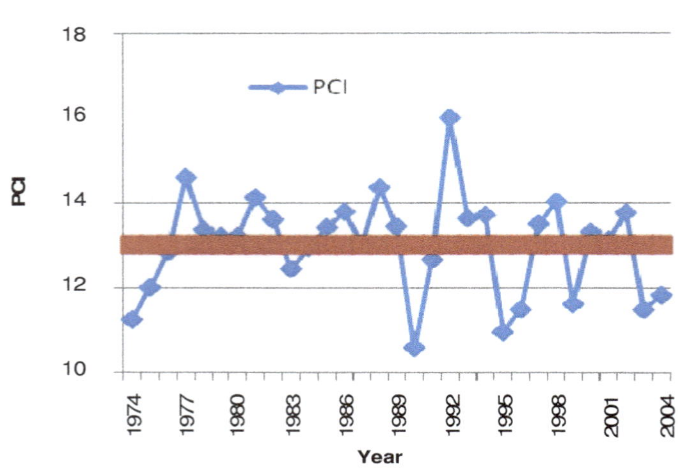

Figure 4. Precipitation concentration for each year Index (PCI) values for each year.

had values above 120 mm. According to the conceptual scale for assessing the MFI proposed by CORNIE (cited by Lujan and Gabriels, 2005), values above 120 show high aggressive rains. Thus, the results obtained indicated that rains in the area are high enough to cause severe soil loss and nutrient transport especially from the upper catchment to the valley bottom.

Figure 4 illustrates the Precipitation Concentration Index (PCI) values for the individual years as well as the mean PCI. The mean value was 13.02. The values ranged from 10.59 (1990) to 16.00 (1992). All the values were greater than 10. According to the scale for assessing PCI proposed by Oliver (1980), values from 11 to 15 denote a moderate seasonal distribution. This could be attributed to the variations in rainfall values. In addition, since precipitation concentration is related to soil erosion, and water management, the results also imply that soil erosion in the area is moderately difficult to control (Lujan and Gabriels, 2005)

Runoff generated by upper catchment

Table 2 shows the average amount of selected soilproperties in the upper catchment. The amount of average annual runoff produced on each plot is presented in Figure 5. The results from this study have shown that soil properties indeed affect the amount of runoff. Low organic matter content, low CEC and high clay content low hydraulic conductivity and low infiltration rates produced the greatest runoff as shown by plot 4 in Figure 5. On the other hand, high organic matter content, high clay content, high hydraulic conductivity and high infiltration rates produced lowest runoff. According to FAO (2005), soil organic matter and CEC improves soil

Table 1. Description of created plots at study site.

Plot	Area (m^2)	Area occupied (%)	Average slope (%)
Plot 1	2488	5.7	20.5
Plot 2	2340	5.4	21.7
Plot 3	2216	5.1	34.7
Plot 4	2123	4.9	30
Valley	20644	47.5	6.3

Table 2. Average values of selected soil properties in the upper catchment.

Land use	Parameter	SOM (%)	CEC	Clay (%)	Ks (mm/h)	If (mm/h)
Plot 1	Mean	2.83	12.01	20.01	31.43	182
	Std. Dev	0.73	1.04	0.47	9.5	89.16
Plot 2	Mean	2.97	13.78	16.04	43.49	419.78
	Std. Dev	1.1	1.4	1.23	4.25	101.67
Plot 3	Mean	4.1	14.24	20.08	60.06	601.78
	Std. Dev	1.35	0.6	0.61	15.65	191.71
Plot 4	Mean	2.48	5.67	22.03	4.11	41.78
	Std. Dev	1.2	1.06	1.11	1.22	14.49

SOM = Soil organic matter, CEC = cation exchange capacity, K_s = hydraulic conductivity, I_f = infiltration rate.

moisture, moisture holding capacity, infiltration capacity, diversity and activity of soil organisms. Also the results showed that runoff generated is highly influenced by land use. Runoff produced under maize cultivation was greater than runoff produced under fallow system. Fallow lands are rich in soil organic matter and therefore form stable aggregate thus reducing runoff. This study has shown that to reduce runoff from the upper catchment there is need to consider fallow period in the cropping calendar.

Sediment yield assessment

Figure 6 presents average annual sediment yield from the hillslope profile under the two land use system (maize and fallow). Sediment yield ranged from a minimum of 2.9 t/ha/year to a maximum of 10.2 t/ha/year. This indicates that for every year between 2.85 to 10.2 tons of sediment per hectare leaving the hillslope are deposited at the 'sawah' fields. The results indicate that sediment yield generation is greatly affected by land use (Figure 6). Average sediment yield leaving the hillslope profile under maize cultivation is more than twice that leaving the profile under fallow system. This shows that landuse in this study site affects the movement of sediment from upper catchment to the lower catchment.

In general, the model results showed a fair estimate as the results conform to the results of sediment flow using cesium- 137 within a footslope at Kotokosu catchment in Ejura (Antwi, 2006). The radio nuclide predicted 4 to 10 tons of sediment per hectare using the Diffusion and Migration model.

Conclusion

Based on the results obtained from the study and the discussions made, the following conclusions can be drawn:

(1) The rains in this study area are quite aggressive to cause detachment and transport of soil particles. Mean Modified Fournier Index (MFI) and mean Precipitation Concentration Index (PCI) were 169.08 and 13.02, respectively.

(2) Land use has great effect on sediment yield and runoff. Sediments leaving hillslope profile under maize hillslope under fallow system. Runoff produced under maize was more than seven times that produced under fallow.

(3) Sediment and runoff are both affected by soil properties such as hydraulic conductivity, soil organic matter, cation exchange capacity and clay content.

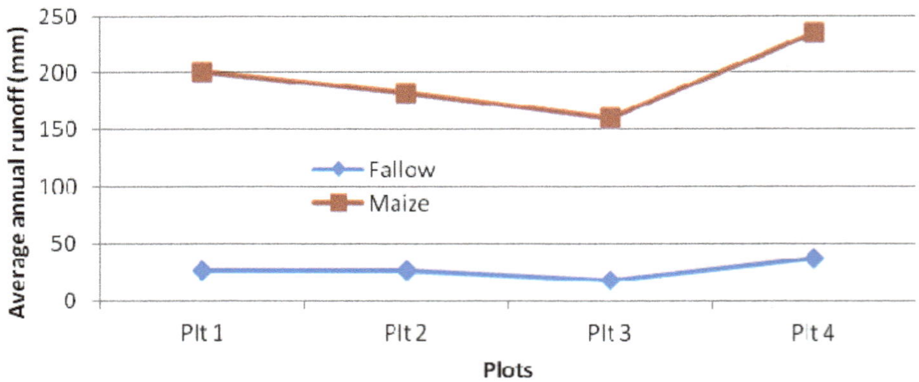

Figure 5. Estimated average annual runoff.

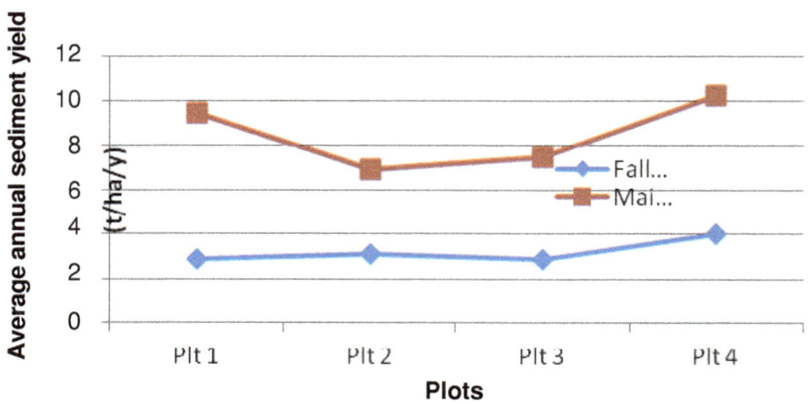

Figure 6. Estimated average sediment yield.

ACKNOWLEDGEMENTS

We are very grateful to Prof. T. Wakatsuiki (Kinki University, Japan) for providing funds for the work. We also thank the Soil Research Institute for providing some field and laboratory assistance.

REFERENCES

Adama AI (2003). The Effect of Different Tillage Practices on Soil Erosion, Water Conservation and Yield of Maize in the Semi-Deciduous Forest Zone of Ghana. M.Sc. Thesis. Department of Crop and Soil Sciences, KNUST, Kumasi, Ghana.

Adu SV (1992). Soils of Kumasi region, Ashanti region, Ghana. Memoir No. 8. Ghana Soil Research Institute. P. 90.

Akyea AO (2009). Effect of Different Land Preparation Methods on Erosion, Growth and Yield of Cassava. Unpublished M.Sc. Thesis. Department of Crop and Soil Sciences, KNUST, Kumasi, Ghana.

Amegashie BK (2009). Assessment of catchment erosion, sedimentation and nutrient export into small reservoirs from their catchments in the Upper East region of Ghana. M.Sc. Thesis. Department of Crop and Soil Sciences, KNUST, Kumasi, Ghana.

Ampofo EA, Muni RK, Bonsu M (2002). Estimation of soil losses within plots as affected by different agricultural management. Hydrol. Sci. J. 47(6):957-967.

Antwi BO (2006). Using Cesium 210 and particle size fraction to model spatial soil erosion rates in the forest Savanna Transition ecology of Volta Basin of Ghana. Unpublished PhD Thesis. KNUST, Kumasi, Ghana. P. 209

Arnoldus HMJ (1980). An approximation of the rainfall factor in the Universal Soil Loss Equation. In: De Boodt M. and D. Gabriels (Eds.) Assessment of Erosion. Wiley, Chichester, West Sussex, UK. pp. 127-132.

Asiamah RD, Quansah C, Dedzoe CD (2000). Soil Degradation: Management and Rehabilitation in Ghana - An Overview Report, In: Proceedings of the FAO/ISCW Expert Consultation on Management of Degraded Soils in Southern and East Africa (MADS - SEA). Pretoria, South Africa. pp. 89-100

Biteete-Tukahirwa JM (1995). Measurement, Prediction and Social Ecology of Accelerated soil erosion in Kibale District of Southwest Uganda. PhD Thesis. Makerere University, Uganda. P. 210.

Bhuyan SJ, Kalita PK, Janssen KA, Barnes PL (2002). Soil loss predictions with three erosion simulation models. Environ. Modell. Software 17:137-146

Black CA (1965). Methods of Soil Analysis. Part 2. Chemical and microbialogical properties. First edition. American Society of Agronomy and Soil Science Society of America. Madison, Wisconsin, USA. Agronomy 9:371-373

Blake GR, Hartge KH (1986). Bulk density. In: Klute, A. (ed.). Methods of soil analysis. Part 1. Physical and mineralogical methods. Second edition. American Society of Agronomy and Soil Science Society of America. Madison, Wisconsin USA. pp. 363-375

Bogardi I, Fogel M, Duckstein L, Bardossy A (1985). In: E. Plate and N. Buras (ed.). Estimating of sediment loading in Agricultural reservoirs with Design and Management of Water Resources Systems. pp. 87-98

Chitrakar S (2004). Testing of the WEPP soil erosion model in the west Usambara mountains of Tanzania. MSc. Thesis. Erosion and Soil and Water Conservation Group. Wageningen Universiteit. The

Netherlands.
FAO (2005). Importance of Soil Organic Matter. FAO Soil Bulletin. P. 94.
Favis Mortlock DT, Quinton JN, Dickson WT (1996). The GCTE Validation of Soil Erosion Model for Global Studies. J. Soil. Water Conserv. 51:397-403.
Flanagan DC, Laflen JM, Lane LG,. Nearing MA (2001). Water Erosion Prediction Project (WEPP) Computer Program (version 2006.5) for Windows. USDA- Water Erosion Prediction Project: Hillsope Profile and Watershed Model Documentation. NSERL Report No. 10. National Soil Erosion Research Laboratory, IN, USA. P. 340.
Flanagan DC, Nearing MA (Eds). 1995. USDA-Water Erosion Prediction Project: Technical documentation. NSERL Report No. 10. National Soil Erosion Research Laboratory, IN, USA. P. 147.
Flanagan DC, Livingston SJ (1995). USDA-Water Erosion Prediction Project: User summary. NSERL Report No. 11. National Soil Erosion Research Laboratory, IN, USA.
Lujan DL, Gabriels D (2005). Assessing the rain erosivity and rain distribution in different agro-climatological zones in Venezuela. *Sociedade and Natureza,* Uberlândia, Special Issue: S&N-2009-632.
Nelson DW, Sommers LE (1996). Total Organic Carbon and Organic Matter. In: Methods. In: Methods of Soil Analysis. Part 3. SSSA Book Series No. 5. SSSA Madison, WI. pp. 214-235.
Obalum SE, Buri MM, Nwite, Hermansah JC, Watanabe CA, Igwe Y Wakastuiki T (2012). Soil Degradation-Induced Decline in Productivity of Sub-Saharan African Soils: The Prospects of Looking Downwards the Lowlands with the Sawah Ecotechnol. Appl. Environ. Soil Sci 2012, Article ID 673926, P. 10 doi:10.1155/2012/673926
Oliver JE (1980). Monthly Precipitation Distribution: A Comparative Index. Professional Geographer. 32(3):300-309.
Peech C (1965). pH. In: Black, C.A. (Ed.). Methods of soil analysis. Part 2: Chemical and microbiological properties. First edition. American Society of Agronomy and Soil Science Society of America, Madison, Wisconsin USA. pp. 371-373.
Quansah C (2001). Integrated soil management for sustainable agriculture and food security. Country case study: Ghana. FAO. Accra. pp.33-75.
Quansah C, Baffoe-Bonnie E, Asiamah RD (1989). Soil erosion in Ghana and strategies for conservation. A paper presented at the African Ministerial Conference on the environment (RCU/SOFERNET) and soil conservation strategy for Africa. Accra, Ghana. 28-30 August, 1989.
Quansah C, Safo EY, Ampontuah EO, Amankwaah AS (2000). Soil Fertility erosion and the associated cost of NPK removal under different soil and residue management in Ghana. Ghana J. Agric. Sci. 33(1):33-42.
Quansah C, Agodzo SK, Gyawu, M, Bonsu J, Dittoh S (2002). National Action programme to Combat Desertification in Ghana. Environmental Protection Agency, Accra, Ghana
Renschler CS, Lee T (2005). Spatially distributed assessment of short- and long-term impacts of multiple best management practices in agricultural watersheds. J. Soil Water Conservat. 60(6):446-456.
Saenyi WW (2002). Sediment Management in Masinga Reseroir, Kenya. PhD Thesis. University of Agricultural Sciences (BOKU), Vienna, Austria. P. 194
Van Reeuwijk LP (2002). Procedure for Soil Analysis. Technical Paper 9. Sixth edition. ISRIC-FAO.
Wakatsuki T, Buri MM, Fashola. OO (2005). Ecological Engineering for Sustainable Rice Production and the Restorattion of Degraded Watershed in West Africa. In: World Rice Research Conference Publication. The International Rice Research Institute (in Press).
Zhang R (1997). Determination of soil sorptivity and hydraulic conductivity from the disc infiltrometer. Determination of soil sorptivity and hydraulic conductivity from the disc infiltrometer. Soil Sci. Soc. Am. J. 61:1024-1030.

Nutrient uptake by *Abelmuscus esculentus* and its effects on changes in soil chemical properties as influenced by residual application of fertilizer

Law-Ogbomo, K. E.

Department of Agriculture, Benson Idahosa University, PMB 1100, Benin City, Nigeria.

A field experiment was conducted as an on-farm trial at Uyinmwendin (Rhodic Paleudult), Edo State, Nigeria, in 2009 and 2010 to investigate the residue of compost manure, compost manure + NPK and NPK fertilizer treatments on nutrient concentration and uptake of *Abelmuscus esculentus* and soil chemical properties. Five treatments were selected viz: untreated (control), compost manure applied at the rate of 20 and 40 t ha^{-1}, NPK applied at the rate of 200 kg ha^{-1} and compost manure + NPK (that is, complementary use of compost manure applied at 20 t ha^{-1} and NPK applied at 100 kg ha^{-1}) using randomized complete block design and replicated three times. Analysis of the soil indicated that the soil is acidic, with low nutrient status. Application of fertilizer influenced the soil nutrient status positively. Optimum fruit yield was obtained at 40 t ha^{-1} compost (49.13 t ha^{-1}). Residually, the N, P and K concentration of okra were significantly higher in compost manure applied at both 20 (2.52% N, 0.43% P and 2.38% K) and 40 t ha^{-1} (2.09% N, 0.41% P and 2.18% K) and compost manure (2.13 % N, 0.43% P and 1.18% K) than NPK (1.22% N, 0.13% P and 0.62% K) and control (0.73% N, 0.11% P and 0.28% K). Compost manure used solely and in complementary quantity with NPK had higher nutrient uptakes than sole NPK use. The soil N, P and K concentrations were reduced by cropping. They were lowest in untreated control and increased with fertilizer application. The compost manure and compost manure + NPK showed a greater potential for increasing plant macronutrients (N, P, K, Ca and Mg) contents. The increases in the soil N, P and K contents after cropping associated with compost manure + NPK treated plots is an indication that compost manure + NPK treatment is a better alternative to inorganic fertilizers because it can sustain continuous cropping.

Key words: Nutrient content and uptake, soil amendments, soil chemical properties.

INTRODUCTION

Okra production in Nigeria is predominantly carried out by the resource-poor farmers. The yield of okra in Nigeria is very low (about 2 t ha^{-1} (FAO, 2007)) owing to low native soil fertility status among other factors. Decline in native fertility status in most humid forest soil is due to intensive weathering and leaching, erosion, run-off, nutrient depletion, nutrient fixation, salinization and water logging (Vanlauwe, 2000). The stability of production depends on replenishing nutrients removed from the soil by crops, maintaining desirable physical condition of the soil, preventing an increase in soil acidity and toxic elements and minimizing or preventing erosion (Sanwal et al., 2007). This emphasizes the importance of fertility restoration in achieving and maintaining high crop productivity and it can be achieved through the use of external fertilizer input.

Inorganic fertilizers have been used as a way of sustaining soil productivity. The addition of inorganic fertilizer is efficient, due to its nutrient concentration and relative ease of transportation. However, the extent to which farmers can depend on this input is constrained by unavailability of the right type of inorganic fertilizer at the right time, high cost, lack of technical know-how, weak extensive service and lack of access to credit (Chude, 1999). Other problems associated with the use of inorganic fertilizers include pollution of ground water after harvest, non-improvement of soil structure (Gordon et al., 1993); increased soil acidity, reduced crop yield, nutrient imbalance and depletion of organic matter. The problems associated with the use of inorganic fertilizers have diverted the attention of agronomists toward making use of organic materials (both organic manures and organic wastes) for improving the soil physical properties that allow profitable crop production (Somani and Totawat, 1996).

Organic manures are rich in nutrient composition and could be easily ploughed back into the field to enhance better performance and sustain agricultural production. They are cheap, easy to come by, always safe to use, non-poisonous to crops and environmentally friendly. In addition, organic manures build healthy soils. However, they have not been fully utilized in the humid tropics due to huge quantities required in order to satisfy the nutritional needs of crops owing to low nutritional concentration compared with inorganic fertilizers. This has resulted in high transportation and handling cost which constitute major constraints.

It is clear that the prospect of obtaining enough inorganic fertilizers to meet the requirement of the teaming population in the tropic is remote. The supply of organic manures and wastes on farms will likely be insufficient to overcome soil nutrient deficiency. Complementary use of organic manures with inorganic fertilizer is widely known to be reliable and sustainable soil fertility management strategy. It minimizes nutrient leaching, particularly in poor sandy soil and subsequent ground water contamination (Manna et al., 1999). This maximizes the use of available organic resources and minimizes the use of costly purchased inorganic fertilizers (Manral and Savena, 2003).

This trial was conducted to evaluate the residual effects of compost manure, compost manure + NPK and NPK fertilizer on nutrient concentration and uptake of *Abelmoschus esculentus* and soil chemical properties in degraded ultisol environment.

MATERIALS AND METHODS

The study was conducted in the forest ecological zone of Edo State, Nigeria in 2009 and 2010. Edo State is located between Latitudes 5° 4″ and 7° 38″ and Longitudes 5° 4″ and 7° 38″ of the equator. The soil used was ultisol of Benin formation (Smith and Montgomery, 1962). Composite soil samples collected from the experimental site before experiment were air-dried at room temperature, crushed and sieved with a 2 mm sieve to remove debris and other materials and prepared for laboratory studies according to Mylavarapus and Kennelley (2002) procedure. The compost manure made up of decomposed plant residues was obtained from Lake Projects Limited, Lagos, Nigeria; while the okra varietal "OCRA" was obtained from Premier Seed Limited, Benin City, Nigeria. The experiment was laid in a randomized complete block design with three replicates. Five treatments were involved viz: untreated (control), compost manure applied at 20 and 40 t ha^{-1}, NPK applied at the rate of 200 (30 kg N + 13.58 kg P + 24.90 kg K), kg NPK ha^{-1} and compost manure + NPK (that is combined compost manure at 20 t ha^{-1}) and NPK at 100 (15 kg + 6.79 kg P + 12.45 kg K) kg ha^{-1}. The compost manure applied was thoroughly mixed with the soil and then left for two weeks to allow for mineralization. NPK was applied one week after transplanting. Each plot size was 3.60 x 3.00 m and consisted of beds.

Three seeds of okra cultivar "OCRA" were sown at 50 x 50 cm which was later thinned to one per stand. At harvest, number of fruits was counted per plant and fresh fruit weights obtained per plot was used to determine yield per hectare. After the field experiment, composite soil samples were taken from individual plots in the field that received different treatment combinations and prepared for laboratory analysis according to the methods described by Mylaravapus and Kennelley (2002). The experiment was repeated in January 2010. This time, no organic fertilizer or inorganic fertilizer was applied. However, at harvest, three fruits were randomly selected from each plot, weighed, sundried, and packed into brown envelopes and oven dried for 2-3 days at 60°C to a constant weight. The dried fruits were ground and passed through a 1mm sieve for N, P and K as per the methods described by Mylavarapus and Kennelley (2002). The nutrient uptakes were computed by multiplying the dried fruit weight with nutrient concentration. Data collected on yield, residual concentration, uptake and soil chemical properties were subjected to statistical analysis using analysis of variance (ANOVA). Mean separation was carried on using Least Significant Difference (LSD) test only where the F-values were significant.

RESULTS AND DISCUSSION

Soil fertility status

The chemical composition of the compost manure used in the trial is shown in Table 1. The compost manure was enriched with nutrients. This implies that compost manure is a source of most essential plant nutrient and thus a complete inorganic fertilizer (NPK) to be used for sustaining crop production. The physical and chemical properties of the soil on which the trial was conducted are presented in Table 2. The soil is slightly acidic (6.00) and the organic carbon is 0.96%. The nutrient contents were low when compared with their critical levels in Southern Nigerian soils (Adeoye and Agboola, 1985). This investigation has revealed that Benin soils are largely deficient in major essential nutrients. Therefore, it is expected that the application of organic and/or inorganic fertilizers such as NPK can increase yield of crops. The various fertilizer types used, raised the fertility status of the soil. NPK fertilizer increased soil acidity through its reduction in soil pH and organic carbon.

Improvement of the soil pH in organic and organomineral fertilizer treated plots might be due to the

Table 1. Chemical properties of compost manure.

Parameter	Value
pH (H_2O 1:1)	7.20
Organic carbon (%)	2.10
Total nitrogen (%)	1.50
Available phosphorus (mg kg^{-1})	92.30
Exchangeable cations (cmol kg^{-1})	
Calcium	20.90
Magnesium	14.10
Potassium	5.70

Table 2. Soil fertility status after treatment with fertilizer before cropping with okra.

Parameter	Treatments				
	Control	20 t ha^{-1} CM	40 t ha^{-1} CM	NPK	CM + N
pH (H_2O) 1:1	6.00	6.10	6.20	5.50	6.10
Organic carbon (%)	0.90	1.10	1.30	0.74	1.30
Total nitrogen (%)	0.12	0.63	0.98	1.05	0.78
Available phosphorus (mg kg^{-1})	9.56	16.13	30.45	96.53	116.72
Exchangeable cations (cmol kg^{-1})					
calcium	1.02	5.15	6.00	9.18	16.10
Magnesium	0.53	1.10	5.50	8.35	16.51
Potassium	1.20	1.70	5,30	8.35	10.51
Particle size analysis (g kg^{-1})					
Clay	250.00	260.00	260.00	250.00	270.00
Silt	100.00	110.00	110.00	90.00	110.00
Sand	650.00	630.00	630.00	660.00	620.00

CM – Compost manure; CM + N – Compost manure + NPK.

possible enrichment of the soil with Ca from the compost manure. The incorporated compost manure into the soil influenced the soil nutrient status. However, the high organic carbon content, with increase in other soil chemical components, is an indication that compost manure has high potential of gradual release of nutrients to the soil that can help to improve the fertility of a degraded ultisol, thereby sustaining yield under intensive cropping system. The higher content of soil total N, available P and exchangeable cations in NPK treated plots compared with compost manure treated plots might be due to fast release of nutrients from it. This however, makes it prone to leaching and percolates deep into the soil, hence not utilized by the crops, it was intended for.

Fruit yield

The effect of fertilizer application on pod yield and its components is shown in Table 3. The yield obtained in the study is a reflection of the application of fertilizer to the soil nutrient. The fruit yield per hectare ranged from 7.92 to 9.13 t ha^{-1} for untreated control and compost at 40 t ha^{-1}, respectively. The observed differences among fertilizer types could be related to nutrient availability to okra plants and release patterns by the fertilizers. The reduced fruit yield produced from the untreated plants could be related to insufficient nutrient uptake as the plants have to rely on the native fertility of the soil. The yield analysis revealed that the fruit yield increase was due to fruit weight and number of fruits per plant. The highest fruit yield produced from 40 t ha^{-1} treated plots might be due to its cumulative effect of nutrients released by the compost manure in the soil.

The average yield of okra in Nigeria was estimated at 5.00 t ha^{-1} (FAO, 2007). But with improved management practice, yield up to 9.13 t ha^{-1} are possible. The observed yield differences among treatments could be related to nutrient availability to crops and release patterns by the organic fertilizer. The fruit yield obtained in the study is a reflection of improvement in nutrient status of the soil as a result of fertilizer application. The highest yield obtained at 40 t ha^{-1} of compost manure + NPK treated plots might be due to the cumulative effect

Table 3. fruit yield and its components.

Treatment	No. of fruits (plant^{-1})	Average fruit wt (g)	Average wt of fruit (g plant^{-1})	Fruit yield (t ha^{-1})
Control	8.33	15.25	126.67	7.92
20 t ha^{-1} Compost manure	10.33	13.57	139.67	8.73
40 t ha^{-1} Compost manure	11.67	12.56	146.00	9.13
NPK	10.67	13.30	141.33	8.84
Compost manure + NPK	11.00	12.73	140.00	8.75
Mean	10.40	13.48	138.73	8.64
LSD (0.05)	0.833	ns	10.113	0.240

ns – Not significant at 5% level of probability.

Table 4. Residual effects of soil amendments on nutrient concentration in okra.

Treatment	Nutrient concentration (%)		
	N	P	K
Control	0.73	0.11	0.28
20 t ha^{-1} Compost manure	2.52	0.43	2.38
40 t ha^{-1} Compost manure	2.09	0.41	2.18
NPK	1.22	0.13	0.62
Compost manure + NPK	2.13	0.43	1.81
Mean	1.74	0.30	1.45
LSD (0.05)	0.581	0.258	0.546

of nutrients released by the compost manure and NPK to the soil.

Nutrient concentration

The residual effect of fertilizer application on nutrient concentration of okra is shown in Table 4. N, P, and K concentration of the plant were dependent on soil nutrient status. N and K were the main nutrients taken from the soil as against the small amount of P. The best residual effect was obtained from the compost manure at 20 t ha^{-1} (2.52%) treated plants. However, it was statistically similar to compost manure at 40 t ha^{-1} and compost manure + NPK treated plants with 2.09 and 2.13%, respectively. Lower N concentration of NPK treated plants compared with compost manure and compost manure + NPK treated plants conformed to the findings of Alabi and Odubena (2001).

Residually, compost manure alone and in complementary quantity with NPK had significant influence on P of the plant. The residual effect of compost manure and compost manure + NPK was not as pronounced on P as N. Compost manure at 40 t ha^{-1} and compost manure + NPK treated plants produced the highest P concentration of 0.43% which was at par with compost manure treated plants of 0.41%. NPK treated plants (0.13%) was not significantly difference from the untreated plants (0.11%). The pattern of distribution of P concentration was similar to K and there was no significant differences between NPK treated plants (0.62 %) and untreated control plants (0.28%). However, other treatments were superior to control. Compost manure at 20 t ha^{-1} produced the highest K concentration (2.38%) which was at par with compost manure applied at 40 t ha^{-1} (2.18%) and compost manure + NPK (1.81%).

Nutrient uptake

The residual effect of compost manure, compost manure + NPK and NPK fertilizer on nutrient uptake of okra is shown in Table 5. The residual effect of compost manure and NPK, used alone and combined had influence on N uptake of okra pod. The application of compost manure at 20 t ha^{-1} residually gave 73.58 kg N ha^{-1} uptake compared to 18.10 kg N ha^{-1} produced from the untreated plants. All treatments were at par but significantly different from untreated plants.

Generally, the nutrient uptakes of P and K under NPK and untreated plants were at par. The uptake of P and K were highest in plots from the residual effect of compost manure and its complementary use with NPK fertilizer. This observation is in agreement with Isitekhale and

Table 5. Residual effects of soil amendments on nutrient uptake of okra.

Treatment	Nutrient uptake (g kg^{-1})		
	N	P	K
Control	18.10	0.28	0.69
20 t ha^{-1} Compost manure	73.58	1.26	69.50
40 t ha^{-1} Compost manure	70.64	1.39	73.68
NPK	38.55	0.41	19.59
Compost manure + NPK	64.11	1.29	54.48
Mean	53.00	0.93	43.59
LSD (0.05)	26.873	0.181	19.519

Table 6. Residual effects of soil amendments application on soil chemical properties.

Parameter	Treatment						
	Control	20 t ha^{-1} CM	40 t ha^{-1} CM	NPK	CM + N	Mean	LSD (0.05)
pH (H_2O 1:1)	5.60	5.90	6.00	5.00	6.00	5.54	0.200
Organic C (%)	0.80	1.00	1.14	0.54	1.35	0.96	0.403
Total N (%)	0.03	0.76	0.99	0.95	1.01	0.71	0.320
Available P (mg kg^{-1})	6.73	18.56	46.17	22.54	101.19	45.19	28.765
Exchangeable cations (cmol kg^{-1})							
Calcium	0.52	5.96	7.40	6.90	13.35	6.63	3.633
Magnesium	0.35	0.95	4.35	6.10	16.00	4.34	3.683
Potassium	0.70	1.50	6.30	5.70	8.90	4.22	1.674

CM - compost manure, CM + N - compost manure + NPK.

Osemwota (2010). Isitekhale and Osemwota (2010) advocated the use of manure since it enhances the release, availability and absorption of nutrients when compared to NPK fertilizer even after the first year of application.

Residual soil chemical properties

The residual effect of fertilizer application on soil chemical properties is shown in Table 6. Residually, compost manure and NPK applied alone and their combinations had significant influence on the soil chemical properties. After cropping with okra, the soil pH was reduced to 5.60 in control, 5.90 in compost manure applied at 20 t ha^{-1}, 6.00 in compost manure applied at 40 t ha^{-1}, 5.00 in NPK and 6.00 in compost manure + NPK. The lowest pH of 5.00 associated with NPK treated plots is in agreement with Ibeawuchi et al. (2006), who observed that continuous use of inorganic fertilizer can lower the soil pH which will result in reduction in crop yield. The least pH recorded in NPK treated plots compared with compost manure treated plots is consistent with acid forming nature of inorganic fertilizer due to its N and P content. Organic carbon also varied significantly among the fertilizer types. Organic carbon was reduced in all the treatments including control except compost manure + NPK treated plots, where it increased from 1.20 to 1.35%. The reduction in organic carbon is owing to complete removal of the biomass from the field (Yihenew, 2002). The low soil organic content has consequently resulted in soil acidity, nutrient imbalance and low crop yield. This observation is in agreement with Agboola and Omueti (1982) who reported that the response of crop to fertilizer application depends on soil organic matter.

Organic carbon contents play a crucial role in sustaining soil fertility, crop production and environmental quality due to their effect on soil physical, chemical and biological properties, such as soil water retention, nutrient cycling, gas flax and plant root growth (Bauer and Black, 1994;Saiju and Kalisz, 1990). The continuous use of NPK alone drastically reduced organic carbon and consequently lowers soil fertility. The total N content increased in the entire treated plots except control and NPK treated plots. In NPK treated plots, the total N content decreased from 1.05% before cropping to 0.95% after cropping. This might be due to leaching and volatilization. This observation is in agreement with Padwick (1983) who reported that many African soils show nutrient deficiency problems after a short period of cultivation, with N being most depleted. Application of compost manure and compost manure + NPK compensated for the low level of total N which is a major defect and characteristic of most tropical soils (Ipinomoroti et al., 2006).

Available P after cropping decreased in control and NPK treated plots and increase in compost manure and compost manure + NPK treated plots. It was highest (101.19 mg kg^{-1}) in compost manure + NPK treated plots. Higher residual available P in plots treated with compost manure than pre-cropping is due to decrease in P sorption.

Exchangeable Ca increased in compost manure and compost manure + NPK treated plots, but decreased in control and NPK treated plots. Exchangeable K varied from 0.52 cmol kg^{-1} in control to 12.35 cmol kg^{-1} in compost manure + NPK treated plots. However, exchangeable Mg was reduced after cropping. It varied from 0.35 cmol kg^{-1} in control to 10.00 cmol kg^{-1} in compost manure + NPK treated plots. The reduction in exchangeable Mg was highest in control plots with the value of 34% followed by NPK (27%) and the least was compost manure + NPK treated plots (5%). Exchangeable K followed the same trend as exchangeable Mg. It ranged from 0.70 cmol kg^{-1}in control to 8.90 cmol kg^{-1} in compost manure + NPK treated soils. The highest reduction was observed in NPK treated plots (46%) and the least was obtained from compost manure applied at 20 t ha^{-1} with a value of 6.30%. The higher mean values in exchangeable cations observed in compost manure and compost manure + NPK were highly influenced by high organic carbon (organic matter) accrued to them.

This study revealed that total N, available P and exchangeable cations were significantly reduced by okra cropping in untreated plots. Higher content of total N, available P, exchangeable Ca and Mg observed in compost manure and compost manure + NPK treated plots can be attributed to the improved organic carbon status of the soil leading to higher organic matter. Organic matter is a natural source of nutrient and exchangeable cations.

This revealed the higher potentials of compost manure and compost manure + NPK for sustained long-term cultivation. However, compost manure + NPK showed a greater potential for increasing soil nutrient contents. This is attributed to the quick release and utilization of inorganic fertilizer by crops to compensate for late release of nutrients from the compost manure which in turn minimizes leaching out of nutrient from inorganic fertilizer. The highest soil nutrient content attributed to compost manure + NPK after cropping is an indication that the NPK requirements of okra can be met by the complimentary use of compost manure and compost manure + NPK as a viable alternative to inorganic fertilizers.

Conclusion

This study has demonstrated the effectiveness of the various fertilizer types in improving the fertility status of degraded ultisol for sustainable okra production. Compost manure when used alone or in combination with NPK residually increased N, P and K concentrations and uptake. The use of the compost manure in combination with NPK best enhanced residual soil fertility and therefore recommended for farmers.

Compost manure + NPK may be a useful management practice to minimize leaching losses, improve soil structure, and reduce production cost by reducing the use of expensive inorganic fertilizer with the additional advantage of a clean environment through the use of compost manure. It also helps to overcome insufficient nutrients in the compost manure.

REFRENCES

Adeoye GO, Agboola AA (1985). The relationship between soil pH, organic matter, available P, exchangeable K, Ca, Mg and nine elements in the maize tissue. Soil Sci. 115 (5):367-375.

Agboola AA, Omueti JAI (1982). Soil fertility problems and its management in tropical Africa. International conference on land clearing and Development. 11 to Ibadan 23-26 November 1982.

Alabi DA, Odubena Y (2001). A preliminary study of the effect of two organic wastes and NPK (15:15:15) on the growth and yield of cowpea. Nig. J. Hort. Sci. 8:19-27.

Chude VO (1999). Perspectives on fertilizer use in the 21st Century. Book of abstracts, Society of Soil Society of Nigeria, 1999. 25th Annual Conference, held at Precious Palm Royal Hotel, Benin City, 21st - 25th November, 1999.

Food and Agriculture Organization (FAO) (2007). FAOSTAT. FAO Statistic Division, Rome.

Gordon WB, Whitney DA, Raney RA (1993). Nitrogen management in furrow irrigated, ridge-tilled corn. J. Prod. Agri. 6:213-217.

Ibeawuchi II, Onweremadu EU, Oti NN (2006). Effects of poultry manure on green (*Amaranthus* cruentus) and7 waterleaf (*Talinum triangulare*) on degraded ultisol of Owerri, Southeastern Nigeria. J. Anim. Vet. Adv. 5(1):53- 56.

Ipinomoroti RR, Daniel MA, Akinlade AM, Makinde EA (2006). Effects of spent cocoa opds on soil properties and plant growth. Nig. J. Soil Sci. 6:78-84.

Isitekhale HHE, Osemwota IO (2010). Residual effects of poultry manure and NPK fertilizer on nutrient contents and uptake by tomato in the forest and derived savanna soils of Edo State. Nig. J. Soil Sci. 20(2):17-25.

Manna MC, Ganguly TK, Ghosy BN (1999). Evaluation of compost maturity and mineral enrichment quality through simple chemical parameters. J. Ind. Soc. Soil Sci. 48:781- 786.

Manral HS, Savena SC (2003). Plant growth, yield attributes and grain yield of soyabean as affected by the application of inorganic and organic sources of nutrients. Bioresour. Technol. 92:110-118.

Mylavarapu RS, Kennelley DE (2002). UF/IFAS extension soil testing laboratory (ESTL): Analytical procedures and training manual. Institute of Food and Agricultural Sciences. University of Florida. Gainesville. U.S.A. p. 28.

Padwick GW (1983). Fifty years of experiment II. The maintenance of soil fertility in tropical Africa: A review. Exp. Agric. 19:283-310.

Saiju UM, Kalisz PJ (1990). Characterization of "coal bloom" horizon in undisturbed forest soils in eastern Kenlucky. Soil Sci. Soc. Am. J. 54:876-882.

Sanwal SK, Lakminarayana K, Yadav RK, Rai N, Yadav DS, Mousumi B (2007). Effect of organic manures on soil fertility, growth, physiology, yield and quality of turmeric. Ind. J. Hort. 64:444-449.

Smith AJ, Montgomery RF (1962). Soil and land use in Central Western Nigeria. Ministry of Agriculture and Natural Resources, Western Nigeria. p. 265.

Somani LL, Totawat KL (1996). Soil conditioners and amendments. Agrotech Publisher Academy, Udaipur, First Edition.

Vanlauwe B (2000). Soil organic matter and crop production in a West African context. In: Agronomy in Nigeria. Agronomy Department. University of Ibadan, Nigeria (Publisher).

Yihenew, G (2002). Selected chemical and physical characteristics of soils. Ade research and its testing sites in North-Western Ethiopia. Ethiop. J. Nat. Resour. 4:199-213.

Modeling of ^{137}Cs and ^{60}Co transport in calcareous soils by groundwater

Ramzi M. Shihab

Soil Science and Water Resource Department, College of Agriculture Tikrit University, Iraq.

The fate of transport of radionuclides is one of the most important factors to be considered for the safety assessment of repositories of radioactive wastes in porous media. Laboratory batch and column experiments were investigated to assess the transport of ^{137}Cs and ^{60}Co in calcareous loam and clay soils leached with groundwater (GW) using convection-dispersion equation model (CDE). Fractionation of Cs and Co in eight sequential fractions in the soils was also measured. Results showed that the distribution coefficient (*Kd*) values for ^{137}Cs found to be much more as compared to ^{60}Co. It was ranged from 20 to 395 ml/g, depending on soil and radionuclide characteristics. The CDE model provided a fairly good fit to the experimental breakthrough curves (BTCs) of both solutes. The retardation factor (*R*) was 821 and 118 for ^{137}Cs, while it was 65 and 88 for ^{60}Co, for both soils, respectively. The dispersion coefficient (*D*) was 2.0 and 2.8 for Cs, and 0.6 and 0.7 cm^2/min for Co for loam and clay soils, respectively. The silt+clay was the major soil fraction in retaining Cs followed by sand fraction, for both soils, while the carbonate was the major soil fraction in retaining Co followed by silt+clay, Fe oxides, and sand for loam soil, and silt+clay, sand and Fe oxides for clay soil. No large change was observed in ^{137}Cs retaining with leaching. The leaching consistently reduced the magnitude of ^{60}Co bound to carbonate and increased fraction bound to silt+clay.

Key words: Transport parameters, ^{137}Cs, ^{60}Co, fractionation, dispersion, calcareous soils.

INTRODUCTION

The release of radioactive materials and fallout to soils has become important task with the advent of nuclear industry. Radionuclides reach groundwater (GW), plants and humans through several paths after their release to the environment. Therefore, the study of radionuclides transport through the soil is an important issue in the safety assessment of nuclear waste repositories and the potential for contamination. These materials can be immobilized in soil components by different mechanisms such as adsorption. It has been recognized that the adsorption process to the soil matrix could limit the transport of radionuclides in the groundwater system (Haque et al., 2011). The mobility of radionuclides is associated with physiochemical characteristics of soil.

The presence of radionuclides like ^{137}Cs and ^{60}Co in the terrestrial environment may be conditioned by nuclear power reactors, nuclear weapon test and nuclear accident. They are part of the major contributors to the radionuclides inventory at low-level liquid waste disposal facilities (Serne et al., 1996). The interaction of radionuclide with the constituents of soil has been recognized as one of the most important processes determining the migration of radionuclide to groundwater. Migration of Co and Cs radionuclides in soil/groundwater system is important for safety disposal radioactive waste containing of these nuclides. Investigations on the behavior of radioactive Cs and Co in soil are addressed in many publications (Xiangke et al., 1999; Hakem et al., 2000; Shihab et al., 2001; Solovitch-Vella and Garnier, 2006; Itakura et al., 2010; Seliman et al., 2010, 2012). Those publications determined some factors affecting the movement of radiocesium or/and radiocobalt in soil.

Mathematical models are integral component of radionuclide transport in soil. An accurate understanding of this transport is required to successfully formulated and use these models. Palagyi and Stamberg (2010) found that the modified convection dispersion equation (CDE) gives rather better results for transport of ^{125}I, ^{137}Cs, and ^{85}Sr in granitoidic rock and soil columns than the classical one. Limited information on the Cs and Co transport parameters such as dispersion coefficient (*D*), retardation factor (*R*), and column Peclet number (*P*) as well as distribution of these radionuclides in soil exists in the literature.

The presence of carbonate in the soil may lead to increase its sorption capacity for radionuclides from radioactive waste and thus decreases their transport. Fahad et al. (1989) had indicated that soil carbonate is an important constituent in the retaining of applied ^{137}Cs. Although, the carbonate fraction of soil plays a key role in the sorption-desorption processes, the effect of carbonate on Cs and Co transport and distribution among soil constituents are still disputable.

The main objective of present study was to investigate the background or feature of calcareous soils in association with groundwater to fit the target radionuclide waste disposal. This was done by (1) evaluating the applicability of convection-dispersion equation model in describing transport of ^{137}Cs and ^{60}Co to understand the impact of leaching with groundwater on radioactive materials disposal under static and dynamic conditions, (2) estimate the transport parameters in undisturbed soil columns, and (3) determine the association of radionuclides with soil constituents using fractionation technique.

MATERIALS AND METHODS

The soils used in this study were calcareous and alluvial loam (soil 1) and clay (soil 2). These soils were taken from two locations surrounding the site of temporary storage of the low level radioactive liquid waste (LW), (formerly, Nuclear Research Center, Iraqi Atomic Energy Commission). Its location is within 33°12′14″N 44°30′49″E. Some characteristics of the soils are given in Table 1, which determined by using standard procedure (Page et al., 1982; Klute, 1986).

Batch experiments

The distribution coefficient (*Kd*) (radionuclide sorbed on soil/radionuclide in solution) of ^{137}Cs and ^{60}Co were determined by the batch technique on samples taken from the surface soil (0-50 cm). The samples were mixed, air dried and ground to pass through a 2 mm sieve. One gram from each sample was equilibrated with the GW (EC= 6.5 dS/m, pH= 7.8 and containing of Ca=275, Mg=314, Na=1067, K=7.0, Cl=1525, SO_4=887, and HCO_3=177 mg/L), for 24 h in a 50 ml centrifuge tubes at 25°C. Borehole was drilled, penetrated the aquifer of the site, and used for obtaining samples of GW. The equilibrating solution was a carrier solution of specified radionuclides in 10:1 liquid to solid ratio. The concentration of nuclide in solution was 10^{-6} mol/L, prepared in GW taken from the aquifer of the same site, and labeled with 0.37 MBq (10 µCi/ml) of CsCl and $CoCl_2$. After equilibrium time, the mixture was then centrifuged for 15 h at 3500 rpm and a aliquot of 10 ml was pipetted from the supernatant solution. Activity of ^{137}Cs and ^{60}Co was measured in the 1 ml solution sample in 2 ml stoppered plastic vials, using a well-type NaI(Tl) detector (LKB,1282 Compugamma, Sweden). The instrument was programmed to count pulses corresponding to energy peak of 0.66 and 1.33 Mev for ^{137}Cs and ^{60}Co, respectively.

Preliminary experiments were used the two soils, two concentrations (10^{-10} and 10^{-6} mol/l), two ratios of liquid to solid (1:1 and 50:1), and four shaking times (2, 15, 24, and 48 h) to select the suitable shaking time. It was found that the sorption equilibrium was attained in 15 h for most cases (data not presented). For this reason a shaking of 15 h was adopted throughout the batch experiments. The amounts of radionuclide retained by sample were determined by subtracting the amounts of radionuclide remaining in the equilibrium solution from the amounts initially added. The Kd was calculated according to the following equation:

$$Kd = \frac{q}{c} = r\left(\frac{Ao}{r - Al}\right)/Al \quad (1)$$

Where: *Kd* = distribution coefficient (L^3 M^{-1}), q = moles of radionuclide/g soil, c = moles of radionuclide/ml liquid, r = liquid to solid ratio, Ao = total radionuclides applied (count rate), Al = radionuclide remaining in solution (count rate).

Column experiments

Undisturbed soil columns were sampled in Perspex tubes 0.5 m long and 0.05 m inner diameter. A steady state saturation water content condition in the soil columns was established before application of radionuclides (the soil columns were saturated with GW from the bottom). Each soil column was spiked with 1.48 MBq (40 µCi) for ^{137}Cs and ^{60}Co (carrier free). The soil columns then were leached with GW until ^{137}Cs and ^{60}Co until very small activity of ^{137}Cs and ^{60}Co was detected in the effluent. The GW was supplied through Marriotte technique maintaining a constant level of 2 cm throughout the leaching period, Table 2 presents physical parameters of soil columns. During the leaching process samples of effluent were daily collected by a fraction collector and measured for ^{137}Cs and ^{60}Co, using a well-type NaI(Tl) detector (LKB, 1282 Compugamma, Sweden) with aid of dual-label program. Data of ^{37}Cs and ^{60}Co from batch and column experiments were corrected for background, dead time, decay and the contribution of high energy ^{60}Co in the low energy ^{137}Cs (spill over). At the end of leaching, the soil columns were sectioned every 2 cm. Soil from each section was air dried, ground and passed through a 2.00 mm

Table 1. Some characteristics of soils.

Character	Soils	
	Soil 1	**Soil 2**
Textural class	loam	clay
Clay (g/kg)	256	393
Silt (g/kg)	463	303
Sand (g/kg)	281	304
ECe[a] (dS/m)	1.0	1.1
pH	7.9	7.8
Equiv.$CaCO_3$ (g/kg)	324	325
OM [b] (g/kg)	14.0	11.4
CEC[c] ($cmol^+$/kg)	13.0	23.4
Soluble ions (mg/L)		
Ca^{2+}	178	108
Mg^{2+}	31.3	31.0
Na^+	13.6	28.3
K^+	17.9	8.6
Cl^-	266	337
SO_4^{2-}	52.9	67.2
Clay minerals (%)		
Dominant 50-90	S(M)[d]	S(M), Ch
Major 20-50	Ch	HM, P
Minor 5-20	P, HM	I
Trace < 5	K, V	V, K

[a]electrical conductivity of soil pest extract, [b]organic matter, [c]cation exchange capacity, [d] S=Smectite; M=Montmorillonite; Ch=Chlorite; P= Palygorskite; HM= Hydrous Mica; I=Illite; K=Kaolinite; V=Vermiculite.

Table 2. Physical parameters of soil columns for ^{137}Cs and ^{60}Co.

Parameter	^{137}Cs		^{60}Co	
	Soil 1	**Soil 2**	**Soil 1**	**Soil 2**
Soil column length, L (m)	0.5	0.5	0.5	0.5
Bulk density, ρ (Mg/m^3)	1.44	1.36	1.44	1.36
Porosity (%)	45.7	48.6	45.6	48.7
Sat.vol.water content, θ (cm^3/cm^3)	0.452	0.481	0.450	0.482
Pore volume, V (cm^3)	443.0	471.4	441.0	472.4
Pore water velocity, v (cm/min)	0.023	0.027	0.022	0.028

sieve. Two grams from each section were taken to measure the activity of ^{137}Cs and ^{60}Co in the same plastic vials in the case of effluent.

Fractionation experiments

The upper 2 cm layer from each soil column was mixed thoroughly for fractionation of the remained unleached ^{137}Cs and ^{60}Co (major part of the two radionuclides had been accumulated in the upper 2 cm soil layer). Ten grams were taken and separated into the following 8 sequential fractions, water soluble, exchangeable, associated with carbonate, organic matter, Mn oxides, Fe oxides, sand, and silt+clay. The first seven fractions except for carbonate were taken following Shuman (1985). The carbonate and silt+clay fractions were taken following the procedure described by Fahad (1988). The carbonate fraction was obtained by washing the sample extracted for exchangeable form with 40 ml deionized water for 3 min and 150 ml 1.0 M NaOAc of pH 5 was added. The soil sample was shaken for 24 h centrifuged at 6000 rpm, and measured for Cs and Co. The silt+clay fraction was separated by sedimentation. The residual sample (sand + silt + clay) was treated with 10 ml of 5% sodium hexametaphosphate and 140 ml of deionized water, shaken for 24 h and sieved through 50 µm sieve to separate sand from silt+clay. The silt+clay suspension was then transferred to 1.0 L cylinder and filled to mark. Based on Stocks' law of particle sedimentation in fluids, sample of 25 ml was pipetted at depth of 10 cm at time zero for silt+clay. The same procedure of fractionation was repeated on soil samples received same activity of ^{137}Cs and ^{60}Co and incubated for the same time of leaching near saturation water content at 25±1°C. All results reported for batch, column and fractionation experiments are average of triplicate determinations expressed on a moisture free basis.

Theory of transport

The basic transport equation is called convection dispersion equation (CDE) (Skagges and Leij, 2002).

$$R\left(\frac{\partial c}{\partial t}\right) = D\left(\frac{\partial^2 c}{\partial x^2}\right) - v\left(\frac{\partial c}{\partial x}\right) \quad (2)$$

Where R is the retardation factor, c is the volume averaged solute concentration, t is time, x is distance, and v is water velocity, approximated by ratio q/Θ, where q is the volumetric flux density and Θ is volumetric water content. Equation 2 can be solved analytically for initial boundary conditions:

$$c(0,t) = c_o \text{ and } \frac{\partial c}{\partial x}(\infty, t) = 0 \quad (3)$$

By assuming that solute distribution inside the column is not affected by an outflow boundary or effluent collection system, and considering the column to be part of an effectively semi-infinite system. Also, by assuming that the concentration is continuous at $x=L$ (length of soil column ($0 \leq x \geq L$), the following equation can be used for solute transport in soil column in one direction for the relative concentration of effluent (c/c_o) in terms of pore volumes, V leached through the column and column Peclet number, P (van Genuchten and Wierenga, 1986):

$$\frac{c}{c_o} = \frac{1}{2} erfc[(\frac{P}{4RV})^{\frac{1}{2}}(R-V)] + 1/2 \exp(P) erfc[(\frac{P}{4RV})^{\frac{1}{2}}(R+V)] \quad (4)$$

$$and \quad P = vL/D \quad (5)$$

where c/c_o is the relative concentration and $erfc$ (x) is the complementary error function (1-erf (x)). In general D depends on pore structure, the solute velocity, and water content. The velocity dependence is often as:

$$D = \lambda v \quad (6)$$

Where λ is the dispersivity (units of length).

Parameters determination

Experimental retardation factor (R_{exp})

R_{exp} was calculated, by using Kd values (determined from batch experiments Equation 1), from the following equation:

$$R_{exp} = 1 + \left(\frac{\rho Kd}{\theta}\right) \quad (7)$$

where ρ is bulk density of soil column. The conditions for batch experiments were similar as possible to those of the column experiments.

Peclet number and dispersion coefficient

Breakthrough curves (BTCs) of ^{137}Cs and ^{60}Co were constructed in which the relative concentration (c/c_o) was plotted as a function of GW pore volumes (V), one pore volume equals to liquid capacity of soil column (Table 2), passed through soil columns. Estimation for P and hence for D can be obtained by matching a series of theoretical curves calculated from Equation (4) with the experimental one. If a reasonably good agreement is achieved (higher correlation coefficient, r and root mean square error (RMSE) the curve is the best fit and parameter is based on that curve (trial and error). Then D was calculated from Equation (5). Data were analyzed by Mathcad software.

RESULTS AND DISCUSSION

Figures 1 and 2 presented observed and predicted BTCs as a function of pore volumes of ^{137}Cs and ^{60}Co of the two soils, respectively. There was good agreement between experimental and theoretical curves (best fit) with significant correlation coefficient (r) greater than 0.9 in all cases studied, and the RMSE was ranged from 0.035 to 0.058. Such fairly well fit indicates the suitability of Equation (4) and (5) to evaluate the respective BTCs of ^{137}Cs and ^{60}Co and hence to estimate and determine the relative transport parameters (Table 3). However, there was some scatter in the theoretical BTCs in terms of shape and position for both solutes reflect the influence of soil characteristics, pore structure, and ion type.

The theoretical non symmetrical sigmoid shape BTCs according to Equation (4) has shown that P number has a substantial influence on both position and value of c/c_o is seen in Figure 3. By increasing the P from 1 to 20 at constant R value (R=100), the curve position is shifted to the left. The position of BTCs depends on both P and R values and the BTCs shape, and the R value depends mainly on the P value (Palagyi and Vodickova, 2009). Also, from Figures 1 and 2 it can be seen that the investigated radionuclides are readily retained, and retention depends on type of soil and their chemical properties. The values of retention at the end of leaching were about 17, 25, 6, and 14% for ^{137}Cs and ^{60}Co of soil 1 and 2, respectively. The highest retention for ^{137}Cs and in less degree for ^{60}Co resulted in high values of Kd and R.

It seems that R values of ^{137}Cs are much higher than those for ^{60}Co (Table 3). Values of R were 821 and 118 observed for Cs and 65 and 88 for Co, indicates that

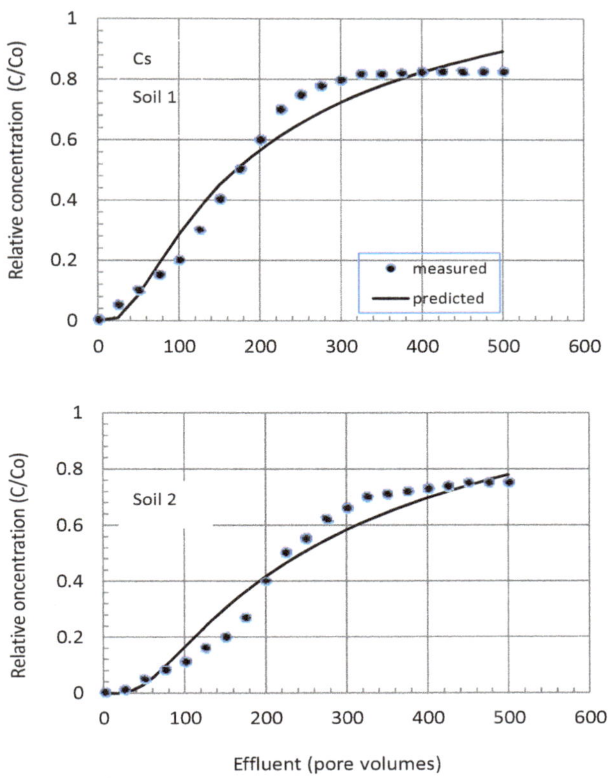

Figure 1. Experimental and theoretical breakthrough curves of ^{137}Cs retained on loam (soil 1) and clay (soil 2).

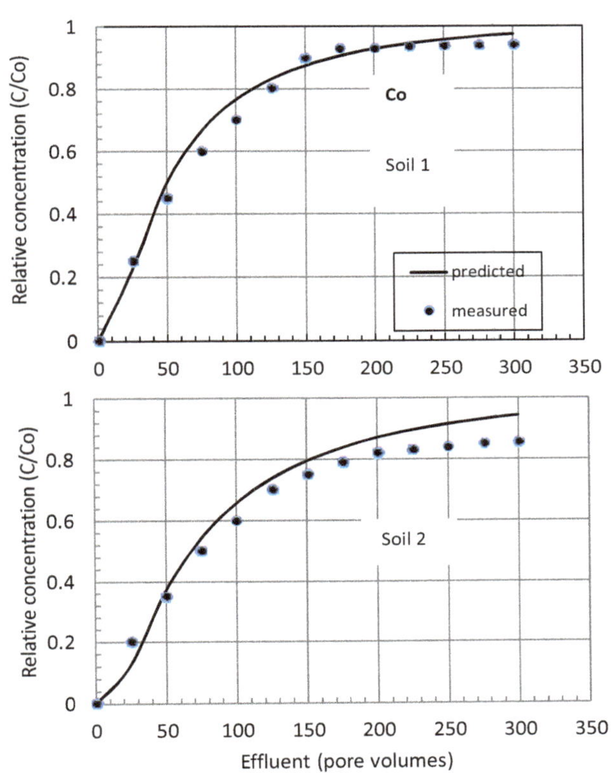

Figure 2. Experimental and theoretical breakthrough curves of ^{60}Co retained on loam (soil 1) and lay (soil 2).

^{137}Cs had been highly retarded comparison with ^{60}Co. Although nearly 500 and 300 pore volumes passed through the soil columns the amounts of Cs and Co retained were very high as indicated by the high retardation factors (*R*). The values of *R* calculated by using data obtained from batch-isotherm experiments were very representative of the observed transport of ^{137}Cs and ^{60}Co in the soil columns experiments. There is close correlation between experimental and theoretical *R* values, which only slightly differ from each other ($R_{theo.}$ was obtained by BTCs fitting as above mentioned). The retention of these radionuclides, in accordance with the values of *Kd* is increasing in the order of $^{137}Cs > ^{60}Co$. The *Kd* values were 257, and 395 for ^{137}Cs and 20, and 31 ml/g for ^{60}Co for soils 1 and 2, respectively (Table 3). It is clear that the *Kd* of ^{137}Cs is about 13 order of magnitude greater than for ^{60}Co in two soils.

The height values of *R* and *Kd* for Cs compared with Co had been obtained for two soils. The silt+clay content play large effect in retention process (Table 4). Clay content and type of clay minerals can play the role of natural barrier in the radionuclides migration (Bucur et al., 2000). Clay and silt have specific sites in their structures that strongly fixed the Cs ions. Cesium has low hydration energy, and if present at the site, it will lose the hydration shell and bind to the negatively charged mineral platelets, resulting in a closing of the site thus fixation the ion (Cornell, 1983). On the contrary, Palagyi et al. (2013) reported that ^{137}Cs sorption is irreversible despite of the fact that ^{137}Cs is prone to ion exchange and practically it forms no complexes.

It is obvious that ^{60}Co showed greater mobility than ^{137}Cs (Figures 1 and 2, Table 3). The ^{60}Co has relatively low sorption affinity to clay fraction. Based on series ofstudies Co was categorized as mobile, exchangeable, or irreversible sorbed in the sand sediment and water system (Mahara and Kudo, 1981). Also, mobile of ^{60}Co was found to consist of non ionic forms which do not contribute to ion exchange absorption onto soil matrix (Hossain et al., 2012). It is well known that cobalt hydrolyzes and form an insoluble hydroxide $Co(OH)_2$ in an alkaline solution. Therefore, under the conditions of the present study the migration rate of ^{60}Co in the soil columns faster than ^{137}Cs.

Radionuclides leached with GW were transported through the soil columns with mechanisms of diffusion. It moved in direction of the higher concentration to lower concentration (Bucur et al., 2006). Dispersion in soil was mainly affected by the geometry of porous structure and by the radionuclide interaction with pore walls (Haque et al., 2011). The relation between *D* and *P* (Equation (5)) is another important factor for recognizing the effect of both diffusion and dispersion on solute mixing. Although, the effects of both diffusion and dispersion on solute mixing are difficult to be separated, Fried and Combarnous (1971) suggested that when *P* is more than 20, the diffusion is at minimum value. Data in Table 3 show that *P* ranged from 0.5 to 2.0. This indicates that diffusion

Table 3. Transport parameters of ^{137}Cs and ^{60}Co in undisturbed soil columns.

Parameter	^{137}Cs		^{60}Co	
	Soil 1	**Soil 2**	**Soil 1**	**Soil 2**
Kd (ml/g)	257.5	394.9	20.2	31.1
R_{exp}	821.4	1117.6	64.4	87.7
R_{theo}	770	1200	65	88
P	0.5	0.5	2.0	2.0
D (cm^2/min)	2.1	2.8	0.6	0.7
λ (cm)	0.01	0.009	0.037	0.039
Correlation coeff., r	0.983	0.985	0.995	0.995
RMSE	0.056	0.0.53	0.035	0.058

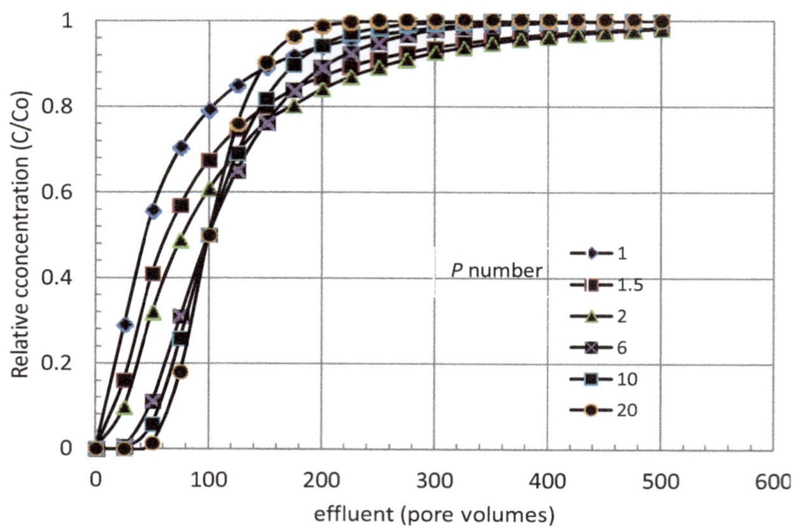

Figure 3. Effect of different values of *P* number on the position and shape of breakthrough curves at value of *R*=100.

may be the dominant process in transport of ^{137}Cs and ^{60}Co. On the other hand, D reaches values 2.1 and 2.8 cm^2/min for ^{137}Cs and 0.6 and 0.7 cm^2/min for ^{60}Co for the loam and clay soils, respectively. Also, λ values were 0.01 and 0.01 cm for ^{137}Cs and 0.04 and 0.04 cm for ^{60}Co for the loam and clay soils, respectively (Table 3). Regardless soil type, the highest value of λ being for ^{60}Co and the lowest for ^{137}Cs. It is evident that the reduction in λ values attributed to high reduction in flow velocities (Table 2).

In the present study the clay content of soils 1 and 2 were 256 and 392 g/kg, respectively (Table 1). Increasing CEC will therefore increase the exchange sites. Comparison of both soils showed that soil 2 had higher CEC, whereas soil 1 had lower CEC. Since clay fraction has a high affinity for sorption of Cs and Co, soil 2 sorbed these two radionuclides higher than soil 1 (Figures 1 and 2). These results are in agreement with the reports of soil samples that contain clay minerals and resulted in large CEC for absorbing a large amount of dissolved ^{137}Cs (Kamel, 2010). When organic matter is higher, and clay minerals are not sufficient, the migration of radionuclide is higher (Dumat and Staunton, 1999; Rosen et al., 2006). This was not so in the present study, because the both soils have low organic matter (Table 1), therefore, the clay fraction plays large role in Cs and Co retention. Effect of length of soil column on travel time (tr) of ^{137}Cs and ^{60}Co are shown in Figure 4. The travel time under dynamic conditions was calculated according to the following equation:

$$tr = \left[\frac{L}{(v/R)}\right] \tag{8}$$

Table 4. Fractionation of Cs and Co remained in the upper layer 2 cm of soil column at the end of leaching.

Soil fraction	% of total added			
	^{137}Cs		^{60}Co	
	Control	Leached soil columns	Control	Leached soil columns
		Soil 1		
Water soluble	0.3	0.3	0.3	2.3
Exchangeable	0.9	0.5	1.7	1.4
Carbonate	1.6	1.5	32.3	12.9
Organic	1.3	2.5	0.5	0.3
Mn oxides	1.0	0.9	4.5	3.0
Fe oxides	4.6	4.3	28.5	34.8
Sand	15.8	14.1	6.1	4.4
Silt+clay	74.5	75.5	26.5	41.0
		Soil 2		
Water soluble	0.2	0.3	0.4	3.7
Exchangeable	1.1	0.4	2.5	2.1
Carbonate	1.9	1.4	47.3	23.7
Organic	1.3	4.4	0.8	0.5
Mn oxides	0.2	0.1	0.7	0.5
Fe oxides	0.6	0.5	5.3	7.6
Sand	15.4	14.2	7.9	7.3
Silt+clay	79.2	78.2	35.2	54.6

The tr represents the theoretical time when the highest activity peak of ^{137}Cs and ^{60}Co appears from soil columns. Because of low GW velocities in the soil columns (Table 2), tr of radionuclides came up very long. Appearing of highest activity peak corresponding to the travel time of ^{137}Cs and ^{60}Co through the soil column, it can be said that the length of soil column affects on tr. However, differences observed in tr of both Cs and Co for a given soil are due to porous media sorption capacity, ionic strength of leaching solution, and nature of undisturbed soil columns. The values of travel time calculated by Equation (8) were 411, 1433 days for ^{137}Cs, and 100, and 112 days for ^{60}Co for loam and clay soils, respectively (Figure 4).

The above results indicate the travel time of Cs was very long compared to Co, and the presence of greater proportion of soil colloids (clay, Fe oxides, and humified organic matter) may increase tr from hundreds to thousands of years for arbitrary distance of 1000 m because ^{137}Cs are more effectively sorbed to the soil colloids. Hossain et al. (2012) found that thickness of soil layer has an effect on tr of ^{137}Cs and ^{60}Co through the soil layer. At a given column length (thickness of soil layer) ^{60}Co travel time was less comparison with ^{137}Cs. The results obtained are consistent with other results given in literature (Meier et al., 2003). This implies that in the presence of soil colloids a large reduction of nuclide travel times must be taken into account. That is the protective effect of porous media like the soil against the migration of radioactive contaminants to the biosphere can be low in the presence of colloids of soil.

Influence of Leaching with GW on fractionation of ^{137}Cs and ^{60}Co in soils

Fractionation of Cs and Co in the control samples as well as in samples of the upper 2 cm of both soils columns leached with GW is given in Table 4. For the control soils, it is clearly evident that major portion, 74 to 79% of applied Cs associated with sil+clay. The sand fraction was the second important soil constituent responsible for Cs retention. About 14% of total Cs was retained by the sand fraction in both soils. The retention percent of the Fe oxides was found to be 4.6% for soil 1 and 0.6 for soil 2. The retention percent of organic fraction was 1.3% for both soils. The rest of Cs added was bound to other fractions of soils.

The leaching increased the amount of Cs bound to organic fractions for both soils 1 and 2. Conversely, the exchangeable, carbonate, Mn oxides, and Fe oxides were found to decrease upon leaching. Comparing the fractionation data of Cs from control and soil columns leached with GW for soil 1 (256 g/kg clay) and soil 2 (392 g/kg clay) one can conclude that both soils appeared similar distribution of retaining ^{137}Cs.

The fact that the retention percent for two soils are

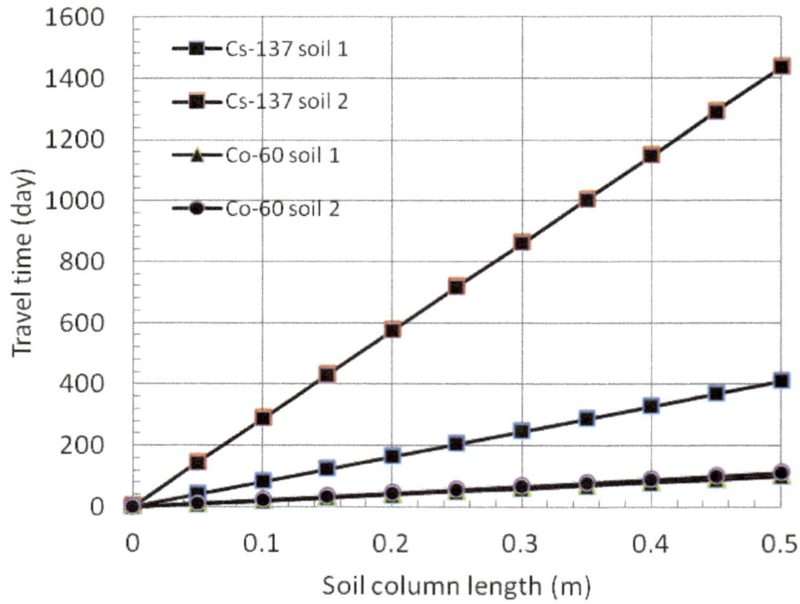

Figure 4. Travel time of ^{137}Cs and ^{60}Co as a function of distance from point of release (soil column length).

similar to each other indicates that the retention of Cs on calcareous soil is mainly determined by the composition of clay minerals. The dominant clay minerals in the two soils are montmorillonite (Table 1), since this type of clay minerals has a high affinity for sorption of Cs. Its affinity for Cs is due to very small hydration energy of the cation and higher electrostatic attraction between Cs and clay particles (Mirkhani et al., 2012). On other hand, the transport of Cs explained the limited replacing ability of the GW because the majority of applied Cs was tightly bound to silt+clay fraction. The amount of Cs released is 75 to 83% over 500 pore volumes using GW as leaching solution (Figures 1 and 2). Also, the *Kd* of Cs was high in the two soils (Table 2). This may also indicate that fixation (irreversible sorption) was the main mechanism responsible for Cs retention in the soils of montmorillonitic clay. These results are essentially consisting with the results discussed above for transport in soil columns. Bellenger and Staunton (2008) recently observed a strong irreversibility for Cs during desorption experiments from different minerals. Giannakopoulou et al. (2007) showed that clay content plays predominant role on sorption of Cs. Also, the results (Table 4) showed that for the control of both soils, carbonate retained 32 to 47% of ^{60}Co added, although substantial amounts were retained by the clay+ silt minerals, Fe oxides, and sand. Different patterns of Co distribution among soil fractions were obtained for the leached soil columns. The leaching consistently increased the amounts bound to silt+clay minerals, Fe oxides, and water soluble in both soils, whereas it either decreased the amount bound to carbonate (13 to 24%), sand, and Mn oxides fractions (in the case of soil 1) or slightly changed in the rest fractions. These results are compared with that obtained by others (Razaq et al., 1993; Kamel, 2002) for the release of ^{60}Co from the soil or from the marine sediments. The ^{60}Co radionuclide was bounded to the exchangeable and carbonate fraction. Also, similar results were reported before for Co desorption from soil constitutions (Backes et al., 1995; McLaren and Backes, 1998; Seliman et al., 2012). The amount of Co released out of both soils columns, respectively is 94 and 86% over 300 pore volumes using GW as a leaching solution (Figures 1 and 2). This indicates a smaller degree of reversibility compared to the amounts of Cs released under the same conditions. Also, this may indicates the highest ability of Co to release out of soil columns when it leached with GW compared with Cs.

The results of the present study (Table 4) showed that the substantial amounts of radionuclides especially for ^{137}Cs were retained by sand fraction. It is unexpected to observe this result since the sand is inert component and has small specific surface area. This is may be attributed to fact that sand in arid and semi-arid soils has considerable amounts of light minerals like calcite, feldspar, mica, and quartz which were responsible for Cs and Co retention. Also, the presence of Fe oxides could enhance the adsorption capacity of some minerals (Sipose et al., 2008). From the above results, one can conclude that the behavior of Co is the combined effect of diffusion and development of bound between the nuclide and adsorption sites.

The low replacing ability of GW for release the two radionuclides may due to the fact that ^{137}Cs and ^{60}Co

were found to be influenced by the (i) composition of the soil, (ii) radionuclide characteristics, and (iii) presence of dissolved cations such as Ca, Mg and K in the soil – groundwater system to form coordination bonds. Some researchers attributed the slow desorption process of trace metals from soil constituents surfaces to different mechanisms such as recrystallization of metals into the mineral structure, diffusion into micropores, surface precipitation, change in the surface complex, and solid state diffusion into the crystal matrix (Glover, 2000).

The positive or negative effects on the obtained retention of ^{137}Cs and ^{60}Co data can be supposed mainly of the silt+clay, and sand content of the soil, as well as the carbonate and Fe oxides which especially in the case of ^{60}Co influence this process. Moreover, the very low concentration of tracer radionuclides and carriers, as well as the competition of major ions in the groundwater, including the isotopic exchange between the stable isotope in soil particles and radioactive isotope in the solution has to be considered in the retention of these radionuclides in soil samples (Soderlund et al., 2011).

Conclusions

The convection-dispersion equation (CDE) model allows predicting the transport of ^{137}Cs and ^{60}Co through soil. Also, CDE model provides estimates of the transport parameters of radionuclides. The transport of both radionuclides were slow in soil columns, in spite of the fact that ^{60}Co showed greater mobility than ^{137}Cs. These findings suggest that Cs and Co are strongly retained on soil fractions. The retention may be influenced by several factors, e.g., quantity and quality of clay minerals and other components of the soil. This study also showed that content of silt+clay was associated with the maximum values of *Kd*, *R*, and the minimum tr, *D*, and *P*.

Most of ^{137}Cs applied was distributed among major soil fractions followed the order silt+clay > sand > organic ≥ Fe oxides, while, ^{60}Co was distributed followed the order carbonate > silt+clay > Fe oxides > sand. Leaching however changed either the order or magnitude of fractions. This study exhibits the significance of radionuclides transport features in soil matrix through groundwater flow from the viewpoints of safety aspects of surface level disposal facility. Current risks appear to be low because such calcareous soils will tend to retain the large amount of radionuclides which might release from the burial site and reduce the potential migration of radionuclides to the environment. More thorough investigations regarding long-term is required to accurately estimate the amount of risk associated with contamination from the site.

Conflict of Interest

The authors have not declared any conflict of interest.

ACKNOWLEDGEMENTS

Author is grateful to Drs, A. A. Fahad AlTaii and H. A. Tawfeek for their assistance and useful comments.

REFERENCES

Backes CA, Mclaren RG, Rate AW (1995). Kinetics of cadmium and cobalt desorption from iron and manganese oxides. Soil Sci. Soc. Am. J. 59:778-785.http://dx.doi.org/10.2136/sssaj1995.03615995005900030021x

Bellenger JP, Staunton S (2008). Adsorption and desorption of 85Sr and 137Cs on reference minerals, with and without inorganic and organic surface coating. J. Environ. Radioact. 99:831-840.http://dx.doi.org/10.1016/j.jenvrad.2007.10.010 PMid:18295381

Bucur C, Popa A, Arsene C, Olteanu M (2000). Diffusion coefficients of critical radionuclides from radioactive waste in geological medium. Proc. WM'00 Conf., Feb. 27-Mar.2, Tucson, AZ.

Bucur C, Olteanu M, Pavelescu M (2006). Radionuclide diffusion in geological media. Rom. J. Phys. 51:469-478.

Cornell RM (1983). Adsorption of cesium on minerals: A review. J. Radioanal. Nucl. Chem. 171:483-500.http://dx.doi.org/10.1007/BF02219872

Dumat C, Staunton S (1999). Reduced adsorption of cesium on clay minerals caused by various humic substances. J. Environ. Radioact. 46:187-200. http://dx.doi.org/10.1016/S0265-931X(98)00125-8

Fahad AA (1988). Fate of zinc applied to calcareous soils using 65Zn as a tracer. 1. Fractionation with time. Arid Soil Res. Rehabil. 2:217-225. http://dx.doi.org/10.1080/15324988809381176

Fahad AA, Ali AW, Shihab RM (1989). Mobilization and fractionation of 137Cs in calcareous soils. J. Radioanal. Nucl. Chem. 130:195-201. http://dx.doi.org/10.1007/BF02037713

Fried JJ, Combarnous MA (1971). Dispersion in porous media. Adv. Hydrosci. 7:665-671.

Gianakopoulou F, Haidouti C, Chronopoulo A, Gasparatos D (2007). Sorption behavior of cesium on various soils under different pH levels. J. Haz. Mater. 149:553-556. http://dx.doi.org/10.1016/j.jhazmat.2007.06.109 PMid:17720309

Glover LJ (2000). The influence of residence time and organic acids on the desorption of trace metals from goethite. M.SC. Thesis, Fac. Virginia Polytech. Instit. State Univ. Virginia.

Haque MM, Ghose S, Islam SMA (2011). A laboratory based study on the movement of radiocesium in some columns by gamma spectrometer. J. Bangladesh Acad. Sci. 35:141-151.

Hakem NL, Al Mahamadi I, Apps JA, Moridis GJ (2000). Sorption of cesium and strontium on Hanford soil. J. Radioanal. Nucl. Chem. 246:275-278. http://dx.doi.org/10.1023/A:1006701902891

Hossain MA, Shamsuzzaman M, Ghose S, Hossain AKM (2012). Characterization of local soils and study the migration behavior of radionuclide form disposal site of LILW. J. Environ. Radioact.105:70-75. http://dx.doi.org/10.1016/j.jenvrad.2011.10.016 PMid:22230023

Itakura T, Airey DW, Leo CJ, Payne T, McOrist GD (2010). Laboratory studies of the diffusive transport of 137Cs and 60Co through potential waste repository soils. J. Environ. Radioact. 101:723-729 http://dx.doi.org/10.1016/j.jenvrad.2010.04.015 PMid:20554096

Kamel NHM (2002). The behavior of 137Cs, 60Co and 85 Sr radionuclides in marine environmental sediment. Sci. World J. 2:1514-1526. http://dx.doi.org/10.1100/tsw.2002.290 PMid:12806132

Kamel NHM (2010). Adsorption models of 137Cs radionuclide and Sr(II) on some Egyptian soils. J. Environ. Radioact. 101:297-303. http://dx.doi.org/10.1016/j.jenvrad.2010.01.001 PMid:20167404

Klute A (ed.) (1986). Methods of Soil Analysis, Part 1, Physical and Mineralogical Methods, Agron. No. 9, 2nd edition, ASA, SSSA, Madison WI, USA. PMid:12267821

Mahara Y, Kudo A (1981). Fixation and mobilization of 60Co on sediments in coastal environments. Health Phys. 41:645-655. http://dx.doi.org/10.1097/00004032-198110000-00006 PMid:7309521

McLaren RG, Backes SA (1998). Cadmium and cobalt desorption kinetics from soil clays: effect of sorption period. Soil Sci. Soc. Am. J. 62:332-337.

http://dx.doi.org/10.2136/sssaj1998.03615995006200020006x

Meier M, Zimmerhackle E, Zeitler G (2003). Modeling of colloid-associated radionuclide transport in porous groundwater aquifer of the Gorlleben site, Germany. Geochem. J. 37:325-350. http://dx.doi.org/10.2343/geochemj.37.325

Mirkhani R, Roozitalab MH, Khaleghpanah N, Majdabadi A (2012). Sorption behavior of cesium and strontium in selected soils of semiarid and arid regions of Iran. J. Radioanal Nucl. Chem. 293:587-594. http://dx.doi.org/10.1007/s10967-012-1779-x

Page AL, Miller RH, Keeny Dr (eds.) (1982). Methods of Soil Analysis, Part 2, Chemical and Microbiological Properties, Agron. No. 9, 2nd edition, ASA, SSSA, Madison WI, USA.

Palagyi S, Vodickova H (2009). Sorption and desorption of 125I-, 137Cs+, 85Sr2+ and 152,154Eu3+ on disturbed soils under dynamic flow and static batch conditions. J. Radioanal. Nucl. Chem. 280:3-14. http://dx.doi.org/10.1007/s10967-008-7436-8

Palagyi S, Stamberg K (2010). Transport of 125I-, 137Cs+,and 85Sr2+ in granitoidic rock and soil columns. J. Radioanal. Nucl. Chem. 286:309-316. http://dx.doi.org/10.1007/s10967-010-0719-x

Palagyi S, Stamberg K, Vodickova H, Herick M (2013). Sorption of 125I, 137Cs+, 85Sr2+ and 152,154Eu3+ during their transport in undisturbed vertical and horizontal soil cores under dynamic flow conditions. J. Radioanal. Nucl. Chem. 395:1447-1458. http://dx.doi.org/10.1007/s10967-012-1924-6

Razaq IB, Fahad AA, Al-Hadethi AA, Ali AW, Tawfeek HA (1993). Uptake of gamma-emmitting activation products,65Co and 65Zn by successive cropping. Basrah J. Agric. Sci. 6:47-61.

Rosen K, Shand CA, Haak E, Cheshire MV (2006). Effects of clay content and wetting-and-drying on radiocesium behavior in a peat and a peaty podzol. Sci. Total Environ. 182:117-133.

Seliman AF, Borai EH, Lasheen YF, Abo-Aly MM, DeVol TA, Bowell BA (2010). Mobility of radionuclides in soil / groundwater system: comparing the influence of EDTA and four of its degradation products. Environ. Pollut. 158:3077-3084. http://dx.doi.org/10.1016/j.envpol.2010.06.035 PMid:20656386

Seliman AF, Borai EH, Lasheen YF, DeVol TA (2012). Remobilization of 60Co, 89Sr, 137Cs, 152Eu, and 241Am from a contaminated soil column by groundwater and organic ligands. Transp. Porous Med. 93:799-813. http://dx.doi.org/10.1007/s11242-012-9983-2

Serne RJ, Felmy AR, Cantrell KJ, Krupka KM, Campbell JA, Bolton H, Fredrickson JK (1996). Characterization of radionuclides-chelating agent complexes found in low-level radioactive decontamination waste. NUREG/CR-6124, U.S. Nuclear Regularity Commission, Washington. http://dx.doi.org/10.2172/219284 PMid:8772355

Shihab RM, Fahad AA, Tawfeek HA, Al-Hassani AA (2001). Movement of radioisotopes Cs and Co in calcareous soil as affected by organic matter and moisture content. Sci. J. Iraqi Atom. Energy Comm. 3:1-9.

Shuman LM (1985). Fractionation method for soil microelements. Soil Sci. 140:11-22. http://dx.doi.org/10.1097/00010694-198507000-00003

Sipos P, Nemeth T, Kovacs Kis V, Mohai I (2008). Sorption of copper, zinc and lead on soil mineral phases. Chemosphere 73:461-469. http://dx.doi.org/10.1016/j.chemosphere.2008.06.046 PMid:18674797

Skagges TH, Leij FJ (2002). Solute transport: Theoretical background. In: Dane J and Topp C (eds.), Methods of Soil Analysis, Part 4, ASA, SSSA, Madison WI, USA.

Soderlund M, Lusa M, Lehto J, Hakanen M, Vaaramaa K, Lahdenpera AM (2011). Sorption of iodine, chlorine, technetium, and cesium in soil. Working Report 2011-04 Posiva Oy. Olkiluoto P. 130.

Solovitch-Vella N, Garnier JM (2006). Comparative kinetics desorption of 60Co, 85Sr, and 137Cs from a contaminated natural silica and columns: influence of varying physicochemical conditions and dissolved organic matter. Environ. Pollut. 141:98-106. http://dx.doi.org/10.1016/j.envpol.2005.08.012 PMid:16198464

Van Genuchten MTh, Wierenga PJ (1986). Solute dispersion coefficients and retardation factor. In: Klute A (ed.), Methods of Soil Analysis, Part 1, Physical and Mineralogical Methods, Agron. No. 9, 2nd edition, ASA, SSSA, Madison WI, USA, pp. 1025-1054.

Xiangke W, Wenmeng D, Jinzhaou D, Zuyi T (1999). Sorption and desorption of radiocesium on calcareous soil: Results from batch and column investigations. J. Radioanal. Nucl. Chem. 240:783-787. http://dx.doi.org/10.1007/BF02349852

Effect of kitchen wastewater irrigation on soil properties and growth of cucumber (*Cucumis sativus*)

Abegunrin, T. P.[1] **Awe, G. O.**[2], **Idowu, D. O.**[1], **Onigbogi, O. O.**[1] **and Onofua, O. E.**[1]

[1]Department of Agricultural Engineering, Ladoke Akintola University of Technology, Ogbomoso, Oyo State, Nigeria.
[2]Department of Crop, Soil and Environmental Sciences, Faculty of Agricultural Sciences, Ekiti State University, Ado Ekiti, Ekiti State Nigeria.

Scarcity of freshwater is becoming an increasing problem primarily in the arid and semi-arid regions of the world. The study was designed to investigate the effects of wastewater irrigation on soil properties and growth parameters of cucumber (*Cucumis sativus*) in southwestern Nigeria. The irrigation treatments consisted of kitchen wastewater, rainwater and groundwater in a randomized complete block design (RCBD) in three replications. Soil samples were collected in the 0 to 20 cm range, analyzed before irrigation application and after harvest according to USDA standard procedures while crop growth parameters were recorded 3, 5 and 7 weeks after planting. Soil pH decreased in all the soil samples after the different water applications, the lowest pH from kitchen wastewater. Electrical conductivity (EC) increased slightly in all soil samples, the highest EC value of 8.59 × 10^{-4} dS/m from groundwater irrigation. The sodium adsorption ratio (SAR) increased in all the irrigation treatments but was below the critical value of 13. The stem girth and leaf area were not significantly affected by water type (WT) during the three periods of evaluation but leaf numbers and plant length differ significantly at seven week of growth. Soil condition was not significantly impacted by the use of kitchen wastewater, however, appropriate wastewater treatment and water management practices have to be followed to remove the toxic elements which could be hazards to vegetable production and the environment.

Key words: Wastewater, groundwater, electrical conductivity, sodium adsorption ratio.

INTRODUCTION

Scarcity of freshwater is becoming an increasing problem primarily in the arid and semi-arid regions of the world (El Youssfi et al., 2012). This trend is however extending to other regions of the world. The rapid growth of the world's urban population continues to place increased demands on limited fresh water supplies. There is a serious struggle to balance water use among domestic, industrial and agricultural users (Darvishi et al., 2010). This rapid increase in population has not only caused an increase in the demand for the limited available freshwater but has also caused an increase in the volume of wastewater generated year by year (Thapliyal et al., 2011). The untreated water generated can find its way into water systems such as rivers, lakes, groundwater and coastal waters with the potential to cause serious pollution (GOI, 2002). Wastewater may contain undesirable chemical constituents and pathogens that pose negative environmental and health impacts (Papadopoulos, 1995).

On the other hand, large amounts of water are needed for irrigation in agriculture. If the wastewater can be used as an alternative water source for irrigation, the dual problems of negative environmental effects and huge water demand for agricultural irrigation would be solved (Darvishi et al., 2010). The reuse of wastewaters for purposes such as agricultural irrigation can reduce the amount of water that needs to be extracted from environmental water sources (Heidarpour et al., 2007). The reuse of wastewater for irrigation purposes gives it a different fate as agricultural crops can make use of the extra water and nutrients. Groundwater recharge is another positive outcome of wastewater irrigation (Walker and Lin, 2008). Wastewater has a potential to supply carbon nutrients (NPK) and micro nutrients to support crop/plant growth (Singh et al., 2011). It serves as a valuable source of plant nutrients and organic matter needed for maintaining fertility and productivity levels of the soil (Rusan et al., 2007). Wastewater can have a positive effect on soil and eventually plant growth, due to its being rich in organic matter and nutrients (Ghanbari et al., 2007; Mohammad and Ayadi, 2004). When wastewater will be used continuously as the sole source of irrigation water for field crops in arid regions, excessive amounts of nutrients and toxic chemical substances could simultaneously be applied to the soil-plant system. This would cause unfavorable effects on productivity and quality parameters of the crops and the soil (Vazquezmontiel et al., 1996).

In agricultural practices, irrigation water quality is believed to affect the soil characteristics, crops production and management of water (Shainberg and Oster, 1978). For instance, the application of saline/sodic water results in the reduction of crop yield and deterioration of the physical/chemical properties of soil (Singh, 2011). Therefore, it is of

concern to the people/farmers if an irrigant is used, which may contain constituents capable of creating adverse effects on the soil and on agricultural produce. In this context, many studies have been conducted to evaluate the effects of treated and untreated wastewater irrigation on soil physical and chemical properties and crop performance and nutrient status across climates, soils and diverse crops, however results vary between different settings. Tabari et al. (2008) reported that wastewater irrigation could enrich soils with heavy metals to levels that may pose potential risk to the environment and human health while Tabatabaei (2007) observed that continuous wastewater application to the soil could alter soil infiltration characteristics. Mojiri et al. (2013) working on the impact of urban wastewater on soil properties and *Lepidium sativum* in an arid region found increased electrical conductivity (EC), organic matter (OM), total N, Na and heavy metals due to wastewater irrigation. However, they reported an increase in root and shoot length. In contrast, Singh et al. (2012) found that domestic wastewater irrigation had no significant effects on properties of a clay soil apart from slight changes in salt solubility and alkalinity. Despite this volume of studies, there is a dearth of information on the impact of household wastewater irrigation on soil properties and quality of irrigated vegetables applicable to southwestern Nigeria where there is an increased popularity in the application of domestic wastewater for irrigation. Thus, we formulated the hypothesis that irrigation using kitchen wastewater will reduce soil quality and negatively impact cucumber growth and yield. Therefore, the objective of this study was to investigate the effects of wastewater irrigation on soil properties and growth parameters of cucumber (*Cucumis sativus*) in southwestern Nigeria.

MATERIALS AND METHODS

Study area

The study was conducted in a screenhouse in the Department of Agricultural Engineering, Ladoke Akintola University of Technology (LAUTECH), Ogbomoso (8°10^{1}N; 4°10^{1}E); in Southwestern Nigeria. The mean annual rainfall is about 1200 mm while the mean annual maximum and minimum temperatures are 33 and 28°C, respectively. The relative humidity of the area is relatively high (74%) throughout the year except in January when the dry wind blows from the North (Olaniyi et al., 2010) while the soil of the area is classified as Hapludalf (SSS, 2006).

Soil sampling and analyses

Deformed soil samples of the soil to be used were collected from soil depth, 0-20 cm. Samples were bulked and passed through a 2-mm sieve for the analysis of particle size (used to classify the soil), P, pH, exchangeable bases (K^+, Ca^{2+}, Mg^{2+}, Na^{2+}) and electrical conductivity (EC). The Na^+, Ca^{2+} and Mg^{2+} values were used to determine the soil sodium adsorption ratio (SAR) using the equation:

$$SAR = \frac{Na^+}{\sqrt{\frac{Ca^{2+}+Mg^{2+}}{2}}}$$

Particle size analysis was done by the Bouyoucos or hydrometer method (Gee and Bauder, 1986), and the corresponding textural class was determined from the USDA textural class triangle. The soil pH was determined with pH-Meter E520 using 1:2.5 soil to water ratio. The EC was determined using a Wilson Grimes conductivity meter. Exchangeable bases were determined by extraction with ammonium acetate, after which potassium (K^+) and sodium (Na^+) were determined by flame photometry while calcium (Ca^{2+}) and magnesium (Mg^{2+}) were determined using atomic absorption spectrophotometry (AAS) as detailed by Anderson and Ingram (1993). Sampling and analysis are carried out before planting and after harvesting for each of the different water types used for irrigating the crops during the course of the experiment.

Water sample collection and analysis

Water samples were collected from the three types of water: rainwater, kitchen wastewater and groundwater, using sterilized sampling bottles. The samples were put on ice and transported to the laboratory within 24 h of collection. In the laboratory, they were kept in a refrigerator until they were analyzed. The sample were

Table 1. Properties of the different water types before planting.

Parameter	Groundwater	Rainwater	Wastewater
HCO_3^- (mg/L)	109.8	85.4	244.0
Na^+ (mg/L)	15.0	5.0	10.0
Ca^{2+} (mg/L)	115.2	25.6	35.2
Mg^{2+} (mg/L)	42.0	5.0	15.0
TSS (mg/L)	188.0	140.0	668.0
SAR	1.69	1.28	2.0
pH	6.0	6.4	6.0

TSS, Total suspended solids; SAR, Sodium adsorption ratio; HCO_3^-, hydrogen carbonate, Ca^{2+}, Calcium, Mg^{2+}, magnesium and Na^+, sodium concentration;

analyzed for parameters such as concentrations of bicarbonate (HCO_3^-), Na^+, Mg^{2+}, Ca^{2+}, TSS as well as pH and EC according to APHA (1998) standard methods.

Experimental design, field procedures and management

The experimental design was a one-factor 3 × 8 experiment arranged in a randomized block design (RBD) with three replications consisting twenty four (24) micro-lysimeters. The treatments consisted of water types (WT): Kitchen wastewater, rainwater and groundwater. Three seeds of Cucumber were planted in each micro-lysimeter and were irrigated sufficiently in the first few days to ensure germination and establishment. After crop establishment, equal amount of groundwater, rainwater and wastewater was applied based on cucumber irrigation water requirements, with water applied using surface irrigation method. Apart from the irrigation treatments, soil management such as fertilizer application, weed removal and crop protection were the same for all treatments.

Data collection

The crop was monitored and the growth parameters (plant length, leaf area, stem girth and number of leaves) were measured and recorded under the different irrigation water type. Growth parameters were measured 3, 5 and 7 weeks after planting. Plant length (PL) was determined by measuring the length of the plant from the ground level to the tip of top-most leaf using a flexible tape. The number of leaves (NOL) per plant was made by a visual count of the green leaves while the stem girth (SG) was measured with the aid of a vernier caliper. The green leaf area was estimated as a product of the leaf length (L) and the widest middle portion of the leaf, the width (W) corrected to 0.85, as described by Blanco and Folegatti (2003) according to the equation:

$$LA = 0.85\,(L \times W)$$

Where: L = leaf length, cm; W = width of widest portion of leaf, cm, and LA =leaf area, cm^2.

Statistical analysis

Data was subjected to analysis of variance (ANOVA) procedure and comparison of the soil chemical characteristics before and after the experiment was made using the Tukey Test (DMRT) at 5% level of significance using the ASSISTAT statistical package (ASSISTAT version 7.6 beta).

RESULTS AND DISCUSSION

Water quality

The results of the water (groundwater, rainwater and wastewater) analyses are shown in Table 1. Of the three water samples, rainwater had the lowest values for all the parameters considered except pH with a value of 6.4. Generally the pH values of the three waters fall within the acidic range. Groundwater had the highest values of Na^+, Ca^{2+} and Mg^{2+}. This may be connected with the rock formation of the area where the well was located. Kitchen wastewater however, had the highest value of 668 mg/L and 2.0 for TSS and SAR, respectively. It also had the highest HCO_3^- concentration. The soap chemicals used and suspended particles of food remnants from kitchen waste may have contributed to the highest values of TSS, SAR and HCO_3^-

Soil properties

The soil texture consisted of 80% sand, 16% silt and 4% clay, classified as sandy loam, according to the USDA classification. The soil properties prior to planting are shown in Table 2. The initial value for electrical conductivity (EC) of the soil was 0.0 dS/m while the pH was 7.2, slightly within the basic range. Figure 1a to g shows the values of the different soil parameters before planting and after the application of the three water treatments. The pH of the soil decreased with the three water treatments (Figure 1a). Soil irrigated with kitchen wastewater had the lowest pH value of 6.0 while soil irrigated with rainwater had the highest pH of 6.9. A pH of 6.7 was obtained for soil irrigated with groundwater (Figure 1a). This is likely due to the decomposition of organic matter and production of organic acids in soils which is in line with the findings of Vaseghi et al. (2005) and Khai et al. (2008). Another explanation for this reduction in pH may be as a result of nitrification of N-NH_4^+ from the wastewater as observed by Stamatiadis et al. (1999). Similar result was reported by Hussein (2009). However, other works (Rattan et al., 2005; Rusan et al.,

Table 2. Soil physical and chemical properties before planting.

Soil properties	pH	Na^+	Ca^{2+}	Mg^{2+}	K^+	SAR	EC	Sand	Silt	Clay	Soil type
		ppm					dS/cm	%			
Value	7.2	110.4	500	380	663	5.26	0	80	16	4	SL

SAR, Sodium adsorption ratio; HCO_3^-, hydrogen carbonate; Ca^{2+}, Calcium; Mg^{2+}, magnesium; Na^+, sodium concentration; EC, electrical conductivity; SL, sandy loam.

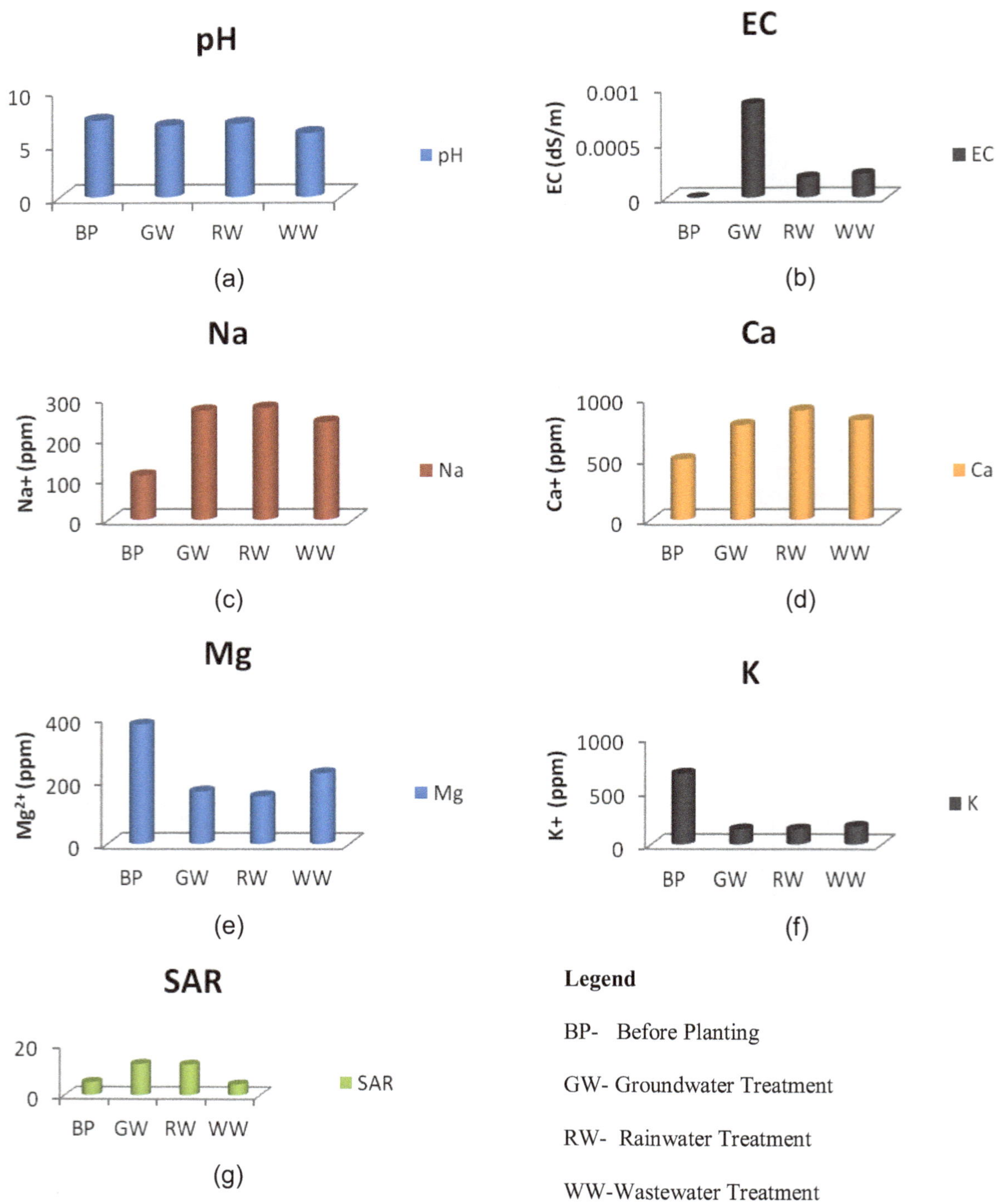

Figure 1. Effects of irrigation water on soil cultivated with cucumber.

Table 3. Soil classification based on electrical conductivity (EC), pH and sodium adsorption ratio.

Soil classification	EC*	pH	SAR
Normal soil	<4	<8.5	<13
Saline soil	>4	<8.5	<13
Sodic soil	<4	8.5-10	>13
Saline-sodic soil	>4	<8.5	>13

*EC, Electrical conductivity; SAR, sodium adsorption ration; pH, level of alkalinity/acidity. Source: Scherer (1996).

2007) have shown that soil pH increased over a long period of irrigation with sewage and wastewater effluents. These investigations described the long term effect while our study was short term and considered kitchen wastewater. According to Mojiri (2011) an initial decrease in pH may be observed in soil irrigated with kitchen wastewater but after a while it may cause an increase of soil pH. SAR increased in soil solution with the three water treatments as maximum value of 12.46 was observed in soil irrigated with groundwater. Soil irrigated with wastewater gave the lowest SAR value of 10.14 while soil irrigated with rainwater gave a value of 12.15 (Figure 1g). Of the treatments, kitchen wastewater was farthest from the critical SAR value of 13 and hence less likely to cause sodicity hazards in the soil. The EC of soil samples treated with groundwater, rainwater and wastewater treatments were 8.59×10^{-4}, 1.85×10^{-4} and 2.17×10^{-4} dS/m, respectively (Figure 1b). At the end of the experiment, the resulted EC value of the treated soil samples remained below the limit of the EC value for agricultural soil, 4 dS/m (Christiansen, 1977). Though EC was not the lowest for wastewater, its value was not far above that for rainwater which contributed the least to soil salinity. The groundwater EC value was the highest. Excess sodium can be detrimental to soil structure and can significantly reduce the soils ability to transmit water (Mace and Amrhein, 2001). The three water treatments added sodium (Na) to the soil (Figure 1c), but Na added by wastewater was not more in comparison to that added by groundwater and rainwater. Therefore, the use of kitchen wastewater is not assessed as leading to an enhanced EC or sodium hazard compared to the other treatments.

Comparing the EC, SAR and pH values of the soils after harvest with threshold values set by Scherer (1996), the soil samples were classified as not posing any threat with respect to salinity and sodicity hazards as well as pH, since EC, SAR and pH values were below 4 dS/m, 13 and 8.5, respectively (Table 3).

Effect of water type irrigation on cucumber growth parameters

For cucumber the number of leaves was not significantly affected by the different water types used for irrigation in the first five weeks of growth, but there was a significant effect at the seventh week of growth (Figure 2a). At 3 weeks after emergence, the average number of leaves for the three water types was about five and there was an increase in the number of leaves between the 3rd and 5th week for all treatments. Plants treated with rainwater had an average of about nine leaves, those with groundwater had about eight, while the kitchen wastewater pots had an average of about seven leaves. Between the 5th and 7th week, the number of leaves from cucumber treated with rainwater increased significantly to about ten compared with kitchen wastewater plots had about eight while pots under groundwater irrigation treatment had the number reduced to about seven (Figure 2a).

The water-type irrigation systems had no significant effect on cucumber average leaf area (Figure 2b). At 3 WAE, the average leaf area for the three water types was between 60 and 65 cm^2 as shown in Figure 2b. At the 5th week, the average leaf area increased to about 114, 90.2 and 92.1 cm^2 for rainwater, groundwater and wastewater irrigation treatments, respectively. From the 5th week to 7th week, different results were obtained for the average leaf area. While average leaf area from groundwater irrigation slightly increased to about 91 cm^2, that of rainwater and wastewater irrigation reduced to about 112.55 and 78.92 cm^2, respectively (Figure 2b). The plant length of cucumber did not differ significantly in the first five weeks of growth, however significant effect was observed at seventh week of growth as shown in Figure 2c. At 3 weeks of growth (WAE), the average plant length was between 25 and 27 cm for the three water types.

At week 5, there was a significant increase in plant length of cucumber treated with rainwater to about 74 cm. Cucumber treated with groundwater and wastewater has approximately equal plant length of about 50 cm. Between 5 and 7 weeks of growth, rainwater treated cucumber increased in length to about 100 cm while groundwater treated cucumber increased significantly to about 58 cm at a decreasing rate and for the wastewater irrigation, it decreased to about 45 cm.

The water-type irrigation systems had no significant effect on cucumber stem girth. From emergence to the 3rd week of cultivation, the average stem girth of cucumber in

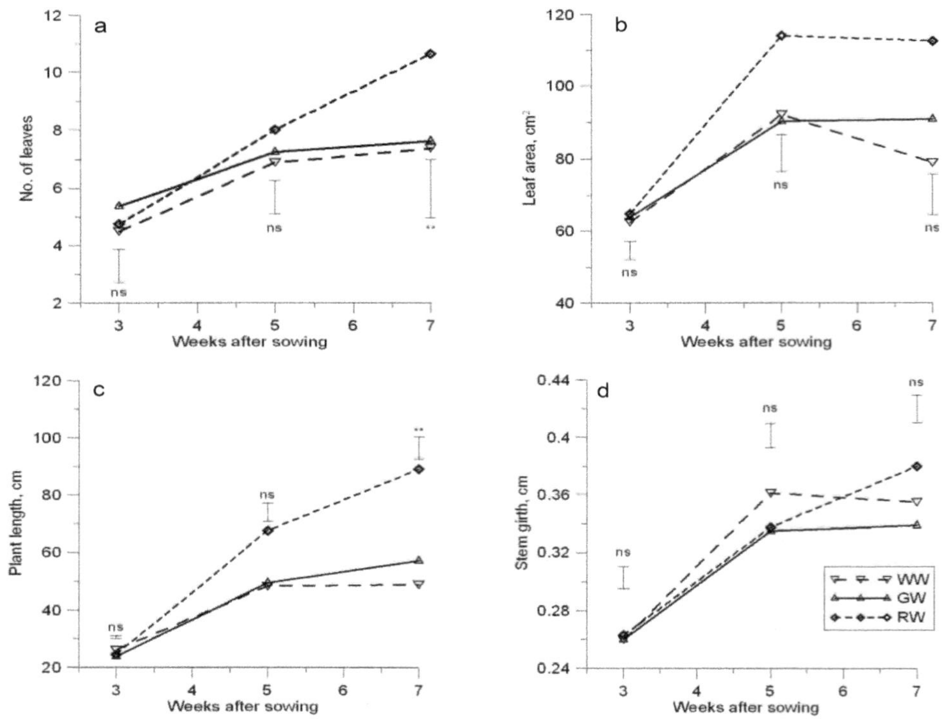

Figure 2. Cucumber growth parameters under the different water types: (a) number of leaves; (b) leaf area; (c) plant length, and (d) stem girth.

Table 4. Summary of the ANOVA for cucumber growth parameters.

Growth parameter	F-values		
	Water-type (WT)	Duration (D)	Interaction
Leaf area	7.61(3.35)	9(3.35)	1.51(2.73)
Stem thickness	2.13(3.35)	5.37(3.35)	0.37(2.73)
No of leaves	2.83(3.35)	14.05(3.35)	0.97(2.73)
Plant length	17.36(3.35)	44.08(3.35)	5.14(2.73)

The data in brackets are the tabulated F-values at 5% level of significance. A significant effect is indicated by lower tabulated F-values.

the three water-types was about 0.26 cm (Figure 2d). Between 5 and 7 weeks of growth, there was almost no change in stem thickness under groundwater (about 0.34 cm) and kitchen wastewater (about 0.36 cm) irrigation treatments whereas cucumber cultivated with rainwater increased from about 0.34 to 0.38 cm in stem girth. These findings are contrary to the results of Hussein (2009) who recorded increase in cucumber growth and yield with increase in sludge application. Mojiri et al. (2013) also found increase in *Lepidium sativum* root and shoot length when irrigated with wastewater. The authors attributed the increase to increased soil aggregation and contribution of organic matter to chemical properties and nutritional status of the soil, such as decrease in soil pH which enhances solubilization of nutrients and increase their availability and supply for root uptake. The reduction in leaf area and static plant length stem girth with wastewater may be attributed to anti-growth elements which the soil may have absorbed from the wastewater and thus influenced crop growth. The negative effect from the wastewater irrigation as observed is a result of formation of inner-sphere complexes, whereby the toxic elements are strongly adsorbed on soil colloids, form insoluble precipitates easily, thus preventing the exchange of cations necessary for crop growth.

The summary of the ANOVA on the effect of water-type and duration on the growth parameters of cucumber is shown in Table 4. From the table, the main effect of water-type and duration significantly affected the leaf area and plant height of cucumber at 5% level of significance. In each case, the calculated F-value was greater than the tabulated F-value. Water-type had no significant effect on the stem girth and number of leaves while duration had significant effect on both. The interaction between water-type and duration had significant effect only on the plant height.

Conclusions

The use of kitchen wastewater for irrigation of cucumbers did not adversely affect the soil as compared to rainwater or groundwater. The soil salinity level remained normal and the sodium levels indicated by the sodium adsorption ratio (SAR) were below the critical level of 13, indicating no threat to soil quality and hence could be suitable for crop production. There were some negative effects on plant growth and leaf area and number, not directly related to the inorganic parameters measured during this study, and ascribed to other wastewater components. Appropriate wastewater treatment and water management practices will have to be followed to allow the reuse of untreated kitchen wastewater for irrigation.

REFERENCES

APHA (1998). Standard methods for examination of water and wastewater, 20th ed. American Public Health Association, Washington, DC, USA.

Anderson JM, Ingram SJ (1993). Tropical biology and fertility: A handbook of methods. England, Wallingford: CAB Int.

Blanco FF, Folegatti MV (2003). A new method for estimating the leaf area index of cucumber and tomato plants. Hortic. Bras. 2(4):666-669.

Christiansen (1977). Irrigation water quality evaluation. J. of Irrig and Drainage, ASCE, 103:155–169.

Darvishi HH, Manshouri M, Farahani HA (2010). The effect of irrigation by domestic wastewater on soil properties. J. Soil Sci. Environ. Manag. 1(2):030-033.

El Youssfi L, Choukr-Allah R, Zaafrani M, Mediouni T, Sarr F, Hirich A (2012). Effect of domestic treated wastewater use on three varieties of amaranth (Amaranthus spp.) Under Semi-Arid Conditions. World Academy of Science, Eng. Technol. P. 62.

Ghanbari A, Abedikoupai J, TaieSemiromi J (2007). Effect of municipal wastewater irrigation on yield and quality of wheat and some soil properties in sistan zone. J. Sci. Technol. Agric. Natural Recou. 10:59-74.

Gee GW, Bauder JW (1986). Paricle size analysis. in aklute ed methods of soil analysis (2nd Ed.) No. 9 ASA Inc. SSA. Inc. Madison, Washington, DC. pp. 383-409

Government of India (GOI) (2002): India assessment water supply and sanitation. a who sponsored study. Planning Commission Government of India, Spectra Visual Word Pvt. Ltd. P. 9.

Heidarpour M, Mostafazadeh-Fard B, Abedi Koupai J, Malekian R (2007). The effects of treated wastewater on soil chemical properties using subsurface and surface irrigation methods. Agric. Water Manag. 90:87-94.

Hussein AHA (2009). Impact of sewage sludge as organic manure on some soil properties, growth, yield and nutrient contents of cucumber crop. J. Appl. Sci. 9(8):1401-1411.

Khai NM, Tuan PT, Vinh CN, Oborn, T (2008). Effects of using wastewater as nutrient sources on soil chemical properties in peri-urban agricultural systems. VNU J. Sci. Earth Sci. 24:87-95.

Mohammad MJ, Ayadi M (2004). Forage yield and nutrient uptake as influenced by secondary treated wastewater. J. Plant Nut. 27(2):351-365.

Mojiri A (2011). Effects of municipal wastewater on physical and chemical properties of saline soil. J. Biol. Environ. Sci. 5(14):71-76

Mojiri A, Aziz HA, Aziz SQ, Gholami A, Aboutorab M (2013). Impact of urban wastewater on soil properties and *Lepidium sativum* in an arid region. Intern. J. Sci. Res. Environ. Sci. 1(1):1-9.

Olaniyi JO, Akanbi WB, Adejumo TA, Akande OG (2010). Growth, fruit yield and nutritional quality of tomato varieties. Afri. J. Food Sci. 4(6):398-402.

Papadopoulos I (1995). Wastewater management for agriculture protection in the near east region, Technical Bulletin, FAO, Regional Office for the Near East, Cairo, Egypt.

Rattan RK, Datta SP, Chhonkar PK, Suribabu K, Singh AK (2005). Long-Term impact of irrigation with sewage effluents on heavy metal content in soils, crops and groundwater – a case study. Agric. Ecosyst. Environ. 109:310-322.

Rusan MJ, Hinnawi S, Rousan L (2007). Long term effect of wastewater irrigation of forage crops on soil and plant quality parameters. Desalination, 215:143-152.

Saxena MC, Singh Y (1965). A note on leaf area estimation of intact maize leaves. Indian J. Agro. 10:437-439 .(Not Cited In The Work).

Shainberg I, Oster JD (1978). Quality of irrigation water. Pergamon Press, London.

Singh PK, Deshbhratar PB, Ramteke DS (2011). Effects of sewage wastewater irrigation on soil properties, crop yield and environment. Agric. Water Manag. 103:100–104.

Soil Survey Staff (SSS), (2006). Keys to soil taxonomy. 10th ed. Washington, USDA-SCS, P. 332.

Stamatiadis S, Doran JW, Kettler T (1999). Field and laboratory evaluation of injection of liquid sewage sludge. Appl. Soil Ecol.12:263-272.

Tabari M, Salehi A, Ali-Arab AR (2008). Effects of Waste Water Application on Heavy Metals(Mn, Fe, Cr and Cd) Contamination in a Black Locust Stand in Semi-Arid Zone of Iran. Asian J. Plant Sci. 7:382-388.

Tabatabaei SH, Najafi P, Amini H (2007). Assessmentof Change in Soil Water Content PropertiesIrrigated with Industrial Sugar BeetWastewater. Pakistan J. Biol. Sci. 10:1649-1654.

Thapliyal A, Vasudevan P, Dastidar MG, Tandon M, Mishra S (2011). Irrigation with domestic wastewater: responses on growth and yield of ladyfinger *(Abelmoschus esculentus)* and on Soil Nutrients. J. Environ. Biol. 32:645-651.

Vaseghi S, Afyuni M, Shariatmadari H, Mobli M (2005). Effect of sewage sludge on some nutrients concentration and soil chemical properties. J. Isfahan Water. Wastewater. (in Persian). 53:15-19.

Vazquezmontiel O, Horan NJ, Mara DD (1996). Management of domestic wastewater for reuse in irrigation, Water Sci. Technol. 33(10–11):355–362.

Walker C, Lin HS (2008). Soil property changes after four decades of wastewater irrigation: a landscape perspective. Catena 73:63-74.

Mapping of limestone deposits and determination of quality of locally available limestone in Rwanda

Nduwumuremyi A.[1], Mugwe N. J.[2], Rusanganwa C. A.[1] and Mupenzi J.[3]

[1]Soil Quality and Plant Nutrition Program, Department of Natural Resources Management, Rwanda Agriculture Board (RAB), P. O. Box 5016, Kigali, Rwanda.
[2]Department of Agricultural Resource Management, School of Agriculture and Enterprise Development, Kenyatta University, P.O. Box 43844-00100, Nairobi, Kenya.
[3]Department of Land Survey, Institute of Applied Sciences (INES-Ruhengeri), Musanze, P. O. Box 155, Rwanda.

Soil erosion, soil nutrient depletion and soil acidity associated to Aluminium (Al) toxicities are the main soil related constraints to agricultural development in many countries of sub-Sahara Africa. The lime plays an important role in improving soil acidity and hence favours plant nutrition. The aim of this study was to map out the available limestone deposits and to determine their selected quality. The selected quality of limestone (CaO, MgO and fineness factor) was analyzed using ethylene diamine tetraacetate (EDTA) method. The global positioning system (GPS) coordinates of limestone deposits were taken. The collected data were processed using Arc GIS 9. The fineness factor was determined by mechanically sieving the lime through a stag of 4 sieves of various aperture dimensions (2, 1, 0.5 and 0.2 mm). The map showed that the limestone deposits are scattered in Northern and Western provinces with more concentration in Northern region. Mpenge limestone had an average of 40% ±4.2 of CaO and 3% ±0.9 of MgO. However, its fineness factor of 63% ±2.01 found to be in good range. Gishyita deposit was heterogeneous in $CaCO_3$, its CaO was 50% ± 20.02 indicating high variability within mines where CaO reached < 30%. Mashyuza deposit in Rusizi region had 50% ±3.4 of CaO and moderate MgO of 3% ±0.6. However, its fineness factor was 55.6% ±1.6 .The difference in lime quality observed in this study was possibly due to the origin of limestone rocks. However, the fineness factors were below the acceptable level because of non-additional grinding and lack of technological skills in producing lime. Therefore, this study recommends the use of grounded limestone after fine grinding.

Key words: Limestone deposit sites, CaO, MgO, fineness factor.

INTRODUCTION

Agriculture is the most important sector of the Rwandan economy, supporting 82% of the population (NISR, 2009). However, agricultural productivity remains critically low. The low productivity of the agricultural sector is largely attributed to low and decreasing soil fertility due to many factors such as soil erosion, continuous cropping, soil acidity and inadequate sustainable soil fertility management. For instance, the acidity affects the fertility of soils through nutrient deficiencies (P, Ca and Mg) and the presence of phytotoxic nutrients such as soluble Al and Mn (Awad et al., 1976). Application of lime in acid soils reduces Al toxicity, improves pH, Ca, Mg and

Table 1. Particle size and efficiency factors of limes.

Particle size(mesh sieving size)	Opening size (mm)	Efficiency factor
>8	>2.36	0
8 - 60	2.36 - 0.25	0.5
<60	<0.25	1.0

Source: Havlin et al. (2005).

increases both P uptake in high P fixing soil and plant rooting system (Black, 1993). The potentials for soils sustainable management are among the options to explore in restoring soil fertility. In agriculture, lime has great importance in improving soil acidity and hence favours plant nutrition. It is observed in Rwanda that even small locally produced limes and liming materials can be used to increase crop production on acidic soils (Yamoah et al., 1992).

The lime is known as a material originated from rocks which can have multiple purposes (construction, cements, water purification, disinfectant, agricultural amendments, etc). Locally available carbonates are relatively common in many countries of sub-Saharan Africa and are well suited for small-scale mining and processing (Van Straaten, 2002). In Rwanda, there is relatively good sources of limestone (travertine) mainly in the Western (Karongi and Rusizi districts) and the Northern (Musanze district) provinces and other small deposits in Gakenke and Rulindo districts. The government of Rwanda through the crop intensification program has singled out the lime use given its impact on crop productivity. Currently, the government is providing some lime to smallholder farmers through the districts administration and local non-governmental organizations (NGOs). However, there is limited information in terms of illustrating the exact location of limestone deposits in Rwanda which implies the insufficient use of lime in many parts of Rwanda. Therefore, this study aimed to map out the lime deposits and to determine the selected quality of locally available limestone in Rwanda.

MATERIALS AND METHODS

This study was conducted in Northern Province: Musanze, Burera, Rulindo and Gakenke districts, and in Western province: Karongi, Nyamasheke and Rusizi districts). The study consisted by laboratory analysis of selected quality (CaO, MgO and fineness factor) of limestones and taking global positioning system (GPS) coordinates of limestone deposits. The collected data were processed using Arc GIS 9. The collected data were supplemented by exploring previous studies on the same matter.

The Ca (%) and Mg (%) were analyzed by ethylene diamine tetraacetate (EDTA) method (Hesse, 1971), the titration was done with EDTA 0.01N. The CaO (%) and MgO (%) were determined by multiplying Ca with conversion factor of 1.3992 and Mg with conversion factor of 1.6578 (Marcus, 2009). Fineness factor (%) was determined by mechanically sieving the lime through a stag of 4 sieves of various aperture dimensions (2, 1, 0.5 and 0.2 mm) resulting to 5 classes of lime. The particles obtained were multiplied by efficiency factor of 0, 0.5 and 1 (very less degradable, less degradable and quick degradable), respectively (Table 1).

$$\text{Fineness factor (\%)} = \sum \left(\frac{\text{Particle * Respective efficiency factor}}{\text{Sample quantity}} * 100 \right) \quad (1)$$

RESULTS AND DISCUSSION

Lime deposits in Rwanda

The deposits of limestone in Rwanda are scattered in Northern and Western provinces with more concentration in Northern region (Figure 1).

Several studies (Verhaeghe, 1963; Grigoriev, 1981; Beernaert, 1999; SOFRECO, 2001; Van Straaten, 2002) indicated that Rwanda has relatively good sources of agricultural lime mainly in Musanze, Karongi and Rusizi districts with small deposits in Gakenke, Rulindo and Nyamasheke districts. Almost all deposits of limestone in Rwanda are located in axis zones of N 35°E, ranging from Mashyuza (south part of Western province) to Ruhengeri (Northern province), this axis is the one of recent major fracture related to the genesis of African rift valley (SOFRECO, 2001).

The study carried out by SOFRECO (2001) found out that the main lime deposits ready to produce lime for agricultural purpose estimated at ±18,700,000 m^3 while the deposits not for agriculture was estimated at ±57,550 m^3, which may be suited for construction purpose. However, current research demonstrates that the required quantity for addressing all acidic soils in Rwanda, estimated at 2,220,480 t. This quantity is quite low compared to the available lime deposits identified by SOFRECO (2001).

Lime deposits in Northern region

Lime deposits in northern region are composed by several sites such as Cyiri (Kinoni sector) in Burera District, Kiryi (Kigombe, Mpenge) and Rwaza in Musanze distrcit, Masangano in Gakenke district and Gitagata in Rulindo district (Figure 2).

These lime deposits are being exploited for many years by cooperatives, groups, privates and companies. However, some were discontinued from operating due to

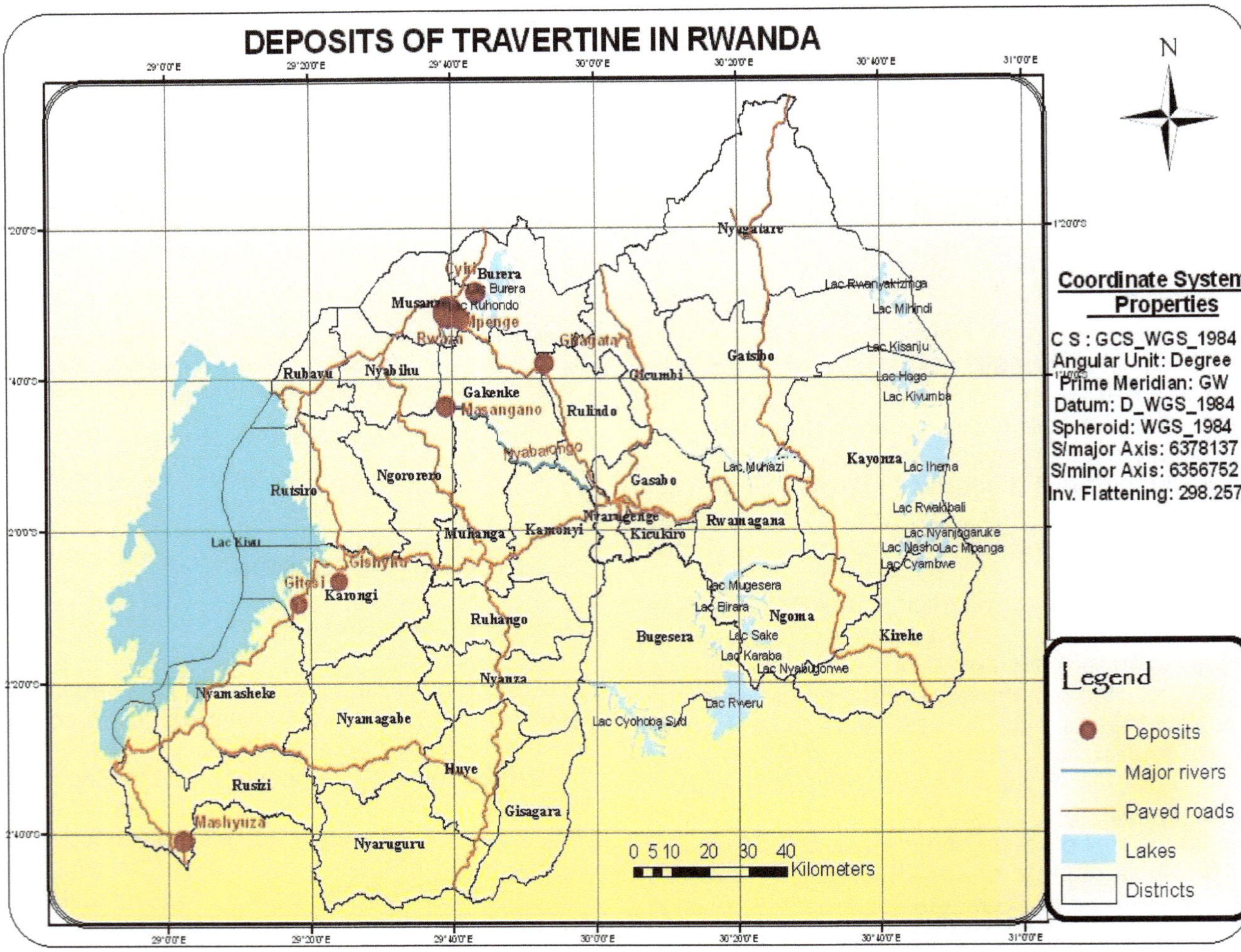

Figure 1. Map illustrating limestone deposits in Rwanda.

1994 Genocide. Majority of product is in the form of burned lime for construction and not for agricultural purpose. Among all sites, Mpenge site has the appropriate lime for agricultural practice, because of hardness, the rest (Cyiri, Kigombe, Rwaza, Gitagata and Masangano) are suitable for construction and other purposes. The studies of Beernaert (1999) found that the grounded Mpenge limestone is suitable for agriculture and could replace burned lime (Nduwumuremyi et al., 2013). All lime producers in this region are relying on traditional techniques and they do not possess machines for finely grinding limestone, consequently, the produced products are less effective and therefore, are very expensive as they are needed in bulk quantity (Coventry et al., 1989) to meet the requirement in the acidic soils.

Selected quality (CaO, MgO and fineness factor) of Mpenge limestone: This study showed that in Northern region, Mpenge limestone (travertine) is suited for agricultural practices with an average of 40% ±4.2 of CaO and 3% ±0.9 of MgO. However, its fineness factor of 63% ±2.01 found to be in good range, but to improve the reactivity (solubility) for reasonable period after its application, it is still to be increased by finely crashing limestone by appropriate machine. The good range of fineness factor was reported to be at least >60% (Crawford et al., 2008). Several studies (Rwagashayija, 1990; Rukeratabaro, 1991; Beernaert, 1999; SOFRECO, 2001) reported that the grounded limestone (travertine) from Mpenge site was ±50% of fineness which is very low to the recommended fineness level and can compromise the effectiveness of lime in soils.

Lime deposits in Karongi region

Lime deposits in Karongi region are composed by two main sites, Ruhundo/Gitesi in Karongi district and Gishyita site, the last borders two districts (Karongi and Nyamasheke). The deposit of Gitesi was discarded from exploitation due to the inaccessibility, poor liming effect or quality (low in CaO) and inconsistency in quality within

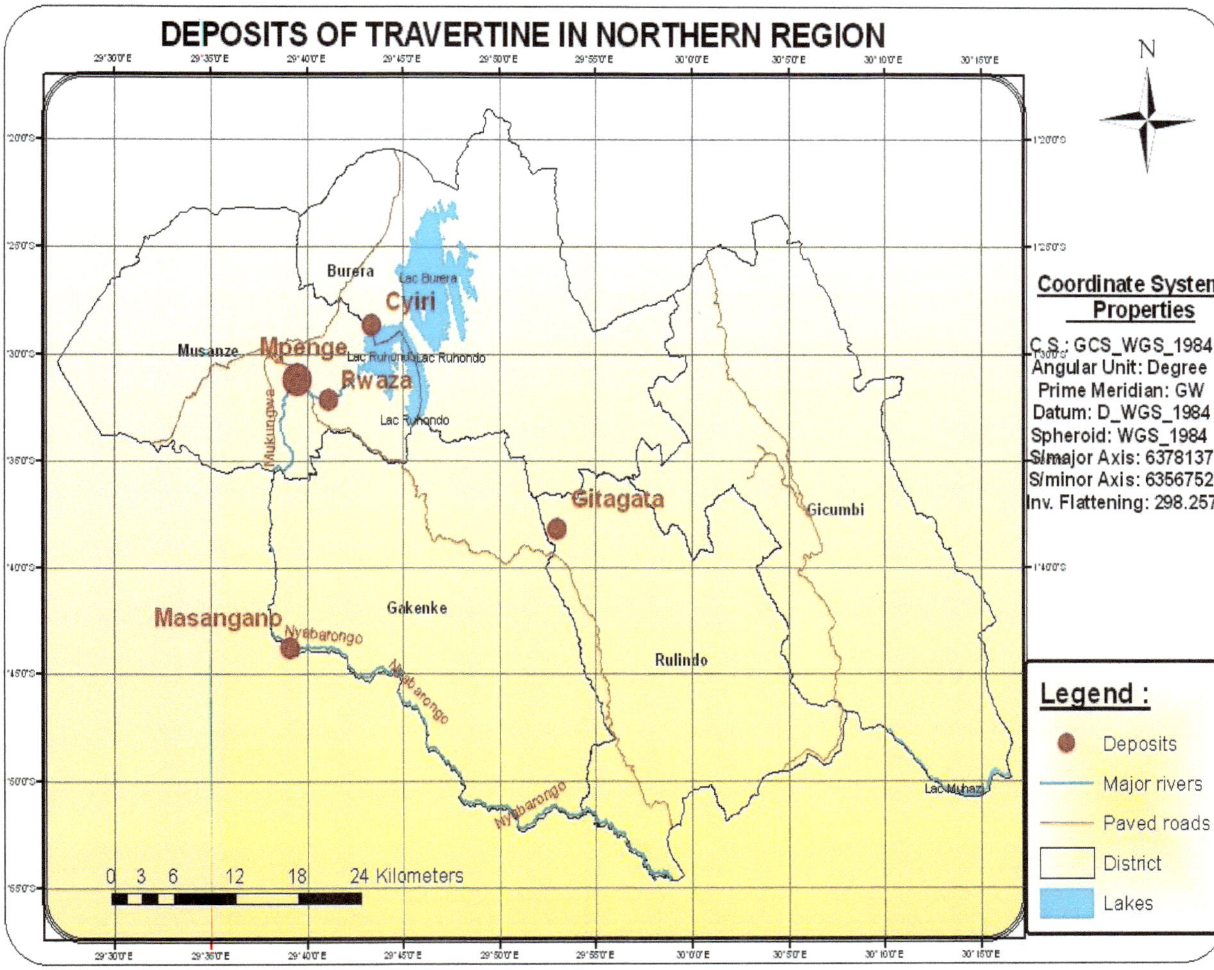

Figure 2. Map illustrating limestone deposits in Northern region.

the deposit. In additional, the high slope of the deposit associated with traditional techniques (manually) without appropriate equipments and safety materials render the deposit dangerous for mine workers. The lime deposit of Gishyita was reported to be approximately 1,000,000t of heterogeneous $CaCO_3$ (SOFRECO, 2001) because of its variability in $CaCO_3$. Additionally, this deposit is geographically accessible by good road on axis Rusizi-Karongi (Figure 3).

However, despite its accessibility, Gishyita deposit is barely used in agriculture probably due to its high variability in CaO.

Quality (CaO, MgO and fineness factor) of Gishyita limestone: Despite the accessibility of the deposit, the lime deposit of Gishyita was found to be heterogeneous in $CaCO_3$. The level of CaO was 50% ±20.02 with high variability within mines where CaO can even be <30% and MgO was 3% ±11.1, but its variability is also a constraint where it can reach 15% of MgO. These are corroborated to the findings of Beernaert (1999) and SOFRECO (2001) who reported Gitesi deposit to be enriched in MgO (3 to 16%) and poor in CaO (<40%).

Lime deposits in Rusizi district

The size of this deposit is approximately one km^2 scattered with small deposits which were not evaluated and measured in this study. Only the site of Mashyuza (Figure 4) was analyzed.

Recently, the CIMERWA (Cemetrie du Rwanda), COCOCHAUMA (Coopérative de Production de la Chaux de Mashyuza) and APC Mashyuza Ltd are the three major enterprises for exploitation of Mashyuza limestone deposits. The CIMERWA uses Mashyuza limestones as one of the materials of mixture for production of Portland cement. The clinker is made by 78% of limestones (travertine), 20% of quartzite, clay and pozzolan. However, COCOCHAUMA and APC Mashyuza Ltd are exploiting limestone for production of quicklime reserved for construction. The production of grounded limestone

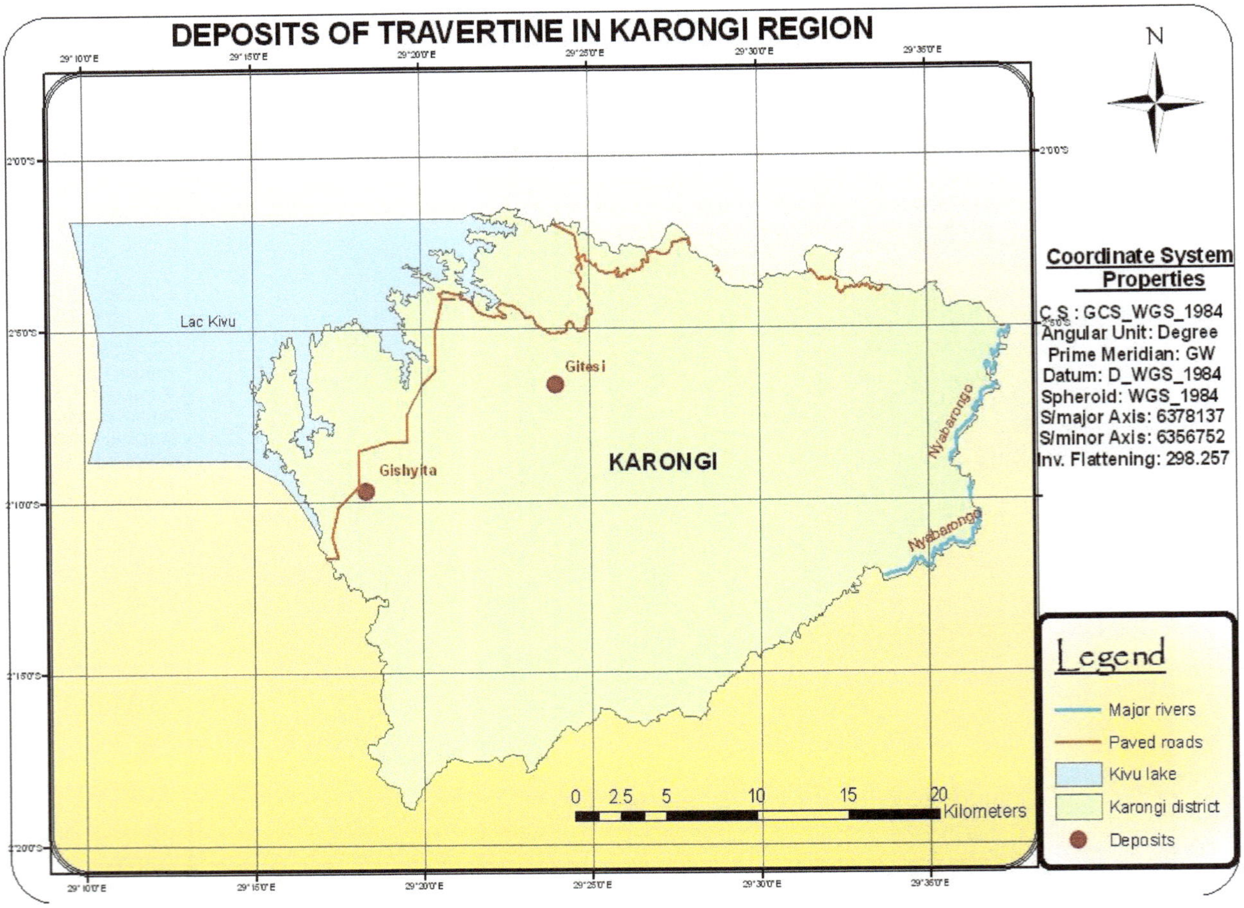

Figure 3. Map illustrating limestone deposits in Karongi region.

for agriculture purposes is limited due to the production technology, storage, and small size of market. Although, CIMERWA is well equipped with skilled personnel and modern technology, it is not involved in grounded limestone production for agricultural purpose. Its mission is to produce cement for construction and it seems to be difficult to deviating them for producing lime for agricultural purpose.

Quality (CaO, MgO and fineness factor) of Mashyuza limestone: The limestones deposits in Rusizi region found to be enriched in terms of CaO estimated at 50% ±3.4 with moderate MgO of 3% ±0.6. These are in agreement with findings of several authors (Verhaeghe, 1963; Beernaert, 1999; SOFRECO, 2001) who reported that non calcined lime of Mashyuza and calcined lime (pure lime) had the similar level of CaO. Like all other limestone sources in Rwanda, its fineness factor was 55.6% ±1.6 which was below the acceptable level (60%) as reported by Crawford et al. (2008).

Conclusion

Rwanda has relatively good sources of agricultural lime mainly in Musanze, Karongi and Rusizi districts with small deposits in Gakenke, Rulindo and Nyamasheke districts. Some of these deposits are suited to agriculture practices; others are suitable for houses and roads construction, and sanitary treatment. Mashyuza limestone and burned lime had the same level of CaO. Mpenge and Mashyuza were reported to be suitable for agriculture practices. Despite good qualities, their fineness factors are still below acceptable level. The total size of deposits is not known and the production is hampered by many factors including technological and financial constraints. Therefore, the study advocates to increasing fineness factor through technical and financial support to the cooperatives engaged in lime production in Rwanda. The further analysis of impurities and total reserves and new deposits of limestone in Rwanda are needed to be explored.

ACKNOWLEDGEMENTS

Special thanks are extended to Rwanda Agriculture Board (RAB), High Institute of Agriculture and Animal Husbandry (ISAE) and Institute of Applied Sciences

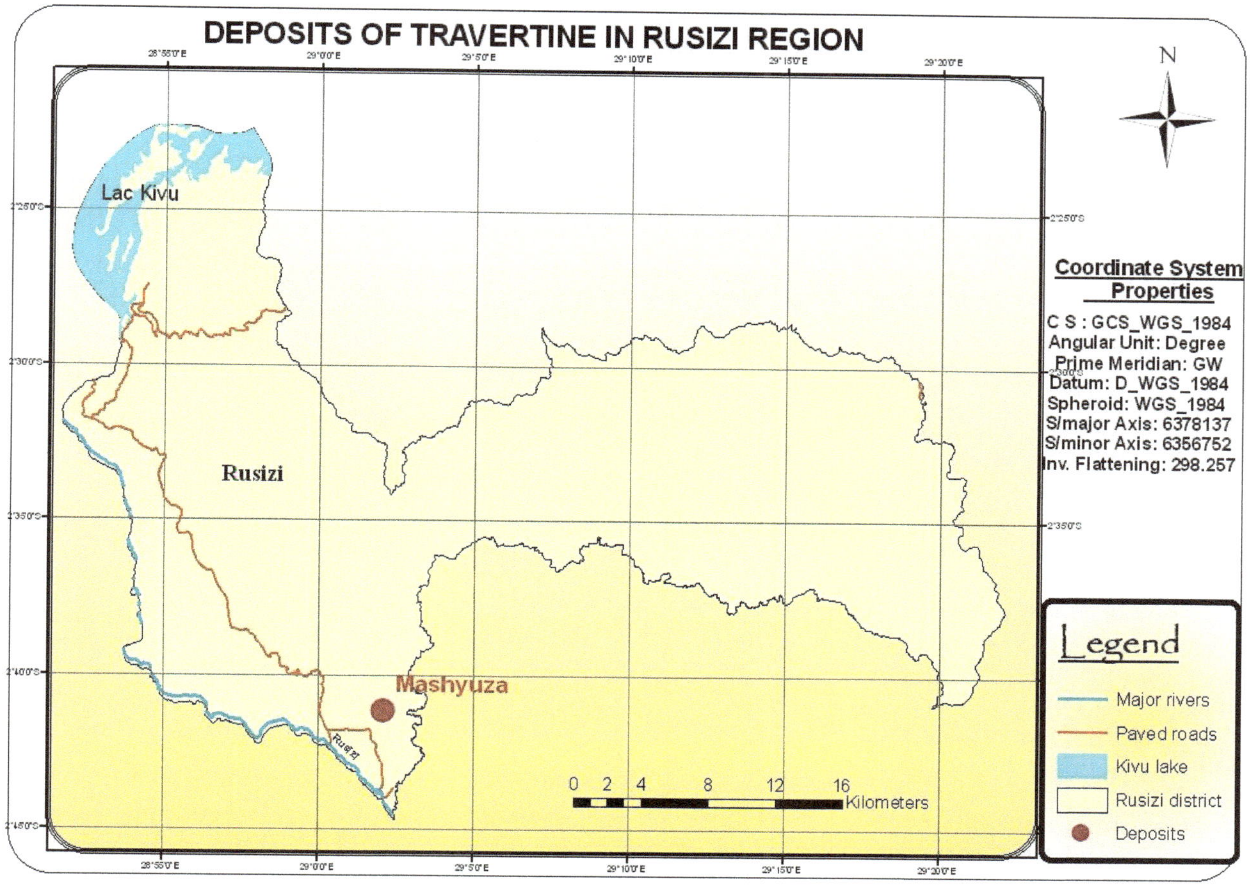

Figure 4. Map illustrating limestone deposit in Rusizi district.

(INES-Ruhengeri) for logistic and laboratory facilities.

REFERENCES

Awad AS, Edwards DG, Milhan PJ (1976). Effect of pH on soluble soil aluminium and on growth and composition of Kikuyu grass. J. Plant Soil Sci. 45:531-542.

Beernaert F (1999). Feasibility study of production of lime and or ground travertine for manegement of acidic soils in Rwanda. Brussels: Pro-Inter Project Consultants.

Black A (1993). Soil fertility evaluation and control. London: Lewis Publisher.

Coventry DR, Walker BR, Morrison GR, Hyland MT, Averz JC, Maden JL, Bartram DC (1989). Yield response to lime of wheat and barley on acidic soils in north-eastern Victoria. Aust. J. Exp. Agric. 29(2):209-214.

Crawford TW, Singh U, Breman H (2008). Solving agricultural problems related to soil acidity in central Africa's great lakes region. Alabama: International center for soil fertility and agriculture development.

Grigoriev S (1981). Travertines of Rwanda, Project of mining research, RWA80/001. Kigali: UNDP.

Havlin JL, Beaton JD, Tisdale SL, Nelson WL (2005). Soil fertility and fertilizers: An introduction to nutrient managemnt (7 ed.). New Jersey: Pearson prentice hall.

Hesse J (1971). A textbook of soil chemical analysis. London: John Murray.

Marcus MA (2009). Fertilizers, Agronomic handbook. Virginia, USA: Virgina Tech University.

Nduwumuremyi A, Mugwe JN, Ruganzu V, Cyamweshi K R, Nyirinkwaya B (2013). Effects of Limestone (travertine) in Improving Selected Soil Properties and Yield of Irish Potato (Solumun tuberosum L.) in Acidic Soils. J. Agric. Sci. Technol. A. 3:175-182

NISR (2009). National census. Kigali, Rwanda: National Statistical office of Rwanda.

Rukeratabaro D (1991). Lime production for soils amendments, Mpnge and Mpatsi limestone exploitation . Kigali: MINIMART.

Rwagashayija T (1990). Travertines of Ruhengeri, prospectives mining of Mpenge. Kigali: MINIMART.

SOFRECO (2001). Evaluation and classification of travertines deposits. Kigali, Rwanda: Ministry of Industries and tourism.

Van Straaten P (2002). Rocks for crops: Agrominerals of sub-saharan Africa. Nairobi, Kenya: ICRAF.

Verhaeghe M (1963). Inventory of deposit of limestone, dolomite and travertine in Kivu, Rwanda and Burundi. Ruhengeri, Rwanda: Ministry of agriculture and economic affairs, Geology service.

Yamoah CF, Burleigh JR, Eylands VJ (1992). Correction of acid infertility in Rwandan oxisols with lime from an indigenous source for sustainable cropping. Exploratory Agric. 28:417-424.

Effects of different land use systems on selected soil properties in South Ethiopia

Alemayehu Kiflu and Sheleme Beyene

Soil Science Department, Hawassa University, Ethiopia

A study was conducted to determine the effect of different land use systems on soil physical and chemical properties in Sodo Zuria Woreda of Wolaita zone Southern Ethiopia. Differences in soil properties in different land use types at two depths (0-15 and 15-30 cm) were observed on many soil properties important to crop growth. Enset (*Ensete Ventricosum*) fields had higher pH, electrical conductivity (EC), available P and Zn, exchangeable Ca and K which is attributed due to the addition of manure, whereas maize fields had lowest average K and Mg, cation exchange capacity (CEC), percentage of base saturation (PBS), total N and OC. These results suggest that land use has persistent, multi-decadal effects on the spatial heterogeneity of soil resources and also the need of land use and landscape research to determine ecologically sound and sustainable land use and management strategies.

Key words: Land use, Enset, grass land, manure, continuous cultivation, soil properties.

INTRODUCTION

Successful agriculture requires the sustainable use of soil resource, because soils can easily lose their quality and quantity within a short period of time for many reasons. Agricultural practice therefore requires basic knowledge of sustainable use of the land. A success in soil management to maintain the soil quality depends on the understanding of how the soil responds to agricultural practices over time. Recent interest in evaluating the quality of our soil resource has therefore been simulated by increasing awareness that soil is critically important component of the earths biosphere, functioning not only in the production of food and fiber, but also in the maintenance of local, regional, and worldwide environmental quality (Negassa, 2001). On the other hand, feeding the ever-increasing human population is most challenging in areas like southern Ethiopia, where there is a very high population density. In addition, the topography has also a great impact on the soil quality and depth due to the interaction impact of cultivation practices and slope. Revising these trends lies in the enhancement of sustainable development of the agricultural sector. However, the basis of this sustainable agricultural development is good quality of soil, since maintenance of soil quality is an integral part of sustainable agriculture.

According to Wang et al. (2001) climate and geological history are importance factors to affecting soil properties on regional and continental scales. However, land use may be the dominant factors of soil properties under small catchment scale. Land use and soil management practices influence the soil nutrients and related soil processes, such as erosion, oxidation, mineralization, and leaching, etc (Celik, 2005; Liu et al., 2010). As a result, it can modify the processes of transport and re-distribution of nutrients. In non-cultivated land, the type of vegetative cover is a factor influencing the soil organic carbon content (Liu et al., 2010). Moreover, soils through land use change also produce considerable alterations (Fu et al., 2000), and usually soil quality diminishes after the cultivation of previously untilled soils (Neris et al.,

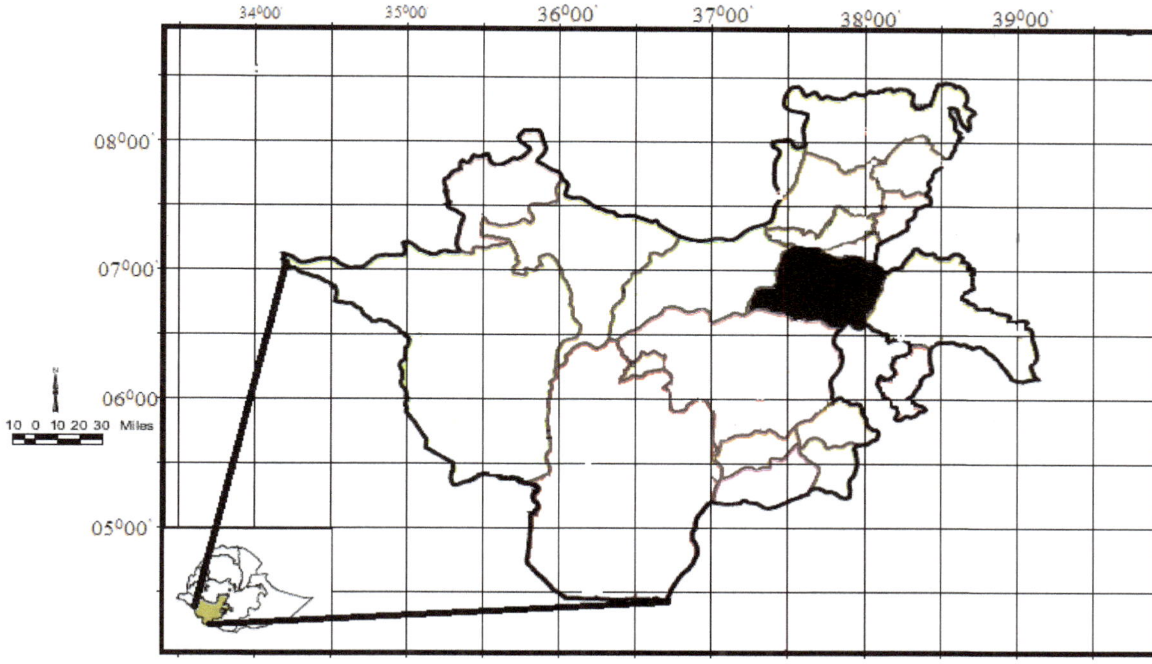

Figure 1. Location map of the study area (Sodo Zuria, Wolayita Southern Ethiopia).

2012). Thus, land use and type of vegetation must be taken into account when relating soil nutrients with environmental conditions (Liu et al., 2010). The particular nature of the typical rugged relief with slopes subjected to cultivation for many years in the study area had lead to decline in soil fertility. Therefore, there is special need for the analysis of soil nutrients in relation to land use. Such a local analysis is necessary to estimate nutrient storage in semi-natural and cultivated ecosystems (Wang et al., 2001), therefore this research was initiated to investigate the influence of different land use types on selected properties of the soil in southern Ethiopia.

MATERIALS AND METHODS

Description of the study area

The study was conducted at Delbo Atwaro watershed, which is located in Sodo Zuria Woreda of Wolayita zone, SNNP Regional State, Ethiopia. Delbo Atwaro is bounded by Damot Woyde on the east, Kindo Koysha on the west, and Bolosso Sore on the north and Humbo on the south. The altitude of Sodo Zuria Woreda ranges from 1500 to 3500 m above sea level, having a "Woinadega to Dega" climatic characteristic. The study area was selected as a representative watershed for high productivity potential in the Wolayita area to use as experimental site for the Sustainable Rural Agricultural Development project funded by CIDA/UPCD in 2003 (Figure 1).

Sodo Zuria Woreda has a total area of 48,125 ha, out of which 27,687 ha are being cultivated, and 4,020 ha are used for grazing and the rest is occupied by other land use types. The study area has a bimodal rainfall with peak rainy months in April and August. The mean seasonal temperature and the annual rainfall of the study area (Delbo watershed) are 20°C and 1296.6 mm, respectively. Soils of Wolayita area are grouped as Nitosols (Ethiopia Mapping Authority, 1998).

Land use and toposequence selection

Enset (*Enset ventricosum*) was selected as one of the land uses because it is as old as agriculture in the study area and used as a staple food. Moreover, maize is grown everywhere in the area and selected as the second land use, whereas grassland is selected as third land use system for the sake of comparison. To take a representative samples, a landscape facing west was selected in the watershed. The toposequence was divided into three slope positions: Upper N 06°54' and E 37°50.589' with 2161 m a.s.l, middle topographic position N 06°54.522' and E 37°50.437 with altitude 2110 m a.s.l and lower topographic position N 06°54.628' and E 37°50.388 and 2087 m a.s.l.

Soil sampling and preparation

All soil samples were air-dried, ground and passed through 2 mm sieve at the Soil Laboratory of the Awassa College of Agriculture. The physicochemical analyses of the soil samples were conducted at National Soil Research Center following standard laboratory procedures. The bulk density determinations were done at Soil Laboratory in Awassa collage of Agriculture. Duplicate soil samples from each topography (Upper, Middle and lower) and land use types (Enset, maize and Grass land) were collected.

Laboratory analyses

Total nitrogen content was determined following the Kjeldahl method (Jackson, 1958). The available phosphorus content of the

Table 1. EC, pH, sand, silt and clay properties of the soil at 0 to15 and 15 to 30 cm depth under different land use systems.

Land use	pH	EC (dS/m)	Sand (%)	Silt (%)	Clay (%)
		0-15 cm			
Enset	6.83	0.15	33.66	48.00^{a}	18.33
Grass	5.80	0.09	44.33	40.66^{b}	15.00
Maize	6.03	0.08	31.66	47.33^{a}	21.00
		15-30 cm			
Enset	6.4	0.10^{a}	29.66	46.00^{a}	24.33^{b}
Grass	5.63	0.06^{ab}	34.33	35.33^{b}	30.33^{a}
Maize	6.2	0.04^{b}	31.00	42.00^{a}	27.00^{b}

Mean values within a column followed by the same letter(s) are not significantly different at $p \leq 0.05$.

soil was analyzed using 0.5 M sodium bicarbonate extraction solution (pH=8.5) following the method of Olsen et al. (1954). The organic carbon determinations were made following the wet oxidation method of Walkley and Black (1934). The exchangeable basic cations (K^+, Ca^{2+}, Mg^{2+}, and Na^+) were extracted with 1 M ammonium acetate at pH=7.0. The CEC of the soil was determined from ammonium acetate saturated sample. The excess ammonium acetate was removed by washing with ethanol. Finally, the exchangeable Ca^{2+} and Mg^{2+} in the ammonium acetate leachate were measured by atomic absorption spectrophotometry (AAS), and K^+ and Na^+ were determined by flame photometer. Hydrometer method was used for the determination of soil particle size distribution. The soil pH was measured using a glass combination pH meter in the supernatant solution of 1:2.5 soil to water solution ratio. The EC was measured by taking 10 g of soil in 25 ml of water with a conductivity meter. A commonly used procedure called diethylenetriaminepenta -acetic acid or DTPA (Lindsay and Norvell, 1978) extraction was used to extract Cu, Fe, Mn and Zn from the samples. The micronutrients extracted with this method were measured by atomic absorption spectrophotometer at 248.3, 279.5, 324.7 and 213.9 nm wavelength for Fe, Mn, Cu, and Zn, respectively.

Statistical analysis

Mean comparisons (LSD) were calculated for the different land uses systems and correlation analysis has been done for the different topographic positions and land uses systems using the SAS software (SAS, 1997), to see the relationship between parameters.

RESULTS AND DISCUSSION

Physico-chemical properties of the soil as influenced by different land use systems

Relatively higher sand content was recorded in grass land soils followed by that of enset and maize fields in the upper 0 to15 cm depth, whereas in the 15 to 30 cm depth silt was found to be higher in grass land soils followed by maize and enset fields (Table 1). On the other hand, higher content of clay was recorded in 0 to 15 cm depth of maize farms. Although texture is inherent property, this might be attributed to accelerated weathering as a result of disturbance during continuous cultivation, as was also concluded by Boke (2004) from the result obtained from the nearby site. The soil texture of the different land use types and the upper layers of the different horizons were found to be the same except for that of grassland soil (15 to 30 cm depth), which was clay loam. This suggests that the different land use types did not have effect on the soil texture of the study area, since texture is an inherent soil property that not influenced in short period of time.

The pH value under enset was found to be the highest followed by maize in both sampling depths. The soil pH could be categorized as slightly acidic under enset and maize fields whereas that of grassland was moderately acidic, following the classification described by Brady and Weil (2002). The higher values of pH under enset in both depths could be due to higher values of exchangeable bases as a result of application of house refuses and wood ash to enset fields. This is evident from the positively correlation between pH and the exchangeable bases in both depths. Relatively higher EC values at both depths were recorded in the enset farms followed by grassland soil. Although the EC values were negligible significant differences ($p \leq 0.05$) were obtained under the different land uses.

The organic carbon content of the soils varied from 2.08 to 3.25% for 0 to 15 cm depth. In 15 to 30 cm depth it ranged from 1.79 to 2.11% (Table 2). Significant differences ($p \leq 0.05$) in OC content of soils were observed among the different land use systems. The average content of soil OC along slope positions, were lower in maize and enset land use systems as compared to that of grassland. The difference could be attributed to the effect of continuous cultivation that aggravates organic matter oxidation. The roots of the grass and fungial hyphae in the grassland soils are probably responsible for the higher amount of total organic matter (Urioste et al., 2006). The results were in agreement with the findings of Negassa (2001) and Malo et al. (2005), who reported less organic carbon in the cultivated soils than grassed soils. At both depths enset farms had higher OC content as compared with the maize farms.

Table 2. Some chemical properties of the soil at 0-15 and 15-30 cm depth under different land use systems.

Land use	TN	OC	Av. P	Fe	Mn	Zn	Cu
	%		mg/kg				
			0-15 cm				
Enset	0.20[b]	2.44[b]	36.35[a]	58.82	21.87[b]	8.63	0.323
Grass	0.29[a]	3.25[a]	3.68[b]	45.40	44.95[a]	8.61	0.443
Maize	0.18[b]	2.08[b]	14.44[b]	21.87	27.70[b]	7.94	0.350
			15-30 cm				
Enset	0.18a[b]	1.95	9.12[a]	18.13[b]	23.58	8.45	0.30
Grass	0.19[a]	2.11	1.16[b]	35.95[a]	29.32	6.30	0.31
Maize	0.15[b]	1.79	5.06[ab]	26.91[ab]	24.14	8.24	0.31

Mean values within a column followed by the same letter(s) are not significantly different at $p \leq 0.05$.

This could be due to the application of house refuse, which also increased total N content of enset farms. Zeleke et al. (2004) reported an increase in OC by 11 and 67% to incorporation of crop residues in Humbo and Alaba.

The mean available P content was significantly ($P \leq 0.05$) different among the land use systems. In all topographic positions and both depths, highest value of available P was found under enset farms followed by maize and grassland soils (Table 2). The higher available P content at both depths under enset is likely the consequence of long-term manure and house refuse applications and the associated increase in microbial activity. Materechera and Mkhabela (2001) have also reported that Organic matter influence P in soil solution by complexing P from adsorption site in ligand exchange and increase the mobility of inorganic P, particularly in acid soils, by decreasing chemical activity of iron and aluminum. The results were also in agreement with that of Boke (2004) who found high available P under enset in soils of Kokate and Adilo and concluded that transformation of organic P to available P through mineralization, addition of manure and crop residue to enset crop may have coated the reaction surfaces of the soil particles and prevent or delayed P sorption, and thereby increased P solubility. Relatively higher content of available P found under maize farms than that of grass land soils could be due to the continuous application of phosphorus fertilizer as was also reported by van der Eijk et al. (2006).

Available P content in all land use types was found to be higher than that of the surface horizons of the respective pedons (Table 2). This could be due to the application of manure, P fertilizers in the case of enset and maize farms, respectively. The difference between available P of grassland soils and surface soil pedons might be due to high number of samples taken from surface as compared to the single point sampling in the case of the pedons and depth of sampling.

Higher total nitrogen (TN) was observed in grassland fields followed by that of enset at both depths. This could be related to the higher organic matter content in the soils of grassland. There was also significant correlation (r=0.94***) and (r=0.50*) between organic carbon and total nitrogen at 0 to 15 cm and 15 to 30 cm depth, respectively. This shows that the contribution of OC to total N is high (Urioste et al, 2006). The results of the present study were in contrast to that of Boke (2004) who obtained relatively higher content of both organic carbon and total nitrogen from enset land use systems than that of the grasslands. The author attributed the higher concentration of total nitrogen under enset land use systems to the higher organic matter content as a result of manure addition.

The values of exchangeable Na were found to be highest under enset followed by grass in 0-15 cm depth (Table 3), while in 15 to 30 cm depth higher available Na was recorded in grass land soils followed by that of maize fields. Although there were differences in available Na concentration of the soils of the different land use systems, their ESP values were below the critical level (15%). For the three land uses and both depths the concentration of exchangeable potassium (K) followed trend of being enset field > grass land >maize farms. High exchangeable K in 0 to 15 cm under enset fields is likely the result of the addition of house refuse and wood ash supply, whereas low exchangeable K concentration in maize farms could be due to the effect of continuous cultivation and crop removal. This result is supported by previous findings that indicate intensity of weathering, cultivation and use of acid forming inorganic fertilizers affect the distribution of K in the soil system and enhance its depletion (Malo et al., 2005).

The exchangeable calcium (Ca) was highest under enset field (Table 3), whereas lower concentrations of exchangeable Ca were found under maize and grass in 0 to 15 cm and 15 to 30 cm depths, respectively. The exchangeable magnesium (Mg) concentrations followed similar trend as that of Ca under different land use systems. Higher and lower values of exchangeable Mg were found under enset and maize fields, respectively. The low exchangeable Ca and Mg observed under maize

Table 3. Some chemical properties of the soil at 0-15 and 15-30 cm depth under different land use systems.

Land use	Exchangeable cations				CEC	PBS
	Na	K	Ca	Mg		
	Cmol(+/kg)					%
			0-15 cm			
Enset	0.173	3.16	13.82	3.12a	23.73	84.66
Grass	0.133	2.35	10.56	3.33a	27.53	65.66
Maize	0.093	1.93	10.09	2.22b	21.80	59.00
			15-30 cm			
Enset	0.11	2.66	11.07	2.68	23.20	71.33[a]
Grass	0.18	1.58	8.10	2.61	24.86	50.0[b]
Maize	0.16	1.45	9.70	2.42	21.33	64.66[a]

Means value within a column followed by the same letter(s) are not significantly different at $p \leq 0.05$

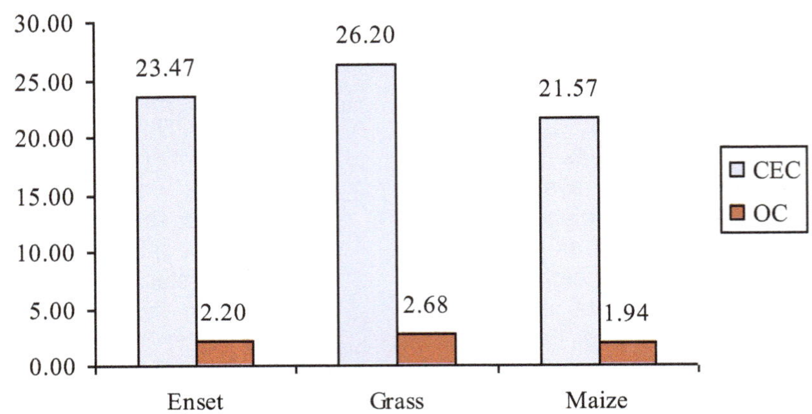

Figure 2. Organic carbon and CEC contents of soil under different land use systems

farms might be due to leaching, soil erosion and crop harvest, as was also reported by Negassa (2001).The percentage of base saturation (PBS) of soils under different land use systems showed significant differences. In 0 to 15 cm depth, the highest PBS was recorded under enset fields (84.66%) followed by grassland (65.66%), whereas in 15 to 30 cm depth, the highest value was found under enset (71.33%) followed by that of maize (64.66%) (Table 3). The highest PBS found under enset field at both depths indicates that the fertility status of enset field is higher as compared to that of the other land use systems. According to Urioste et al (2006), addition of organic matter increases the amount of exchangeable bases. Moreover, intensive cultivation and continuous use of inorganic fertilizers in the cultivated fields that enhance loss of base cations through leaching, erosion and crop harvest (Negassa, 2001).

Relatively, highest cation exchange capacity (CEC) values were observed under grassland (27.53 cmol (+)/kg) followed by that of enset (23.73 cmol (+)/kg) at both sampling depth (Table 3). In accordance with the organic carbon content, CEC values of the soil decreased consistently from grassland to enset and maize (Figure 2). This was also evident from the positively and highly correlation (r=0.91***) and (r =0.41) of CEC with organic carbon for 0 to 15 and 15 to 30 cm depths, respectively. The depletion of organic carbon as a result of intensive cultivation had, therefore, reduced the CEC of the soils under maize land use. These results were in agreement with pervious findings of Boke (2004) and Negassa (2001).

The micronutrients status of the soils was influenced by different land use systems (Table 2). Significant variations ($p \leq 0.05$) in available iron (Fe) at the 15 to 30 cm depth and manganese in 0 to 15 cm depth were observed among different land use systems. The highest available Fe was measured under enset (58.82 mg/kg) followed by grassland (45.40 mg/kg) at 0 to 15 cm depth. In 15 to 30 cm depth, highest Fe was obtained in grassland soils (35.95 mg/kg) followed by that of maize (26.91 mg/kg). In spite of the significant variation observed, available Fe was in a sufficient level for plant growth under all land use systems based on the Fe rating established by Havlin et al. (1999).

At both depths, available Mn concentrations were higher in grassland soils followed by that of maize fields (Appendix Table 1). According to the nutrient toxicity level suggested by Lindsay and Norvell (1978), the concentration of Mn was in the toxic level in all land use systems, as the concentrations of Mn in both layers were greater than 21.87 mg/kg compared to the critical level of 5 mg/kg. This higher content of Mn could be attributed due to pH of the soil where manganese becomes more available in acidic soils. Highest concentration of available Zinc (Zn) was found on the surface layer (0 to 15 cm depth) of enset field (8.63 mg/kg) followed by grassland soils (8.61 mg/kg), whereas in the lower depth (15 to 30 cm depth), the highest available Zn concentration under enset (8.45 mg/kg) was followed by maize land use (8.24 mg/kg). The concentrations of available Zn in all land use systems were within adequate level as indicated by Havlin et al. (1999). In the 0 to 15 cm depth the concentration of available Zn in the maize farms was the lowest as compared to the other land use systems. According to Negassa (2001), low Zn concentration in farm fields might be due to continuous harvesting of crop, organic matter oxidation, removal of the topsoil by sheet and rill erosion that is aggravated by tillage activities. Available cupper (Cu) values were not affected by the different land use systems. In 0 to 15 cm depth, however relatively highest available Cu content was observed in grassland soils (Appendix Table 2). This could be due to the relation of copper with organic carbon. According to Havlin et al. (1999) the concentrations of available Cu in all land use systems were in deficient range, except for grassland soils in depth of 0 to 15, where concentration falls under medium level.

Conclusion

In the southern region of Ethiopia, the growing population and more intensive land use have resulted in sharp declines of soil fertility by continuous removal of grain and vegetation, and erosion. Characterizing the spatial variability of soil nutrients in relation to site properties such as climate, land use, topography and other variables is important for understanding how the ecosystem works and assessing the effects of further land use change on soil properties (Wang et al., 2000). Findings of this study suggest that many soil properties are influenced by land use. At both depths, higher values of soil pH, EC, Av. P, Zn, K, Ca and PBS were obtained under enset fields as compared to the grassland and maize fields. Under maize fields the contents of Av. K, Mg, CEC, PBS, total N and OM were significantly lowest. These effects are attributed to the higher organic matter management in the enset fields and crop removal coupled with frequent disturbance of soil due to tillage practice under maize fields. The information generated from the present study will assist in developing sustainable and ecologically stable land use management strategies for the study area. Moreover, organizations, people living in the study area and others who intend to invest, transfer or introduce new agricultural technologies in the area will benefit from the study.

REFERENCES

Boke S (2004). Soil phosphorus fractions as influenced by different cropping systems in Andosols and Nitosols in Kembata-Tembaro and Wolayta Zones, SNNPRS. MSc Thesis. Alemaya University, (AU). P. 131.

Brady NC, Weil RR (2002). The nature and properties of soils. 13th ed. The Iowa state PVT. Ltd., India.

Celik I (2005). Land-use effects on organic matter and physical properties of soil in a southern Mediterranean highland of Turkey. Soil Tillage Res. 83:270–277.

Ethiopian Mapping Authority (1998). National Atlas of Ethiopia, Addis Abeba, Ethiopia.

Fu B, Chen L, Ma K, Zhou H, Wang J (2000). The relationships between land use and soil conditions in the hilly area of the loess plateau in northern Shaanxi, China. Catena 39:69-78.

Havlin JL, Beaton JD, Tisdale SL, Nilson WL (1999). Soil fertility and fertilizers: an introduction to nutrient management. 6th ed. Prentice Hall. Upper Saddle River, New Jersey. P. 499.

Jackson ML (1958). Soil Chemical Analysis. Prentice Hall, Inc., Engle Wood Cliffs. New Jersey. pp. 183-204.

Lindsay WL, Norvell WA (1978). Development of a DTPA soil test for zinc, iron, manganese, and copper. Soil Sci. Soc. Am. J. 42:421-428.

Liu XL, He YQ, Zhang HL, Schroder JK, Li CL, Zhou J, Zhang ZY (2010). Impact of land use and soil fertility on distributions of soil aggregate fractions and some nutrients. Pedosphere **20**(5):666–673.

Malo DD, Schumacher TE, Doolittle JJ (2005). Long-term cultivation impacts on selected soil properties in the northern Great Plains. Soil. Tillage Res. 81:277–291

Materechera SA, Mkhabela TS (2001). Influence of land-use on properties of a ferralitic soil under low external input farming in southern Swaziland. Soil Tillage Res. 62:15-25.

Negassa W (2001). Assessment of Important Physicochemical Properties of Nitosols under Different Management Systems in Bako Area, Western Ethiopia. M.Sc. Thesis, Alemaya University, Alemaya. p. 109.

Neris J, Jiménez C, Fuentes J, Morillas G, Tejedor M (2012). Vegetation and land-use effects on soil properties and water infiltration of Andisols in Tenerife (Canary Islands, Spain). Catena 98:55–62.

Olsen SR, Cole CV, Watanabe FS, Dean LA (1954). Estimation of available phosphorus in soils by extraction with sodium bicarbonate. USDA Circular 939:1-19.

SAS (1997). SAS Institute Inc., Cary. NC. USA.

Urioste AM, Hevia a GG, Hepper EN. Anton LE, Bono AA, Buschiazzo DE (2006). Cultivation effects on the distribution of organic carbon, total nitrogen and phosphorus in soils of the semiarid region of Argentinian Pampas. Geoderma 136:621–630.

van der Eijk D, Janssen BH, Oenema O (2006) Initial and residual effects of fertilizer phosphorus on soil phosphorus and maize yields on phosphorus fixing soils A case study in south-west Kenya. Agric. Ecosys. Environ. 116:104–120

Wakene Negassa, (2001). Assessment of Important Physicochemical Properties of Nitosols under Different Management Systems in Bako Area, Western Ethiopia. M.Sc. Thesis, Alemaya University, Alemaya. pp 109.

Walkley A, Black CA (1934). An examination of Degtjareff method for determining soil organic matter and proposed modification of the proposed modification of the chromic acid titration method. Soil Sci, 37:29-38.

Wang J, Fu B, Qiu Y, Chen L (2001). Soil nutrients in relation to land use and landscape position in the semi-arid small catchment on the loess plateau in China J. Arid. Environ. 48:537–550.

Zeleke TB, Grevers MCJ, Si BC, Mermut AR, Beyene S (2004). Effect of residue incorporation on physical properties of the surface soi1 in the South Central Rift Valley of Ethiopia. Soil Till.Res. 77:35-46.

APPENDIX

Table 1. Correlation matrix of soil properties under different land uses of 0 to 15 cm depth at Delbo Watershed, 2006.

	pH	Clay	Na	K	Ca	Mg	CEC	TN	OC	Av. P	Fe	Mn	Zn	Cu
pH	1.0	-0.20	0.29	0.79*	0.85**	0.14	0.02	-0.28	-0.22	0.89**	0.02	-0.78*	0.10	-0.21
Clay		1.0	-0.16	-0.66*	-0.45	-0.49	-0.75*	-0.77*	-0.71*	0.05	0.073	-0.18	-0.70*	-0.34
Na			1.00	0.45	0.06	0.49	0.08	0.01	0.19	0.41	0.14	-0.31	0.33	-0.42
K				1.0	0.82**	0.42	0.49	0.24	0.30	0.57	0.07	-0.47	0.45	-0.10
Ca					1.0	0.33	0.45	0.12	0.16	0.55	0.24	-0.46	0.11	-0.09
Mg						1.0	0.75*	0.51	0.73*	-0.01	0.35	0.24	0.14	0.11
CEC							1.0	0.83**	0.91***	-0.33	0.23	0.44	0.24	0.20
TN								1.0	0.94***	-0.52	0.24	0.68*	0.58	0.36
OC									1.0	-0.46	0.32	0.64	0.43	0.25
Av. P										1.0	-0.01	-0.83**	0.13	-0.18
Fe											1.0	0.10	0.06	-0.19
Mn												1.0	0.11	0.54
Zn													1.0	0.31
Cu														1.0

*Significant at the < 0.05, ** < 0.01, and ***< 0.001 levels.

Table 2. Correlation matrix of soil properties under different land uses of 15 to 30 cm depth at Delbo Watershed, 2006.

	pH	Clay	Na	K	Ca	Mg	CEC	TN	OC	Av. P	Fe	Mn	Zn	Cu
pH	1.0	-0.62	0.002	0.80**	0.75*	0.25	-0.04	-0.57	-0.24	0.81**	-0.37	-0.84**	0.22	-0.21
Clay		1.0	0.37	-0.60	-0.42	-0.11	0.23	0.48	0.78*	-0.47	0.13	0.84**	-0.72	0.11
Na			1.0	0.10	-0.21	-0.45	0.19	-0.14	0.42	-0.39	-0.43	-0.03	-0.63	-0.59
K				1.0	0.47	0.30	0.17	-0.17	-0.04	0.60	-0.44	-0.73*	0.05	-0.34
Ca					1.0	0.03	0.17	-0.43	-0.13	0.73*	-0.43	-0.50	0.22	-0.22
Mg						1.0	0.21	0.17	0.10	0.39	0.53	-0.05	-0.06	0.52
CEC							1.0	0.33	0.41	-0.19	0.11	0.25	-0.47	-0.17
TN								1.0	0.50	-0.28	0.19	0.74*	-0.32	0.17
OC									1.0	-0.11	-0.21	0.55	-0.82**	-0.08
Av. P										1.0	-0.23	-0.54	0.38	0.20
Fe											1.0	0.35	0.18	0.78*
Mn												1.0	-0.41	0.30
Zn													1.0	0.38
Cu														1.0

*Significant at the < 0.05, ** < 0.01, and ***< 0.001 levels.

Re-defining and quantifying inorganic phosphate pools in the soil and water assessment tool

Richard L. Haney[1], Michael J. White[1], Jeffrey G. Arnold[1], Elizabeth B. Haney[2] and R. Daren Harmel[1]

[1]USDA-ARS, Grassland, Soil and Water Research Laboratory, 808 East Blackland Road, Temple, TX 76502, USA.
[2]Texas A&M University, Texas AgriLife Research and Extension Center, 720 E. Blackland Road, Temple, TX 76502, USA.

The soil and water assessment tool (SWAT), a large-scale hydrologic model, is used to estimate phosphate (P) loading in streams and water bodies. The labile, active, and stable P pools are currently used to represent P cycling in SWAT; however, these pools are conceptual without any chemical basis. The current structure allows SWAT to reasonably represent P cycling; however, restructuring and incorporation of recent research results should produce more accurate P simulations. This paper presents a redefined SWAT structure using four inorganic soil P pools (water soluble, labile, active, and stable) combined with soil extraction methods to determine initial values for each pool. The redefined structure was selected by examining the chemical characteristics of laboratory tests relative to soil P pools and analyzing the relationships between test results. Water soluble P, labile P and active P are determined using water, H3A, Mehlich 3 extraction, respectively. Stable P is determined with an acid digestion. Redefining SWAT's inorganic soil P pools based on soil and extractant chemistry will improve the defensibility, credibility, and the accuracy of SWAT P routines in water resource planning, management, and decision making.

Key words: Soil and water assessment tool (SWAT), phosphate, soil extractants, H3A, Olsen, Mehlich 3.

INTRODUCTION

Phosphate (P) is an essential nutrient for plant growth and is applied to agricultural and urban lands in inorganic (fertilizer) and organic forms (manure). Excessive application of P to the landscape increases P losses to surface waters, which may contribute to eutrophication of fresh water systems and coastal estuaries (Sharpley et al., 1994). Intensive cropping systems can produce soils with elevated nutrient contents, and the associated tillage increases sediment losses from these enriched fields and streams banks to lakes and streams.

While the application of inorganic P is relatively uniform on a regional basis on urban and agricultural lands, manure produced by animal feeding operations tends to be a regional issue. Animal feeding operations tend to cluster around existing animal support and processing infrastructures. Intensive manure applications in these areas often exceed the capacity of the soils to retain nutrients. Repeated applications of excess manure can increase P losses in runoff (Sharpley et al., 2001; Pierson et al., 2001) to stream and river waters, with significant consequences for downstream water quality.

Phosphate concentration in surface water may be tied

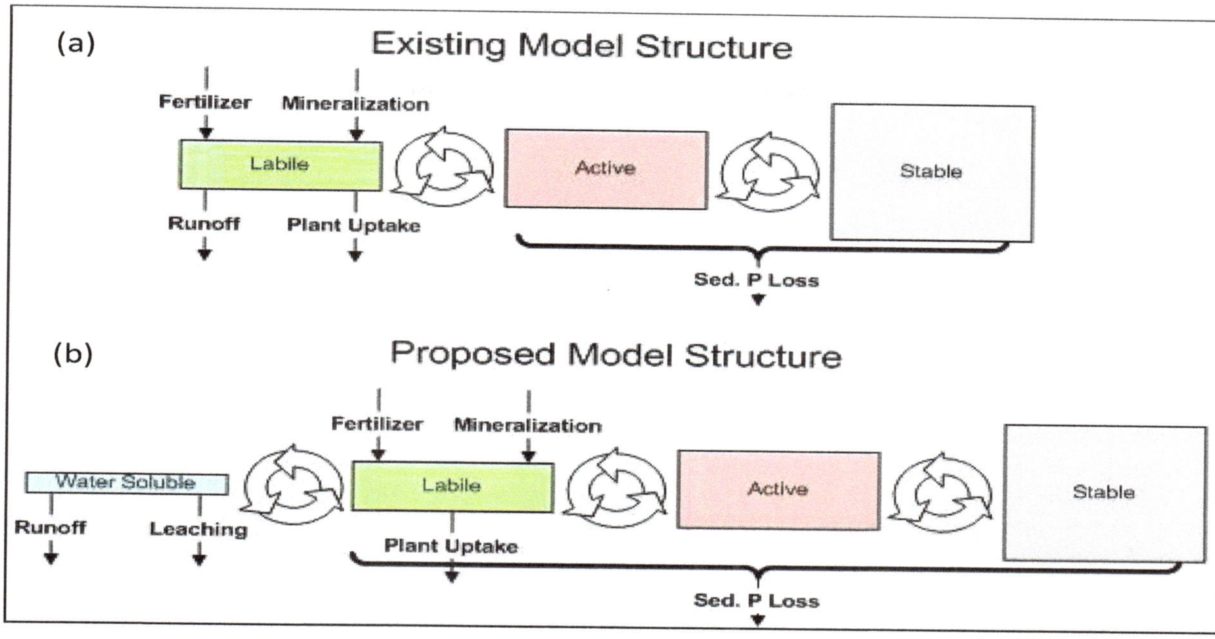

Figure 1. Soil inorganic phosphorus pools and processes that move P in and out of pools: a) as currently defined in SWAT (Neitsch et al. 2005), and b) as redfined.

directly to water quality degradation. Total P concentrations of 50 µg L^{-1} are sufficient to sustain chlorophyll concentrations of 10 µg L^{-1}, the lower boundary for a trophic state in temperate streams (Dodds et al., 1998) and a eutrophic state in lakes (Smith, 2003). Using the Redfield (1958) ratio P concentrations exceeding 65 µg L^{-1} are sufficient to support algal growth in many water bodies when nitrate-N concentrations exceed 10 mg L^{-1}. It has been shown that these concentrations of total P can be from ground water in intensive agricultural regions (Burkart et al., 2004). Dissolved P concentrations from compiled peer-reviewed nutrient export studies on agricultural lands averaged 1.38 mg L^{-1}; however, the median concentration was 0.40 mg L^{-1} since most concentrations were well below the average value (Harmel et al., 2008).

Tools to estimate P runoff for assessment of conservation effects or land management planning range from simple export coefficients (Reckhow et al., 1980) to process based models such as the soil water assessment tool (SWAT) (Arnold et al., 1998) and PPM Plus (White et al., 2010). SWAT is a river basin or watershed scale model developed to predict the impact of land management practices on water, sediment, and agricultural chemical yields over time (Neitsch et al., 2005). SWAT predicts the impact of topography, soils, land use, management, and weather on water, sediment, nutrient (N and P), and agricultural chemical yields from the landscape at a variety of spatial scales (Gassman et al., 2007). SWAT simulates the fate and transport of water, sediment, nutrients, and pesticides from upland areas by routing these constituents through a channel network where they are subjected to various in stream processes.

SWAT has been extensively used to study land management impacts on water quantity and quality. For example, SWAT has been used in to evaluate precision feed management programs for controlling P losses and soil-P build-up at field and watershed scales (Ghebremichael et al., 2008), to quantify P sources throughout the Rock River basin in Wisconsin and quantify impacts from the application of best managements practices (Kirsch et al., 2002), and to evaluate best management practices related to dairy manure management (Santhi et al., 2001).SWAT also has been integrated into USEPA's modeling framework, Better Assessment Science Integrating Point and Nonpoint Sources (BASINS).

The current version of SWAT (SWAT, 2009) contains three inorganic soil P pools (labile, active, and stable) as shown in Figure 1a. SWAT contains a full P balance; P may be added to the soil by fertilizer, manure, or crop residues, and lost from the soil through plant uptake and runoff (sediment bound and soluble P). In addition, organic P forms transform into inorganic P forms through mineralization. In SWAT and as in natural soils, most of the inorganic P resides in the stable pool. In SWAT, inorganic P in the labile pool is assumed to be in rapid equilibrium with the active pool, while the active pool is in slow equilibrium with the stable pool. Users may define the initial labile P value for various soils. If the user does not input initial labile P, labile P in all soil layers is initially set to 5 mg P kg^{-1}. Active and stable inorganic P pools are not input by the user but are initially set using the following equations:

$$Active\ P = Labile\ P \times \frac{1-psp}{psp} \tag{1}$$

$$Stabile\ \ P = 4 \times Active\ P \tag{2}$$

Where *psp* is the phosphorus availability index (PAI), which can be user input (default value is 0.4).

A PAI of 0.4 indicates that 40% of P added to the soil remains as labile P and 60% becomes active P (Vadas and White, 2010). For user input, the PAI may be determined experimentally using a 6 month long incubation study (Vadas and White, 2010) or estimated from soil texture, pH, and base saturation (Sharpley et al., 1984, 1989). Total P is defined in SWAT as the sum of the labile, active, and stable inorganic P values plus the fresh and humic organic P pools.

SWAT's soil P routines were originally developed for the EPIC model (Jones et al., 1984), where labile P is synonymous with solution P and was quantified through extraction with anion exchange resin (Sharpley et al., 1984). Vadas and White (2010) indicate that any P extraction method that is comparable to the anion exchange resin can be used to quantify solution P and that P extraction results from routine soil tests, such as Mehlich 3, Bray 1, or Olsen, are sometimes well related to anion exchange resin results. Therefore, it is possible to use routine soil test P data to initialize solution P. Since the default values for the active, stable, and labile P pools are arbitrary and because guidance is not provided on methods for estimating reasonable initial input values (Vadas and White, 2010), the initial P pool values may not reflect actual soil conditions. It is more accurate for users to input the initial labile P concentration (Vadas and White, 2010), and the same should be true for the active and stable P pools as well. Therefore, our research objectives were to develop a new structure for inorganic soil P pools in SWAT and to identify readily available laboratory methods to determine initial values for each pool. This structure redefines inorganic soil P pools from conceptual to chemically based and allows site-specific initial value determination; both of which will improve the accuracy and defensibility of SWAT P simulation.

MATERIALS AND METHODS

The methods used in this study consist of two major steps. First, soil samples were analyzed with numerous common soil extractants to determine soil P concentrations. Then, results from the various extractants were all compared to each other to examine potential correlation, and the chemical "appropriateness" for each extractant in terms of representing soil P pools was examined.

Soil testing

Sixty-two soils of varying type from the North American Proficiency Testing (NAPT) program were used to evaluate various soil test methods for determining initial values for soil P pools. NAPT is a voluntary proficiency testing program of the Soil Science Society of America for laboratories throughout the country. Soil samples are sent to laboratories quarterly for testing, and laboratory results are statistically analyzed. Samples are also available for sale for use in research and as laboratory standards. Analytical data from each quarter are available at www.naptprogram.org. Seven of the 62 soils used in this study were from 2003 (laboratory identification numbers 103, 105, 112, 113, 114, 117, and 119). The remaining samples were from 2005 (laboratory identification numbers 101-104 and 106-120), 2006 (laboratory identification numbers 101-114 and 116-120), and 2007 (laboratory identification numbers 101, 103-104, 106-107, and 109-120). Each soil sample received from the NAPT program were previously dried and ground to 2 mm. NAPT reports the following extractable P results: water; Olsen et al. (1954); Bray (1:7); Bray (1:10), Bray and Kurtzb (1945); AB-DTPA, Modified Morgan, Modified Kewlona, Strong Bray, Mehlich 1, and Mehlich 3 (Mehlich, 1984).

In our laboratory, the samples were extracted with a 10:1 extractant to soil ratio with H3A (Haney et al., 2010), shaken for 10 min at 180 rev/min, centrifuged at 3500 rpm, and filtered through Whatman 2V filter paper. Filtered extracts were then analyzed for PO_4-P using a Rapid Flow Analyzer (Flow Solution IV, OI Analytical; College Station, TX). For quality control, twenty soil samples were selected randomly and extracted using Mehlich 3, Olsen, Bray (1:7), and water for P determination. The results were compared with the NAPT results for the same samples, and no significant differences were detected (data not shown).

Selecting extractants to represent inorganic P pools

The results from the various soil extractants were examined in terms of their chemical "appropriateness." In other words, extractants were examined based on their design intent and more importantly their chemical composition to determine whether they could realistically represent the inorganic soil P pools. Since water is the universal extractant and the key to natural rainfall, runoff, and leaching processes, the examination started with water. All possible combinations of soil P data from the ten extractants used in NAPT were compared to determine significant correlations (α = 0.05). Specifically, correlation between paired extractant results indicated that both extractants measured some portion of the same soil P pool or at least that a consistent relationship existed between the two.

RESULTS AND DISCUSSION

Soil testing

Results of soil tests conducted by NAPT and by our laboratory are shown in Table 1. As shown, soils varied widely in terms of relevant physical and chemical properties, particularly pH, organic C, % clay, and extractable P; therefore, the redefined pool structure should be applicable for a majority of soil types.

Selecting extractants to represent inorganic P pools

When the correlation between soil P levels for the 10 extractants was examined, it was interesting that the correlations were much better for the commonly applied

Table 1. Descriptive statistics of relevant soil properties, particularly extractable P, for the 62 samples used in this study.

Soil property	Max	Min	Median	25%	75%
pH (1:1)w	8.4	4.3	7.3	6.0	7.9
Total organic C (combustion)	7.510	0.378	1.795	1.020	2.690
Clay (0-2 um)	33	2	17	9	23
Extractable P (mg P kg^{-1} soil)					
Water	20.3	0.2	2.9	1.9	6.6
H3A	135.0	2.0	23.7	9.2	40.7
Mehlich 3	223.0	4.0	50.9	30.5	100.5
Olsen	94.7	1.0	23.7	10.8	38.4
Bray (1:7)	165.9	1.0	37.0	15.0	65.0

Table 2. Correlation coefficients comparing water, Bray, Olsen, Mehlich 3, and H3A extractable P (mg P kg^{-1} soil). All correlations were significant ($P \leq 0.05$; $n = 62$).

Parameter	Water	Bray	Olsen	Mehlich 3	H3A
Water	.	0.46	0.68	0.66	0.84
Bray	.	.	0.82	0.90	0.74
Olsen	.	.	.	0.90	0.86
Mehlich 3	.	.	.	.	0.88
H3A	.	.	.	.	.

extractants (Mehlich 3, Bray 1, Olsen, and H3A). In fact, the correlation between each of these extractants was significant at α = 0.05 (Table 2). These results indirectly support the appropriateness and the common use of Mehlich 3, Bray 1, Olsen, and H3A extractants.

Water soluble pool

In redefining the inorganic soil P pools, it was important to start with water extraction (Figure 2) because dissolution and transport of soil P by rainfall, leaching, and runoff is a fundamental natural process and a fundamental modeling need in SWAT. As such, it is more appropriate to define water soluble P as its own pool, which is fed by the labile pool and diminished by runoff and leaching (Figure 1b), instead of simply including it as a component of labile P (Figure 1a). In addition, anion exchange can overestimate this value relative to water since loosely bound water extractable P is generally a very small fraction of the total P in soils or sediments (Kuo, 1996).

Labile pool

The labile pool, which includes the water soluble pool, is fed by fertilizer addition and mineralization of organic P and diminished by plant uptake, represents the readily plant available pool in SWAT (Figure 1b). The H3A extractant (Haney et al., 2010) was chosen to quantify the labile pool (Figure 2) because H3A extractable P was more strongly correlated with water extractable P than any of the other extractants (Figure 3, Table 2) and because of the chemical nature and design of H3A. For soil tests to accurately represent P availability, they should respond to soil characteristics in manner similar to that of plants (Kuo, 1996); therefore, H3A utilizes organic acids that mimic plant root exudates in the extraction of soil nutrients (Haney et al., 2010). H3A extracts P at or near soil pH which is important in simulating natural soil conditions since pH and P solubility are highly correlated (Golterman, 1998; Sharpley, 1993). H3A extractable P is comparable to resin exchangeable P upon which the original EPIC labile pool was based, yet is a much simpler laboratory analysis (unpublished results). In addition, H3A extractable P was strongly correlated with Olsen, Bray, and Mehlich 3 extractable P (Table 2), which are known to be highly correlated with resin-extractable P.

Active pool

Whereas the water soluble P pool (runoff, leaching) and labile P pool (plant available) are easily visualized, the

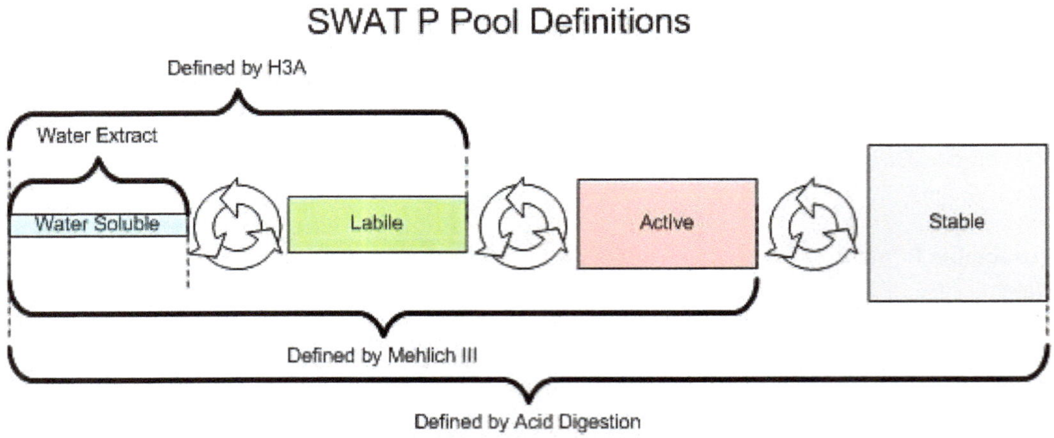

Figure 2. Redefined SWAT inorganic P pools presented with the soil test method used to quantify initial values for each pool.

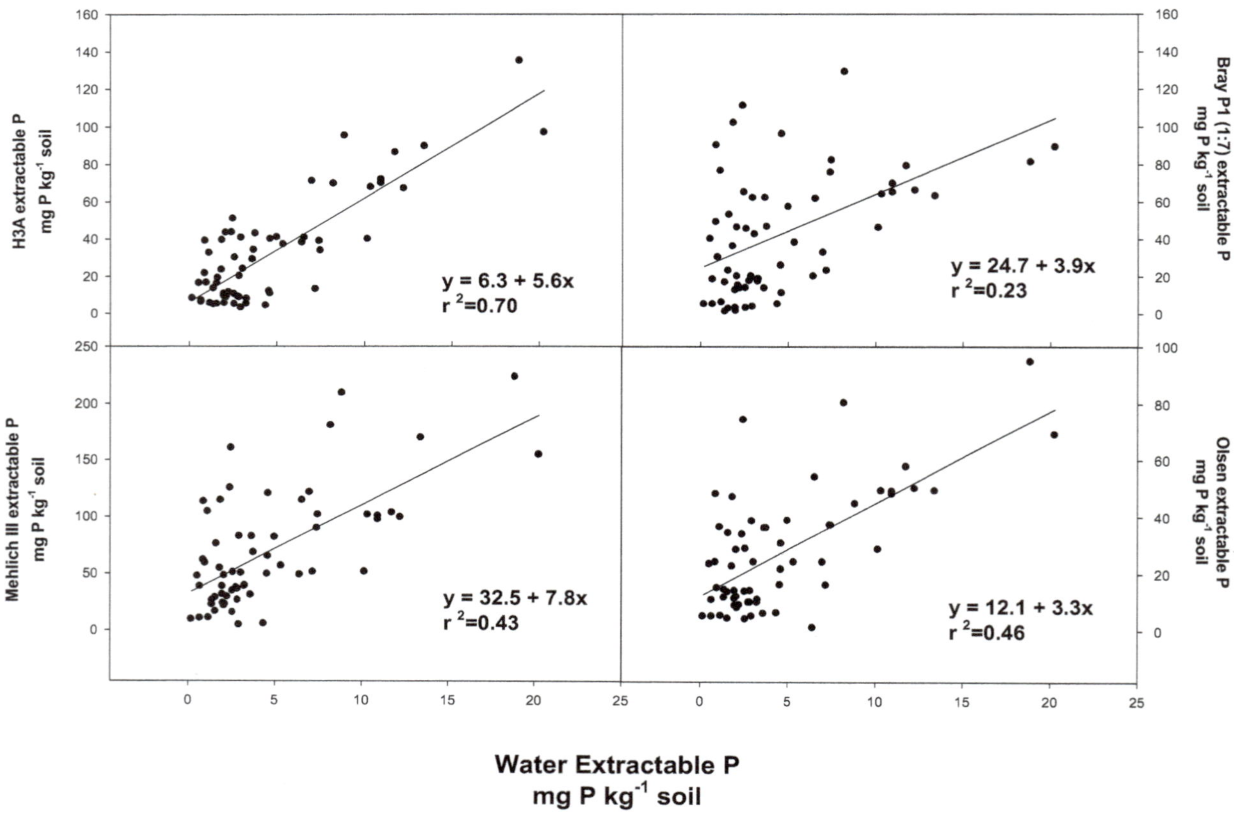

Figure 3. Relationship between water extractable P and Olsen, Bray, Mehlich 3, and H3A extractable P.

active pool is a bit more nebulous (Figure 1b). The active pool is in essence a reserve P supply pool that is relatively stable and composed of slowly desorbable P (Vadas et al., 2006).

Mehlich 3 was chosen to quantify the active pool (Figure 2) because H3A and Mehlich 3 extractable P are strongly correlated (r = 0.88, Table 2) and Mehlich 3 extracted more P than H3A (Figure 4). Mehlich 3 extracted more P than the other extractants studied (Table 1) because it dissolves iron (Fe-), aluminum (Al-) and calcium (Ca-) phosphates (Nelson et al., 1953) with its strongly acidic solution (buffered to pH 2.6). The Fe-, Al-, and Ca- phosphates dissolved by Mehlich 3 extraction are relatively insoluble P forms in the soil

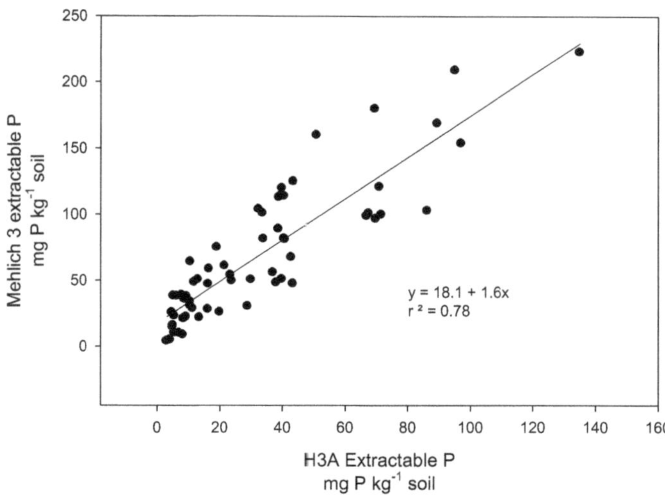

Figure 4. Relationship between H3A and Mehlich 3 extractable P.

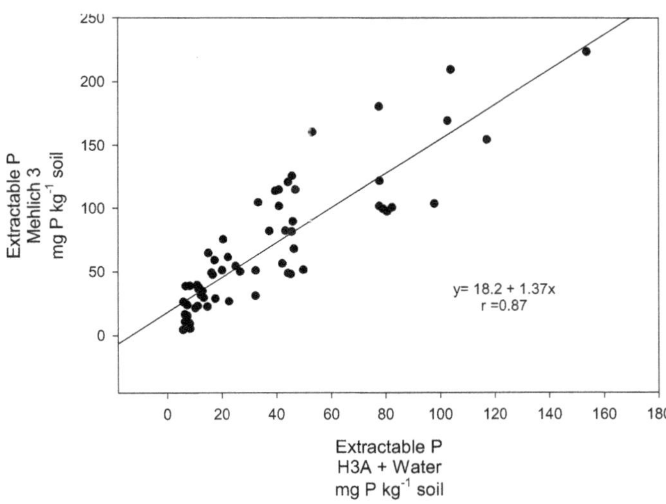

Figure 5. Relationship between Mehlich 3 and H3A plus water extractable P.

(Ketterings et al., 2002), which are more representative of relatively slow reacting pools than labile or water soluble P pools. Mehlich 3 is poorly correlated with water extractable P; however, the relationship is significantly improved when water and H3A extractable P are summed (Figure 5), which further supports the redefined pool structure.

Stable pool

Similar to active P, stable P is more difficult to visualize than water soluble and labile P (Figure 1b). P in the stable pool is strongly adhered to mineral particles or occurs in low solubility particles; therefore, a small portion of P in the stable pool becomes available at an extremely slow rate. Quantifying this soil P pool is best accomplished by determining the total soil P through acid digestion (Figure 2).

Redefined inorganic soil p pool structure

The correlation between and increasing magnitude of water, H3A, and Mehlich 3 extractable P (Tables 1 and 2), along with the design intent and chemical basis of these extractants, provided support for a redefined inorganic soil P structure in SWAT. The correlation between water and H3A extractable P (r=0.84) and between H3A and Mehlich 3 extractable P (r=0.88) indicates that the extractants measured some portion of the same soil P pool or at least that consistent relationships exist between the pools. Although many fractionation processes exist for determining the inorganic P status and availability in soil, the alternatives are time consuming and not practical for most SWAT users; therefore, the soil test methods listed in Table 3 corresponding to each pool are recommended.

The soil P pools as shown in Figures 1b and 3 are subset of the next larger set (Table 3). For the soils analyzed in this dataset, the water soluble pool comprised 14% of the labile pool (H3A extractable) which comprised 47% of the active pool (Mehlich 3 extractable). Similarly, the active pool comprised 14% of the total P pool. These defined pools and relative sizes provide an initial basis for estimating the relative speed of transfer between the soil P pools. Qualitatively, when water soluble pool is diminished by runoff or leaching, it is fed by the labile P pool. Similarly, when the labile pool is diminished by plant uptake, it is rapidly fed by fertilizer addition and more slowly fed by mineralization and the active pool, which in turn is even more slowly replenished by the stable pool.

Incorporating this redefined inorganic soil P structure into SWAT will require a relatively simple modification of the code. The model is not dissimilar to the current approach. To fully realize an improved mineral soil P model, additional research will be needed to further refine transformation rates between the proposed pools. The transformation rates between these pools dictate how long it takes applied P to move from highly available pools (water, labile) to less available forms (active, stable). The definition of these pools by soil extractant allows these rate coefficients to be estimated from laboratory data. A soil sample can be spiked with mineral P and tested over a period of time to quantify P movement from pool to pool. With a sufficient sample population, pool to pool rate transformation coefficients can be estimated. Both the relative pool sizes derived from this research and transformation coefficients are needed to accurately simulate the fate of P fertilizer applied to soils.

Table 3. Mean and standard deviation of soil P levels averaged across all soils.

Soil P test	Inorganic P pool	P concentration (mg P kg^{-1} soil)	Percentage of next higher pool
Water	Water soluble	4.6 (4.3)	14% of the labile pool
H3A	Labile	32.3 (28.6)	47% of the active pool
Mehlich 3	Active	69.2 (50.8)	14% of the stable pool
Digestion	Stable	505.9 (451.0)	

Conclusions

Redefining the inorganic soil P pool structure in SWAT will allow users to rapidly and accurately estimate initial values for each pool with commonly applied soil test methods. The new soil P pool structure defines water soluble P as water extractable P, labile P (plant available) as H3A extractable, and active P (reserve supply pool) as Mehlich 3 extractable. The stable P pool, a majority of which is only very slowly transformed to active P, is determined by acid digestion. This redefined soil P pool structure based on commonly used soil test methods has several advantages. SWAT users will be able to rapidly and inexpensively determine initial values and relative magnitude of soil P pools from soil samples collected from watersheds of interest instead of using SWAT default ratios. This will produce more accurate soil P values, which will directly improve simulation of soluble P in runoff and other P processes, since calculations are directly dependent on soil P pool values. Further improvement in P simulation will also result from more accurate estimation of parameters (e.g. phosphate soil partitioning coefficient (Kd)) that are affected by soil P values. The improved simulations will be especially valuable in P threatened or impaired watersheds. Another benefit is the ability to better characterize P transformation rates between soil pools in quasi-equilibrium, which is a critical aspect of the model (as important as the relative magnitudes of the pools themselves). With soil pools defined with available laboratory tests, the transformation rates can be evaluated under differing environmental conditions and pool imbalances due to fertilization and crop uptake. By testing soils over time, the dynamics between the pools can more readily be identified. Finally, the redefined inorganic soil P pool structure will improve the defensibility of model predictions, since soil P pools will be definable and reproducible and not simply conceptual.

REFERENCES

Arnold JG, Srinivasan R, Muttiah S, Williams JR (1998). Large area hydrologic modeling and assessment. Part I: Model development. J. Am. Water Resources Assoc. 34(1):73-89.

Bray RH, Kurtz LT (1945). Determination of total, organic and available form of phosphorus in soil. Soil Soc. 59:39-45.

Burkart MR, Simpkins WW, Morrow AJ, Gannon JM (2004). Occurrence of total dissolved phosphorus in surficial aquifers and aquitards in Iowa. J. Am. Water Resour. Assoc. 40(3):827-834.

Dodds WK, Jones JR, Welch EB (1998). Suggested classification of stream trophic state: distributions of temperate stream types by chlorophyll, total nitrogen, and phosphorus. Water Resourc. 32(5):1455-1462.

Gassman PW, Reyes MR, Green CH, Arnold JG (2007). The Soil and Water Assessment Tool: Historical development, applications and future research directions. Trans. ASABE. 50(4):1211-1250.

Ghebremichael LT, Veith TL, Hamlett JM, Gburek WJ (2008). Precision feeding and forage management effects on phosphorus loss modeled at a watershed scale. J. Soil Water. Conserv. 63(5):280-291.

Golterman, HL (1998). Reflections on fractionation and bioavailability of sediment bound phosphate. Archive Hydrobiology 30:1-4.

Haney RL, Haney EB, Hossner LR, Arnold JG (2010). Modifications to the new soil extractant H3A-2-1: A multinutrient extractant. Commun. Soil Sci. Plant Analysis. 41(12):1513-1523.

Harmel RD, Qian SS, Reckhow KH, and Casebolt P (2008). The MANAGE database: Nutrient load and site characteristic updates and runoff concentration data. J. Environ. Qual. 37(6):2403-2406.

Jones CA, Cole CV, Sharpley AN, Williams JR (1984). A simplified soil and plant phosphorus model: 1. Documentation. Soil Sci. Soc. Amer. J. 48(4):800-805.

Ketterings QM, Czymmek KJ, Reid WS, Wildman RF (2002). Conversion of modified Morgan and Mehlich 3 soil tests to modified Morgan soil test values. Soil Sci. 167:830-837.

Kirsch K, Kirsch A, Arnold JG (2002). Predicting sediment and phosphorus loads in the Rock River basin using SWAT. Trans. ASAE. 45(6):1757-1769.

Kuo S (1996). Phosphorus. In Methods of Soil Analysis, Part 3 – Chemical Method, ed. Sparks DL, Page AL, Helmke PA, Loeppert RH, Soltanpour PN, Tabatabai MA, Johnston CT, Sumner ME. 869-919. Madison, WI, SSSA.

Mehlich A (1984). Mehlich 3 soil test extractant: A modification of the Mehlich 2 extractant. Commun.Soil Sci. Plant Analysis 15:1409-1416.

Neitsch SL, Arnold JG, Kiniry JR, Williams JR, King KW (2005). Soil and Water Assessment Tool (Version 2000). Theoretical documentation. GSWRL 02-01, BRC 02-05, TR-191. College Station, Texas: Texas Water Resources Institute.

Nelson WL, Mehlich A, Winters E (1953). The development, evaluation, and use of soil tests for phosphorus ability. In Soil and Fertilizer Phosphorus in Crop Nutrition, ed. Pierre WH Norman AG, 153-188. New York: Academic Press.

Olsen SR, Cole CV, Watanabe FS, Dean LA (1954). Estimation of available phosphorus in soils by extraction with sodium bicarbonate (USDA Cir. 939). Washington, D.C.: U.S. Government Printing Office.

Pierson ST, Cabrera ML, Evanylo GK, Schroeder PD, Radcliffe DE, Kuykendall HA, Benson VW, Williams JR, Hoveland CS, McCann MA (2001). Phosphorus losses from grasslands fertilized with broiler litter: EPIC simulations. J. Environ. Qual. 30:1790-1795.

Reckhow KH., Beaulac M, and Simpson J (1980). Modeling phosphorus loading and lake response under uncertainty: A manual and compilation of export coefficients. U.S. Environmental Protection Agency, EPA 440/5-80-011, 214 pg.

Redfield AC (1958). The biological control of chemical factors in the environment. Am. Sci. 46:205-221.

Santhi C, Arnold JG, Williams JR, Hauck LM, Dugas WA (2001). Application of a watershed model to evaluate management effects on point and nonpoint source pollution. Trans. ASAE 44(6):1559-1570.

Sharpley AN (1993) An innovative approach to estimate bioavailable phosphorus in agricultural runoff by Fe oxide-impregnated paper. J. Environ. Qual. 22:597-601.

Sharpley AN, Jones CA, Gray C, Cole CV (1984). A simplified soil and plant phosphorus model: 2. Prediction of labile, organic, and sorbed phosphorus. Soi. Sci. Soc. Amer. J. 48(4):805-809.

Sharpley AN, Singh U, Uehara G, Kimble J (1989). Modeling soil and plant phosphorus dynamics in calcareous and highly weathered soils. Soil Sci. Soc. Amer. J. 53(1):153-158.

Sharpley AN, Chapra SC, Wedepohl R., Sims JT, Daniel TC, Reddy KR (1994). Managing agricultural phosphorus for protection of surface waters: Issues and options. J. Environ. Qual. 23:437-451.

Sharpley AN, McDowell RW, Weld JL, Kleinman PJA (2001). Assessing site vulnerability to phosphorus loss in an agricultural watershed. J. Environ. Qual. 30:2026-2036.

Smith VH (2003). Eutrophication of freshwater and coastal marine ecosystems: A global problem. Environ. Sci. Pollut. Res. 10:1-13.

Vadas PA, White MJ (2010). Validating soil phosphorus routines in the SWAT model. Trans. ASABE. 53(5): 1469-1476.

Vadas PA, Krogstad J, Sharpley AN (2006). Modeling phosphorus transfer between labile and nonlabile soil pools: Updating the EPIC model. Soil Sci. Soc. Am. J. 70:736–743.

White MJ, Storm DE, Busteed PR, Smolen MD, Zhang H, Fox GA (2010). A quantitative phosphorus loss assessment tool for agricultural fields. Environ. Modelling Software, 25(10):1121-1129.

29

Nutrient budget analysis under smallholder farming systems and implications on agricultural sustainability in degraded environments of semiarid central Tanzania

Richard Y. M. Kangalawe

Institute of Resource Assessment, University of Dar es Salaam; P. O. Box 35097, Dar es Salaam, Tanzania.

Soil nutrient budget is one of the critical concerns in most farming systems, especially in degraded environments. This paper reveals a net nutrient removal from the agricultural system that resulted in negative budgets, particularly for N. Large variations existed between villages, with N budget ranging from -9.6 to -252.4 kg/ha/year, with a mean budget of -91.9 kg/ha/year. Various factors seem to have contributed to the negative budgets, including soil erosion, crop harvest and residue removals. The latter two contributed significantly less to nutrient losses compared to soil erosion. The main input sources for nutrients into the system were found to be manure, crop residues, atmospheric deposition and to a limited extent inorganic fertilisers. Manure application added into the soil about 13, 1.5, and 0.04 kg/ha/year of N, P and K respectively. Its use was, however, limited by the inadequate amounts available. Crop residues added about 3.9, 1.9 and 11 kg/ha/year of N, P and K respectively. The use of crop residues for nutrient recycling and soil maintenance is however, constrained by its competitive use for animal feed. Generally, this farming system seems to be characterised by negative nutrient budgets, limiting agricultural productivity and sustainability.

Key words: Nutrient budget, smallholder farming systems, degraded environments, agricultural sustainability, semiarid Tanzania.

INTRODUCTION

Farming in the tropical environments is largely on upland soils under rainfed agriculture. These soils easily degrade when subjected to over-exploitation, for example through cultivation (Kang, 1993). In the past the traditional bush and fallow system with short cropping cycles and long fallow periods allowed for regeneration of adequate soil productivity. The system was of high biological stability and sustained agricultural production for many generations. In recent years various socio-economic factors have caused the fallow periods to be shortened or abolished as observed in many parts of the tropics (Kang, 1993; Mohamed, 1989; Adesina, 1994; Scholes et al., 1994). The shortened fallow periods have resulted in increased soil degradation in farmlands, more weed infestation, declining crop yields, and reduced production of both food and cash crops. Increasing and sustaining food and cash crop production on these inherently erodible and low fertility tropical soils is an elusive goal.

One of the reasons for low agricultural production is the gradual process of soil nutrient depletion in many of the sub-Saharan land-use systems (Smaling, 1993). Among the processes responsible for this nutrient depletion are soil erosion, crop harvesting, residue removals and leaching, phenomena that have received less public attention compared to aspects like erratic rainfall, crop failures, and civil wars which are directly related to famine and starvation (Smaling, 1993). The supply of nutrients dictates the productivity of land and dwindling soil fertility has become one of the major constraints to crop production, especially in sub-Saharan Africa (Smaling, 1993; CWFS, 1985). In most land-use systems, the amount of nutrients leaving the soil in crops, by leaching and erosion, is grossly exceeding the nutrient input to the system by natural processes such as atmospheric deposition and biological nitrogen fixation, or artificially by organic manure and mineral fertilisers (Smaling, 1993; CWFS, 1985). This has resulted into reduced land productivity and agricultural sustainability (Lynam and Herdt, 1989; Spenser and Swift, 1992; Rockström, 1997).

This paper analyses nutrient budgets in the Irangi Hills (Figure 1) located in Kondoa District, which forms an area commonly known as the Kondoa Eroded Area (KEA). It presents findings from three villages namely, Baura, Bolisa and Mafai. The main objective of the study was to determine the nutrient budget for the crop/livestock and soil management systems in the area. Specifically, the objectives were to determine input sources and pathways for nutrient loss, and to compute the difference between nutrient input and output into and from the agricultural system. The KEA is a badly degraded area in the hills where comprehensive land-rehabilitation activities have been tried out since the 1930s, including the eviction of grazing animals implemented in 1979. The studied villages are characterised by gullies as deep as 20 m.

The climate of the study area is semiarid with an average annual precipitation of 600 mm. The dry season normally lasts from June through October and the rainy season from November to May. The rainfall distribution pattern is influenced by elevation (Ngana, 1992). Mafai village, which is much higher in altitude (ranging from 1620 to 2010 m), compared to the other two villages (ranging from 1200 to 1545 m) is sub-humid, while Baura and Bolisa are semiarid. The long-term rainfall pattern for the study area shows that rainfall is highly variable between years and is also very erratic. Temperature varies from 16 to 23°C, the coolest period being in June and July.

MATERIALS AND METHODS

Laboratory analyses of soil, plant and farmyard manure samples

Plant samples (grain and stover) from major crops grown in the area were analysed for chemical nutrient content, mainly N, P, and K. The objective was to determine the amounts of these nutrients that are removed from the soil through crop harvest and residue removals, and that returned back into the soil by the residues retained in the farm. Farmyard manure was also analysed with the aim of establishing the amounts of same nutrients that it contributes to the soil when applied. Soil samples were collected from soil profiles dug to 1 to 1.5 m depth from representative locations for major soil groups identified by the farmers during transect walks. Soil profiles were described to at least 80 cm depth, according to the internationally accepted guidelines for soil-profile description (FAO, 1990). Visual observation showed no distinct horizons; as such samples were taken at specified depth ranges as described by (Hodgson, 1985) and (FAO, 1990), at 20 cm intervals, starting with the upper 0 to 20 cm. The soil samples were then analysed for particle-size distribution, pH, organic carbon, total nitrogen, extractable phosphorus, exchangeable bases (Ca^{2+}, Mg^{2+}, K^+ and Na^+), and cation exchange capacity (CEC), in order to obtain information important for estimating nutrient availability and soil productivity. This determination essentially aimed at estimating the inherent soil fertility status and, subsequently used the findings in computing the quantities of nutrients (N, P, and K) lost from the soil through soil erosion and other pathways.

Household interviews

Interviews were also conducted to 10% (147 households) of the village households so as to solicit information on crop yield and fertiliser application whose data were later used in computing the nutrient budget.

Inputs

Nutrient input to the agricultural system included inorganic fertilisers, farm yard manure, incorporation of crop residues and atmospheric deposition, as described in the following sections.

Inorganic fertilisers (Input 1)

Information on this input based on data obtained from household interviews conducted to the farmers on the type and quantities of fertilisers they used in the preceding growing seasons. The major inorganic fertilisers used by the respondent households were Calcium Ammonium Nitrate (CAN), Nitrogen-Phosphorus-Potassium (NPK), Sulphate of Ammonia (SA), UREA and Triple Supper Phosphate (TSP). However, the amounts of fertilisers used by respective households varied considerable.

Manure (Input 2)

The quantities of manure used by the farmers based on responses regarding the number of livestock units that are kept by the interviewed farmers. One livestock unit (250 kg live weight) requires a total of about 3000 kg of dry matter per year, of which about 40% is given out as manure (Mero et al., 1988). This gives an average of 1200 kg of manure being produced by one livestock unit in a year.

Crop residues (Input 3)

This refers to that portion of crop plants which is retained in the field after harvesting the main crop and later incorporated in the soil. Nutrients returned back into the soil through this pathway was determined by using the percentage nutrient content in the stover and the proportion of the total residue that is retained in the field. Responses from the survey in Mafai village indicated that of the

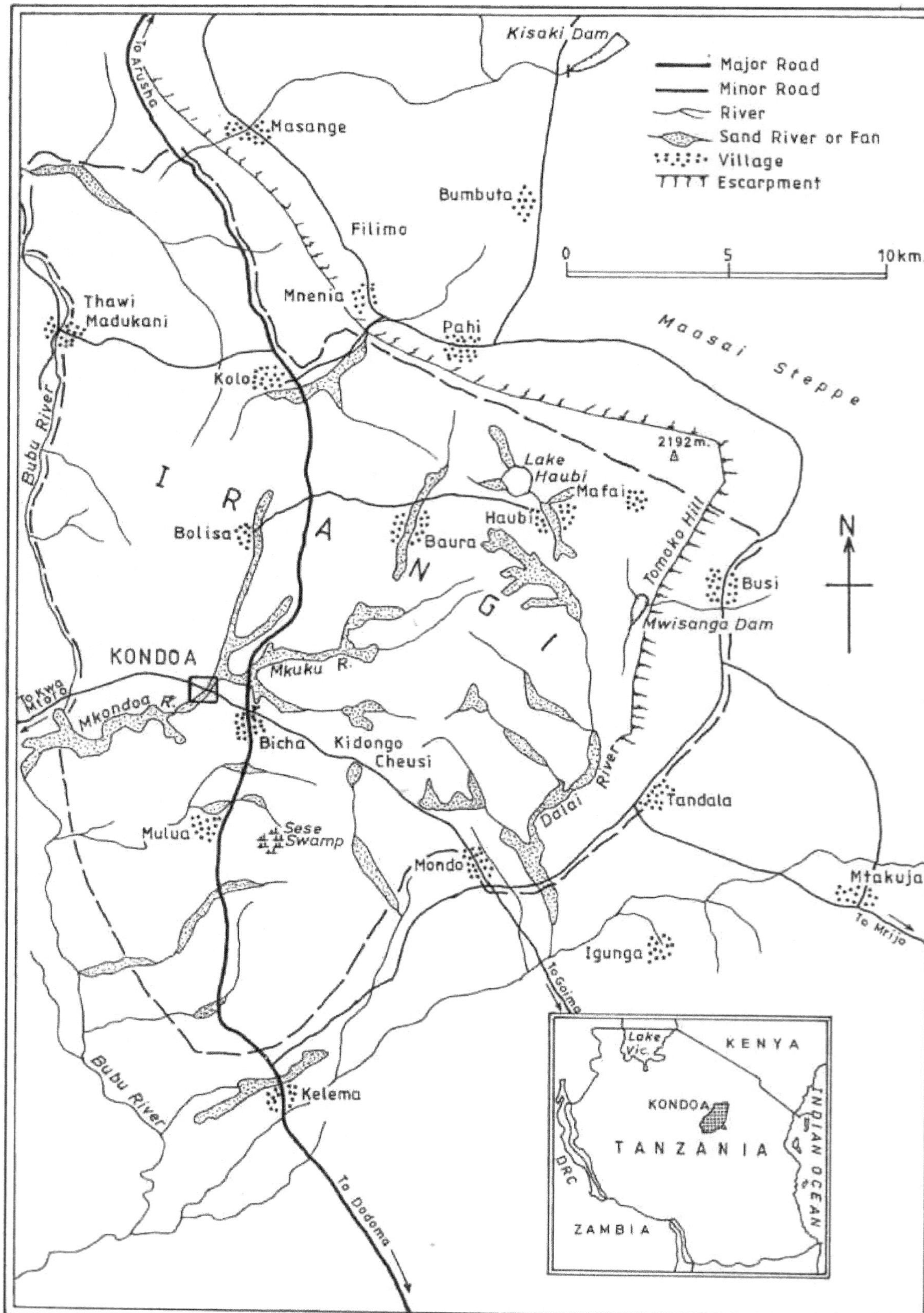

Figure 1. Map of the Irangi Hills. The mapped area forms the Kondoa Eroded Area (KEA), an area recognised for its severe soil erosion with extensive gullies. The insert shows the location of the KEA within Tanzania. Source: Kangalawe (2001).

Table 1. Land characteristics used in the calculation of soil loss by erosion in the survey villages.

Property	Mafai	Baura	Bolisa
Slope gradient in degrees (percent)	11 (20)	6 (10)	5 (8)
Slope length in metres	100	100	100
Rainfall factor (R)	200	200	200
Soil erodibility factor (K)	0.15	0.15	0.15
Slope length factor (L)	2.126	2.126	2.126
Slope steepness factor (S)	4.106	1.579	1.187
Crop cover factor (C)	*	*	*
Land management practice factor (P)	0.8	0.5	0.44

* Refer to Table 12 for crop cover factor (C).

total crop residue produced in the farm 55.4% is incorporated into the soil, 25.1% used as feed to livestock or other purposes and 19.5% is burnt. In Baura the corresponding values were 27, 25 and 48% respectively. Those for Bolisa village were 29.4, 44.4 and 26.2%. The residue yield per hectare was calculated based on harvest indices of the different crops.

Atmospheric deposition (Input 4)

Nutrient deposition from the atmosphere (wet and dry) was determined basing on the expressions below (Stoorvogel and Smaling, 1990; Smaling and Fresco, 1993).

$N = 0.14 * [\text{mean annual rainfall (mm)}]^{0.5}$
$P = 0.123 * [\text{mean annual rainfall (mm)}]^{0.5}$
$K = 0.092 * [\text{mean annual rainfall (mm)}]^{0.5}$

Where = N, P and K are kilograms of nitrogen, phosphorus and potassium deposited per hectare per year.

Outputs

Various pathways can be discerned regarding nutrient removal from the soil. The major pathways analysed during this study were nutrient removals (mainly N, P and K) from harvested crops plants, residue removals and loss through soil erosion. The determination of these nutrient output pathways from the farming system is described below. The landscape characteristics used to calculate soil and nutrient losses through erosion summarised in Table 1.

Harvested products (Output 1)

Nutrient output in harvested products was determined from crop harvest data reported by interviewed households. Crop yield data was then multiplied by the nutrient content in the product (e.g. grain, tubers and roots) so as to obtain the total amount of nutrients (N, P and K) that are removed from the soil through crop harvest.

Residue removals (Output 2)

The quantity of residue produced per unit area was determined based on the yield of harvested products and the respective harvest index for each crop. The term harvest index is used here referring to the proportion of the total crop biomass that is harvested. What remains is considered as residue. The total residue removal was derived from the interviews on the proportion taken away as animal feed, burnt or for other purposes. The amount of residue removed was multiplied by nutrient content in the residue to estimate the amounts of N, P and K taken away from the farming system.

Soil erosion (Output 3)

Nutrient loss in eroded soil material was determined by quantifying soil loss under the different crops grown in the area and later multiplying by the nutrient content in the soil samples. Quantities of soil loss by erosion were determined using the Universal Soil Loss Equation (USLE), that is, $A = R * K * L * S * C * P$. Where A = predicted soil loss in tonnes/hectare/year; R = rainfall factor (half the mean annual rainfall (Roose, 1976; Babu et al.,1978; FAO, 1979; Singh et al., 1981); K = soil erodibility factor; L = slope length factor; S = slope steepness factor; C = crop cover and management factor (shortly, crop cover); P = land management (erosion control) practice factor (Young, 1991; El-Swaify and Fownes, 1992). The values of L and S in the equation above were computed based on Smaling et al. (1993), while P was computed based on Wenner (1981) as follows.

$L = (d/22.13)^{0.5}$
$S = (0.43 + 0.30 * s + 0.043 * s^2)/6.613$
$P = 0.2 + 0.03 * s$

Where: d = slope length in meters; s = slope gradient in percent. Inherent soil fertility was then used to translate the soil loss into N, P and K losses. Soil erosion is a selective process, taking away nutrients and fine soil particles first. The eroded material is thus richer in nutrients than soil it originates from (Stocking, 1984; Gachene, 1987; Mnkeni, 1992). The amount by which the eroded material is richer in nutrients than the soil, from which it is taken, is referred to as the '*enrichment factor*'. The calculated nutrient loss in the above procedure was then multiplied by an enrichment factor of 2.
It should be noted that although leaching and denitrification are among the contributors to soil nutrient loss, they were not considered in this study because of the aridity of the climate in the study area. Due to low precipitation and high evapotranspiration of the area, no effective leaching or denitrification takes place (Young, 1991; FAO, 1978). As such, their role in nutrient loss from the surface soil layers in this area is considerably small.

RESULTS

Landscape and soil characteristics

Most of the soils in the study area are severely eroded

Table 2. Some characteristics from composite soils samples (0-20 cm depth) in the study area.

Soil characteristic	Mafai	Baura	Bolisa	Mean	NSS Classification
pH (H_2O)	4.80	5.20	5.40	5.10	Strongly acid
Total N (%)	0.29	0.07	0.08	0.14	Low
Available P (ppm)	4.20	7.20	3.20	4.87	Low
CEC me/100g	17.12	3.84	13.89	11.62	Low
Exchangeable K (me/100g)	0.61	0.61	0.52	0.58	Medium
Exchangeable Ca (me/100g)	4.39	1.83	2.74	2.99	Medium
Exchangeable Mg (me/100g)	1.61	0.40	1.20	1.07	High
Exchangeable Na (me/100g)	1.86	0.92	0.80	1.19	High
Organic carbon (%)	2.62	0.31	0.54	1.16	Low
Particle size distribution					
% Sand	54.83	80.27	69.20	68.10	na
% Silt	14.57	10.60	17.63	14.27	na
% Clay	30.60	9.13	13.17	17.63	na

na = not applicable; NSS = National Soil Service.

exposing the red argilic layer. A few exceptions exist in some sites in the conserved forest areas (e.g. in Mafai). Soils of this area are classified as ferric acrisols (Tosi et al., 1982), ferralic cambisols (Mowo et al., 1993) and luvisols (Payton et al., 1992). According to FAO/UNESCO (1974), the three soil types belong to the low erodibility class. The mean texture of the sampled soil profiles is medium, thus qualifying for the soil erodibility factor (K) of 0.15 in the USLE equation. Chemical soil analysis based on the top 20 cm soil layer because root systems of most annual crops grown in the area occupy this plough layer. Table 2 presents characteristics of this soil layer.

The organic carbon content ranged from low (0.31 and 0.54% in Baura and Bolisa) to high (2.62% in Mafai). The average for the three villages was classified as low (Table 2). The texture of the soils was sandy clay in Mafai, loamy sand in Baura and sandy loam in Bolisa. Based on the National Soil Service (NSS, 1994) classification, the soils in this area are strongly acid. However, the recorded pH values are higher than the critical pH value of 5 for a maize/sorghum, cowpea and cassava based cropping system of the area (Aune and Lal, 1995) and four for general plant and microbial growth (Logan, 1990). These soils also had low N, P, CEC and low to medium values of other exchangeable cations. These low figures suggest that soil fertility may be exhausted by continuous cultivation without adequate replacement.

Calculating inputs

Inorganic fertilisers (Input 1)

The inorganic fertilisers used in the studied villages were N-P-K (20:10:10), UREA (46% N), Sulphate of ammonia (SA-21% N), Calcium ammonium nitrate (CAN-26% N) and triple super phosphate (TSP-46% P_2O_5). Among the respondents who reported to have used inorganic fertilisers were 8.9% in Mafai village, while in Baura and Bolisa the percentages were 2.05 and 7.53% respectively. The remaining households did not use any fertiliser during the study period, and some had never ever used inorganic fertilisers before. The total amounts of N, P, and K nutrients applied through the use of these fertilisers are presented in Table 3. Considering the total area cultivated by respondents in these villages, it has been found that on average, the application of inorganic fertilisers on a hectare basis is about 2.60, 0.65 and 0.65 kg of nitrogen, phosphorus and potassium respectively (Figure 2).

Manure (Input 2)

Laboratory analysis for nutrients in farmyard manure indicated that the manure samples contained 1.56% N, 0.18% P and 0.005% K (on a dry matter basis). The N content of the manure samples compares well with the 1.3% reported by Smaling et al. (1993). However, the values for the other two nutrients, P and K are smaller. Smaling et al. (1993) used values as high as 0.5% for P and 1.6% for K. The availability and rate of manure application are influenced by the number of households keeping livestock and number of livestock units. Among the interviewed households only 40% are keeping livestock at present. The number of livestock units reported by these households was 42 in Mafai, 92 in Baura and 180 in Bolisa. These figures correspond to a total of 54.4, 110.4, and 216 tonnes of manure per year in Mafai, Baura and Bolisa respectively. Of all respondents,

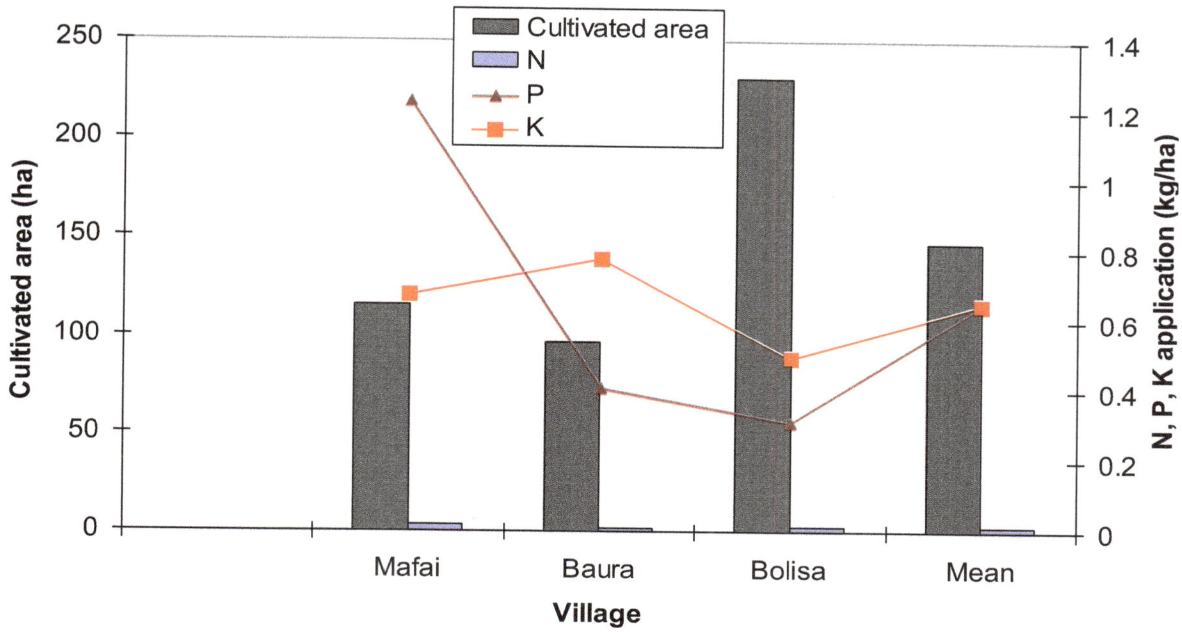

Figure 2. Area cultivated and average N, P and K application (kg/ha) from inorganic fertilisers.

67.32% had access to farmyard manure. It can be observed that the percentage of households using manure is larger than that of households keeping livestock. This disparity is due to the fact that the non-livestock keepers obtain some manure from relatives and/or neighbours who keep livestock. Among those using manure 37.4% are residents of Bolisa, whereas Mafai and Baura had 16.32 and 13.6% respectively. The average manure application rate for all respondent households have been found to be 1.5, 3.2, and 2.9 tonnes/household per year for Mafai, Baura and Bolisa, corresponding to an application rates are 0.44, 1.15 and 0.94 tonnes/ha/year respectively. The average application rate for the three villages is 0.84 tonnes/ha/year. Table 4 shows the application rates for various nutrients.

Crop residue additions (Input 3)

Determination of N, P and K nutrients added to the soil through crop residues was based on the percentages of crop residue retained in the field after harvesting the main crop and the nutrient concentration in the residues. The harvest indices (HI), area under each crop and amount of harvested product are presented in Table 5. For P and K the analysis included the percent of the total residue that was burned, because they are not removed from the arable land by burning (Smaling et al., 1993). The nutrient contents in the crop residues used in this computation are presented in Table 6. Figure 3 shows the total residue yield that was used in nutrient calculation for both inputs and outputs, whereas the quantities of residue-nutrients added into the soil by each crop are given in Table 7.

The largest N-input per hectare came from pigeon peas and sunflower stover which contributed 8.6 kg/ha each. Bean's residues ranked second in the quantity of N added per unit area, providing 3.4 kg/ha, followed by finger millet (3.3 kg/ha). Other crops with significant residue-N input were maize, cowpea, and sorghum (Table 7). The average N additions from all crop residue are 5.9 kg/ha in Mafai, 3.2 kg/ha in Baura and 2.7 kg/ha in Bolisa. Significant P additions per hectare were contributed by sunflower (8.6 kg) followed by cassava (2.2 kg) (Table 7). Lowest values are found in Irish and sweet potato because of the low inherent quantities of this nutrient in the residue. Pigeon peas residue provided the largest share of K additions from residues, amounting to 20.1 kg/ha, followed by maize (16.2 kg) and sunflower (15 kg/ha). The average K addition per hectare ranged between 8.3 and 13.9 kg with considerable variations between villages. Some of the crops such as beans, cowpeas and groundnuts had high nutrient concentrations in their residue, giving a high nutrient input per hectare.

However, because of the small farm areas they occupy, their contribution to total residue input has been relatively small (Table 6).

Atmospheric deposition (Input 4)

The mean annual rainfall for Kondoa district for the past 63 years (1931-1993) has been 643 mm (Mbegu, 1988;

Table 3. N, P and K nutrients (in kg) application from fertilisers by the surveyed households.

Village	Nitrogen (N)				Phosphorus (P)		Potassium (K)	All sources		
	CAN	NPK	SA	UREA	NPK	TSP	NPK	N	P	K
Mafai	39.0	190	120	69	41.4	100.2	78.9	418.0	141.6	78.9
Baura	0	180	0	0	39.2	0	74.7	180.0	39.2	74.7
Bolisa	0	280	0	253	61.0	10.0	116.2	533.0	71.0	116.2
Total	39.0	650	120	322	269.7	110.2	269.8	1131	251.9	269.8

Table 4. Area cultivated and average N, P and K application (kg/ha) from inorganic fertilisers.

Village	Cultivated area (ha)	N	P	K
Mafai	115.3	3.63	1.23	0.68
Baura	96.2	1.87	0.41	0.78
Bolisa	230.7	2.31	0.31	0.50
Mean	147.4	2.60	0.65	0.65

Table 5. Nutrient addition (kg/ha) from farmyard manure in the survey village.

Village	Farm area (ha)	FYM (kg/ha)	N	P	K
Mafai	115.3	437.12	6.82	0.77	0.022
Baura	96.2	1147.61	17.90	2.02	0.057
Bolisa	230.7	936.28	14.61	1.65	0.047
Mean	147.4	840.34	13.11	1.48	0.042

Table 6. Crop, harvest index (HI), cultivated area (ha) and yield of harvested product (kg/ha).

Crop	HI	Mafai		Baura		Bolisa		Total area cropped	Mean Yield/ha
		Area	Yield	Area	Yield	Area	Yield		
Maize	0.30	53.1	827	29.1	565	60.5	919	142.8	770
Sorghum	0.30	16.8	604	21.7	490	34.9	720	73.4	605
Pearl millet	0.30	0	0	23.1	413	55.0	727	78.1	569
Finger millet	0.30	26.3	786	11.5	497	33.7	439	71.5	574
Cowpeas	0.40	0	0	1.62	550	0.5	570	2.1	560
Beans	0.40	5.5	600	0.2	500	0.4	650	6.2	583
Pigeon peas	0.30	0.1	750	4.2	800	1.7	540	6.0	560
Groundnut	0.40	0	0	0	0	5.3	500	5.3	500
Cassava	0.70	0	0	0.9	3000	15.5	4100	16.4	3550
Sweet potato	0.85	6.6	3000	1.0	3000	1.3	3000	8.9	3000
Irish potato	0.73	3.7	2700	0	0	0	0	3.7	2700
Sunflower	0.30	1.6	600	2.8	520	21.9	750	26.3	623
Sugarcane	0.78	1.5	3500	0	0	0	0	1.5	3500
Total		115.3		96.2		230.7		442.2	

HI=harvest index (Agata, 1982; Cock, 1985; Victorio et al., 1986).

Ngana, 1990). Using the theoretical expressions for atmospheric depositions described above, the following results were obtained. The amount of N deposited in the study area is 3.55 kg/ha/year (that is, 0.14 * 25.36) and that of P is 3.119 kg/ha/year (that is, 0.123 * 25.36). The amount of K deposited from atmospheric sources is 2.333

Table 7. Nutrient content in harvested products and in residues (in percent).

Crop	Harvested product			Crop residue		
	N	P	K	N	P	K
Maize	1.48	0.28	0.24	0.30	0.09	1.34
Sorghum	1.06	0.34	0.34	0.30	0.07	1.15
Pearl millet	1.92	0.43	0.43	0.30	0.08	1.05
Finger millet	1.43	0.19	0.34	0.60	0.12	1.24
Cowpea	2.40	0.32	1.05	1.04	0.08	1.27
Beans	3.80	0.33	1.05	1.04	0.08	1.28
Pigeon pea	2.40	0.32	1.41	1.40	0.11	1.77
Groundnut	3.73	0.60	0.83	0.58	0.10	0.31
Cassava[1]	0.42	0.05	0.43	0.05	0.22	0.22
Sweet potato[1]	0.27	0.05	0.24	0.20	0.06	0.34
Irish potato[1]	0.34	0.05	0.41	0.20	0.06	0.34
Sunflower	3.89	0.32	0.52	1.98	0.59	1.03
Sugarcane[2]	0.61	0.09	1.47	0.31	0.12	1.15

Source: [1]Mnkeni (1992); [2]Smaling et al. (1993).

Table 8. Total crop residue yield (kg/ha) used in the nutrient input and output calculations.

Crop	Mafai	Baura	Bolisa	Mean
Maize	1930	1317	2143	1815
Sorghum	1410	1144	1679	1411
Pearl millet	na	963	1697	1329
Finger millet	1835	1160	1023	1339
Cowpeas	na	1155	975	1065
Beans	900	750	1005	885
Pigeon peas	2333	2030	1260	1597
Groundnut	na	na	750	750
Cassava	na	7000	9567	8283
Sweet potato	529	529	529	529
Irish potato	999	na	na	999
sunflower	1400	1213	1750	1454
sugarcane	778	na	na	778

na = not grown by the respondents during the period under study.

kg/ha/year (that is, 0.092 * 25.36). Similar findings for N deposition are reported for Northern Nigeria (Wilkes et al., 1984).

Calculating outputs

Nutrient removals in harvested products (Output 1)

Crop harvesting entails removal of nutrients from the arable land in the harvested products. The total quantity of the nutrients removed is a function of the area cultivated, yield per unit area and the nutrient content in the product. Grain is the major product harvested. Table 5 shows the land area that was grown with the different crops and yield of harvested products involved in the analysis, while Table 6 presented the nutrient content in the harvested products. The calculated overall average nutrient removals through harvested products from the respondents' crop fields were 14.0, 1.9 and 7.2 kg/ha for N, P and K respectively. At the village level, the nutrient removals were 14.7 kg N/ha in Mafai, 12.1 kg N/ha in Baura and 15.1 kg N/ha in Bolisa. Removals of P amounted to 2, 1.6 and 2.1 kg/ha respectively. In the same order, the values for K removed per hectare were 10.7, 5.2 and 5.7 kg (Table 8).

It can be seen from Table 8 that among the crops involved in the nutrient budget analysis, sunflower had

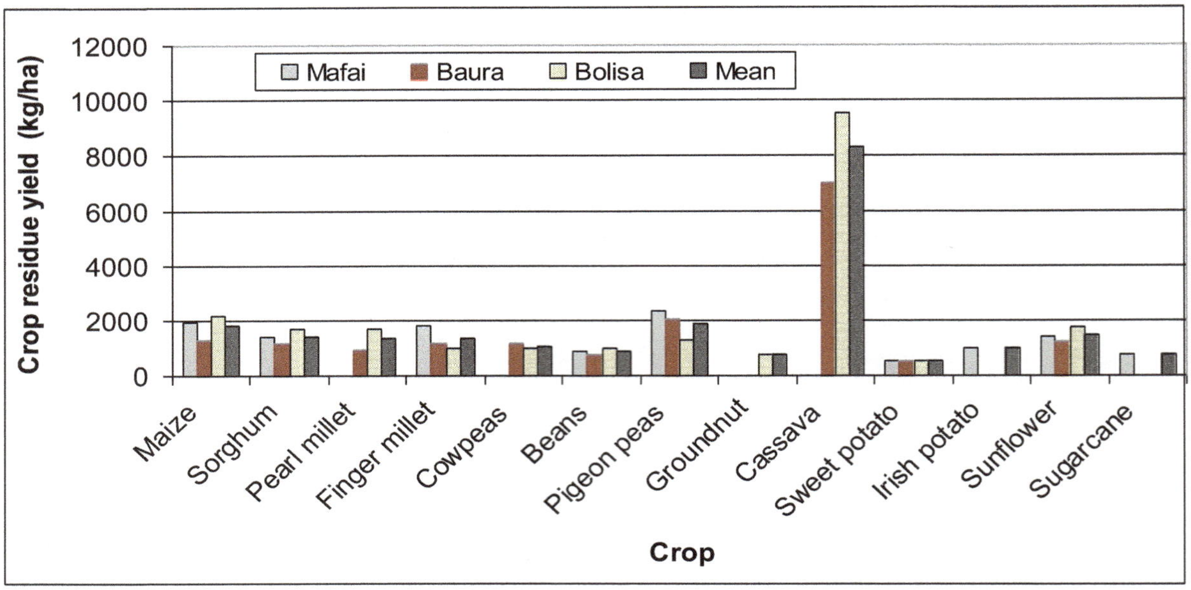

Figure 3. Total crop residue yield (kg/ha) used in the nutrient input and output calculations. Some crops were not grown by all the respondents during the period under study (e.g. cassava, cowpeas, groundnut, Irish potato, pearl millet, sugarcane), which raise their mean values.

the largest N removal (at a rate of 24.2 kg/ha), followed by beans (22.2 kg/ha), sugarcane (21.4 kg/ha) and groundnut (18.6 kg/ha). Lowest removals were observed under sorghum (6.4 kg/ha). This may be explained at least in part by the relatively low content of N in its (sorghum) grain compared to other cereals involved in the analysis (Table 5). The largest P removals were observed under sugarcane and groundnut that extracted from the soil 3.2 and 3.0 kg/ha respectively. Other crops with large removals were pearl millet (2.5 kg/ha) and pigeon peas (2.2 kg/ha). Lowest quantities of P were removed by finger millet grain at a rate of 1.1 kg/ha. Sugarcane featured highest in terms of extracting K from the soil (Table 8). However, because of the small area planted with this crop its total contribution to the overall nutrient loss through crop harvest was relatively small. Other crops removing large amounts of K from the arable land are cassava, Irish potato and Pigeon peas, taking away 15.3, 11.0 and 9.8 kg/ha respectively. The remaining crops had smaller rates of K nutrient extraction from the soil. The overall removals by these crops were however large because of the large land area they occupy, especially for cereals.

Crop residue removals (Output 2)

The removal of nutrients from the arable land in form of crop residues was calculated in a similar way as calculations for harvested products (Output 1). The calculation based on the proportion of the residue that is taken away from the arable land and the nutrient content in this residue. The average crop residue yield used in the analysis is presented in Figure 3. The proportion of crop residue considered here is that taken away as animal feed and for other purposes, which was 25.1% in Mafai, 25% in Baura and 44.4% in Bolisa. For nitrogen, the proportion of residue burned was also taken into consideration since N is lost by burning. The latter constituted 19.5, 48 and 26.2% of total residue in Mafai, Baura and Bolisa respectively. P and K are not taken away from the system by burning (Smaling et al., 1993). Quantities of nutrient removed through this pathway are given in Table 9. The average residue nutrient removal from the agricultural system was established to be 5.7, 0.4 and 4.3 kg K/ha. Baura and Bolisa villages featured high in the removal of nutrients (especially N) in crop residues (Table 9). The explanation to the observed trends is the larger proportion of the total residue that is taken away from the arable land by burning, for animal feed and other purposes. The two types of residue removals in Baura and Bolisa village accounted for 73% and nearly 71% of the residue produced in the farm respectively.

Regarding individual crops, sunflower, pigeon peas, cowpea and bean residues had the highest N removals (18.1 and 14.1 kg/ha respectively). Other crops removed relatively small quantities of N (Table 9). Cassava removed largest quantities of P (1.2 kg/ha), followed by finger millet (0.6 kg/ha), maize and Pigeon peas (0.5 kg/ha each). Other crop residue removals accounted for limited quantities, ranging from 0.1 to 0.4 kg P/ha. As for K, largest removals in crop residues were observed in pigeon peas whose residue removed 8.7 kg/ha. Maize was the next highest ranking crop in terms of removing K

Table 9. Nutrient addition (kg/ha) into the soil through incorporation of crop residues.

Crop[1]	Mafai			Baura			Bolisa			Mean values		
	N	P	K	N	P	K	N	P	K	N	P	K
Maize	3.2	1.3	19.3	1.1	0.9	13.2	1.9	1.0	16.0	2.1	1.1	16.2
Sorghum	2.3	0.7	12.1	0.9	0.6	9.9	1.5	0.7	10.7	1.6	0.7	10.9
Pearl millet	na	na	na	0.8	0.6	7.6	1.5	0.7	10.0	1.2	0.7	8.8
Finger millet	6.1	1.7	17.0	1.9	1.1	10.7	1.8	0.7	7.0	3.3	1.2	11.6
Cowpeas	na	na	na	2.3	0.5	7.9	2.6	0.4	6.0	2.5	0.5	8.5
Beans	5.2	0.5	8.6	2.1	0.5	7.2	3.0	0.4	6.9	3.4	0.5	7.6
Pigeon peas	13.5	1.4	23.2	7.0	1.5	24.8	5.2	0.8	12.4	8.6	1.2	20.1
Groundnut	na	na	na	na	na	na	1.3	0.4	1.3	1.3	0.4	1.3
Cassava	na	na	na	0.2	2.1	2.1	0.3	2.2	2.2	0.3	2.2	2.2
Sweet potato	0.6	0.2	1.3	0.3	0.2	1.3	0.3	0.2	1.0	0.4	0.2	1.2
Irish potato	1.1	0.4	2.5	na	na	na	na	na	na	1.1	0.4	2.5
sunflower	15.4	8.3	14.4	6.5	7.2	12.5	10.2	10.3	18.0	8.6	8.6	15.0
sugarcane	0	0.9	8.5	na	na	na	na	na	na	0	0.9	8.5
Total	5.9	1.7	11.9	3.2	2.3	13.9	2.7	1.6	8.3	3.9	1.9	11.4

[1]na = not grown by the respondents during the period under study.

(7.9 kg/ha) from the agricultural land.

Nutrient loss through soil erosion (Output 3)

The quantities of soil loss under each crop in respective villages were obtained by multiplying the factors in the USLE for each crop so as to get respective quantities of eroded soil material (Figure 4). Let us look at an example of maize crop in Mafai village and how soil loss from erosion was determined. The factors considered here are rainfall factor (R = 200), soil erodibility factor under ferric acrisols (K = 0.15), slope length factor on 100m slope (L = 2.126), slope gradient factor, with slope gradient of 20% (S = 4.11), and land management (erosion control) practice factor (P = 0.8). Multiplication of these factors gives a soil loss 83.8 tonnes/ha/year. The calculated average soil losses from cropland owned by the respondent households are 62.4, 16.5 and 11.2 tonnes/ha/year in Mafai, Baura and Bolisa respectively. The average soil loss on a hectare basis was highest under sunflower, maize, sorghum, finger millet and pigeon pea crop covers (Figure 4). The erosion loss was used in the determination of nutrient losses through erosion presented in Table 10. Soil erosion contributed greatest to the N loss. The calculated figures are in the magnitude of 253.0 kg/ha/year in Mafai, 17.8 kg/ha/year in Baura and 15.2 kg/ha/year in Bolisa.

Nutrient budgets

The term nutrient budget is used here to refer to the difference between total quantities of input to and outputs from the agricultural system. In all the three villages there were negative nitrogen budgets to the magnitudes of -252.4, -9.58 and -13.53 kg/ha/year in Mafai, Baura and Bolisa respectively (Table 11). Corresponding budgets for P were 4.16, 5.67 and 4.02 kg/ha/yr, while those for K were 0.07, 8.75 and -0.33 kg/ha/yr in the same order. The mean difference between the four inputs and three output factors for the three villages has indicated values of -91.85 kg N, 4.62 kg P, and 2.82 kg K/ha/yr. Among the four input factors, application of inorganic fertilisers has shown to play a small role as nutrient input source in the study area, contributing for only 2.6 kg N, 0.65 kg P and 0.65 kg K/ha/yr (Table 11), equivalent to 11.2, 9.1 and 4.5% of total N, P and K input to the system. Manure ranked highest as N input source, third as P source and fourth as K source, providing about 56.5, 20.8 and 0.3% of input N, P and K, in the same order. Incorporation of crop residues into the soil was the major contributor to K input and second as P and N source. Residues added into the system accounted for nearly 17.0, 26.3 and 79.0% of input N, P & K respectively. Atmospheric deposition contributed significant quantities of all the three elements, N, P, and K, ranking highest as input source for P, third N source and second K source, providing about 15.3, 43.8 and 16.2% of input N, P and K respectively (Table 11).

Removal of nutrients in eroded soil material has been found to be the major pathway for N losses. This was followed by nutrient removal in harvested products. Residue removals had the lowest contribution to the overall N losses (Tables 10 and 11). The mean values for the three output factors in the same order are 95.33, 13.97 and 5.73 kg/ha respectively. As for P and K budget, largest removals appeared to be in form of

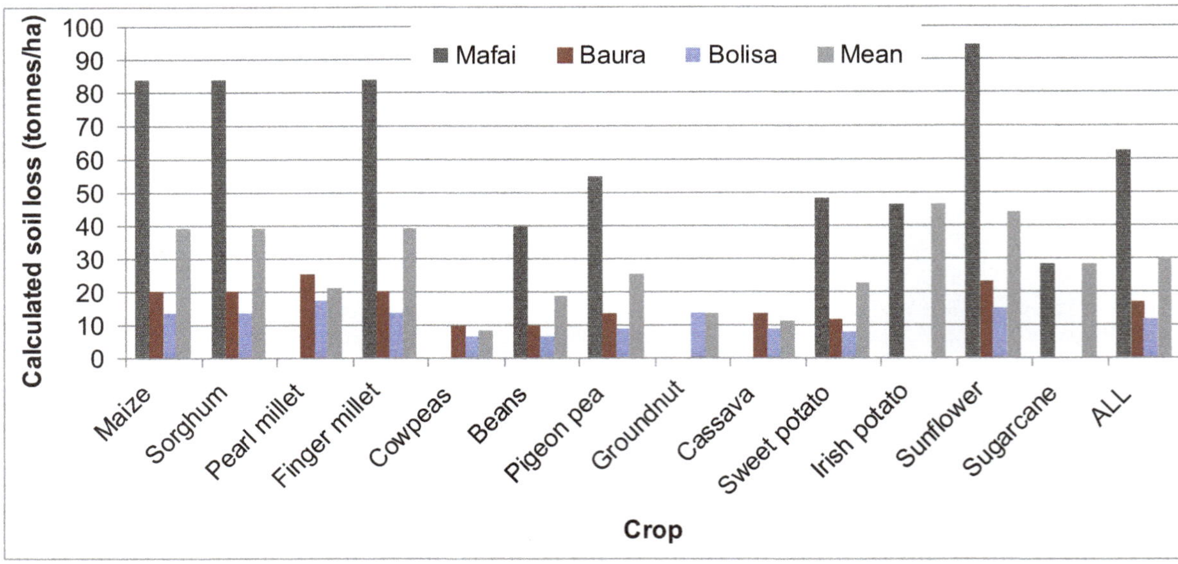

Figure 4. Calculated soil loss (tonnes/ha) by erosion under different crop covers. [1]For crop cover factor (C) refer to Smaling et al. (1993), Roose (1986) and Lewis (1987).

harvested products, withdrawing 2.0, 1.6 and 2.1 kg P/ha/yr, and 10.7, 5.2 and 5.7 kg K/ha/yr in Mafai, Baura and Bolisa respectively. The overall average P and K removals in harvested products were 1.90 and 7.2 kg/ha/yr, in that order (Table 11).

DISCUSSION

Landscape and soil features

The various landscape and soil characteristics presented in Table 2 reveal that in of the study villages the soil are generally of low organic carbon content. Such soils are characterised by low nutrient and water holding capacity which makes these soils of limited productivity compared to the sandy clay and loamy textures. The low N, P, CEC and low to medium values of other exchangeable cations that these soils had suggests that soil fertility may be exhausted by continuous cultivation without adequate nutrient replacement.

Major sources of soil nutrient inputs

The major sources of nutrient supply to the agricultural system were found to be inorganic fertilisers (Input 1), manure (Input 2), crop residue additions (Input 3) and atmospheric deposition (Input 4). Although no fertiliser recommendations exist for study area, the whole of Dodoma (Mowo et al., 1993), the quantities of inorganic fertiliser used by respondent farmers fall far below the recommended rates for other places in Tanzania. In areas with similar soil types and agro-climatic conditions such as Shinyanga, the recommended rates ranges from 12-80 kg N/ha and 10-27 kg P/ha (Tosi et al., 1982). Limited access to fertilisers and high prices were among the reasons put forward by farmers for the low fertiliser application rates. The low rates of soil nutrient replacement from inorganic fertilisers may lead to decreased crop productivity if other nutrient sources are also limiting.

Manure is another important sources of soil fertility I the study area. However, the average application rate for the three study villages was very low, ranging from 0.44 to 0.94 tonnes/ha/yr) with an average of about 0.84 tonnes/ha/year. The observed manure application rate falls far below the 5 tonnes per hectare recommended for a sorghum/millet based farming system of Dodoma region, and 7-10 tonnes/ha necessary to meet the requirements of most crops (Agata, 1982).

This situation indicates the inadequate supply of nutrients required to boost modest crop production. This nutrient application rates from manure does not compare favourably with rates applied elsewhere. In Kisii district is Kenya, for example, nutrient application rates from manure are as high as 24 kg N, 5 kg P, and 25 kg K/ha/year (Smaling and Fresco, 1993). The obvious explanation is low manure application rates in the study area attributed to the fact that only few households keep livestock.

Crop residue addition is another important source of the major soil nutrients, N, P and K. However, it contribution to the soil nutrients pool is influenced by the proportion of the plant material that is retained in the field as crop residue after harvesting the main crop and the quality (nutrient concentration) of the residues. The proportion of the crop removed from the field is referred

Table 10. Nutrient removals (kg/ha) in harvested products in the three villages.

Crop	Mafai			Baura			Bolisa			Mean		
	N	P	K	N	P	K	N	P	K	N	P	K
Maize	12.2	2.3	2.0	8.4	1.6	1.4	13.6	2.5	2.2	11.4	2.1	1.9
Sorghum	6.4	2.1	2.0	5.2	1.7	1.6	7.6	2.5	2.4	6.4	2.1	2.0
Pearl millet	na	na	na	7.9	1.8	1.8	14.0	3.1	3.1	11.0	2.5	2.5
Finger millet	11.2	1.5	2.6	7.1	0.9	1.7	6.3	0.8	1.5	8.2	1.1	1.9
Cowpeas	na	na	na	13.2	1.8	5.8	13.7	1.8	6.0	13.5	1.8	5.9
Beans	22.8	2.0	6.3	19.0	1.6	5.3	24.7	2.1	6.8	22.2	1.9	6.1
Pigeon peas	18.0	2.4	10.6	19.2	2.6	11.3	13.0	1.7	7.6	16.7	2.2	9.8
Groundnut	na	na	na	na	na	na	18.6	3.0	4.1	18.6	3.0	4.1
Cassava	na	na	na	12.5	1.4	12.9	17.1	1.9	17.6	14.8	1.7	15.3
Sweet potato	8.2	1.4	7.3	8.2	1.4	7.3	8.2	1.4	7.3	8.2	1.4	7.3
Irish potato	9.1	1.4	11.0	na	na	na	na	na	na	9.1	1.4	11.0
sunflower	23.3	1.9	3.1	20.2	1.6	2.7	29.2	2.4	3.9	24.2	2.0	3.2
sugarcane	21.4	3.2	51.5	na	na	na	na	na	na	21.4	3.2	51.5
All	14.7	2.0	10.7	12.1	1.6	5.2	15.1	2.1	5.7	14.0	1.9	7.2

Table 11. Nutrient loss (kg/ha) from agricultural land through removals of crop residues.

Crop	Mafai			Baura			Bolisa			Mean		
	N	P	K	N	P	K	N	P	K	N	P	K
Maize	2.6	0.4	6.5	2.9	0.3	4.4	4.5	0.8	12.7	3.3	0.5	7.9
Sorghum	1.9	0.2	4.1	2.5	0.2	3.3	3.6	0.5	8.6	2.7	0.3	5.3
Pearl millet	na	na	na	2.1	0.2	2.5	3.6	0.6	7.9	2.9	0.4	4.4
Finger millet	4.9	0.6	5.7	5.1	0.4	3.6	4.3	0.6	5.6	4.8	0.6	5.0
Cowpeas	na	na	na	6.3	0.2	2.6	6.3	0.3	4.8	6.3	0.3	3.7
Beans	4.2	0.2	2.9	5.7	0.2	2.4	7.2	0.3	5.5	5.9	0.2	3.6
Pigeon peas	10.9	0.5	7.8	19.0	0.5	8.3	12.4	0.6	9.9	14.1	0.5	8.7
Groundnut	na	na	na	na	na	na	3.1	0.3	1.0	3.1	0.3	1.0
Cassava	na	na	na	0.5	0.7	0.7	0.6	1.7	1.7	0.6	1.2	1.2
Sweet potato	0.5	0.1	0.5	0.8	0.1	0.4	0.7	0.1	0.8	0.7	0.1	0.6
Irish potato	0.9	0.2	0.9	na	na	na	na	na	na	0.9	0.2	0.9
sunflower	12.4	0	0	17.5	0	0	24.4	0	0	18.1	0	0
sugarcane	3.1	0	0	na	na	na	na	na	na	3.1	0	0
ALL	4.6	0.3	4.1	6.2	0.4	3.1	6.4	0.5	5.8	5.7	0.4	4.3

[1]Note: sugarcane and sunflower residues are burned, thus only N gets lost from the system. P and K are not removed (Smaling et al. 1993).

to as harvest indices (HI), as described by Agata (1982), Cock (1985) and Victorio et al. (1986). Harvest index depends on the genetic traits of the cultivar and the conditions of the environment, such as photoperiod, air temperature, solar radiation, water supply, and level nitrogen fertilisation (Mazurczyk et al., 2009). Findings presented in Table 5 revealed that the largest N-input per hectare came from pigeon peas and sunflower stover which contributed 8.6 kg/ha each. Bean's residues ranked second in the quantity of N added per unit area, providing for 3.4 kg/ha, followed by finger millet (3.3 kg/ha). These observations demonstrate that integration of leguminous crop plants in the farming system may partly help alleviate the soil fertility problems. Increased growing of these crops may enhance soil N-status taking advantage of their N-fixing properties. Pigeon peas residue also provided the largest share of K additions from residues, amounting to 20.1 kg/ha/year, followed by maize (16.2 kg) and sunflower (15 kg/ha). Although there were considerable variations between the villages regarding the nutrient input from crop residue additions, this source of soil nutrient appears to be particularly important given that many farmers cannot afford inorganic fertiliser, and have limited access to manure. Soil fertility improvement in the field would thus depend on careful management of crop residues. Atmospheric deposition (Input 4) is another source of nutrient input to the soil. However, since this input is dependent on the amount of rainfall, its contribution to the Kondoa environment is small due to the semiarid conditions. Similar findings were reported for Northern Nigeria (Wilkes et al., 1984).

Mechanisms for soil nutrient removals (outputs)

Three major outlets contribute to the net removals of soil nutrients from the agroecosystems, namely nutrient removals in harvested products (Output 1), crop residue removals (Output 2), and nutrient loss through soil erosion (Output 3). Crop harvesting entails removal of nutrients from the arable land in form of harvested products. The total quantity of the nutrients removed is a function of the area cultivated, yield per unit area and the nutrient content in the product. Grain is the major product harvested. Crops such as sunflower, beans, sugarcane, groundnut were found to be among the crops that had the largest N removal, at a rate of 24.2, 22.2, 21.4 and 18.6 kg/ha, respectively. Other crops removed large amounts of other nutrients, particularly K from the arable land. These included cassava, Irish potato and pigeon peas, taking away 15.3, 11.0 and 9.8 kg/ha respectively. The possible implication of the observed trends in nutrient removal through crop harvests is that continuous growing of these crops that extract large quantities of nutrients from the soil may lead to continuous deterioration of the soil if equal or larger quantities of nutrients are not replaced into the soil through the various input sources. The ultimate impact is continued soil degradation and lowered productivity of the land and limited future sustainability.

Regarding crop residue removals (Output 2), the larger proportion of the total residue appears to be that taken away from the arable land by burning and for animal feed. The two types of residue removals in Baura and Bolisa village accounted for 73% and nearly 71% of the residue produced in the farm respectively, and contributing to considerable removals as presented in crop residue removals (Output 2), accounting for 5.7 kg N/ha, 0.4 kg P/ha and 4.3 kg K/ha being removed from the agricultural system through crop residue removals. It has been established from this study that among the crops grown, maize, sorghum, finger millet and pearl millet straws removed the largest amounts of all N, P and K, partly because of the high nutrient concentration in their residues and in part because of the larger land area under these crops. Reducing the quantities of residues removed from the cropland would thus significantly help maintain the status of soil fertility in the existing farming system.

Figure 4 presented the quantities of eroded soil material by estimating the quantities of soil loss under each crop in respective villages. Respective quantities of eroded soil material were obtained multiplying the crop factors in the USLE. At village level, the calculated average soil losses from croplands owned by the respondent households were 62.4, 16.5 and 11.2 tonnes/ha/year in Mafai, Baura and Bolisa respectively; the highest soil loss being under sunflower, maize, sorghum, finger millet and pigeon pea crop covers. The figures for eroded soil material are slightly smaller than those reported for some parts of Dodoma region. In Mpwapwa for example, Christiansson (1981), reported yearly values as high as 105 tonnes/ha/yr (that is, 70 m^3/ha/year) under mixed native food crops in flat cultivation. In maize alone values of soil loss were 96.2 tonnes/ha/yr (that is, 64 m^3/ha/yr), while in sorghum and pearl millet it was 85.1 tonnes/ha (that is, 56.7 m^3/ha/yr) and 99 tonnes/ha/yr (that is, 66.6 m^3/ha/yr) respectively. Higher figures are also possible under similar rainfall conditions. In Kenya for example, figures recorded are as high as 350 tonnes/ha/year (Bekele and Thomas, 1992). In all the three villages erosion removed smaller quantities of both available P and exchangeable K. This may be explained by the fact that often the available P and exchangeable K in the soil are in smaller proportions compared to quantities of total N. Mafai village contributed a larger proportion, over 80% of the total nitrogen loss in eroded soil material (Table 10). This is attributed to the steeper slopes of the terrain in this village that resulted in larger values of both S and P factors in the USLE (Table 1) with a consequent effect of compounding the calculated soil loss compared to Baura and Bolisa. Another explanation to this observation is that Mafai soils are nearly four times richer in topsoil N than both Baura and Bolisa (Table 2). It is evident from Figure 4 and Table 10, and studies elsewhere (Christiansson, 1981) that land-use systems under cereal crops such as maize, finger millet and sorghum, and root and tuber crops (such as sweet and round potato) grown on steep slopes could be highly erosive, leading to large quantities of valuable nutrients being transported away in eroded materials, particularly if no sound soil conservation measures are practised.

Predominantly nutrient mining agricultural system

In all the three villages there were negative nitrogen budgets to the magnitudes of -252.4, -9.58 and -13.53 kg/ha/year in Mafai, Baura and Bolisa respectively (Table 11). Implicitly this gives reflections on the prevalence of a general nutrient mining from the cropping system in the study area, which leads to a net negative nutrient budget, particularly for N and marginal positive budgets for P and K. The removals of nutrients is from eroded soil material and nutrient removals in harvested products have been found to be the major pathway for N losses, accounting for 95.33 and 13.97 kg/ha respectively. As for P and K budget, largest removals appeared to be in form of harvested products. Thus careful management of these output pathways is likely to enhance soil fertility and

Table 12. Crop cover factors and the calculated soil loss (tonnes/ha) by erosion under different crop covers.

Crop	Cover factor[1]	Calculated soil loss by erosion			
		Mafai	Baura	Bolisa	Mean
Maize	0.4	83.8	20.1	13.3	39.1
Sorghum	0.4	83.8	20.1	13.3	39.1
Pearl millet	0.50	na	25.1	17.0	14.1
Finger millet	0.40	83.8	20.1	13.3	39.1
Cowpeas	0.19	na	9.7	6.3	5.3
Beans	0.19	39.5	9.7	6.3	18.5
Pigeon pea	0.26	54.6	13.1	8.7	25.5
Groundnut	0.40	na	na	13.3	4.4
Cassava	0.26	na	13.1	8.7	7.3
Sweet potato	0.23	48.0	11.6	7.6	22.4
Irish potato	0.22	46.1	na	na	15.4
Sunflower	0.45	94.1	22.7	15.0	43.9
Sugarcane	0.20	27.9	na	na	9.3
ALL	na	62.4	16.5	11.2	30.0

[1]Reference for crop cover factor (C) are found in Smaling et al. (1993), Roose (1986) and Lewis (1987).

Table 13. Nutrient loss (kg/ha) in eroded soil material under different cropping systems in the survey villages.

Crop	Mafai			Baura			Bolisa			Mean		
	N	P	K	N	P	K	N	P	K	N	P	K
Maize	486.0	0.70	0.10	28.2	0.28	0.02	21.2	0.08	0.02	178	0.36	0.17
Sorghum	486.0	0.70	0.10	28.2	0.28	0.02	21.2	0.08	0.02	178	0.36	0.17
Pearl millet	na	na	na	35.2	0.36	0.04	27.2	0.01	0.02	20.8	0.16	0.02
Finger millet	486.0	0.70	0.10	28.2	0.28	0.02	21.2	0.08	0.02	178	0.36	0.17
Cowpeas	na	na	na	13.6	0.14	0.02	10.0	0.04	0.01	7.8	0.06	0.01
Beans	229.2	0.34	0.04	13.6	0.14	0.02	10.0	0.04	0.01	84.2	0.18	0.02
Pigeon peas	316.6	0.46	0.06	18.4	0.18	0.02	14.0	0.06	0.01	116	0.24	0.03
Groundnut	na	na	na	na	na	na	21.2	0.08	0.02	7.1	0.02	0.01
Cassava	na	na	na	18.4	0.18	0.02	14.0	0.06	0.01	10.8	0.08	0.01
Sweet potato	278.4	0.40	0.06	16.2	0.16	0.02	12.2	0.04	0.01	102	0.20	0.03
Irish potato	275.4	0.38	0.06	na	na	na	na	na	na	91.8	0.13	0.02
sunflower	574.8	0.80	0.12	31.8	0.32	0.02	24.0	0.10	0.02	201	0.27	0.05
sugarcane	161.8	0.24	0.04	na	na	na	na	na	na	53.9	0.08	0.01
All	253.0	0.36	0.06	17.8	0.18	0.02	15.2	0.06	0.02	95.3	0.20	0.03

ensure the sustainability of agricultural system.

What is definitely worrying from the findings of this study is the large deficit of the most growth limiting nitrogen. The point of concern here is how long the nutrient resource base is going to sustain production given the present soil management system and extraction rates. Both P and K are in marginal positive nutrient budgets. Such marginal positive trends do not rule out the possibility of these nutrients becoming limiting in the near future, or in some parts of the landscape, especially if nutrient uptakes in harvested products, residue removals or erosion loss are not replenished by equal or larger quantities.

The use of inorganic fertilisers (Input 1) remains the major source of nutrient supplement into the soil (at least in the short-run) to offset the continuous depletion that has resulted into the observed negative nutrient budgets. No fertiliser recommendations exist for the study area, and for Dodoma region as a whole. However, based on the calculated nutrient budget, 5 bags (1 bag = 50 kg) of UREA (46%N) or 9 bags of N-P-K (20-10-10) compound fertiliser would be required to offset the calculated negative N-budget. Another approach to replenish fertility in the soil would be increased planting of leguminous

Table 14. Summary on nutrient input, output (in kg/ha/yr) and nutrient budget[1] from the different sources.

Source	Mafai			Baura			Bolisa			Mean		
	N	P	K	N	P	K	N	P	K	N	P	K
Input sources												
(1) Fertilisers	3.63	1.23	0.68	1.87	0.41	0.78	2.31	0.31	0.51	2.60	0.65	0.65
(2) Manure	6.82	0.77	0.02	17.90	2.02	0.06	14.61	1.65	0.05	13.1	1.48	0.04
(3) Crop residue	5.90	1.70	11.90	3.20	2.30	13.90	2.70	1.60	8.30	3.93	1.87	11.4
(4) Atmospheric deposition	3.55	3.12	2.33	3.55	3.12	2.33	3.55	3.12	2.33	3.55	3.12	2.33
Total input	19.90	6.82	14.93	26.52	7.85	17.07	23.17	6.68	11.19	23.18	7.12	14.4
Output sources												
(1)Harvested products	14.70	2.00	10.70	12.10	1.60	5.20	15.10	2.10	5.70	13.97	1.90	7.20
(2) Residue removals	4.60	0.30	4.10	6.20	0.40	3.10	6.40	0.50	5.80	5.73	0.40	4.33
(3) Erosion	253.0	0.36	0.06	17.80	0.18	0.02	15.20	0.06	0.02	95.33	0.20	0.03
Total output	272.30	2.66	14.86	36.10	2.18	8.32	36.70	2.66	11.52	115.03	2.50	11.6
Nutrient Budget	-252.40	4.16	0.07	-9.58	5.67	8.75	-13.53	4.02	-0.33	-91.85	4.62	2.82

[1]Nutrient budget = Total input - Total output in kg/ha/yr.

crops and tree species which fix nitrogen and add organic matter to the soil and reduce soil erosion. Leguminous trees incorporated in the farm are reported to fix up to 50 kg N/ha/yr and raise the productivity of semiarid soils (Mortimore, 1989). This can be adopted in areas with similar climatic conditions such as Kondoa. Trees incorporated in crop fields not only improve the fertility status of the soil but also reduces soil erosion through reduction of runoff (Young, 1991; Liniger, 1992).

Implication of the nutrient budget on agricultural sustainability

The agricultural system of the study area is characterised by declining crop productivity, mainly associated with negative balances for major nutrients like nitrogen. The amount of nutrients leaving the soil in harvested crops, by erosion and by residue removals, is grossly exceeding the nutrient input to the system by natural processes such as biological nitrogen fixation and atmospheric deposition, or artificially by organic manure and inorganic fertilisers.

Increasing and sustaining agricultural production in the inherently erodible and low fertility soils of the study area is a major challenge. This has resulted from among others factors, soil erosion and increased soil degradation in farmland, continuous harvesting without proportionate nutrient replenishment and residue removals.

One of the reasons for low agricultural production is the gradual process of soil nutrient depletion in many of the sub-Saharan land-use systems (CWFS, 1985; Smaling, 1993). Among the processes responsible for this nutrient depletion are soil erosion, crop harvesting, residue removals and leaching, phenomena that have received less public attention compared to aspects like crop failures due to erratic rainfall and climatic and non-climatic factors. The nutrient status of the soil dictates the productivity of land and dwindling soil fertility has become one of the major constraints to crop production, in the Kondoa area as well other parts of Tanzania and Africa in general. This has resulted into reduced agricultural sustainability (Lynam and Herdt, 1989; Spenser and Swift, 1992; Rockström, 1997). Thus, in order to ensure agricultural sustainability there is need to enhance the management of all the nutrient input pathways while minimising the nutrient output pathways.

Conclusions

It has been established in the foregoing discussion that the agricultural system in the study area faces extremely large nutrient deficit, particularly for N, as expressed by the enormous negative nutrient budgets. The major concern among the nutrient output factors from the agricultural land is soil erosion. This nutrient loss pathway accounted for more than 80% of the N losses in eroded material. Thus efforts to control erosion are necessary to reduce the amount of soil removed from the arable land, and as such help to halt the negative nutrient budget trends in the farming system. Increased effort on on-farm soil conservation is an important undertaking. Among the on-farm conservation strategies could be the increased incorporation of crop residues into the soil (Input 2). For ensured sustainability of the agricultural system protection of the land against soil erosion has to take a long-term perspective. In the short run, one of the important factors in the efforts to raise agricultural productivity is increased availability and application of inorganic fertilisers, especially under conditions of heavy population pressures and dwindling reserves of uncultivated arable land. Given the growing land pressure in the study area and elsewhere in the country and rising fertiliser prices, a subsidy on fertilisers is critical especially to the resource-poor farmers to ensure food security and soil productivity. However, further research has to be conducted on crop responses to fertilisers in this area so as to come out with more feasible recommendations bearing in mind the climatic and socio-economic conditions that prevail in the area.

Manure application (Input 2) provides nutrient and additional benefits such as increased water storage and nutrient retention in the soil. However, since not all households have got access to farmyard manure following expulsion of livestock from the area in the late 1970s, increased use of this input is unlikely to be achieved in the short-run. Present average manure application rate has been observed to be far below the recommended rates for the sorghum, pearl millet, and maize based farming system of study area. Such low rates pose a particular challenge given shortages of inorganic fertilisers. Incorporation of crop residues into the soil would, apart from contributing significantly to the maintenance of high nutrient levels in the soil, is also necessary to provide surface cover to the ground that is required so as to reduce surface runoff and control soil loss by erosion. Other conservation efforts such as including agroforestry trees in croplands and making contour ridges to control soil erosion need to be encouraged. Finally, a well-planned long term conservation strategy becomes an absolute necessity if agricultural productivity is to be made sustainable under such fragile semiarid environments.

Conflict of Interests

The author(s) have not declared any conflict of interests.

ACKNOWLEDGEMENTS

This paper is based on a study undertaken as part of my MSc study, which was undertaken with support from a NORAD fellowship through the Management of Natural Resource and Sustainable Agriculture programme. Am grateful to my mentor Jens Aune of NORAGRIC and to the farmers in the Kondoa Eroded Area for their participation in the study.

REFERENCES

Adesina FA (1994). A preliminary investigation into agroforestry practices in savannah belt of Western Nigeria. Agrofor. Syst. 27(3):197-206, http://dx.doi.org/10.1007/BF00705056

Agata W (1982). The characteristics of dry matter and yield production in sweet potato under field conditions. In: Villareal RL, Grigs TD (eds). Sweet Potato. Proceedings of the first international symposium, Asian Vegetable Research and Development Centre. Shanhua, Tainan, Taiwan, China, pp. 119-127.

Aune JB, Lal R (1995). The tropical soil productivity calculator - A model for assessing effects of soil management on productivity. In: Lal R and Stewart BA (eds). Soil management experimental basis for sustainability and environmental quality. CRC Press, Lewis Publishers, Coca Raton, USA, pp. 499-520.

Babu R, Tejwani KG, Agrawal ML, Bhushan LS (1978). Distribution of erosion index and iso-erodent map of India. Indian J. Soil Conserv. 6:1-12.

Bekele MW, Thomas DB (1992). The influence of surface residue on soil loss and runoff. In: Hurni H and Tato K (eds). Erosion, conservation and small-scale farmers, Geographica Bernsia, International Soil Conservation Organisation (ISCO), World Association of Soil and Water Conservation (WASWC), pp. 439-452.

Centre for World Food Studies – CWFS (1985). Potential food production increases from fertiliser aid. A case study of Burkina Faso, Ghana and Kenya, Vol. 1. Centre for World Food Studies, Wageningen.

Christiansson C (1981). Soil erosion and sedimentation in semi-arid Tanzania. Studies on environmental changes and ecological imbalance. Scandinavian Institute of African Studies, Uppsala.

Cock JH (1985). Cassava - new potential for a neglected crop. Westview Press, Boulder & London.PMCid:PMC1432425.

El-Swaify SA, Fownes JH (1992). Erosion processes and models: Applications in the tropics. In: Hurni H and Tato K (eds). Erosion, conservation, and small scale farming. GB, ISCO, WASWC, pp. 135-149.

FAO (1978). Report on the FAO/UNEP expert consultation on Methodology for assessing soil degradation. FAO/UNEP, Rome.

FAO (1979). A provisional methodology for soil degradation assessment. FAO, Rome.

FAO (1990). Guidelines for soil profile description, 3rd Edition, FAO, Rome.

FAO/UNESCO (1974). FAO/UNESCO soil map of the world 1:5000000 – Legend. P. 1, UNESCO, Paris.

Gachene CKK (1987). Nutrient losses in eroded material from some Kenyan soils. In: Thomas DB, Biamah EK, Kilewe AM, Lundgren L and Machonge BO (eds). Soil and water conservation in Kenya. Proceedings of Third National Workshop, University of Nairobi/SIDA, pp. 34-37.

Hodgson JM (1985). Soil survey field handbook: Describing and sampling soil profiles. Technical Monograph No. 5. Soil Survey, Rothamsted Experimental Station, Harpenden, Herts.

Kang BJ (1993). Alley cropping: Past achievements and future directions", Agrofor. Syst. 23(2-3):141-155. http://dx.doi.org/10.1007/BF00704912

Kangalawe RYM (2001). Changing land-use patterns in the Irangi Hills, central Tanzania: A study of soil degradation and adaptive farming strategies. Dissertation, Department of Physical Geography and Quaternary Geology, Stockholm University P. 22.

Lewis LA (1987). Predicting soil loss in Rwanda. In: Beek KJ, Burrough PA and McCormack DEM (eds). Quantified land evaluation procedures. ITC Publication 6, ITC, Enschede, The Netherlands, pp 137-139.

Liniger H (1992). Soil cover and management - Attractive water soil conservation for drylands in Kenya. In: Haskins PG and Murphy BM (eds). Proceedings of the 7th International Soil Conservation Organisation (ISCO) Conference, 1. Sydney, pp. 130-139.

Logan TJ (1990). Chemical degradation of soil. Adv. Soil Sci. 11:186-221.

Lynam JK, Herdt RW (1989). Sense and sustainability: Sustainability as an objective in international agricultural research. Agric. Econ. 3:381-398. http://dx.doi.org/10.1016/0169-5150(89)90010-8

Mbegu AC (1988). The HADO Project: What, Where, Why, How? Forestry and Beekeeping Division, Ministry of Lands, Natural Resources and Tourism, Dar es Salaam.

Mero RN, Urio NA, Ekern A. (1988). The effect of supplementing Rhodes grass (*Chloris gayana*) with Siratro (*Macroptilium*

atropurpereum) on dry matter digestibility and voluntary intake. In: Improved dairy production from cattle and goats in Tanzania, NORAGRIC Occasional. Series B, 8:42-53.
Mnkeni PNS (1992). Role of soil management in enhancing sustainability of small-holder cropping systems in some agro-ecosystems of Tanzania: A review. Ecology and Development Paper No. 3. Ecology and Development Programme, Agricultural University of Norway.
Mohamed SA (1989). The effect of short fallowing on agricultural productivity in Mgololo, Mufindi district, Tanzania. Research Institute of Resource Assessment, University of Dar es Salaam. P. 22.
Mortimore M (1989). The causes, nature and rate of soil degradation in the northernmost states of Nigeria and An assessment of the role of fertiliser in counteracting the processes of degradation. Environment Development Working Paper The World Bank, Washington DC. P. 17.
Mowo JG, Floor J, Kaihura FBS, Magoggo JP (1993). Review of fertiliser recommendations in Tanzania. Part 2 - Revised fertiliser recommendations for Tanzania. Soil Fertility Report No. F6. National Soil Service, Agricultural Research Institution Mlingano, Tanga, Tanzania.
National Soil Service – NSS (1994). NSS Classification for soil fertility evaluation. National Soil service, Mlingano, Tanga.
Ngana JO (1990). Modelling for periodic features in seasonal rainfall and its implications to water resources and agricultural planning. Research Report . Institute of Resource Assessment, University of Dar es Salaam. P. 27.
Ngana JO (1992). Climatic assessment of Kondoa Eroded Area. Research Report Institute of Resource Assessment, University of Dar es Salaam. P. 80.
Payton RW, Christiansson C, Shishira EK, Yanda P and Erickson MG (1992). Landform, Soils and Erosion in the north-eastern Irangi Hills, Kondoa, Tanzania. Geografiska Annaler 74(A241):65-79 http://dx.doi.org/10.2307/521285
Rockström J (1997). On-farm agrohydrological analysis of the Sahelian yield crisis: Rainfall partitioning, soil nutrients and water use efficiency of pearl millet. PhD Thesis. Department of Systems Ecology, Stockholm University.
Roose EJ (1976). Use of universal soil loss equation to predict erosion in West Africa. In: Soil erosion: Prediction and control. Soil Conservation Society of America. Ankey, Iowa, USA, pp. 60-74.
Roose EJ (1986). Runoff and erosion before and after clearing depending on the type of crop in West Africa. In: Lal R, Sanchez PA and Cummings Jr. RW (eds). Land clearing and development in the tropics. Rotterdam, Balkema, The Netherlands, pp. 317-330.
Scholes MC, Swift MJ, Heal OW, Sanchez PA, Ingram JSI, Dalal R (1994). Soil fertility research in response to the demand for sustainability. In: Woomer PL and Swift MJ (eds). Biological management of tropical soil fertility. John Wiley & Sons Inc, pp. 1-14.
Singh G, Babu R, Chandra S (1981). Soil loss prediction research in India. ICAR Bulletin T-12/D-9, Central Soil and Water Conservation Research and Training Institute, Dehra Dun, India.
Smaling E (1993). An Agro-ecological Framework for Integrated Nutrient Management with special reference to Kenya. PhD Thesis. Wageningen Agricultural University, The Netherlands.
Smaling EMA, Fresco LO (1993). A decision support model for monitoring nutrient balances under agricultural land use (NUTMON). In: An Agro-ecological Framework for Integrated Nutrient Management with special reference to Kenya, Smaling E (ed). Ph.D. Thesis. Wageningen Agricultural University, The Netherlands, pp. 123-149.
Smaling EMA, Stoorvogel JJ, Windmeijer PN (1993). Calculating Soil Nutrient Balance in Africa at Different Scales II. District Scale. In: Smaling EMA (Ed). An Agro-ecological Framework for Integrated Nutrient Management with special reference to Kenya, PhD Thesis. Wageningen Agricultural University, The Netherlands, pp. 93-120.
Spenser DSC, Swift MJ (1992). Sustainable agriculture: Definition and Measurement. In: Mulongoy K, Gueye M and Spenser DSC. (eds). Biological nitrogen fixation and sustainability of tropical agriculture, John Wiley & Sons, Chichester.
Stocking M (1984). Erosion and soil productivity: A review. Consultants' Working Paper No. 1. FAO, Rome.
Stoorvogel JJ, Smaling EMA (1990). Assessment of nutrient depletion in Sub-Saharan Africa, 1983 -2000. Report The Winand Staring Centre for Integrated Land, Soil and Water Research, Wageningen, The Netherlands. P. 28.
Tosi JA, Hartshorn GS, Quesada CA (1982). HADO Project Development Study and Status of the Catchment Forestry. A Report to the Ministry of Natural Resources and Tourism, Tanzania, Tropical Science Centre, San Jones.
Victorio RG, Moreno U, Black JR CC (1986). Growth, partitioning and harvest index of tuber bearing Solanum genotypes grown in two contrasting Peruvian environment. Plant Physiol. 82:103-108. http://dx.doi.org/10.1104/pp.82.1.103, PMid:16664974 PMCid:PMC1056074
Wenner CG (1981). Soil conservation in Kenya. Soil Conservation Extension Unit, Ministry of Agriculture, Nairobi.
Wilkes BM, Duke BJ, Jimoh WLO (1984). Mineralogy and chemistry of Harmattan dust in Northern Nigeria. Catena 11:91-96. http://dx.doi.org/10.1016/S0341-8162(84)80009-0
Young A (1991). Agroforestry for Soil Conservation. CAB International, Wallington, UK.

30

The influences of forest fire on the vegetation and some soil properties of a savanna ecosystem in Nigeria

Fatubarin, A. R.[1] and Olojugba, M. R.[2]

[1]Department of Biological Sciences, Joseph Ayo Babalola University, Ikeji-Arakeji, Osun State, Nigeria.
[2]Department of Soil Science and Management, Joseph Ayo Babalola University, Ikeji-Arakeji, Osun State, Nigeria.

The effect of forest fire on natural forest in Southern Guinea Savanna in Nigeria was investigated. The study was carried out in Oro forest reserve in Kwara State, Nigeria. The study site was located in the north-eastern and south-western portions of the reserve. In each location in the forest, one hectare (100 m × 100 m) was divided into 100 plots of 10 m × 10 m. Twenty plots were randomly selected for determining the frequency of burned trees, re-sprouted trees and the numbers of seedling/ha. Soil samples were also collected at three depths: 0 - 5, 6 - 10, 11 - 15 and 16 - 20 cm. Soil samples were taken before burning and one year after burning. The effect of fire on tree species recovery showed that regeneration of *Byrsicarpus coccineus*, *Grewia mollis* and *Butyrospermum Paradoxium* were very encouraging in the burned area. In contrast, *Adenodolichos peniculatus, Fadogia pobegunii* and *Terminalia aviceniodes* were very sensitive to fire as they failed to regenerate or poorly reproduced in the burned area. Fire had no effect on soil texture except 0 - 5 cm. Soil pH significantly increased available phosphorus increased significantly, whereas, soil organic matter, available acidity and total nitrogen significantly decreased in the burned area. Also, metallic cations (Ca, Mg, K) and cation exchange capacity increased in the burned area. These changes were related primarily to oxidation of the organic matter layer during fire and concurrent changes in the soil environment following fire (e.g. a reduction in organic matter content of the soil, and increased soil pH).

Key words: Guinea Savanna, soil properties, forest fire, Regeneration, *Byrsicarpus cocineus, Adenodolichos peniculatus.*

INTRODUCTION

Fire is an important natural disturbance in most forest ecosystems and can lead to rapid changes in soil and biogeochemical cycling, which, in turn can have important implications for long-term ecosystem dynamics (Schmoldt et al., 1999). Fire resulted in disturbance of many forest lands depending on its severity and forest composition also; severity and extent of forest fires had a significant effect on seeds regeneration (Lecomte et al., 2005). On the other hand, fire modifies the ecosystem for short, long term and some biota may be adapted to these changes in the ecosystem (Rydgren et al., 2004). In many instances, fire may used for forest management purposes e.g.

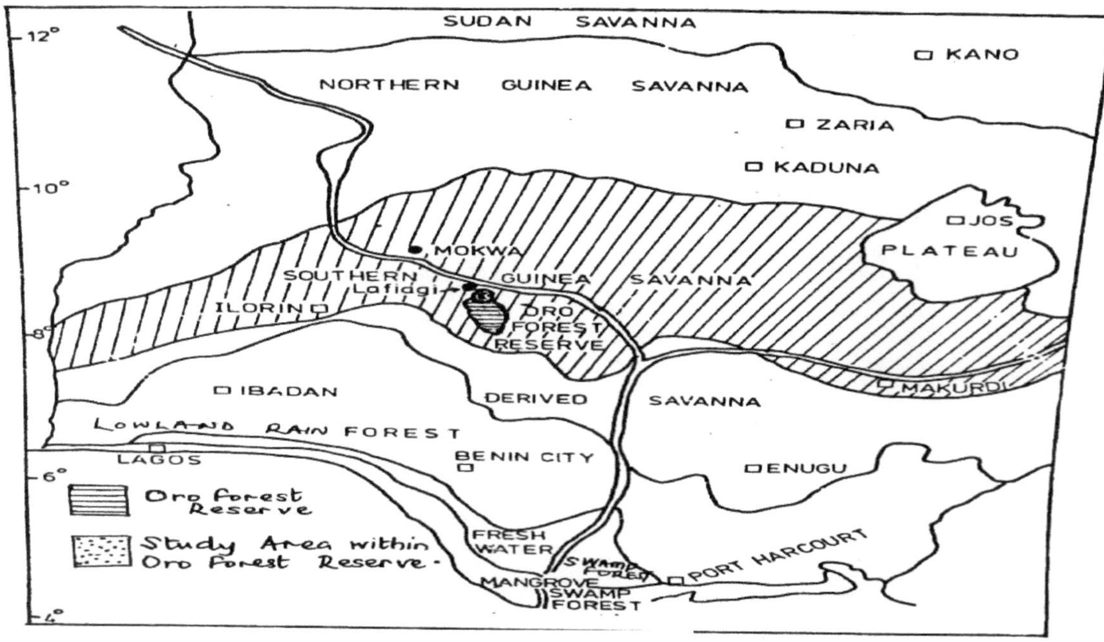

Figure 1. The location of Oro Forest Reserve within the Southern Guinea Savana of Nigeria.

prescribed burning (Bergeron et al., 2002). Basically, there are two types of forest fires, prescribed (controlled) fires and wildfires. Prescribed burning of naturally accumulated forest floor or slash following tree harvest is a standard practiced to reduce fuel levels with the intention of minimizing the extent and severity of wildfire or facilitating germination and growth of desired forest species. They are carried out when soil is moderatelymoist and thereby show a low level of severity (Wasterd et al., 1990).

It has been stated that logging after fire might have a considerable long term effect on vegetation and diversity in natural forest (Fraser et al., 2004). Spatial fire distribution has been widely used to categorize fire regime (Cui and Perera, 2008). Savanna ecosystems are characterized by the co-existence of carbon rich woody and carbon poor herbaceous plants dominated by grasses. Savannas occur in over 20 countries mostly in the seasonal tropics Nigerian inclusive with a limited distribution in temperate regions (Hutley and Setterfield, 2007).

Frequent and severe fires commonly result in degradation of soil (Anderson et al., 1981) forest fires threatened wild endangered species in Indonesia (Whitehouse and Mulyana, 2004). Fire is capable of exerting serious effects on soil properties (Giacomo, 2005). However, scarcity information is available on the influences of savanna burning on the vegetation and soil properties of southern Guinea savanna zone in Nigeria.

The objective of this study was to investigate the impact of forest fire on tree diversity and its effects on some soil physicochemical properties in the southern Guinea savanna vegetation of Oro forest reserve located in the southern Guinea savanna zone of Nigeria.

MATERIALS AND METHODS

The field studies were carried out in 2010 at Oro Forest Reserve in Kwara State Nigeria at longitude 08°53[1]E and latitude 05°22[1]N. The reserve is a wood land Savanna of about 5413 ha. The study site was located in the north-eastern and south-western portion of the reserve (Figure 1).

Plot description and samples collection

In each location in the forest, one hectare (100 m × 100 m) was divided into 100 plots of 10 m × 10 m. Twenty plots were chosen for determining the frequency of burned trees, re-sprouted trees and the numbers of seedling/ha. The frequency and numbers of trees before and after were determined in each sample plot using quadrant. Soil samples were also collected at three depths: 0-5, 6-10 and 11-15 and 16-20 cm. Soil samples were taking before burning and one year after burning by wild fire.

A Dutch auger was used to collect soil sample which were carefully kept in well labeled plastic bags and sent immediately to Laboratory for analysis.

Particle size analysis

This was done by hydrometer method (Gee and Bauder, 1986) using sodium haxametaphosphate (calgon) as dispensing agent.

Chemical analysis

The soil samples were dried for few days sieved to pass through 2 mm mesh and chemically analysed. The pH (in water) was

Table 1. Site plant community in the study area.

Species	Density/ha	Rank	Relative density (%)
Acadia nigrescan	17.25	13	1.5
Annona Senegalensis	5.75	19	0.5
Bridelia scleroneura	17.25	13	1.5
Bridelia stenocarpa	46.01	8	4.00
Butyrospermum paradoxum	143.77	3	12.50
Combretum collinumfresen	80.51	4	7.00
Daniellia oliveri	11.50	16	1.00
Detarium microcarpum	230.03	1	20.00
Ficus ingnens	5.75	19	0.50
Grewia mollis	184.03	2	16.00
Lannea kerstingii	5.75	19	0.5
Maytenus senegalensis	11.50	16	1.00
Prosopis africana	5.75	19	0.50

Table 2. Population dynamics of selected species of regenerating plants in permanent quadrant.

Species	Population sizes												
	J	F	M	A	M	J	J	A	S	O	N	D	J
Adenodolichos Paniculatus	25	26	35	30	30	28	26	26	22	20	16	10	0
B. paradoxum	89	97	109	105	75	75	75	24	72	72	70	68	4
Byrscarpus coccineus	115	123	122	101	166	154	142	35	25	15	12	7	0
D. microcapum	40	39	42	33	39	33	33	36	23	20	15	10	5
Fadogia pobeguinii	15	18	20	20	22	18	18	18	14	14	14	6	4
G. mollis	80	82	79	57	68	60	56	58	56	50	40	38	5
M. seregalansis	54	58	56	41	55	44	42	45	30	25	22	10	2
Terminalia aricenioides	48	51	57	44	60	54	52	51	52	42	40	28	2
Piliostigma thonningii	48	46	48	43	42	40	35	33	32	25	22	20	4
Total species	514	540	568	474	556	512	474	362	323	278	251	177	26

determined in a 1:2.5 solution (soil: distilled water) and was measured with a standard glass electrode. The organic carbon content of the soil was determined according to Walkley and Black (1965) dichromate oxidation method. The percentage organic matter content in the samples was calculated by multiplying the values of organic carbon by the conventional Van Bammeller factor of 1.724. Total soil nitrogen was determined by Macro kjeldahl methods (Bremner, 1965). Available phosphorus was extracted using Bray II method (Bray and Kurtz, 1965) and determined by spectrophotometer.

Exchangeable Na, K, Ca and Mg were extracted with $BaCl_2$ 0.1 M (Hendershot, et al., 1993) and analysed by atomic absorption. Exchangeable acidity was determined from 0.1 NaCl extracts and titrated with 1.0 N HCl.

Cation exchangeable capacity (CEC) was determined by summing up total exchangeable bases (TEBS) and total exchangeable acidity (TEA), which the base saturation = TEB/CEC × 100, where TEB = Total exchangeable bases, and CEC = Cations Exchangeable capacity.

Statistical analysis

The data were analyzed using two way analysis of variance (ANOVA) and means were separated by Duncan New Multiple Range Test and student t- test was used to test the level of significance of some properties in the pre and post burn periods at 5% level.

RESULTS

Burning and regeneration

In the study area, Detarium microcarpum had the highest density/ha of 230.03 and relative density of 20.00, while Annona selegalensis, Ficus ingnens, Lannea kerstingii and Prosopis africana had the least density/ha with relative density of 0.5 (Table 1).

It is apparent that *Adenodolrchios peniculatus* and *Byrsicarpus coccinetts* were most susceptible to forest fire. This is evident in the number of re-spouting of trees which were reduced to 0.0% (Table 2). However, fire also triggered–off regeneration more in *Brysicarpus coccineus*, followed by *Butyrospermum paradoxum* than any other species in the in the study area (Table 2).

The forest fire annually occurs between November and

Table 3. Effect of fire on soil texture.

Vegetation	Soil depth (cm)	Sand (%)	Silt (%)	Clay (%)
Southern guinea savanna	0 -5	82.30[a]	9.36[a]	8.35[a]
	6 - 10	84.90[a]	7.30[a]	7.80[ab]
Before burning	11 -15	82.75[a]	9.73[b]	7.53[ab]
	16 - 20	85.43[a]	7.40[a]	7.53[ab]
After burning	0 - 5	83.31[b]	10.37[ab]	6.26[b]c
	6 - 10	84.90[a]	8.37[ab]	6.73[b]c
	11 - 15	82.75[a]	9.73[b]	7.53[ab]
	16 - 20	85.43[a]	7.80[a]	7.53[ab]

Means followed by the same letter in a column in a forest are not significantly different at P ≤ 0.05.

January. The most resistant tree species are *Detarium microcarpum and Fodogia probeguinic* which reduced from 10 to 5 and 6 to 4 respectively. It is also noted that fire did not affect them in term of regeneration.

Soil texture

Fire did not affect soil texture at the level below 5 cm. However, effect of fire was found in the top layer (0-5 cm). This was indicated by the fact that sand, silt and clay (%) were not different when compared together at the sampling level below 5-10, 10-15 and 15–20 cm in unburned and burned sites (Table 3)

Soil pH, soil organic matter and soil N

Soil pH increased significantly ($p \leq 0.05$) from 5.0 to 6.3 at 0 - 10 cm depth and 5.7 to 6.2 to 6.3 at 11 - 20 cm depth for both pre-burn and post – burned plots, soil pH was lower at 10 - 20 cm depth. The values were statistically different at both sampling depths (Table 5). The soil nitrogen was lower in the post-burned plots with the values decreased from 0.09 to 0.05% at 0 - 10 cm depths. While the values decreased from 0.07 to 0.04% at 10 - 20 cm sampling depths ($P \leq 0.05$). It was also noted that, the values of soil nitrogen was lower at 10 – 20 cm depth.

Available phosphorus

Available phosphorus was higher in post-burned plots at all sampling depths and increased from 12.01 to 22.50 mg/kg at 0 - 10 cm depth. At 10 - 20 cm depth, it increased from 11.4 to 21.5 mg/kg, (Tables 4 and 5).

Exchanged cations, aluminum and cation exchange capacity

Exchangeable calcium was significantly ($P<0.05$) higher in the post-burned plots at all sampling depths and increased from 3.90 cmol/ kg of soil to 4.20 cmol/kg of soil at 0 - 10 cm depth, the values increased from 1.41 cmol/kg of soil to 2.42 cmol/kg of soil at 10 - 20 cm depths and was statistically different (Tables 4 and 5). Exchangeable magnesium was also higher in the post burned plots in all sampling depths and not significantly different and increased from 1.20 cmol/kg of soil to 2.95 cmol/kg of soil at 0-10 cm depth. While at 10-20 cm depth, it increased from 1.0 cmol/kg of soil to 1.80 cmol/kg of soil. Exchangeable potassium values did not show any significant different in all the sampling depths and but higher in the post-burned plots. Exchangeable aluminum decreased from 2.5 cmol/kg of soil to 1.5 cmol/kg of soil at 0 - 10 cm depth and decreased from 0.30 cmol/kg of soil to 0.15 cmol/kg of soil at 10 - 20 cm depth and did not statistically different (Table 5).

The cation exchangeable capacity increased from 11.10 cmol/kg of soil to 12.65 cmol/kg of soil after fire at 0 - 10 cm depth as well as, increased from 3.16 cmol/kg of soil to 4.97 cmol/kg of soil after fire at 10 - 20 cm depth.

DISCUSSION

Very sketchy studies have investigated factors that affect tropical savanna forest (Sukumar et al., 2005). Climate change, animals and fire are some factors that affected vegetation dynamics in those ecosystems (Silori and Mishra, 2001). As forest fire could be useful or destructive to the forest, while tree species vary in their responses to forest fires. Prescribed fire may play an important role in sustainable management of forest subjected to wild fires (Matthias et al., 2009). *B. coccineus* benefited from fire, as was indicated by the substantial increase of regeneration and the fact that considerable number of B. *coccineus, Grawia mollis* as well as *B. paradoxium* responded after fire (Table 2). Hein et al. (2010) mentioned that in mixed forests, the probability is low for succession after fire. In hard seeded legume, heat from fire may break the hard seed coat and

Table 4. Comparison of the pre and post burning soil chemical properties at 0-10cm depth by student's t-test.

Properties	Pre-burning	Post-burning	t-test values
pH	5.9	6.3	2.94*
OC	0.8	0.58	2.34*
OM	1.38	0.8	2.36*
N	0.09	0.05	2.44*
C/N	0.6	0.69	2.31*
Ex. Ca	3.9	4.2	2.80*
Ex. Mg	1.20	2.95	1.02^{ns}
Ex. K	3.5	4.00	1.20^{ns}
Ex.A1	2.5	1.5	0.67^{ns}
CEC	11.10	12.65	0.91^{ns}
Av. P	12.01	22.5	2.12*

*Significant at ($p \leq 0.05$), ns = not significant at ($p \leq 0.05$), OC = organic carbon, OM = organic matter, Ex. Ca = exchange calcium, Ex. Mg = exchange magnesium, Ex. K = Exchange potassium and CEC = cation exchange capacity.

Table 5. Comparison of the pre and post burning soil chemical properties at 0-20 cm depth by student's t-test.

Properties	Pre-burning	Post-burning	t-test values
pH	5.2	6.2	2.55*
OC	0.7	0.4	2.4*
OM	1.21	0.69	2.28*
N	0.07	0.04	2.24*
C/N	6.2	5.5	2.15*
Ex. Ca	1.41	2.42	2.11*
Ex. Mg	1.0	1.8	1.15^{ns}
Ex. K	0.45	0.60	0.91^{ns}
Ex.A1	0.30	0.15	1.11^{ns}
CEC	3.16	4.97	1.04^{ns}
Av. p	11.4	21.5	2.24*

*Significant at ($p \leq 0.05$), ns = not significant at ($p \leq 0.05$), OC = organic carbon, OM = organic matter, Ex. Ca = exchange calcium, Ex. Mg = exchange magnesium, Ex. K = Exchange potassium and CEC = cation exchange capacity.

allow water absorption by the seeds and hence dormancy is broken (Mbalo and Witkowski, 1997). Removal of stem tips and young leaves by fires, for instance, may allow dormant buds to develop into new shoots. In appears that the presence of *B. coccineus, G. mollis, B. paradixium* and *Maytenus senegalensis* in fire prone ecosystem, as the case in the present study, facilitated its adaptation to fire in various ways including thickening of bark, ability to respond and dispersing of seeds. Hare (1965) mentioned that some trees have thick insulating bark, which protected them from the scorching heat of surface fires. In contrast, *Adenodolichos peniculatus*, *Fadogia pobeguinii* and *Terminalia avicenioides* in the present study, were very sensitive to forest fire since they failed to regenerate or poorly recovered in the burnt soils, *Detarium microcaupum* and *Piliostigma thonningii* were more or less negatively sensitive to fire in terms of regeneration and recovery after forest fire. Species diversity was considerably affected by forest fires, although this effect dependent on the severity of fire (Kodandapani et al., 2008).

Fire has no significant effect on soil texture in the studied locations. However sand content increased in the post-burned plots at the 0-5cm depth which might be due to destruction of soil organic matter. Soil pH increased significantly after fire in all locations studied. This finding is in line with the earlier studies (Creighton and Santelies, 2003; Ekinci, 2006). Giacomo (2005) reported that soil pH in non-calcareous soils increased after fire because of the release of the alkaline cations (Ca,Mg,K and Na) bound to the organic matter sites and a decreased in organic acids produced during the biological oxidation

of organic matter (Wells, 1971).One pertinent implication of increased pH in the burned soils is the possibility of increasing nitrogen fixation under
environmental conditions favouring both symbiotic and free-living nitrogen fixers.

The loss of soil organic matter represent one of the more obvious alteration to soil physical properties after a fire, significant reductions in soil organic matter were recorded in all locations studied during post-burned. The effect of fire on organic matter comprises volatilization, charring or oxidation (Giovannini et al., 1988). Fire causes burning and or total removal of organic matter (Simard et al., 2001). The decreased levels of soil organic matter observed in the burned plots in this study suggest important ramifications for other soil biological, chemical and physical properties. However, it is not likely that litter fall rates in post-burned forests will approximate those on pre-burned sites for sometimes and thus, soil organic matter content in post-burned plots will more than likely remain low for sometimes period. There was a significant reduction in total soil nitrogen after fire at all sampling depths in the study locations, which is in agreement with Neff et al. (2005). Forest floor layers are a major reservoir of soil N and their removal during forest fires can cause significant reductions in it (Driscoll et al., 1999). A likely explanation for decreased soil nitrogen in the post burning plots is through increased leaching of NO_2 during the rainy season that is characteristic of this area. Consumption of the litter layer during fire likely leads to increased infiltration rates for burned soils in rainy season, than in turn could lead to increased leaching of nitrogen in the form of NO_2.

Significant increase in phosphorus might be attributed to the fact that burning coverts some of the organic pool of soil P to orthophosphate (Cade-Menun et al., 2000). Also, at a large portion of the nutrient reserve in most forest ecosystems is contained in the organic material on the forest floor and that distributed throughout the profile (Wagner and Wolf, 1998). Nutrients contained in these organic materials are slowly released into the soil though biological oxidation is decomposition (Fuhrmann, 1998).

Exchangeable cation such as Ca, Mg and K were higher after fire, which is in agreement with Adams and Boyle (1980) who reported increase in alkaline cations after fire. Exchangeable aluminum was lower in the post-burned soil due to the conversion of aluminum into complex oxides, thereby reduces soil acidity after fire.

Conclusion

The present study revealed that fire encourages fire tolerant tree species and discouraged fire sensitive species as reported by Ivanauskas et al. (2003). *D. microcarpum* and *F. pobeguinic* have relatively thick and fissured barks and this might explain their resistance to fire relatively to *A. peniculatus* and *B. caccineus* which have relatively thin barks, probably the reason for their susceptible to fire. Effects of wild fire on soil properties in this highly fragmented ecosystem are not well conceived. Data from this study suggest soil properties vary in their response to fire; soil pH and available phosphorus were higher following fire. In the same vein, exchangeable cations such as Ca, Mg and K were more at post fire while the exchangeable acidity became low during post-fire period, changes in soil properties were more than likely a result of the oxidation of the litter layer during the fire and concurrent changes in the soil environment following the fire. Nitrogen content decrease was due to leaching of NO_3 while phosphorus increase might be due to changing of organic phosphorus to the available form. Organic matter content of the soils was lower in post-fire plots due to the destruction of organic layer in the forest. The savanna burning in Nigeria southern Guinea savanna is annual one, which is used as a cheaper means of land clearing for planting seasons. However, we found that fire can significantly affect some soil chemical and physical properties as well as plant species composition (Litton and Santelices, 2002), in these native forests. In light of these discoveries, we recommend that more emphasis should be placed on preventing and fighting human caused fires in native vegetation, aggressive public awareness on the effect of forest fire and fire breaks should be established around native forest to prevent further degradation of this ecosystem. In addition, further research is needed on long term effects of fire on Guinea savanna in Nigeria.

REFERENCES

Adams PW, Boyle JR (1980). Effects of fire on soil nutrients in clear cut and whole-tree harvest sites in central Michigan. Soil Sci. Soc. Am. J. 44:847-850. http://dx.doi.org/10.2136/sssaj1980.03615995004400040038x

Anderson DW, Bettany TR, Stewart JWB (1981). Particle size fractions and their use in studies of soil organic matter. I. The nature and distribution of forms of Carbon. nitrogen and sulfur. Am. J. Soil Sci. 45:767-772. http://dx.doi.org/10.2136/sssaj1981.03615995004500040018x

Bergeron Y, Leduc A, Harvey B, Gauthier S (2002). Natural fire regime: a guide for sustainable management of the Canadian boreal forest. Silva Fennica 36:81-95.

Bray RH, Kurtz LT (1965). Determination of total organic carbon and available form of phosphorous in soil. J. Soil Sci. 59:39-45. http://dx.doi.org/10.1097/00010694-194501000-00006

Bremner JM (1965). Nitrogen method of soil analysis, part one (ed) C.C. Black American Society of Agro forestry. Madison. Wisconsin, USA. Pp. 1149-1178.

Cui W, Perera AH (2008). What we do know about forest fire size distribution, and why is this knowledge useful for forest management. Int. J. Wildland Fire 17:234-244.http://dx.doi.org/10.1071/WF06145

Creighton ML, Santelices R (2003). Effect of wildfire on soil physicaland chemical properties in a Nothofagus glauca forest. Chile Rev. Chil. Hist. Nat. Santiago. 76:16.

Cade-Menun BJ, Berch SM, Preston CM, Lavkulich LM (2000). Phosphorus forms and related soil chemistry of Podzolic soils on northern Vancouver Island. II. The effects of clear-cutting and burning. Can. J. For. Res. 30:1726-1741.http://dx.doi.org/10.1139/x00-098 http://dx.doi.org/10.1139/x00-099

Driscoll KG, Arocena JM, Massicotte HB (1999). Post-fire

nitrogencontent and vegetation composition in Sub-Boreal spruce forests of British Columbia's central interior. Can. For. Ecol. Manage. 121:227-237.http://dx.doi.org/10.1016/S0378-1127(99)00003-1

Ekinci H (2006). Effect of forest fire on some physical, chemical andbiological properties of soil in Canakkale, Turkey. Int. J. Agric. Biol. 8:102-106.

Fraser E, Landhausser S, Lleffers V (2004). The effect of fire severity and salvage logging traffic on regeneration and early growth of aspen suckers in north-central Alberta. For. Chronicle 80:251-256.http://dx.doi.org/10.5558/tfc80251-2

Fuhrmann JJ (1998). Microbial metabolism. In: Sylvia DM, JJ Fuhrmann, PG Hartel & DA Zuberer (eds) Principles and applications of soil microbiology: Prentice-Hall, Princeton, New Jersey, USA, pp. 189-217.

Gee GW, Bauder JW (1986). Particle-size Analysis. Pages 383-411 in Methods of Soil Analysis Part 1. A. Klute (edi.), Soil Science Society of America Book Series 5, Madison, Wisconsin, USA.

Giacomo C (2005). Effects of fire on properties of forest soils: a review. Oecologia, 143:1-10.http://dx.doi.org/10.1007/s00442-004-1788-8 PMid:15688212

Giovannini G, Lucchesi S, Giachetti M (1988). Effects of heating onsome physical and chemical properties related to soil aggregation and erodibility. Soil Sci. 162:479-486.http://dx.doi.org/10.1097/00010694-199707000-00003

Hare RC (1965). Contribution of bark to fire resistance of southern trees. J. For. 63:248-251.

Hein G, Justice O, Andrisano A (2010). From monospecific to mixed forest after fire? An early forecast for the montane belt of Majella, Italy. For. Ecol. Manage. 259:433-439.http://dx.doi.org/10.1016/j.foreco.2009.10.040

Hendershot WH, Lalande H, Duquette M (1993). Soil reaction and exchangeable acidity. In; Soil Sampling and Methods of Analysis (ed. M.R. Carter). Lewis publishers, USA, pp. 141-145.

Hutley LB, Setterfield SA (2007). Savannas. In: Encyclopaedia of Ecology (ed. Jorgeensen SE), Elsevier, Amsterdam (in press)

Ivanauskas NM, Monteiro R, Rodrigues RR (2003). Alterationsfollowing a fire in a forest community of Alto Rio Xingu. For. Ecol. Manage. 184:239-250.http://dx.doi.org/10.1016/S0378-1127(03)00156-7

Kodandapani N, Cochrane MA, Sukumar R (2008). A comparative analysis of spatial, temporal and ecological characteristics of forest fires in seasonally dry tropical ecosystemsin the Western Ghats, India. For Ecol. Manage. 256:433-439.

Lecomte N, Simard M, Bergeron Y, Larouche A, Asnong H, Richard PJH (2005). Effects of fire severity and initial tree composition on understory vegetation dynamics in a boreal landscape inferred from chronosequence and paleoecological data. J. Veg. Sci. 16:665-674.http://dx.doi.org/10.1111/j.1654-1103.2005.tb02409.x

Litton CM, Santelices R (2002). Early post-firesuccession in a Nothofagus glauca forest in the Coastal Cordillera of south-central Chile. Int. J. Wildland Fire. 11:115-125.http://dx.doi.org/10.1071/WF02039

Mbalo BA, Witkowski ETF (1997). Tolerance to soil temperaturesexperienced during and after the passage of fire in seeds of Acacia karroo, A. nilotica and Chromolaena odorata: a laboratory study. South Afr. J. Bot. 63:421-425.

Neff JC, Harden JW, Gleixner G (2005). Fire effects on soil organic matter content, composition and nutrients in boreal interior Alaska. Can. J. For. Res. 35:2178-2187.http://dx.doi.org/10.1139/x05-154

Rydgren K, Økland RH, Hestmark G (2004). Disturbance severity and community resilience in a boreal forest. Ecology 85:1906-1915.http://dx.doi.org/10.1890/03-0276

Silori C, Mishra BK (2001). Assessment of livestock grazing pressure in and around the elephant corridor in Mudumalai wildlife sanctuary, south India. Biodiver. Conserv. 10:2181-2195.http://dx.doi.org/10.1023/A:1013285910650

Simard DG, Fyles JW, Pare D, Nguyen T (2001). Impacts of clear cutharvesting and wildfire on soil nutrient status in the Quebec boreal forest. Can. J. Soil Sci. 81:229-237.http://dx.doi.org/10.4141/S00-028

Sukumar R, Suresh HS, Dattaraja HS, Srinidhi S, Nath C (2005). The dynamics of a tropical dry forest in India: climate, fire, elephant and the evolution of lifehistory strategies. In: Burslem, D.F.R.P.,M.A. Pinard and S.E. Hartley (eds.), Biotic Interactions in the Tropics: Their role in the maintenance of species diversity. Cambridge University Press, Cambridge.http://dx.doi.org/10.1017/CBO9780511541971.022

Wasterd JD, Radosevich SR, Sanberg DV (1990). Natural and prescribed fire in the pacific Northwest forest Oregon State. University Press, Corvallis.

Wells C (1971). Effects of prescribed burning on soil chemical properties and nutrient availability. Prescribed Burning symposium Proceedings. United States Department of Agriculture, Forest Service, Southern-eastern Forest Experiment Station. Ashevilla, State, USA. P. 20.

Whitehouse E, Mulyana AS (2004). Coal fires in Indonesia. Int. J. Coal. Geol. 59:91-97.http://dx.doi.org/10.1016/j.coal.2003.08.010

Walkley A, Black IA (1965). Determination of organic carbon in soil. Soil Sci. Soc. Am. 37:31-43.

Wagner GH, Wolf DC (1998) Carbon transformations and soil organic matter formation. In: Sylvia DM, JJ Fuhrmann, PG Hartel & DA Zuberer (eds) Principles and applications of soil microbiology: Prentice-Hall, Princeton, New Jersey, USA, pp. 218-258.PMid:9548537

Schmoldt DL, Peterson DL, Keane RE, Jm lenihan D, Mckenzie, DR Weise-Sandberg DV (1999). Assessing the effects of fire disturbance on ecosystems: a scientific agenda for research and management. United States Department of Agriculture, Forest Service, General Technical Report PNW-GTR-455. P. 104.

Permissions

All chapters in this book were first published in JSSEM, by Academic Journals; hereby published with permission under the Creative Commons Attribution License or equivalent. Every chapter published in this book has been scrutinized by our experts. Their significance has been extensively debated. The topics covered herein carry significant findings which will fuel the growth of the discipline. They may even be implemented as practical applications or may be referred to as a beginning point for another development.

The contributors of this book come from diverse backgrounds, making this book a truly international effort. This book will bring forth new frontiers with its revolutionizing research information and detailed analysis of the nascent developments around the world.

We would like to thank all the contributing authors for lending their expertise to make the book truly unique. They have played a crucial role in the development of this book. Without their invaluable contributions this book wouldn't have been possible. They have made vital efforts to compile up to date information on the varied aspects of this subject to make this book a valuable addition to the collection of many professionals and students.

This book was conceptualized with the vision of imparting up-to-date information and advanced data in this field. To ensure the same, a matchless editorial board was set up. Every individual on the board went through rigorous rounds of assessment to prove their worth. After which they invested a large part of their time researching and compiling the most relevant data for our readers.

The editorial board has been involved in producing this book since its inception. They have spent rigorous hours researching and exploring the diverse topics which have resulted in the successful publishing of this book. They have passed on their knowledge of decades through this book. To expedite this challenging task, the publisher supported the team at every step. A small team of assistant editors was also appointed to further simplify the editing procedure and attain best results for the readers.

Apart from the editorial board, the designing team has also invested a significant amount of their time in understanding the subject and creating the most relevant covers. They scrutinized every image to scout for the most suitable representation of the subject and create an appropriate cover for the book.

The publishing team has been an ardent support to the editorial, designing and production team. Their endless efforts to recruit the best for this project, has resulted in the accomplishment of this book. They are a veteran in the field of academics and their pool of knowledge is as vast as their experience in printing. Their expertise and guidance has proved useful at every step. Their uncompromising quality standards have made this book an exceptional effort. Their encouragement from time to time has been an inspiration for everyone.

The publisher and the editorial board hope that this book will prove to be a valuable piece of knowledge for researchers, students, practitioners and scholars across the globe.

List of Contributors

Mojtaba Mehrabanian
Department of Soil Science, Faculty of Agricultural Engineering and Technology, Collage of Agriculture and Natural Resources, University of Tehran, Iran

Agena Anjulo Tanga
Department of Plant Science, College of Agricultural Sciences, Arba Minch University, P. O. Box: 21, Ethiopia

Tilahun Fromssa Erenso
Ziway Soil Laboratory, Oromia Institute of Agricultural Research, Addis Ababa, Ethiopia

Bekele Lemma
Wondo Genet College of Forestry and Natural Resources, Hawassa University, Ethiopia

Bocar Ahamadou
Unit of Soil Science, Department of Agronomy, Katibougou Rural Polytechnical Institute of Training and Applied Research, P. O. Box 06 Koulikoro, Mali

Qiaoyun Huang
Key Laboratory of Subtropical Agriculture and Environment, Ministry of Agriculture, Huazhong Agricultural University, Wuhan, 430070, P.R. China

Lin Yaping
Key Laboratory of Subtropical Agriculture and Environment, Ministry of Agriculture, Huazhong Agricultural University, Wuhan, 430070, P.R. China

Javed Iqbal
Department of Plant and Soil Science, 2207 Agronomy Hall, Iowa State University, Aames IA, 50010, USA

Vernhar Gervais-Beaulac
Département des sciences de l'environnement, 3351, boul. des Forges, C.P. 500, Trois-Rivières, Québec, G9A 5H7, Canada

Diane Saint-Laurent
Département des sciences de l'environnement, 3351, boul. des Forges, C.P. 500, Trois-Rivières, Québec, G9A 5H7, Canada
Géographie et Laboratoire de recherche en géomorphologie fluviale et sols, Université du Québec à Trois-Rivières, 3351, boul. des Forges, C.P. 500, Trois-Rivières, Québec, G9A 5H7, Canada

Jean Sébastien Berthelotand
Département des sciences de l'environnement, 3351, boul. des Forges, C.P. 500, Trois-Rivières, Québec, G9A 5H7, Canada

Mhamed Mesfioui
Département de Mathématique et informatique, Université du Québec à Trois-Rivières, 3351, boul. des Fortes, C.P. 500, Trois-Rivières, Québec, G9A 5H7, Canada

Mewael Kiros Assefa
Department of Plant Sciences, Aksum University, Shire-campus P. O. Box 314, Shire-endaslassie, Ethiopia

S. von Tucher
Plant Nutrition, Technische Universität München, Emil-Ramann-Straße 2, 85350 Freising-Weihenstephan, Germany

Urs Schmidhalter
Plant Nutrition, Technische Universität München, Emil-Ramann-Straße 2, 85350 Freising-Weihenstephan, Germany

Sharma Richa
Department of Biotechnology and Allied Sciences, Jayoti Vidyapeeth Women`s University, Jaipur, Rajasthan, India

Chandra Subhash
Department of Biotechnology and Allied Sciences, Jayoti Vidyapeeth Women`s University, Jaipur, Rajasthan, India

Singh Amrita
Department of Biotechnology and Allied Sciences, Jayoti Vidyapeeth Women`s University, Jaipur, Rajasthan, India

Yacob Alemayehu
Soil and Water Research Section, Gambella Agricultural Research Center, P. O. Box 62, Gambella, Ethiopia

Heluf Gebrekidan
School of Natural Resources Management and Environmental Sciences, Haramaya University, P. O. Box 138, Dire Dawa, Ethiopia

Sheleme Beyene
School of Plant and Horticultural Sciences, Hawassa University, P. O. Box 5, Hawassa, Ethiopia

Gebremedhin Kiros
Department of Soil Resources and Watershed Management, Aksum University, Shire- campus: P. O. Box -314, Shire - Endaselassie, Ethiopia

Mitiku Haile
Department of Land Resources Management and Environmental Protection, Mekelle University: P. O. Box -231, Mekelle, Ethiopia

Girmay Gebresamuel
Department of Land Resources Management and Environmental Protection, Mekelle University: P. O. Box -231, Mekelle, Ethiopia

J. A. Osunbitan
Department of Agricultural Engineering, Obafemi Awolowo University, Ile - Ife, Nigeria

K. O. Adekalu
Department of Agricultural Engineering, Obafemi Awolowo University, Ile - Ife, Nigeria

P. O. Aina
Department of Soil Science, Obafemi Awolowo University, Ile - Ife, Nigeria

Nguyen Huu Thanh
Faculty of Natural Resources and Environment, Hanoi University of Agriculture, Hanoi, Vietnam

Tran Thi Le Ha
Faculty of Natural Resources and Environment, Hanoi University of Agriculture, Hanoi, Vietnam

Cao Viet Ha
Faculty of Natural Resources and Environment, Hanoi University of Agriculture, Hanoi, Vietnam

Nguyen Duc Hung
Faculty of Natural Resources and Environment, Hanoi University of Agriculture, Hanoi, Vietnam

Phan Quoc Hung
Faculty of Natural Resources and Environment, Hanoi University of Agriculture, Hanoi, Vietnam

Kiyoshi Kurosawa
Institute of Tropical Agriculture, Kyushu University, Fukuoka, 812-8581, Japan

Kazuhiko Egashira
Kyushu University, Fukuoka 812-8581, Japan

A. L. Ngala
Department of Soil Science, University of Maiduguri, P. M. B. 1069, Maiduguri, Borno State, Nigeria

H. Yakubu
Department of Soil Science, University of Maiduguri, P. M. B. 1069, Maiduguri, Borno State, Nigeria

I. J. Tekwa
Department of Agricultural Technology, Federal Polytechnic, P. M. B. 35, Mubi, Adamawa State, Nigeria

Lee Fergusson
Prana World Consulting, P. O. Box 1620, Oxenford, Queensland 4210, Australia

G. Y. Jamala
College of Agriculture, Ganye, Adamawa State, Nigeria

D. O. Oke
Federal University of Technology Akure, Ondo State, Nigeria

Imoro Abukari Ziblim
Department of Range and Wildlife Management, Faculty of Renewable Natural Resources, University for Development Studies, Nyankpala Campus. Tamale, Ghana

Gakpo Selorm Paul
Department of Range and Wildlife Management, Faculty of Renewable Natural Resources, University for Development Studies, Nyankpala Campus. Tamale, Ghana

Khan Aikins Timothy
Department of Range and Wildlife Management, Faculty of Renewable Natural Resources, University for Development Studies, Nyankpala Campus. Tamale, Ghana

Aidin Parsakhoo
Department of Forestry, Faculty of Forest Science, Gorgan Agricultural Sciences and Natural Resources University, Golestan Province, Gorgan City, Iran

Majid Lotfalian
Department of Forestry, Faculty of Natural Resources, Sari Agricultural Sciences and Natural Resources University, Mazandaran Province, Sari City, Iran

Hamid Jalilvand
Department of Forestry, Faculty of Natural Resources, Sari Agricultural Sciences and Natural Resources University, Mazandaran Province, Sari City, Iran

S. U Okpamen
Chemistry Division, NIFOR, P.M.B 1030, Benin City, Edo State, Nigeria

E. G Ilori
Edo State Institute of Technology and Management, Usen, Edo State, Nigeria

I Agho
Biochemistry Division, NIFOR, P.M.B 1030, Benin City, Edo State, Nigeria
Analytical Laboratory Division, NIFOR, P.M.B 1030, Benin City, Edo State, Nigeria

A Nkechika
Chemistry Division, NIFOR, P.M.B 1030, Benin City, Edo State, Nigeria

F. U. Maidoh
Chemistry Division, NIFOR, P.M.B 1030, Benin City, Edo State, Nigeria

P. N Okonjo
Analytical Laboratory Division, NIFOR, P.M.B 1030, Benin City, Edo State, Nigeria

Daniel Hagos Berhe
Axum Agricultural Research Centre, P. O. Box 230, Axum, Ethiopia

Agena Anjulo
Wondo Genet College of Forestry and Natural Resources, P. O. Box 128, Shashemene, Ethiopia

Abdu Abdelkadir
Wondo Genet College of Forestry and Natural Resources, P. O. Box 128, Shashemene, Ethiopia

Sue Edwards
Institutefor Sustainable Development (ISD), P. O. Box 171, Code 1110, Addis Ababa, Ethiopia

Godsteven P. Maro
TaCRI, Lyamungu P. O. Box 3004, Moshi, Tanzania

Jerome P. Mrema
Department of Soil Science, Faculty of Agriculture, Sokoine University of Agriculture, P. O, Box 3008, Morogoro, Tanzania

Balthazar M. Msanya
Department of Soil Science, Faculty of Agriculture, Sokoine University of Agriculture, P. O, Box 3008, Morogoro, Tanzania

James M. Teri
TaCRI, Lyamungu P. O. Box 3004, Moshi, Tanzania

J. M. Miriti
Kenya Agricultural Research Institute (KARI), ARC Muguga South Centre, P.O. 30148-00100 Nairobi, Kenya

G. Kironchi
Faculty of Agriculture, University of Nairobi, P.O. Box 30197-00100 Nairobi, Kenya

A.O. Esilaba
Kenya Agricultural Research Institute (KARI) Headquarters, P.O. Box 57811-00200 Nairobi, Kenya

C. K. K. Gachene
Faculty of Agriculture, University of Nairobi, P.O. Box 30197-00100 Nairobi, Kenya

L. K. Heng
International Atomic Energy Agency (IAEA), Soil and Water Management and Crop Nutrition Section, Wagrammer Strasse 5, P.O. Box 100, 1400 Vienna, Austria

D. M. Mwangi
Faculty of Agriculture, University of Nairobi, P.O. Box 30197-00100 Nairobi, Kenya

Drissa DIALLO
Unité des Sciences du Sol, DER STA IPR/IFRA de Katibougou BP6 Koulikoro, Mali

P. N. M Njeru
KARI Muguga South, P.O Box 30148-00100, Nairobi

J Mugwe
Kenyatta University, P.O Box 43844, Nairobi

I Maina
KARI Headquarters, P.O Box 57811-002004, Nairobi

M Mucheru-Muna
Kenyatta University, P.O Box 43844, Nairobi

D Mugendi
Embu University College-Nairobi University, P.O Box 6- 60100, Embu

J. K Lekasi
KARI Muguga South, P.O Box 30148-00100, Nairobi

S. K Kimani
KARI Muguga South, P.O Box 30148-00100, Nairobi

J Miriti
KARI Muguga South, P.O Box 30148-00100, Nairobi

A. O Esilaba
KARI Headquarters, P.O Box 57811-002004, Nairobi

F Muriithi
KARI Headquarters, P.O Box 57811-002004, Nairobi

E. T. Atakora
Department of Agricultural Engineering, Kwame Nkrumah University of Science and Technology, Kumasi, Ghana

N. Kyei-Baffour
Department of Agricultural Engineering, Kwame Nkrumah University of Science and Technology, Kumasi, Ghana

E. Ofori
Department of Agricultural Engineering, Kwame Nkrumah University of Science and Technology, Kumasi, Ghana

B. O. Antwi
Soil and Water Management Division, CSIR- Soil Research Institute of Ghana, Academy Post Office, Kumasi, Ghana

K. E Law-Ogbomo
Department of Agriculture, Benson Idahosa University, PMB 1100, Benin City, Nigeria

Ramzi M. Shihab
Soil Science and Water Resource Department, College of Agriculture Tikrit University, Iraq

T.P. Abegunrin
Department of Agricultural Engineering, Ladoke Akintola University of Technology, Ogbomoso, Oyo State, Nigeria

G. O. Awe
Department of Crop, Soil and Environmental Sciences, Faculty of Agricultural Sciences, Ekiti State University, Ado Ekiti, Ekiti State Nigeria

D. O. Idowu
Department of Agricultural Engineering, Ladoke Akintola University of Technology, Ogbomoso, Oyo State, Nigeria

O. O. Onigbogi
Department of Agricultural Engineering, Ladoke Akintola University of Technology, Ogbomoso, Oyo State, Nigeria

O. E. Onofua
Department of Agricultural Engineering, Ladoke Akintola University of Technology, Ogbomoso, Oyo State, Nigeria

A. Nduwumuremyi
Soil Quality and Plant Nutrition Program, Department of Natural Resources Management, Rwanda Agriculture Board (RAB), P. O. Box 5016, Kigali, Rwanda

N. J. Mugwe
Department of Agricultural Resource Management, School of Agriculture and Enterprise Development, Kenyatta University, P.O. Box 43844-00100, Nairobi, Kenya

C. A. Rusanganwa
Soil Quality and Plant Nutrition Program, Department of Natural Resources Management, Rwanda Agriculture Board (RAB), P. O. Box 5016, Kigali, Rwanda

J. Mupenzi
Department of Land Survey, Institute of Applied Sciences (INES-Ruhengeri), Musanze, P. O. Box 155, Rwanda

Daniel Hagos Berhe
Axum Agricultural Research Centre, P. O. Box 230, Axum, Ethiopia

Agena Anjulo
Wondo Genet College of Forestry and Natural Resources, P. O. Box 128, Shashemene, Ethiopia

Abdu Abdelkadir
Wondo Genet College of Forestry and Natural Resources, P. O. Box 128, Shashemene, Ethiopia

Sue Edwards
Institute for Sustainable Development (ISD), P. O. Box 171, Code 1110, Addis Ababa, Ethiopia

Alemayehu Kiflu
Soil Science Department, Hawassa University, Ethiopia

Sheleme Beyene
Soil Science Department, Hawassa University, Ethiopia

Richard L. Haney
USDA-ARS, Grassland, Soil and Water Research Laboratory, 808 East Blackland Road, Temple, TX 76502, USA

Michael J. White
USDA-ARS, Grassland, Soil and Water Research Laboratory, 808 East Blackland Road, Temple, TX 76502, USA

Jeffrey G. Arnold
USDA-ARS, Grassland, Soil and Water Research Laboratory, 808 East Blackland Road, Temple, TX 76502, USA

Elizabeth B. Haney
Texas A&M University, Texas AgriLife Research and Extension Center, 720 E. Blackland Road, Temple, TX 76502, USA

R. Daren Harmel
USDA-ARS, Grassland, Soil and Water Research Laboratory, 808 East Blackland Road, Temple, TX 76502, USA

Richard Y. M. Kangalawe
Institute of Resource Assessment, University of Dar es Salaam; P. O. Box 35097, Dar es Salaam, Tanzania

A. R. Fatubarin
Department of Biological Sciences, Joseph Ayo Babalola University, Ikeji-Arakeji, Osun State, Nigeria

M. R. Olojugba
Department of Soil Science and Management, Joseph Ayo Babalola University, Ikeji-Arakeji, Osun State, Nigeria